THE ROUTLEDGE COMPANION
TO ADAPTATION

The Routledge Companion to Adaptation offers a broad range of scholarship from this growing, interdisciplinary field. With a basis in source-oriented studies, such as novel-to-stage and stage-to-film adaptations, this volume also seeks to highlight the new and innovative aspects of adaptation studies, ranging from theatre and dance to radio, television and new media. It is divided into five sections:

- **Mapping**, which presents a variety of perspectives on the scope and development of adaptation studies;
- **Historiography**, which investigates the ways in which adaptation engages with – and disrupts – history;
- **Identity**, which considers texts and practices in adaptation as sites of multiple and fluid identity formations;
- **Reception**, which examines the role played by an audience, considering the unpredictable relationships between adaptations and those who experience them;
- **Technology**, which focuses on the effects of ongoing technological advances and shifts on specific adaptations, and on the wider field of adaptation.

An emphasis on adaptation-as-practice establishes methods of investigation that move beyond a purely comparative case study model. *The Routledge Companion to Adaptation* celebrates the complexity and diversity of adaptation studies, mapping the field across genres and disciplines.

Dennis Cutchins is an associate professor of English at Brigham Young University, USA.

Katja Krebs is Senior Lecturer in Theatre and Performance Studies at the University of Bristol, UK.

Eckart Voigts is Professor of English Literature at TU Braunschweig, Germany.

i

THE ROUTLEDGE COMPANION TO ADAPTATION

*Edited by Dennis Cutchins, Katja Krebs
and Eckart Voigts*

Routledge
Taylor & Francis Group

LONDON AND NEW YORK

First published 2018
by Routledge

2 Park Square, Milton Park, Abingdon, Oxfordshire OX14 4RN
52 Vanderbilt Avenue, New York, NY 10017

Routledge is an imprint of the Taylor & Francis Group, an informa business

First issued in paperback 2020

British Library Cataloguing-in-Publication Data
A catalogue record for this book is available from the British Library

Library of Congress Cataloging-in-Publication Data
Names: Cutchins, Dennis R. (Dennis Ray), 1963– editor. | Krebs, Katja, editor. | Voigts-Virchow, Eckart, editor.
Title: The Routledge companion to adaptation / [selected and edited by] Dennis Cutchins, Katja Krebs, and Eckart Voigts.
Description: New York: Routledge, 2017. | Includes index.
Identifiers: LCCN 2017050768| ISBN 9781138915404 (hardback) | ISBN 9781315690254 (ebook)
Subjects: LCSH: Literature–Adaptations–History and criticism. | Film adaptations–History and criticism.
Classification: LCC PN171.A33 R68 2017 | DDC 809–dc23
LC record available at https://lccn.loc.gov/2017050768

ISBN: 978-1-138-91540-4 (hbk)
ISBN: 978-0-367-51781-6 (pbk)

Typeset in Bembo
by Deanta Global Publishing Services, Chennai, India

CONTENTS

CONTRIBUTORS

Anna Blackwell is a lecturer in the Centre for Adaptations, De Montfort University, where she works on the underlying assumptions, associations and cachet that Shakespeare carries in the contemporary moment. Her recent research into Shakespeare and popular digital culture can be seen in *Adaptation, Awards Culture and the Value of Prestige* (Palgrave 2017), the forthcoming *Broadcast Your Shakespeare* (Arden) and her first monograph, *Digital Shakespeareans* (Palgrave).

Shannon Brownlee teaches in the Cinema and Media Studies program at Dalhousie University in Halifax, Canada. In addition to adaptation theory, her research interests include film theory, animated film, and experimental queer and feminist cinemas.

Johan Callens teaches at the Vrije Universiteit Brussel and has published widely on American drama and performance. Essays of his have appeared in, among others, *Text and Performance Quarterly, Theatre Research International, The Journal for Dramatic Theory and Criticism, Modern Drama, The Drama Review, Theatre Journal* and *PAJ: A Journal of Performance & Art*. He recently coedited *Dramaturgies in the New Millennium: Relationality, Performativity, and Potentiality* (Narr, 2014).

Sarah Cardwell is Honorary Fellow in the School of Arts at the University of Kent. She is the author of *Adaptation Revisited* (Manchester University Press 2002), *Andrew Davies* (Manchester University Press 2005) and numerous articles on film and television aesthetics, literary adaptation, contemporary British literature, and British cinema and TV. She is founding co-editor of *The Television Series* (Manchester University Press), and on the editorial boards of *Critical Studies in Television* and the book series *Adaptation and Visual Culture* (Palgrave Macmillan).

Patrick Cattrysse teaches narrative studies and adaptation studies at the Universiteit Antwerpen, Screenwriting Studies at the Université Libre de Bruxelles, and Media Theory and Criticism and Intercultural Communication at Boston's Emerson College European Center. He has published internationally on adaptation and screenwriting studies. He is the author of *Descriptive Adaptation Studies: Epistemological and Methodological Issues* (Garant Publishers, 2014) and the co-editor of *Transcultural Screenwriting: Telling Stories for a Global World* (Cambridge Scholars Publishing, 2017).

FIGURES

Contents

Bernadette Cochrane is a lecturer in drama at the University of Queensland, and her publications include *New Dramaturgy: International Perspectives on Theory and Practice* (co-edited with Katalin Trencsényi, Methuen Drama, 2014) and "Screening from the Met, the NT, or the House: What changes with the live relay" (*Theatre to Screen*, special issue on Adaptation, July 2014, with Frances Bonner). She is currently working on the monograph, *Screened live: A new paradigm for theatrical performance*.

Malcolm Cook is a lecturer in film at the University of Southampton. He has published a number of chapters and articles on animation, early cinema, and their inter-medial relationships. His book *Early British Animation* will be published by Palgrave Macmillan in 2018. His forthcoming work includes research into the use of music in Len Lye's British films, the place of singalong films in early cinema and an edited collection (with Kirsten Thompson) on animated advertising.

Dennis Cutchins is an associate professor of English at Brigham Young University where he regularly teaches courses in adaptation and American literature. In 2000 he won the Carl Bode award for the best article published in the *Journal of American Culture*, and in 2004 received the Charles Redd Center's Butler Award. He is currently working on ways to apply cognitive brain research to adaptation studies.

Pamela Demory is on the faculty of the University of California, Davis, where she teaches writing in film studies, queer cinema, and advanced composition. She has published articles on film adaptation and on queer cinema in *The Journal of Popular Culture*, *Literature/Film Quarterly*, and *New Directions in Adaptation Studies* (2010). She is also the co-editor of *Queer Love in Film and Television* (Palgrave Macmillan, 2013), and is currently editing a collection of critical essays on queer adaptation.

Kamilla Elliott is Professor of Literature and Media in the Department of English and Creative Writing at Lancaster University. Her principal teaching and research interests lie in British literature of the long nineteenth century and literature's relations with other media generally. Author of *Rethinking the Novel/Film Debate* (Cambridge University Press, 2003) and *Portraiture and British Gothic Fiction: The Rise of Picture Identification, 1764–1835* (John Hopkins University Press, 2012), she is currently working on sequels to both: *Rethinking the Adaptation/Theorization Debate* and *British Literature and the Rise of Picture Identification, 1836–1918*.

Rainer Emig is Chair of English Literature and Culture at Johannes Gutenberg University, Mainz, Germany. Among his publications are *Modernism in Poetry* (1995) and *W.H. Auden* (1999) as well as edited collections on *Gender ↔ Religion* (with Sabine Demel, 2008), *Hybrid Humour* (with Graeme Dunphy, 2010), *Performing Masculinity* (with Antony Rowland, 2010), *Commodifying (Post-) Colonialism* (with Oliver Lindner, 2010) and *Treasure in Literature and Culture* (2013).

Riccardo Fassone is a post-doctoral researcher at the University of Torino, Italy. He is the author of two monographs – *Every Game is an Island* (Bloomsbury, 2017) and *Cinema e videogiochi* (Carocci, 2017) – and of a number of articles and chapters in books. His main research interests are the history of the Italian video game industry and the relation between games and other media.

André Gaudreault is a full professor at the Université de Montréal, Canada Research Chair in Cinema and Media Studies, director of the Canadian section of the TECHNÈS international research partnership and the GRAFICS (Research Group on the Emergence and Development of Cinematic and Theatrical Institutions). He cofounded the Observatory of Cinema in Quebec. His publications include many books and more than 120 chapters of books or articles of decisive intellectual importance.

Robert Geal is Lecturer in Film and Television Studies at the University of Wolverhampton. He has published papers on authorial artifice in film adaptation; the historical development of film theory; gender and sexuality in animation; and spectacle in science fiction in journals including *Literature/Film Quarterly*, *New Review of Film and Television Studies*, *Film International*, and *Adaptation*.

Joyce Goggin is Senior Lecturer in Literature at the University of Amsterdam, where she also conducts research in film, TV, and media studies; cultural studies; finance; gambling; computer gaming; and economic history. Her most recent publications include a co-edited volume entitled *The Aesthetics and Affects of Cuteness* (Routledge, 2016), as well as numerous articles, including "Crise et comédie: Le système de John Law au théâtre néerlandais" (*Gagnons sans savoir comment: Représentations du Système de Law du XVIIIe à nos jours*. Ed. Florence Magnot-Ogilvy. Presses Universitaires de Rennes, 2017) and "Trading and Trick-Taking in the Dutch Republic: Pasquin's 'Wind Cards and the South Sea Bubble" (*Playthings in Early Modernity: Party Games, Word Games, Mind Games*. Ed. Alison Levy. Series: *Cultures of Play, 1300–1700*, Western Michigan University Press, 2017). She is currently completing a translation and edition of two of Pieter Langendijk's Dutch theatrical comedies on the market of 1720, and writing essays on gaming, fan labour, and historical card games.

Yvonne Griggs is Lecturer in Media and Communications at the University of New England where she teaches film and television. She specializes in adaptation studies and has published monographs, articles and chapters in edited collections in her specific research area, the most recent being *The Bloomsbury Introduction to Adaptation Studies: Adapting the Canon in Film, TV, Novels and Popular Culture* (2016). Her forthcoming monograph, *Adapting TV: Rewiring the Text* (Palgrave Macmillan), is due for publication in 2018.

Julie Grossman is Professor of English and Communication and Film Studies at Le Moyne College, USA. She is author of *Literature, Film, and Their Hideous Progeny* (Palgrave Macmillan, 2015) and, with Barton Palmer, co-editor of the series *Adaptation and Visual Culture* and the collection *Adaptation in Visual Culture* (Palgrave Macmillan, 2017). Other books include *Rethinking the Femme Fatale in Film Noir* (Palgrave Macmillan, 2009, 2012) and (with Therese Grisham) *Ida Lupino, Director* (Rutgers UP, 2017).

Richard J. Hand is Professor of Media Practice at the University of East Anglia, UK. He is the founding co-editor of the *Journal of Adaptation in Film and Performance*. In addition to adaptation and translation studies, his interests include cross-disciplinarity in performance media, using critical and practical research methodologies. He has published on radio drama, popular horror cultures, Joseph Conrad and Graham Greene. As a practitioner, he has produced stage and radio plays in the UK and US.

Elaine Indrusiak is Professor of English and Literature at Universidade Federal do Rio Grande do Sul, Brazil, where she coordinates the research group *Biopics: Literature in film through historiographic metafiction*. She has co-translated, with Elvio Funck, Christopher Marlowe's *Edward II* into Portuguese (Movimento) and has published articles on adaptation studies, translation studies, transmedia narratology and comparative literature.

Glenn Jellenik is Assistant Professor of English at University of Central Arkansas. His research approaches adaptation from an historical perspective – specifically, long-eighteenth-century adaptation. His recent essay, "The Origins of Adaptation: The Birth of a Simple Abstraction" (*The Oxford Handbook of Adaptation Studies*, Oxford University Press, 2017), traces the rise of current notions of adaptation to the late-eighteenth century. At the centre of all of his work is the desire to complicate our cultural assumptions about adaptation and originality.

Katja Krebs is Senior Lecturer in Theatre and Performance Studies, University of Bristol, and has written, edited and co-edited numerous books and articles on adaptation, translation and performance. She was also founding co-editor of the *Journal of Adaptation in Theatre and Film* from 2007 to 2015.

Keith Lawrence is an associate professor of English at Brigham Young University, where he teaches courses in literature and film, American literature, and world literatures. An adaptation studies enthusiast who researches Asian American literature on the side, Keith is lead editor of *Recovered Legacies* (Temple University Press, 2005) and the upcoming *Asian Images* (Ironweed, 2018). His current project considers the works of Maxine Hong Kingston, David Henry Hwang, and Amy Tan as trans-Pacific adaptations.

Thomas Leitch is Professor of English at the University of Delaware. His most recent books are *Wikipedia U: Knowledge, Authority, and Liberal Education in the Digital Age* (Johns Hopkins University Press, 2014) and *The Oxford Handbook of Adaptation Studies* (Oxford University Press, 2017). He is currently working on *The History of American Literature on Film*.

Claire McCarthy is a PhD candidate at the School of Humanities, University of Tasmania, with an Elite Research Scholarship. Claire's research is in the area of adaptation, cultural and gender studies, with a focus on immigration, popular culture and the adaptation of history. Claire's work has appeared in the *Journal of Ecocriticism*, and she has presented at the Adaptation Studies Conference, Oxford, and at the Menzies Centre for Australian Studies, London.

Brian McFarlane is Adjunct Professor at Swinburne University and Adjunct Associate Professor at Monash University, Melbourne. He is the author/editor of over twenty books and of hundreds of articles and reviews of film and literature, and of their interconnection. He is the editor, compiler and chief author of *The Encyclopedia of British Film* (Manchester University Press, 2014); his last book was *Twenty British Films: A Guided Tour* (Manchester University Press, 2015); and his next is *Making a Meal of It: Writing about Film*.

Philippe Marion is a professor at the Université Catholique de Louvain. His research focuses on the fields of media narratology and the comparative analysis of media and media discourses. He is a founding member of the Media and Journalism Observatory (ORM) and the School of Journalism at Louvain. His publications focus on the study of narratives in images, on contemporary media culture and genres, and the genealogy of media.

Kathryn Meeks is an English MA student at Brigham Young University with an emphasis in rhetoric and American literature. She currently teaches persuasive writing and a specialized writing course for the BYU Honors Program. When Kathryn is not studying, writing, or preparing class, she is most often enjoying the world of adaptation in whatever medium it presents itself.

Kyle Meikle is Assistant Professor of English and Communication at the University of Baltimore. His essays have appeared in *Adaptation*, the *Journal of Adaptation in Film & Performance*, and *Literature/Film Quarterly*. He is currently working on a book about adaptations in the franchise era.

Carol Poole has taken up the role of Edge Hill University Ambassador to Media City (Salford, Manchester), following 15 years as Head of the Media Department. She now represents the university in aspects of the creative and digital industries in the northwest of England. Her research focuses on the production aspects of text to screen adaptations and the role of canonic cultural text in the economy of adaptations.

Ana Iris Ramgrab has received a degree in journalism from Pontifícia Universidade Católica do Rio Grande do Sul in 2002 and a Masters in Literature in English at Universidade Federal do Rio Grande do Sul in 2013. Her independent research interests include, but are not restricted to, adaptation studies, fan studies, and the works of, or based on, Jane Austen.

Amanda Ruud is Provost's Fellow and Ph.D. candidate in English and Visual Studies at the University of Southern California. Her work examines the intersection of poetics, performance, and visual culture in early modern England, and her dissertation, "Shakespeare's Speaking Pictures," focuses on the ethical import of the silences, pauses, and tableaux that punctuate Shakespearean theatre. Amanda's ongoing research extends to the afterlife of those moments in various forms from the photograph to silent film.

Josh Sabey is a film director for goingsane.org and boldrush.org. He won the Award of Merit at the 2017 Impact Doc Awards for his documentary *Going Sane*. He writes and produces films with his wife in Raleigh, North Carolina, and publishes online at *Heterodox Academy*, *The Office for the Study of Christian Values in Literature*, *The Federalist*, and *Pysch Central*.

Gregory Semenza is Professor of English at the University of Connecticut and has published *The English Renaissance in Popular Culture* (Palgrave Macmillan, 2010), *Milton in Popular Culture* (Palgrave Macmillan, 2006), and three other non-adaptation-related books. He recently edited "Shakespeare and the Auteurs" for *Shakespeare Bulletin* (2016). His most recent book, *The History of British Literature on Film: 1895–2015* (with Bob Hasenfratz; Bloomsbury, 2016), is the first volume in the "Bloomsbury History of World Literatures on Film" series, which he is editing.

Max Sexton is a lecturer at the University of Surrey and an independent scholar. After a career in British television, he has published on the links between aesthetics and technology in *Adapting Science Fiction Television: Small Screen, Expanded Universe* (Rowman & Littlefield, 2015) and the de-stabilization of genre, and he has engaged in debates around live performance and special effects. His book *Secular Magic and the Moving Image* (Bloomsbury, 2017) is about the performance, presentation and representation of magic and illusion in film and television.

Suzanne Speidel is Senior Lecturer in Film Studies at Sheffield Hallam University. Her publications include articles and chapters in *The Journal of Adaptation on Film and Performance*; *British Rural Landscapes on Film* (Manchester University Press, 2016); *Introduction to Film Studies* (Routledge, 2011); and *The X-Files and Literature* (Cambridge Scholars Press, 2007). She is currently completing a monograph on adaptations of the novels of E. M. Forster.

Bradley Stephens is Senior Lecturer in French Studies at the University of Bristol. He specialises in French literature and its reception from the nineteenth century onwards, and has published widely in this field. His latest book, *"Les Misérables" and Its Afterlives: Between Page, Stage, and Screen* (Routledge, 2015; co-edited with Kathryn M. Grossman), explores new readings of Victor Hugo's popular nineteenth-century novel and its prolific adaptations. He is currently working on a new biography of Hugo.

Dawn Stobbart completed her doctorate at Lancaster University. She has a BA (Hons) in English Literature and an MA in Contemporary Literature, and is currently focusing on how videogames function as a carrier for horror. She has an interest in contemporary literature, especially the way this translates to videogames. She has published on Gothic fiction, narrative studies, and folklore, primarily focusing on how videogames construct narratives for these genres.

Lissette Lopez Szwydky is Assistant Professor of English at the University of Arkansas. Her research and teaching interests include Romantic and Victorian literature and culture, adaptation studies, gender studies, and professional issues in the humanities. She has published articles on nineteenth-century stage adaptations of Three-Fingered Jack, Mary Shelley's *Frankenstein*, Victor Hugo's *Notre-Dame de Paris*, and others. She is currently working on a book project, titled *The Nineteenth-Century Culture Industry: Literature, Adaptation, History*.

Ruxandra Trandafoiu is Associate Professor of Communication at Edge Hill University. She is the author of *Diaspora Online* (Berghahn, 2013) and co-editor of *The Globalization of Musics in Transit: Music Migration and Tourism* (Routledge, 2013) and *Media and Cosmopolitanism* (Peter Lang, 2014). Her research focuses on migration, digital communication and the social and cultural context of adaptations.

Eckart Voigts is Professor of English Literature at TU Braunschweig, Germany. He has written, edited and co-edited numerous books and articles, predominantly on intermediality, adaptation, neo-Victorianism and theatre/drama. More recently, he has explored the interrelationship of franchise adaptation and participatory culture on social media platforms, with a special focus on mashup aesthetics, paratextual engagement, transmedia modes of reception and reader-response criticism.

ACKNOWLEDGEMENTS

The editors want to express their sincere gratitude to Routledge's Kate Edwards, Harriet Affleck and Ben Piggott, and Lisa Keating at Deanta for their tireless support for this project.

A project of this size brings with it many trials and tribulations, which are by far outweighed by the pleasure of bringing together scholars from thirteen different countries across four continents. We are proud of such breadth of perspective this Companion is able to offer and would like to thank each and every contributor for their investment in and engagement with such an endeavour.

All three editors have, of course, a specific list of colleagues, friends and families who supported this project in more than one enduring way. The following names are probably only the tip of the iceberg, without which we wouldn't have been able to bring it all together. Sincere thanks are due to all the anonymous peer-reviewers, whose comments and suggestions on each and every chapter have been instrumental in developing this Companion. Our gratitude extends to Dr Kenton E. Barnes, as well as Jörn Surborg and Ana Atalaia at Braunschweig University, for all their proofreading. Peter Jachimiak and Millie Krebs-Jachimiak deserve heartfelt thanks for their ongoing patience, support and good humour. And thanks to Wade, who sacrificed many fishing trips while this was underway.

INTRODUCTION TO
THE COMPANION

Dennis Cutchins

Although the term 'adaptation studies' is relatively new, scholarly attention to the varied relationships that exist between the arts in different media is not new at all. In fact, the study of adaptation, by other names, has been integral to the development of Western literature. Both Plato and Aristotle wrote extensively about mimesis, or the representation of ideal forms or aspects of nature in art, and it is not difficult to see that artistic work as kind of adaptation. Although most modern scholars would argue with Plato's conception of an ideal world, all would agree that he was deeply interested in the relationships between the world we live in and the artistic representations of that world.

At least as far back as Aristotle and Phaedrus, scholars have written about *ekphrasis*, or the way one art form may be portrayed by another. John Keats' "Ode on a Grecian Urn" is part of a long line of poems that attempt to recreate in words works of graphic arts (even if those graphic arts are imaginary, as was Keats' urn). That sounds simple enough, but imagine an artist creating a painting that somehow represents a dance. She or he would be forced to find a way to illustrate movement in a static medium, along with rhythm and synchronous action. In each case, the artist would have to adapt these dynamic activities to pigments on a canvas. Edgar Degas's 1874 painting, *Stage Rehearsal*, gives some idea of what this might look like, and suggests both the difficulty and the potential beauty of such an attempt. Certainly this tradition of ekphrastic adaptation, as well as the scholarship surrounding it, is part of the prehistory of adaptations and adaptation studies (Figure I.1).

Something like adaptation studies was also present at the inception of the modern era in literary scholarship. Influence studies, popular in the early to mid-twentieth century, is another antecedent of adaptation studies. Often working to expose unacknowledged sources, influence studies typically involves a kind of literary forensic work. John Livingston Lowes' *The Road to Xanadu: A Study in the Ways of the Imagination* (1927) is a good example of this genre. Samuel Taylor Coleridge claimed that "Kubla Khan; or, A Vision in a Dream: A Fragment" was composed one night after waking from an opium-induced dream, but Lowes argues that the composition of the poem was both more painstaking and more deliberate than that. He uses Coleridge's notes and letters to trace literary and historical sources for "Kubla Khan," as well as for "The Rime of the Ancient Mariner." Lowes' painstaking archival work is still impressive, and is indicative of the kind of source-seeking common in early adaptation studies.

Figure I.1   Edgar Degas, *The Rehearsal of the Ballet Onstage*, ca. 1874. H. O. Havemeyer Collection, Gift of Horace Havemeyer, 1929

Although all of these brands of scholarship have largely fallen out of favour, their focus on the complex relationships between texts has not. Adaptation studies continues to grow as a field, likely driven by the explosion of adaptive texts in practically every kind of art. At least a third of the songs on our playlists are 'covers,' or new recordings of older songs, and even ostensibly new songs often 'sample' elements of older songs. More than half of the plays on Broadway or the West End at any given moment are adaptations of films, novels, and television shows. Marvel's film adaptations of comic book storylines have owned the movie box office for the past ten years and are increasingly popular on television and streaming services. But that economic dominance of adaptations is really nothing new. Since the establishment of the Academy Awards™ in 1927, most of the films nominated for 'Best Picture' have been adaptations, and two of the most popular television shows in recent years, *Game of Thrones* and *The Walking Dead* are both based on literary texts. In fact, perhaps the most popular television show on the planet is Simon Cowell's *X-Factor*, which has been adapted in more than 49 different countries. This is, in short, the golden age of adaptation studies. We live in a world of adaptation, and a failure to study that world means we must ignore an increasingly important part of contemporary culture.

George Bluestone's 1957 *Novels into Film* may have been the first acknowledged text in the field, but in 2017 dozens of new books were published on adaptation studies, and new articles appear each month. Three journals are dedicated to the field: the *Literature/Film Quarterly*; the *Journal of Adaptation Studies*; and the *Journal of Adaptation in Film and Performance*. The *Literature/Film Quarterly*, founded in 1973 by Jim Welsh and Tom Erskine, is the oldest of the three and for years focused more or less exclusively on literature to film adaptations. The *Journal of Adaptation Studies*, founded by Deborah Cartmell and Imelda Whelehan in 2008, and the *Journal of Adaptation in Film and Performance*, founded by Richard Hand and Katja Krebs in 2007, have been wider in their scope since their inception. Anyone planning research in the field of adaptation studies would be well-advised to pay attention to these three journals.

Adaptation studies centers itself not on texts, but on the varied relationships that exist between texts. Understandably, the effort to define 'adaptation' is a regular exercise in books dealing with the subject. At least since Bluestone authors have worked to circumscribe the field of what eventually became known as adaptation studies. What exactly, our students often ask, is adaptation? That is a great question. It is clear that a film like *Mary Shelley's Frankenstein* (1994) should be considered an adaptation. By announcing its relationship to an earlier novel, the film more or less demands that viewers consider it in the context of Shelley's book. But it's less clear that we should also treat a film like Spike Jonez's *Her* (2013) as an adaptation of Shelley's novel (or an adaptation of other adaptations of Shelley's novel). All three of these texts explore the dangers of technical advances, and all three portray a human creation that has a mind of its own and eventually escapes its bounds. But abstract summations like that hide a multitude of differences. At most, *Her* is an unacknowledged adaptation, and would thus fall out of the scope of some definitions of adaptation studies. Studying it as an adaptation, however, despite Jonez's intentions (whatever they were) is certainly interesting, and would likely make the field of adaptation studies richer for the effort.

Because of problems like this, we suggest it may be more fruitful to consider adaptation as a way to think about all sorts of texts, rather than as a particular kind of text. A great many items, from Hollywood blockbusters to fan fiction, and from video games to novelizations, might productively be approached using the comparative tools of adaptation studies, and many of those approaches are, indeed, illustrated in this Companion.

The subject matter of this Companion is eclectic and varied. The entries here focus on dance, film, radio, stage plays, literature, television, and new media from a wide range of theoretical approaches. And they are not presented here in terms of their own forms, but instead the Companion offers a structure which allows readers to see adaptation in terms of history, identity, reception, and technology. No matter which section the essays in the Companion belong to, they have at least one thing in common. All of them engage in one way or another with some kind of comparative method. In other words, to study a text as an adaptation is, always, to study at least two texts. Such a position does not, however, imply a hierarchy, be that temporal or structural. This comparative method is altogether fitting in a poststructuralist scholarly landscape. If, as Mikhail Bakhtin and others have suggested, texts only have meaning in relationship to other texts, then attention to the relationships between texts is centrally important. As a student of ours recently pointed out, other approaches look at texts thorough the lens of theories like Marxism or feminism, but adaptation studies looks at texts through the lens of other texts. In that respect, the 'theory' of adaptation studies may be hiding in plain sight. Our dogged insistence that the texts we study can be understood in the light of other texts must, in the end, be considered more than a method. It at least approaches the status of a theory.

The fact that we have titled the collection you now hold a 'Companion' suggests a few things. It implies, for instance, that there are fellow travellers, and that this book might escort you on your journey as well as become an acquaintance and confidant, offering you useful instructions, or at least hints, on how the journey may be approached. We certainly intend that this book be useful, but it turns out that the tasks related to adaptation are not as well-defined as, say, travelling from A to B, and so the maps and signposts and journeys you find in this Companion are more likely to take the form of examples. Nevertheless, we offer here one piece of advice: learn how to think about multiple things at the same time.

This is quite similar to the advice John Keats offered his brothers, if they would avoid being bores. He suggested that they develop the ability to hold multiple ideas, even contradictory ideas, in their minds at the same time. He wrote:

> several things dovetailed in my mind, & at once it struck me, what quality went to
> form a Man of Achievement especially in Literature & which Shakespeare possessed

so enormously—I mean *Negative Capability*, that is when man is capable of being in uncertainties, Mysteries, doubts, without any irritable reaching after fact & reason.

To study texts as adaptations absolutely requires the scholar to hold multiple things in his or her mind at the same time. Comparing a simple fairy tale to a film, for instance, requires the scholar to imagine the tale, or at least big chunks of it, at the same time he or she studies scenes from the film, or visa-versa. And that's just the beginning. The scholar also may be forced to consider the creative roles played by publishers, reviewers, screenwriters, directors, producers, actors, and cinematographers, not to mention the reading and viewing public. Moreover, cultural, social, economic, racial, or historical factors almost certainly affected the production of both texts. And there's always the possibility that other literary or film texts influenced the process at any stage.

That may sound daunting, but it should also feel exciting and inspiring. Adaptational thinking is definitely a skill that can be learned. The more scholars practice it, the better they get. Moreover, adaptation studies encourages unexpected connections. The pages of this Companion, in fact, are full of connections that most of us have missed, despite the fact that in some cases they have been under our noses for years. Perhaps the most inspirational thing we could say is that newcomers to the field, including students, are often quite adept at making those connections.

PART I

Mapping the field

Katja Krebs

> Maps are often an abstraction of the physical world – a symbolic depiction of a space
> or idea that allows one to understand and navigate an unfamiliar topography or com-
> plex topology. But while most conventional charts, plans and diagrams claim to offer
> an accurate, even objective picture of the world, each one is bound by the specific
> agendas of its creators and users.
>
> *Hans Ulbrich Obrist 2014: 11*

This opening section of the *Routledge Companion to Adaptation* offers four different cartographies
of adaptation studies: "cartographies can be altered endlessly to reflect different priorities, hierarchies, experiences, points of view and destinations" (Obrist 2014: 11). We encounter not only
four different points of views but also priorities, experiences and destinations.

Sarah Cardwell draws the first of the four maps in "Pause, rewind, replay: adaptation, intertexuality and (re)defining adaptation studies". Being concerned with the development of adap
tation studies as quintessentially unmappable in its current form, as it has changed from "an
under-acknowledged, narrowly focused field of study into an all-embracing perspective which
eagerly consumes all intertexts in its path", Cardwell tries to establish much needed focus in
order to halt adaptation studies' "sowing the seeds of its own destruction". Her suggestion to hit
replay and engage with the development of a "more precise vocabulary from which new conceptual insights and debate can evolve" is put into the context of an otherwise unmappable field
where "the study of adaptation mutates into the study of intertextuality" and becomes "logically
unlimited". In other words, adaptation studies needs to (re)define itself, decide what to include
in an otherwise unnavigable sea of texts.

Kamilla Elliott's "The theory of BADaptation" points the compass in a different direction
whereby this chapter uses the analysis of existing maps of the field of adaptation studies in order
to uncover their specific agendas and perspectives. Subsequently, Elliott is then able to propose
that we need to "study adaptations to discover ways in which theorization is lacking, allowing
adaptations to challenge and adapt even our most cherished theoretical beliefs". In other words,
the object of the map has to be allowed to take an active role in its direction. Rather than developing "new theory or more theories" Elliott calls for "a major rethinking of the dysfunctional
and oppositional relationship between adaptation and theorization".

In "Adaptation and the concept of the original", Rainer Emig conceptualises one of the pivotal concerns of adaptation studies, i.e. the relationship between adaptation and the original. Taking fidelity criticism as the start of his discussion, Emig demonstrates the inherent contradiction in notions of fidelity which, at specific moments in history, ranged from "precise copying of models" to "conformity to an underlying theological idea" and are inextricably linked to the concept of "the authenticating creator (usually male)" who "supposedly infused an original work with his ideas and skills". However, rather than offering us a neat solution to the inherent contradictions of notions of the original and authorship, Emig's discussion of fidelity criticism as one of the cornerstones of adaptation studies takes us on a journey from the Romantic ideals of the creative genius via Structuralism, Formalism, Poststructuralism and Deconstruction, to a position which, he argues, is in itself not necessarily any less problematic.

Patrick Cattrysse takes us in a very different direction in his discussion of "An evolutionary view of cultural adaptation: Some considerations". Using adaptation studies' close cousin Translation Studies as his starting point, Cattrysse categorises adaptations as comprising both 'ipsative' and 'additive' processes, whereas translation studies scholars have abandoned the ipsative translation in favour of the additive one. While Cattrysse's starting point may be markedly different, one of his main concerns is also fidelity, yet he shifts the focus from fidelity – or, more broadly, the relationship between source and adaptation – to the relationship "between the adaptation and its new target context". His conclusion is a call to a "descriptive–explanatory approach", very much akin to the approach exemplified by Descriptive Translation Studies, which is in a position to disregard "the (inter)-personal taste of the analyst".

Each of these four chapters is as distinct from one another as if they were points on a compass, taking position at different ends of the spectrum. However, without a North there would be no South, and without an East there would be no West. As such, what each chapter in this mapping section does do is offer a redrawing of established territories in order to "challenge the authority with which maps depict the 'truth' and question the very grounds on which" (Obrist 2014: 11) adaptation studies exists.

Works cited

Obrist, Hans Ulbrich (ed.) (2014), *Mapping It Out: An Alternative Atlas of Contemporary Cartographies*, London: Thames and Hudson.

1

PAUSE, REWIND, REPLAY

Adaptation, intertextuality and (re)defining adaptation studies

Sarah Cardwell

'Adaptation studies' today is a massively expanded and proliferating field. From medium-specific versus comparative approaches, to intertextuality and metatextuality, onwards to intermediality and transmediality, twenty-first-century adaptation studies has broadened its original scope from literary/theatrical adaptations on screen to innumerable permutations and degrees of adaptation and related practices. Eclectic interpretations of individual adaptations have burgeoned. Scholars from far beyond the traditional enclaves of literature or film departments contribute to the subject's growth.

Rejecting the prejudices and hierarchies of pre-1980s adaptation studies, embracing and asserting pluralism, today's scholars positively celebrate adaptation as a diverse cultural practice. Our notion of adaptation has expanded ostensibly in response to contemporary, real-world creative developments, exhibiting openness to many forms of 'adaptiveness', but the field's current relativistic pluralism also conforms to dominant scholarly, theoretical trends. The two seem fruitfully to coincide and drive adaptation studies inevitably into the future, rejecting older conceptions of adaptation (based mostly on literature-screen examples) "in favour of an inclusivist conception of adaptation as a freewheeling cultural process: flagrantly transgressing cultural and media hierarchies, wilfully cross-cultural, and more weblike than straightforwardly linear in its creative dynamic" (Murray 2012: 2).

However, this happy concord conceals a gap at the heart of adaptation studies. Today's apparent eclecticism and openness constitute a breadth that belies a lack of depth. There are many oversights in this new age of adaptation studies: topics that lie neglected, questions that remain buried and unanswered, and alternative approaches not yet adopted. These are the unforeseen drawbacks of the particular nature of expansion the field has undergone since the late 1990s.

It is crucial to recognise that fundamentally underpinning the recent transformation of adaptation studies is a radically amended notion of what (an) adaptation is, and a greater recognition of its connectedness with other cultural practices, such as borrowing, remaking, translating and so on (manifested, for instance, as we shall see later, in collections edited by Robert Stam and Alessandro Raengo [2005], Mireia Aragay [2005] and Ulrike H. Mienhoff and Jonathan Smith [2000]). But the concept of adaptation *per se* has necessarily been diluted as it expands to form a catch-all category for these permutations. The relationship and differences between adaptation and related forms of textual interconnectedness are exceedingly complex, and the focused conceptual work required to elaborate them is currently thin on the ground.

Moreover, and correspondingly, the vocabulary adaptation scholars once relied upon appears to be in crisis, its functional and explanatory powers increasingly undermined. Thoughtful challenges to key concepts which once served to define and delimit adaptation (e.g. original/source/ur-text, interpretation/translation) too often lead to the words being summarily cast aside; alternative, looser and more fashionable terms (e.g. meta- and inter-textuality, and inter- and trans-mediality) proliferate; potentially valuable distinctions between adaptation and related practices become blurred; and adaptation becomes harder and harder to pin down. Indeed, adaptation studies has mutated from an under-acknowledged, narrowly focused field of study into an all-embracing perspective which eagerly consumes all intertexts in its path – and which, by doing so, may be sowing the seeds of its own destruction. Now is the moment to pause, rewind a little and reflect upon the fascinating course of adaptation studies, and its present trajectory and purposes, with the hope of ensuring its future.

Golf studies knows its focus: golf, not tennis. An enterprising scholar of the subject might provocatively conceive a tennis match as golf, but such enterprises would not challenge the core remit of the field. Adaptation studies is different, for every adaptation is always something else too: a film, a TV programme, a book, a play. Therefore, historically, the purpose of adaptation studies was to study an adaptation *as an adaptation*. This worked well. It meant that (for example) a film could be studied as a film, from a range of film studies' perspectives, or as an adaptation, with similarly varied emphases. To approach the film as an adaptation meant prioritising its connections with its source text; to examine it as a film did not. There was something distinctive about an adaptation studies approach: it acknowledged the curious "equivocatory use of the word adaptation, a result of the homonymic verbal conflation of (the process of) adaptation and (the end product) adaptation" (Cardwell 2002: 11). This dual definition of adaptation shaped our field, inspiring and guiding those who would seek out and examine its instances, as well as delimiting their focus.

Adaptation studies today, in contrast, could be characterised as the study not only of adaptations-as-adaptations but also of a wide range of texts-as-adaptations. Scholars pull more and more creative works into the field, some of which are not adaptations – or even texts – in any established sense; a recent book on adaptation includes "remakes, video games, biopics, fan fiction and celebrity culture" (Carroll 2009: cover), for example. This can and does generate exciting new interpretations of individual works and specific practices, but it has consequences for the conceptual coherence of our discipline. The adaptation studies scholar no longer requires an adaptation, but instead needs only take the appropriate attitude to the work under scrutiny. There is an ontological fissure. The performance of adaptation studies has come unfixed from a notion of what (an) adaptation is. Adaptations are no longer logically necessary for adaptation studies. How did this happen?

Intertextuality and the transformation of adaptation studies

Adaptation studies was transformed by the inexorable rise of intertextuality. Of course, intertextuality has always been attended to, whether under that label or not, as the study of textual influence and referentiality, for the practice is ubiquitous and inevitable in art. But in the late 1990s and early 2000s, the striking, special links between intertextuality and adaptation were brought to centre stage.

In 1996, in his important monograph on comparative studies of adaptation, Brian McFarlane argued that "[m]odern critical notions of intertextuality represent a more sophisticated approach [than fidelity criticism], in relation to adaptation" (1996: 10); he reiterated that the source text is only one of an adaptation's intertexts, and not always the most salient. In 2002, Sarah Cardwell argued even more forcefully for the importance of intertextual criticism in her study of the sub-genre of television classic-novel adaptations, emphasising that an adaptation's "intertextual references […] may prove to be even more relevant to [its] meanings and effects than its novelistic source text. A careful and responsive analysis needs to recognise such intertextual nuances in order to comprehend fully the film or programme and all its shades of meaning" (2002: 67). Both McFarlane and Cardwell were concerned with how an awareness of intertextuality could enhance one's appreciation, understanding and interpretations of particular adapted texts. Intertextuality augmented and enhanced an adaptation, and was part of its artistic integrity. Therefore, a recognition of intertextuality, they argued, was crucial to fair, sensitive criticism and interpretation.

But in 2005, intertextuality *per se* gained traction and momentum in the field when Robert Stam placed it at the very centre of his project on adaptation (Stam and Raengo 2005). Stam's work was strongly influenced by cultural studies and its fascination with 1970s literary and critical theory (often characterised, especially by its critics, as Theory with a capital T), which was in turn inspired by continental philosophy.[1] Stam expanded upon and emphasised the broader scholarly context of intertextuality as a theoretical construct, correspondingly nudging adaptation studies further away from literary/film/television studies and embedding it more deeply within cultural studies. The power and colonial expansiveness of this influx of 'Theory' precipitated conceptual and methodological disruption at the heart of adaptation studies. Indeed, intertextuality began to threaten the previously distinct and coherent identity of adaptation studies itself, as it challenged the very existence of the category of adaptation.

As Stam acknowledged (2005: 45), following Barthes, intertextuality is "the condition of any text whatsoever" (1981: 39). Consequently, as the study of adaptation mutates into the study of intertextuality, the range of texts available for our attention increases dramatically – indeed, it is logically unlimited. It is true that intertextuality is a crucial defining feature of adaptation, but it has become hard to see what adaptation is but a special instance of intertextuality. And such a characterisation is uncomfortable, for it contradicts the ideological drive in adaptation studies away from special – or privileged – texts and the hierarchy they imply.

Perhaps this is why many adaptation scholars find it simpler to bury or dismiss questions of what (an) adaptation is; how adaptation differs from intertexuality; and how one might delineate the field of adaptation studies, enjoying instead the supposedly revolutionary impact of reconceiving adaptation studies as intertextual studies, "plac[ing] adaptation as part of the larger phenomenon of rewriting and of a theory of intertextuality" (Aragay 2005: 203). The implication is that intertextuality has made adaptation studies redundant – that we may as well allow it to be subsumed within cultural studies (which is where it seems currently to sit most comfortably), completing the movement away from those awkwardly medium-specific (or, more accurately, art form-specific) collegiate fields of literary, film, television and new media studies. So, is the future of adaptation studies actually intertextual studies? Or would it be reasonable – and valuable – to attempt to demarcate and retain *adaptation studies* as a distinct field? The answer to the latter, I would suggest, is yes.

The continued existence and vitality of the specific field of adaptation studies depends upon establishing a persuasive, if soft-edged, distinction between intertextuality and adaptation. This would conceptually underpin separate and complementary fields of intertextual and adaptation studies, and support a breadth of approaches to adapted texts. Interestingly, a little attention to common language and dialogue reveals that creators and consumers of adaptations outside the academy already instinctively perceive some rough-and-ready delineation. Whilst adaptation studies embraces intertextual equality, rejects old hierarchical distinctions and seeks to blur the boundaries between adaptations and related forms, downplaying the possibility of a specific, discernible category of adaptation, the sheer variety of common language (adaptation, remake, appropriation, copy, spin-off, spoof, parody, etc.) discloses ongoing, obstinate attempts to draw, negotiate and maintain distinctions between something one might term adaptation proper and other kinds of intertextuality.

Moreover, whilst wide-ranging intertextual webs are one key source of pleasure, it is also the case that creators and audiences of adaptation continue to seek joy in the specific and privileged connectedness between one text and another, in reliving and repeating the memory of one cherished work via a new work (cf. Ellis 1982). People on both sides of the creative process continue to intuit a distinction between the pleasures of intertextuality and the pleasures of adaptation – as if the two were distinguishable. Scholarly study could reconnect with pre-theoretical, intuitive perceptions, pleasures and dialogue in pursuit of clearer conceptual answers to the questions above. But first we must recognise the extent of the impact of intertextuality upon adaptation studies – how it has changed not only our notion of adaptation, but also the very way in which adaptation studies is conducted – before engaging in a paradigm shift.

Language, metacriticism and conceptual analysis

Intertextuality as conceived in cultural studies has shaped today's adaptation studies; in return, "adaptation studies feeds cultural studies" (Andrew 2011: 28). Such a powerful relationship is sustained by the repeated reassertion of core principles and language, which inevitably act as a (self-)conservative force. Since the 1990s, the "tide of theory" (Cartmell 2010: 11) within adaptation studies has been dominated by giants of continental philosophy and post-structuralist theory, including those who conceived and championed various versions of intertextuality: Saussure, Genette, Bakhtin, Barthes, Kristeva. And adaptation studies of course sits within a wider context: the lingering domination of 1970s' 'Theory' across many humanities subjects. Given this, it is not surprising that whilst adaptation studies has not wanted for theory, *conceptual* work is still extraordinarily thin on the ground.

Few recent scholars distinguish clearly between the two practices of theorising and conceptualising. Kamilla Elliott's contributions constitute particularly notable exceptions. Her initial work on novel-to-film adaptation (2003) explored existing conceptions of adaptation in an attempt to elucidate common ways in which the category is formulated. Exhibiting an uncommonly keen awareness of the difference between metaphor and model, conceptualisation and theory, Elliott uses language precisely and sensitively, as a tool for metacritical analysis. In her (2013) perceptive analysis of theory within adaptation studies and the shortcomings and blind spots in the field, she displays characteristic openness to alternative approaches drawn from subject areas currently less popular (including cognitive science), and suggests that theories must adapt to (our study of) adaptations, rather than the other way around. Elliott proffers the vision of rejuvenated, more rigorous and more varied adaptation studies.

Generally, though, adaptation scholars manifest the current predilection in UK/European humanities for theory over conceptual analysis. Interestingly, adaptation studies in the US exhibits a wider range of approaches and emphases, incorporating more traditional comparative work on literature–screen adaptation alongside the work of scholars such as Stam. The different perspectives exhibited by the key British journal *Adaptation* and long-established US journal *Literature Film Quarterly*, for example, are indicative. Yet even in the US, where greater methodological amenability is apparent, conceptual analysis is thin on the ground.

That is not to deny an enduring interest from adaptation scholars in the attempt to formulate more clearly the terms we use to define and discuss adaptation – which is, at heart, an interest in conceptualisation. Unfortunately, currently dominant theoretical paradigms inadvertently undercut such good intentions. Attempts to tackle the giant of intertextuality by defining types and subtypes are liable simultaneously to generate yet more technical vocabulary. Gérard Genette's important work (1982) promised to provide the tools for detailed analyses of varying intertextual relationships between a text and its source(s), but also left scholars to navigate and deploy not only the intertextual but also the transtextual, incorporating the intertextual as well as the paratextual, metatextual, architextual and hypertextual. Patrick Cattrysse's recent (2014), laudable attempt to distinguish adaptation, intertextuality and intermediality, though eschewing jargon or wordplay, nevertheless resulted in a terminologically dense and abstract account. Thus, rapid and ongoing changes in those key concepts which should help us define and delimit adaptation – the original/source/ur-text, interpretation/translation, meta- and inter- textuality and now inter- and trans- mediality – rather than clarifying distinctions or settling helpful vocabulary for debate, tend instead to end up providing scholars with a "candy store of available approaches" (Westbrook 2010: 43), dissuading them from spending extended time subjecting each concept to rigorous critical analysis, or sorting the wheat from the chaff.

The idea and study of intertextuality are valuable and here to stay. But there is space also for adaptation – and for adaptation studies. Adaptation scholars are well placed to return to the project of defining adaptation for this new era, cognisant of the unexpected challenge posed by intertextuality, but working hopefully towards making fine discriminations between closely related but distinct creative practices, works and concepts. This requires conceptual spadework, which has been sorely neglected: a focus not on 'Theory' or theories, but on concepts and language. Some of those knotty, awkward, unfashionable terms which once defined our field, and which were hastily discarded in the early twenty-first century, are worthy of renewed attention. If we subject them to rigorous critical analysis, we may find ourselves working towards a deeper understanding of adaptation.

One way to rebalance the emphasis, and rejuvenate focus on conceptual work, is to broaden our scope and look to new possibilities and influences. Adaptation studies has maintained strong connections with cultural studies, and has similarly preferred theories and approaches inspired by the continental philosophical tradition. Yet the Anglo-American analytic tradition of philosophical aesthetics[2] offers us an alternative route, focusing on conceptual – including ontological and epistemological – questions, not by expanding our terminological vocabulary still further but by returning to and drilling down into common, pre-theoretical, shared language. It offers the opportunity for metacriticism via a reflection upon the tools of our trade: our words.

We could seek a working vocabulary to delimit and explore adaptation from a conceptual point of view, simultaneously valuing the process, the journey, the discussions, rather than expecting or hoping that we will ultimately concur on a fixed set of terms. An agreed-upon definition of adaptation is neither a realistic aim nor necessary to justify a delimited field of adaptation studies. just as analytic aestheticians have never settled on an unequivocal definition of their primary focus: art. And yet, analytic scholars never abandon the question 'what is art?' – incorporating, necessarily, 'what is *not* art?' (note that both questions are categorical, not evaluative, in tone). Instead, they recognise that it is a vital part of aesthetics to attempt, constantly, to define art, leaving the category flexible, fluid, debated, contentious but nevertheless central (see Carroll 2000). In the same way, keeping sight of the core question "What is (an) adaptation?" (Cardwell 2002: 9–30) – and grasping that some closely related, apparently homologous practices/texts are *not* adaptation(s) – is key to the future of adaptation studies.

(Re)defining adaptation distinct from intertextuality – some tentative demarcations

The query "What is (an) adaptation?", which pertains to the very definition and delineation of our field, lies neglected, and yet is more pressing now than ever, when much recent work blurs the boundaries between adaptation and intertextuality, or allows the latter to subsume or supplant the former. Some scholars have pinpointed this dilemma: as Nico Dicecco encapsulates, "[o]n what conceptual grounds do adaptation scholars draw the distinction between adaptation and other kinds of intertextuality?" (2015: 161). Most notably, Thomas Leitch has pursued relentlessly the need to distinguish intertextuality and adaptation (2003, 2008), but although he often poses the question, he too finds answering it a challenge. He has attempted to posit a genre of adaptation via textual markers, though the resulting category is not epistemologically sound (for example, he acknowledges that audiences might misread a film as being an adaptation).

Importantly, Leitch takes a non-relativistic approach to an adaptation's intertextuality, prioritising textual features over "extra textual signalling" (2008: 108), and noting that the nature and extent of a film's intertextuality determine whether the film constitutes a special instance of adaptation: "comparisons that are discretionary in all texts, because they are all intertexts, become foundational to the extent that any audience experiences an adaptation as an adaptation" (2008: 117). There are the beginnings of a valid and helpful distinction here between intertextuality and adaptation.

Perhaps the first distinction to make is that *intertextuality is a necessary but not sufficient condition for adaptation.* That is, an adaptation is necessarily intertextual, but a text's being intertextual does not mean that it is necessarily an adaptation. Intertextuality is the larger category which includes the smaller subset, or special case, of adaptation.

This requires a little disambiguation, however. Adaptation is employed in a slippery fashion here, in both its nominative (noun) and verbal (verb) form, to refer to both process and end-product. Intertextuality, on the other hand, seems to refer to a state or quality of a text. So the creator of a television adaptation, for example, undertakes adaptation-as-process to form an adaptation-as-product, one inevitable quality of which is intertextuality. The major part of the adaptation's intertextuality arises because of the intentional actions of its creator, and because those actions were in pursuit of adaptation. Other intertextual qualities in the adaptation may be intended (e.g. generic references, situating the work in the context of a TV genre) or unintended by the creator (e.g. qualities that evoke other works, despite the creator's unawareness of this).

Thus, we affirm that indeed all texts are intertextual. But we also see clearly that we can distinguish conceptually between at least three common types of intertextuality within adaptation (as above) – and that one of those, that is, elements arising intentionally and in pursuit of the practice of adaptation, is unique to adaptation. The adaptive element of intertextuality is also temporally distinctive: whilst an intertextual connection can exist between contemporaneous, earlier or later texts in any combination, adaptation requires the existence of a prior text, and cannot precede its source.

We also see that for both creator and viewer, adaptation exists in two senses: literally, in terms of an adaptation-as-end-product, and also implicitly, in the sense that, generally speaking, both perceive that adaptation-as-process has occurred in order for the end-product to exist. Yet intertextuality exhibits different qualities: it is a state of being, rather than a process/product, to which creator and viewer bear different relationships. The creator inevitably contributes to the text's intertextuality, but in the case of adaptation, his/her primary concern is to adapt. The viewer perceives intertextuality to varying degrees, dependent upon his/her skills, knowledge and enthusiasm. Whilst adaptation is a creative process and an end-product which are central to the creator's practice and perceived by the viewer, intertextuality, on the other hand, is a textually based feature to which both creator and viewer contribute on a more equal footing. The viewer can augment the text's intertextuality, but not its adaptive qualities (i.e. its qualities as an adaptation), for the viewer does not undertake the creative process of adaptation.

Perhaps this tells us why adaptation is losing ground to intertextuality. While the first distinction we found was categorical, it leads us to a second, which is ontological. *Adaptation and intertextuality are ontologically distinct.* They come into being via different means. As John Frow astutely summarises, "[t]he identification of an intertext is an act of interpretation. The intertext is not a real and causative source but a theoretical construct formed by and serving the purposes of a reading" (1990: 46). Whilst Frow considers the perspective of intertextual criticism to be transformative, he wisely notes that there are ontological implications. The study of intertextuality is the study of a world of interconnected texts undertaken by an all-important consumer. Adaptation studies as originally conceived is, in contrast, an exploration of consciously referential, creative practice and the works arising from it. This latter perspective challenges the current hegemony which places the all-knowing consumer in the most powerful position of determining the value and meaning of texts.

And yet, the brief conceptual analysis above clearly validates the option of taking at least two distinct perspectives on an adapted text. It is a legitimate and conceptually grounded position to shift the balance back towards attention to integrous, individual works and the creative ventures that bring them into being – to undertake a conscious act of humility and place another's artistic practice and achievements ahead of our attempts to own the work ourselves. This is not to kowtow to an old hierarchy in which the reader attempts to uncover the artist's genius, or assumes that a text's meanings are all the result of the creator's intentions. It is instead an act of recognition and of choice: we recognise what it means to call a text an adaptation, and choose to let our interpretation be guided by the creator's confident embrace of that term.

Adaptation can also be regarded as the purposeful 're-fitting' of material from one artistic context to another. But can we specify more closely what is meant by artistic context? It is interesting that the focus on intertextuality has been lately supplemented with attention to intermediality and transmediality (see Bruhn *et al.*, 2013). This revealing development acknowledges one of the original insights and specific strengths of adaptation studies: an appreciation of the complexities of movement from one medium (art form) to another. Renewed emphasis on mediality is promising for our conceptual project of delimiting adaptation studies, suggesting a further means by which we may be able to distinguish adaptation – from one medium (art form) to another – from similar practices such as translation and remaking, which take place within one medium (art form).

Of course, distinctions between media/art forms are notoriously tricky, and today's scholars are anxious not to be suspected of harbouring medium-specific tendencies; thus, adaptation studies continues to neglect enduring questions of medium specificity which are actually more, not less, important in today's multi-media, intermedial context. The consequent dearth of conceptual grounding within this area only increases unease. The vagueness with which the term 'medium' is employed across the humanities exacerbates the problem. In most cases, when adaptation is cursorily described in terms of the movement from one medium to another, what is really meant by medium is art form (see Cardwell 2014). The simplistic conflation and confusion of these two categories has profound implications for conceptual exploration, and inhibits the development of more precise conceptual, ontological and technical delineations.

In Bruhn *et al.*'s (2013) engaging collection of essays on adaptation and intermediality studies, for example, several contributors acknowledge the need for greater precision in our exploration of adaptation as an intermedial practice. Regina Schrober raises the "problem of transmediality" (2013: 91) which fails to recognise medial boundaries and assumes that attributes can be genuinely shared across different media (*sic*). She notes that in fact even simple, apparently non-medium-specific elements such as rhythm and sound are manifested and contextualised so differently within different media (*sic*) that they cannot be considered to be elementarily identical. Her broad argument is persuasive, but the absence of art-level categories (including art forms, works and practices) necessitates her adoption of the terms 'medium' to refer to large-scale artistic forms, such as poetry and music, and 'modalities' for smaller, non-medium-specific, ostensibly transmedial elements. Again, vocabulary is multiplied without a proportionate increase in conceptual clarity.

In the same volume, Lars Ellestrøm recognises the central problem: the distinction between art forms and media is not properly understood within adaptation studies (2013: 113). He avers that adaptation scholars should therefore take up existing terminology available within intermedial studies. Unfortunately, his essay suggests that intermedial studies has not yet clearly established its own viable working definitions. Ellestrøm offers the example that a book page is a technical medium that can mediate various kinds of media content, such as poems or drawings. Whilst this seems successfully to distinguish medium from art form, it begs its own questions: What is included in this technical medium, and how can it be distinguished from other, very similar media? How is a book page a different medium from a piece of paper? Would a different kind of paper, ink or font constitute a change of (technical) medium? How can we divide media content from the technical medium which mediates it? What about in the case of a painting: what is the media content here? Is it the image – and if so, how can this be reasonably separated from its presentation in paint (does it exist without its mediator)?

Throughout intermedial studies such difficulties persist. Media refers sometimes to art practices (e.g. literature, film); sometimes to component traditions and forms (poetry, documentary); sometimes to small constituent elements (the written word, the image); sometimes to variations either cultural (different languages, genres) or technical (fonts and inks, digital film or 35mm). How exactly is a medium to be conceptualised? These problems are fascinating and extraordinarily complex. Much more metacritical and analytical work is needed to explore, as a matter of priority, a rudimentary demarcation of medium from art form. Intermedial and cognitive studies have been proposed as routes to clearer and more precise categorisation and conceptualisation, but there is a more obvious and potentially helpful alternative in analytic philosophical aesthetics and art history, whose very focus has always been these kinds of questions as they pertain to all arts. The work of analytic aesthetician Noël Carroll (see 1996) offers a model for the exploration of medium specificity from a sceptical perspective; his ideas are being developed from an alternative perspective within film and television aesthetics in ways which are readily applicable to adaptation studies (see Cardwell 2014).[3]

Whilst it can be argued that all media products are open to intermedial studies (Ellestrøm 2013: 115), the same distinctions apply as those made above regarding intertextuality. Intermediality is a necessary but not sufficient condition for adaptation, that is, all adaptation exhibits intermediality, but not all instances of intermediality constitute adaptation(s). Adaptation exists as a special instance of consciously intended and temporally specific intermediality. Adaptation is a special case of both intertextuality and intermediality.

Pause, rewind, replay: an argument for adaptation studies

Adaptation studies appears to be flourishing. International academic interest in the field is lively. Yet, although it is tempting to press ever onwards in the current path, it is often valuable briefly to pause and reflect.

The boundaries between adaptation and intertextual (and more recently, intermedial) studies are often elided. Yet by definition, intertextual studies are virtually limitless in terms of their remit: they are, for their adherents, merely the study of art today, encompassing everything from literary works to television to film to theatre and beyond. Within that broader field, adaptation is rarely acknowledged as a specific, differentiated practice (see Meinhof and Smith, 2000).

Intertextuality is most helpful when regarded as a tool, a perspective, not an overarching ideology. It is not a replacement for or competitor with adaptation studies; as we have seen, adaptation is a sub-category of intertextuality. The redefinition and reassertion of adaptation studies as a distinct field is not only plausible but also valuable. It does not depend upon establishing a concrete taxonomy of texts and specifying what an adaptation is, once and for all. Instead it requires an acknowledgment that there exists within common and creative understanding a certain category called adaptations. The texts included therein will change and will be contended, but in weighing a text to ascertain its inclusion or not, we undertake a task that matters not only to the delineation of our discipline, but also to creative artists, their works and their audiences.

Often, our currently preferred model of adaptation represents not so much facts about the creative works before us, but rather a critical route that we choose to take into those works. If our preferred critical/theoretical perspective is an eager embrace of interconnectedness, we are likely to suppress awkward matters of difference and incompatibility; the task of discriminating clearly between kinds and degrees of adaptation (and non-adaptation); the disambiguation of terms; and conceptual and metacritical debates which might enable – or encourage – us to change direction.

Today, we find ourselves in a strange position in which the very subject of our attention – adaptation – is feared by some in the field to be "inherently conservative" (Sanders 2006: 9). Intertextuality feels a more comfortable incarnation, delivering a somewhat misleading impression of breadth and openness while actually dominating and constraining our remit. Other influences, other foci, other perspectives are too often relegated, or dismissed as out of date, as the reaction to Colin MacCabe *et al.*'s 2011 provocative, contrapuntal collection on fidelity criticism revealed (see Voigts and Nicklas 2013). And yet, *rewinding* can reveal rich possibilities, as Timothy Corrigan (2010) notes: "evolutionary progress can also be a return to positions that we may have archived too quickly." Many important questions raised by writers from the 1930s to the 1970s, such as those around medium specificity, remain unanswered; our moving on from a debate does not mean it is thereby settled.

Adaptations have always been deployed to fight ideological battles (Cardwell 2002. 69), this requires that we remain alert. Whilst political sensitivity will always be important in the study of the arts, ideological concerns can sometimes hamper the free thought and debate necessary for progress and development. As Elliott sagely observes, adaptation studies "is an especially divided field whose polarizations have perhaps precluded new theories from developing" (2013: 23), and we must have the confidence to "challenge theoretical as well as textual canonicity" (2013: 33). Adaptation studies would benefit from greater open-mindedness to new influences such as Anglo-American analytic aesthetics, or cognitive studies, just as our field is enriched by recent scholarship which exploits the correspondences and crossovers between adaptation and translation studies (see Raw 2012, Krebs 2013).

R. Barton Palmer celebrates (appropriates) adaptation studies as the 'most exciting anti-discipline on the current cultural studies scene' (2009: 89), but the flipside is that the subject is at risk of losing a discernible identity entirely. Adaptation studies would benefit from hitting 'replay' and re-instituting a vital, energetic and ongoing engagement with its key terms and concepts – including adaptation itself, for the word describes a socially grounded artistic practice which changes over time. Old definitions and debates will no longer do, but we must avoid throwing the baby out with the bathwater. Instead, via the metacriticism of common, shared language, adaptation scholars could seek to develop a more precise vocabulary from which new conceptual insights and debate can evolve. This would include working towards a more particular conceptualisation of adaptation, overlapping with but disambiguated from related forms and practices.

How might the day-to-day practice of adaptation studies look different following these interventions? Whilst the concept of adaptation might be more narrowly focused, the wealth of possible approaches would broaden. Adaptations could still be celebrated as instances of intertextuality and/or intermediality; they could also be examined as independent artworks within the fields of film, television, theatre, new media studies, etc. The comparative approach, which is logically supported by the adaptation/intertextuality distinction explored herein, can uniquely open up conceptual questions (Cardwell 2007), especially around medium specificity; even fidelity might prove valuable as an interpretative (if not evaluative) tool in some instances. With the terms of our dialogue clearly on the table for discussion, we could initiate greater and deeper reflexivity. Finally, the acts of metacriticism and conceptualisation combined could forge a new area of adaptation studies, inspired by the analytic alternative, to sit alongside and debate with existing perspectives. We might begin slowly to build a new, analytical aesthetics of adaptation.

Notes

1 There are two recognised, overarching philosophical traditions: (Anglo-American) analytic and continental (the implicit continent is Europe). Of course, both branches embrace scholars from all over the world; the terms are not intended to imply geographical specificity but rather to delineate two very different approaches to philosophy and – more specifically for our purposes here – aesthetics. I employ this established distinction herein. For a valuable introductory collection which presents both traditions, highlighting their differences, see H. Gene Blocker and Jennifer M. Jeffers (eds.) (1999) *Contextualizing Aesthetics*.

2 Interested readers who are less acquainted with the analytic (rather than continental) tradition can find an engaging introduction to classic Anglo-American analytic aesthetics in Oswald Hanfling's (ed.) (1992) *Philosophical Aesthetics*.

3 Cardwell (2014) recommends the term 'art form' (or simply 'art') to refer to the principal, top-level categories such as film and television, often labelled by scholars as media. I propose that each art form is constituted by many media, and that the particular combination and number of those media are always changing and developing with each new art work. This initial, simple distinction between art form and medium chimes with established, common usage within (fine) art practice, history and criticism.

Works cited

Andrew, Dudley (2011) "The economies of adaptation," in Colin MacCabe, Kathleen Murray and Rick Warner (eds.) *True to the Spirit*, 27–39.

Aragay, Mireia (ed.) (2005) *Books in Motion: Adaptation, Intertextuality, Authorship*, Amsterdam: Rodopi.

Barthes, Roland (1981) "Theory of the text," in Robert Young (ed.) *Untying the Text*, London: Routledge, 31–47.

Blocker, H. Gene and Jennifer M. Jeffers (eds.) (1999) *Contextualizing Aesthetics: From Plato to Lyotard*, Belmont, CA: Wadsworth Publishing.

Bruhn, Jørgen, Anne Gjelsvik and Eirik Frisvold Hanssen (eds.) (2013) *Adaptation Studies: New Challenges, New Directions*, London: Bloomsbury.

Cardwell, Sarah (2002) *Adaptation Revisited: Television and the Classic Novel*, Manchester: MUP.

Cardwell, Sarah (2007) "Adaptation studies revisited: Purposes, perspectives, and inspiration," in James M. Welsh and Peter Lev (eds.) *The Literature/Film Reader: Issues of Adaptation*, Lanham, UK: Scarecrow Press, 51–64.

Cardwell, Sarah (2014) "Television amongst friends: Medium, art, media," *Critical Studies in Television* 9(3): 6–21.

Carroll, Noël (1996) *Theorizing the Moving Image*, Cambridge: Cambridge University Press.

Carroll, Noël (ed.) (2000) *Theories of Art Today*, Madison, WI: University of Wisconsin Press.

Carroll, Rachel (ed.) (2009) *Adaptation in Contemporary Culture: Textual Infidelities*, London: Continuum.

Cartmell, Deborah (2010) *Screen Adaptation: Impure Cinema*, Basingstoke, UK: Palgrave.

Cattrysse, Patrick (2014) *Descriptive Adaptation Studies: Epistemological and Methodological Issues*, Antwerp: Garant.

Corrigan, Timothy (2010) "Adaptations, refractions, and obstructions: the prophecies of Andre Bazin," http://revistafalsomovimento.com/adaptations-refractions-and-obstructions-the-prophecies-of-andre-bazin-timothy-corrigan. [1 Nov 2015]

Dicecco, Nico (2015) "State of the conversation: The obscene underside of fidelity," *Adaptation* 8(2): 161–175.

Ellestrøm, Lars (2013) "Adaptation within the field of media transformations," in Jorgen Bruhn, Anne Gjelsvik and Eirik Frisvold Hanssen (eds.) *Adaptation Studies: New Challenges, New Directions*, 113–132.

Elliott, Kamilla (2003) *Rethinking the Novel/Film Debate*, Cambridge: Cambridge University Press.

Elliott, Kamilla (2013) "Theorizing adaptations/adapting theories," in Jorgen Bruhn, Anne Gjelsvik and Eirik Frisvold Hanssen (eds.) *Adaptation Studies: New Challenges, New Directions*, 19–45.

Ellis, John (1982) "The Literary Adaptation," *Screen* 23(1): 3–5.

Frow, John (1990) "Intertextuality and ontology," in Michael Worton and Judith Still (eds.) *Intertextuality: Theories and Practices*, Manchester: MUP, 45–55.

Genette, Gérard (1982) *Palimpsests: Literature in the Second Degree*. Trans. Channa Newman and Claude Doubinsky, 1997, Lincoln, NE: University of Nebraska Press.

Hanfling, Oswald (ed.) (1992) *Philosophical Aesthetics: An Introduction*, Oxford: Open University Press.

Krebs, Katja (ed.) (2013) *Translation and Adaptation in Theatre and Film*, London: Routledge.

Leitch, Thomas (2003) "Twelve fallacies in contemporary adaptation theory," *Criticism* 45(2): 149–71

Leitch, Thomas (2008) "Adaptation, the genre," *Adaptation* 1(2): 106–120.

MacCabe, Colin, Kathleen Murray, and Rick Warner (eds.) (2011) *True to the Spirit: Film Adaptation and the Question of Fidelity*, Oxford: Oxford University Press.

McFarlane, Brian (1996) *Novel to Film: An Introduction to the Theory of Adaptation*, Oxford: Clarendon Press.

Meinhof, Ulrike H. and Jonathan Smith (eds.) (2000) *Intertextuality and the Media: From Genre to Everyday Life*, Manchester: Manchester University Press.

Murray, Simone (2012) *The Adaptation Industry: The Cultural Economy of Contemporary Literary Adaptation*, London: Routledge.

Palmer, R. Barton (2009) "Review of *Journal of Adaptation in Film and Performance* 1.1 and 1.2," *Adaptation* 2(1): 87–9.

Raw, Lawrence (ed.) (2012) *Translation, Adaptation and Transformation*, London: Bloomsbury

Sanders, Julie (2006) *Adaptation and Appropriation*, London: Routledge.

Schrober, Regina (2013) "Adaptation as connection – Transmediality reconsidered," in Jorgen Bruhn, Anne Gjelsvik and Eirik Frisvold Hanssen (eds.) *Adaptation Studies: New Challenges, New Directions*, 89–112.

Stam, Robert (2005), 'Introduction: The theory and practice of adaptation', in Stam and Raengo (eds.) *Literature and Film: A Guide to the Theory and Practice of Film Adaptation*, Malden, MA: Blackwell Publishing, pp. 1–52.

Stam, Robert and Alessandro Raengo (eds.) (2005) *Literature and Film: A Guide to the Theory and Practice of Film Adaptation*, Malden, MA: Blackwell Publishing.

Voigts, Eckart and Pascal Nicklas (2013) "Introduction: Adaptation, Transmedia Storytelling and Participatory Culture," *Adaptation* 6(2): 139–42.

Westbrook, Brett (2010) "Being adaptation: The resistance to theory," in Christa Albrecht-Crane and Dennis Cutchins (eds.) *Adaptation Studies: New Approaches*, Cranbury, NJ: Rosemont Publishing, 25–45.

2

THE THEORY OF *BAD*APTATION

Kamilla Elliott

BADaptation—a term coined by J. Kraus (2012: 258) and developed by Constantine Verevis (Verevis 2014: 216)—is a resonant portmanteau in adaptation studies. In 2010, Deborah Cartmell and Imelda Whelehan subtitled a book *Impure Cinema* "to call attention to the bad press that adaptations have received since the beginning of film's history" (Cartmell and Whelehan 2010: 127). The rhetoric of BADaptation precedes cinema: describing an 1838 stage play of *Oliver Twist*, Charles Dickens the Younger decrees it the worst in "the very long list of bad adaptations of popular stories" (Dickens 1892: xxvii); decades earlier, a periodical reviewer addresses "the bad adaptation of hymns to tunes" (anon 1856: 98) and a letter to *The Players*, a nineteenth-century penny British theatrical journal, declares: "that our stage should become the receptacle for bad adaptations of immoral French buffoonery, we feel a national degradation" (anon. 1860: 2). While Verevis defines "'BADaptation' [as] a concept employed to engage with and challenge those approaches to adaptation and remaking that routinely employ a rhetoric of betrayal and degradation, of 'infidelity' to some idealized original" (Verevis 2014: 216), these examples make clear that adaptations have been dubbed bad (as well as many synonyms for bad) for violating moral and national ideologies as well as theories of ideal originals.

Early and mid-twentieth-century critics predicate their discourses of BADaptation on what they perceive to be irreconcilable differences between media forms. A letter to the *Scenario Bulletin Digest* in 1922 is unilateral in its insistence that "The reason an 'adaptation' is bad is because a good one cannot be made. This has been proven, and when a thing is proven it ought to settle all arguments" (M. G. 1922: 28). The first academic monograph on literary film adaptation contends similarly that adaptation is always essentially bad translation and bad art, because it violates formalist semiotic and medium specificity theories (Bluestone 1957).

While there were occasional sociological studies of film adaptation in the first half of the twentieth century—and Bluestone also considered industry and social aspects mid-century—following the cultural revolution of the 1960s in the West, critics began to examine art and media to locate their cultural ideologies. In 1985, the next book-length treatment of literary film adaptation[1] explains the fact that "nobody loves an adaptation" (Boyum 1985: 17) in terms of aesthetic rivalries between literature and film, as well as competing aesthetic theories: "it finds itself in a no-man's land, caught somewhere between a series of conflicting aesthetic claims and rivalries" (Boyum 1985: 17). In 2005, Robert Stam identifies theoretical, disciplinary, and cultural reasons for the discourse of BADaptations: theories of seniority as priority, iconophobia and logophilia, anti-corporeality, beliefs concerning the facility and parasitism of adaptations, disciplinary rivalries, and class prejudices (Stam 2005: 1–52). Surveying such dynamics in 2010, Cartmell concludes that "'bad' adaptations receive more coverage than good ones" (Cartmell 2010: 126–28).

While George Bluestone proposes an end to the practice of literary film adaptation ("The film and the novel [should] remain separate institutions, each achieving its best results by exploring unique and specific properties", Bluestone 1957: 218), later critics recommend new theories to solve the problems of theoretically BADaptations: narratology to offset the problems of linguistic translation for semiotic theories, dialogics and poststructuralist inter-textuality to combat hierarchical binaries of originals and copies, and postmodern cultural studies to defy the aesthetics of high-art humanism. However, these new theories create new discourses of BADaptations, in which adaptations continue to fail new theories. The transtheoretical and transhistorical discourse of BADaptation therefore suggests a more systemic impasse between the *processes* of adaptation and theorization *themselves* and a need to reconsider their dysfunctional relationship. Simply applying more theories or new theories to adaptation has not redressed their dysfunctional relationship; indeed, from one angle of view, it has intensified it.

There are several aspects to this dysfunctional relationship. First, adaptation and theorization are rival, overlapping processes, each seeking to rework cultural products and each other in its image, and each resisting the other's reworking. Second, adaptations have rarely been theorized as adaptations, but instead as books, films, art, music, theatre (etc.); as rhetoric, translation, or narrative; as vehicles of historical, political, cultural, national, and philosophical ideologies; as manifestations of industry or mind, the latter viewed in existential, psychoanalytic, phenom-enological, and cognitive terms. While all of these theories, deriving from many humanities disciplines, indubitably illuminate adaptation, none theorizes it on its own terms—as adaptation. The final part of this essay ponders what a reciprocal discourse and practice of "adapting theorization" to and through adaptation might bring to the hitherto one-way rhetoric and practice of 'theorizing adaptation'.

From George Bluestone's "Word to Image: The Problem of the Filmed Novel" (1956) to Fredric Jameson's "Adaptation as a Philosophical Problem" (2011) and beyond, scholars have identified adaptation as a theoretical problem to be resolved by theorization. The rhetoric features regularly and prominently in the titles, subtitles, and sub-headings of books, essays, chapters, articles, blogs, and reviews:

1967: Donald W. McCaffrey, "Adaptation Problems in Two Unique Media: The Novel and the Film"

1971: Jean Mitry, "Remarks on the Problem of Adaptation"

1973: Edward Murray, "*In Cold Blood*: The Filmic Novel and the Problem of Adaptation"

1979: Roger Manvell, *Theater and Film: A Comparative Study of the Two Forms of Dramatic Art and of the Problems of Adaptation of Stage Plays into Films*

1987: Thomas Sobchack and Vivian Carol Sobchack, "Film and Literature: The Problem of Adaptation"

1990: J. B. Bullen, "Is Hardy a 'Cinematic Novelist'? The Problem of Adaptation"

1997: Greg Jenkins, "The Problem of Adaptation"

2007: Marc DiPaolo, "The Problem of Adaptation"

2008: Joseph M. Boggs and Dennis W. Petrie, "The Problems of Adaptation"

2009: M. J. Kidnie, *Shakespeare and the Problem of Adaptation*
2010: Robert McKee, 'The Problem of Adaptation'
2011: Federico Pagello, 'True to *The Spirit*? Film, Comics and the Problem of Adaptation'
2014: Anthony Friedmann, 'The Problem of Adaptation'

While adaptation scholars, myself included (Elliott 2017), have focused on the theoretical problems of adaptation *scholarship*, what is striking for this discussion is that these titles figure *adaptation itself* as the theoretical problem.

Since definition is the first stage of theorization (de Man 1986: 5), and the definition of adaptation is hotly contested by critics from various disciplines and camps, and since adaptation's definition is often limited to literary film adaptation or otherwise constrained (Hutcheon 2006: 15–22), obscuring it as a more general process, in this discussion of adaptation as a process akin to theorization, it is essential to define adaptation more generally and fundamentally. I know nowhere better to do so than via the *Oxford English Dictionary*, which defines adaptation as "The action or process of adapting one thing to fit with another, or suit specified conditions, esp. a new or changed environment" and also as the product of such a process (*OED* 1989). If this is an adaptation, then a BADaptation is one that does not suit its new environment or changed conditions. Most studies of adaptation have focused on new media, cultural, or historical environments: my chapter considers the failure of adaptations to adjust to theoretical environments. I argue that this is a transtheoretical, transhistorical dynamic that cannot be explained locally using only some theories or overcome by progressivist theories of theorization; it must be explained through a larger purview of adaptation as a more general process at odds with the more general process of humanities theorization. If, at the turn of the twenty-first century, adaptations and their scholars were attacked for resisting the theoretical turn in the humanities (Elliott 2013), prior to that time, adaptations were denounced for violating the very theories they were subsequently charged with favouring: formalist medium specificity, New Critical organic unity, high-art aestheticism, Romantic theories of authorship, logocentrism, metaphysical theories of representation, right-wing patriarchy, and humanism. Scholars have been so concerned with theorizing what was wrong with adaptations that it has not occured to us to ask, what is wrong with theorization in the humanities? In 1957, rather than challenge the medium specificity theory that had dominated interart theory for nearly two centuries, as adaptation itself was doing, Bluestone called for an end to the *practice* of literary film adaptation (Bluestone 1957: 210). Prior to film, eighteenth-century theorists addressing poetry and painting and early twentieth-century critics addressing book illustration in relation to prose argued similarly (Elliott 2003: 13; 79). Hybrid art forms and transmedial adaptations of all kinds violate neoclassical, Russian formalist, and structuralist medium specificity theories, which hold that each art or medium must maintain essential, ontological, materially based properties in order to distinguish it from other arts and media, and that these principles prohibit adaptation from being aesthetically effective or viable. Medium specificity theorists argue further that certain forms are better suited to representing certain subjects, and that each medium should be restricted to representing subjects best suited to its form. Since transmedial adaptations cross media forms and seek to represent subject matter represented by other media, they are doubly prone to being condemned as theoretically and aesthetically bad.

Yet for all their protests against adaptation's aesthetic and representational hybridity, advocates of aesthetic purity, Romantic originality, and formalist semiotics have allied to create *theoretical* hybrids to oppose media hybrids such as adaptation, yoking neoclassical theories proclaiming the essentialist separatism of the arts to Romantic theories proclaiming their organic unity. While a full discussion of these and other theoretical hybrids is beyond the scope of this chapter, and will be pursued in a forthcoming book, their alliances recall how biological theories of separate species and religious theories of original creation once combined to oppose biological adaptation. Academic disciplines are built upon theories of medium specificity; they are economic and professional institutions that cannot be easily dismantled on purely intellectual grounds. This is in part why, in spite of poststructuralist, postmodern, and posthuman theories that have effectively dismantled medium specificity theory, humanities scholars, being housed in and employed by and dependent upon separate media spheres for their professional existence, continue to champion and support it. Humanities scholars are further subdivided professionally according to national boundaries and historical periods. Adaptations, straddling many disciplines and leaping across the globe and centuries, are disciplinary, national, and historical bastards. According to William Blackstone's *Commentaries of the Laws of England* (1765–69), a bastard is both nobody's child (*filius nullius*) and the people's child (*filius populi*) (Blackstone 1839 edition: 495). Adaptations are simultaneously nobody's children and everybody's children, belonging to no field, nation, or era, yet claimed by all. Their position as everybody's child allows for their universal theoretical use and abuse in disciplinary, national, cultural, political, philosophical, and psychological theoretical wars, as scholars wrangle over adaptation's custody; their position as nobody's child allows for their universal neglect and marginalization by all. Literary scholars, for example, theorize literary film adaptations as literary translations (Cattrysse 2014: 47 ff.) or criticism (Sinyard 1986: 117 ff.); film scholars seeking to establish film as an independent art form according to medium specificity theory, which requires it to be original and distinct from other arts and media, theorize adaptations as "films in their own right" (Phillips 2006: 130) rather than as aids to literary study. Geoffrey Wagner's hierarchized adaptation taxonomy, for example, champions film's independence from literature: translation (transposition) is ranked bottom, because good translations are inevitably bad art, since they lack the uniqueness and originality required by both medium specificity theory and Romantic originality. Slightly above transposition comes commentary, in which film adaptation undertakes literary criticism—in a critical, not just interpretive, sense. At the top are works that pay almost no attention to their sources (analogies); only these can be good art (Wagner 1975: 222–26). While these modes of study indubitably illuminate single disciplines, they illustrate the limitations of single discipline adaptation study. Film scholars understandably object to limiting film adaptation to a mode of literary criticism; yet it is problematic that film adaptation can only be valued when it casts off its identity *as* adaptation and emerges as independent, original art. Wagner's rejection of fidelity in adaptation furthermore sits uneasily with his insistence on adaptations' fidelity to dominant humanities theories of the day.

In the final quarter of the twentieth century, as a majority of humanities scholars began to turn from conservative aesthetic and formal theories to radical political, philosophical, and cultural ones, adaptations continued to be judged good or bad for their adherence to or departure from mainstream humanities theories. Although Julie Sanders voices a growing critical consensus that academic studies should not be "aimed at identifying 'good' or 'bad' adaptations" or about "making polarized value judgments" (Sanders 2005: 20), polarized judgments persist when adaptations are judged good for supporting left-wing political theories and bad for supporting right-wing ones or good for supporting postmodern philosophies and bad for supporting modernist ideologies. Scholars in recent decades have redirected the rhetoric of aesthetically and theoretically nonconforming BADaptation towards theoretically 'bad' racist, sexist, heterosexist, capitalist, nationalistic, colonizing adaptations (Sanders 2005: 45–62). Heritage criticism, which synthesizes many of these theories, offers a particularly marked discourse of such BADaptations (Vincendeau 2001).

New theories have not only failed to eradicate the discourse of BADaptation, they have further multiplied the ways in which adaptations can be bad, as older theories persist alongside newer ones and as postmodern inclusivity has expanded the media, disciplines, and aspects of adaptations that can be judged bad for failing theoretical tenets. The continuity of BADaptations amidst theoretical revolution is thus a transhistorical phenomenon. Just as the neoclassical imitations revered by Augustan critics were condemned by Romantic theorists (Del Villano 2012: 179), so too the adaptations deemed national treasures and civilizing forces of colonization by the Victorians are now condemned by postcolonial theorists as nationalist propaganda and oppressive politics (Ponzanesi 2014: 109–55).

Yet there are differences: while adaptations were seen to fail older aesthetic and formal theories unilaterally, they fail newer political and philosophical theories selectively. Adaptations are theoretically promiscuous, as likely to support political conservatism as left-wing identity politics, as prone to valorize high art as low, as resistant to poststructuralist semiotics as to medium specificity, as likely to engage in fidelity as infidelity to theories. Adaptations thus resist new theories as well as old: they flout their materiality in the face of the abstracting philosophical turn; their both/and dynamics challenge deconstruction's neither/nor; their flaunting of deliberate agency affronts both psychoanalysis and poststructuralist intertextuality; their ready collusion with corporate capitalism and right-wing nostalgia offends radical political ideologies.

The multidisciplinary nature of adaptation and theorization in the humanities has led to a panoply of theories being applied to them. Postmodern scholars welcome theoretical pluralism; many see it as the solution to the theoretical problem of adaptation. Rainer Emig rightly argues that "adaptation cannot and must not rely on one theory or even one clearly prescribed set of theories only … its multi- and interdisciplinary status also determines its multi-, inter- and transtheoretical attachments" (Emig 2012: 14). Brett Westbrook recommends a "glorious plurality" of diverse, even dissonant, theories to theorize the field, calling critics to "choose from a candy store of available approaches: semiotics, feminist criticism, Russian Formalism, media studies – the whole menu" (2010: 43–44). Linda Hutcheon's *A Theory of Adaptation* (2006) is far from being a single theory, combining many without reconciling their ideological differences in a postmodern theoretical pastiche. Yet given the anti-adaptation stances of so many humanities theories, applying more theories deriving from more disciplines that deem adaptation to be bad and that do not theorize adaptation as adaptation has equally led to a greater variety of theoretical discipline, punishment, and abuse.

Scholars have rendered adaptations weapons and casualties alike in theoretical and disciplinary wars that have failed to theorize them *as adaptations*: instead, they have been defined, taxonomized, and theorized as other things—as books, films, arts, and other media; as signs, narratives, and translations; as aesthetics, culture, politics, geography, and history; as sociology, economics, psychology, philosophy, technology, industry, and more. While adaptations participate in all of these, they do not reduce to any: neither does their sum total amount to 'adaptation' or clarify processes of adaptation.

Prior scholars have recommended theorizing "adaptations *as adaptations*" (Hutcheon 2006: 4, original emphasis). Yet for Hutcheon, such theorization is optional and limited to the lowest common denominator of baseline audience response:

> In the book, I thought I had to treat adaptations as adaptations. The reality is that as readers or moviegoers, if we don't know it's an adaptation, or we know it is but don't know the adapted text, then it doesn't matter, we simply experience it the way we would any other work of art. So, I don't think we have to treat adaptations as adaptations.
>
> *Hutcheon in Zaiontz 2009: 1*

Even so, Hutcheon's expansion of adaptation studies beyond literature and film has invaluably contributed to treating adaptations as adaptations, as their wider disciplinary expanse highlights their common features as adaptations. Prior to Hutcheon, Sarah Cardwell recommended theorizing adaptations as adaptations by studying them as a genre with shared conventions (Cardwell 2002: 67). Subsequently, leading adaptation scholars have opposed adaptation taxonomies on postmodern theoretical grounds (Cartmell and Whelehan 2007: 2). Yet if we are to theorize adaptations as adaptations, we have to consider them in relation to each other. Such groupings need not amount to universal claims or hierarchical rankings, as in older modes of taxonomization; they can be dialogically or dialectically or intertextually negotiated; they can dissipate along poststructuralist axes of random differentiation or proliferate as postmodern pastiche.

Unless we discuss adaptations in relation to each other, adaptation studies will continue to be seen as a subset of other fields, disciplines, topics, and theories, informing them while offering little insight into adaptation and adaptive processes and products. Amid their protests against adaptation taxonomies, Cartmell and Whelehan understand that "the will to taxonomize" is "symptomatic of how the field has tried to mark out its own territory" (Cartmell and Whelehan 2007: 2), indicating "more than anything its need to establish a critical perspective of its own" (Cartmell and Whelehan 2010: 6). Grouping adaptations and setting those groups in dialogue with each other has been and will continue to be essential to theorizing adaptations as adaptations.

Mieke Bal has argued, "No concept is meaningful for cultural analysis unless it helps us to understand the object better on its – the object's – own terms' (Bal 2007: 8). In his essay 'The Resistance to Theory", Paul de Man insists that "Literary theory can be said to come into being when the approach to literary texts is no longer based on non-linguistic, that is to say historical and aesthetic, considerations" (De Man 1986: 10). In the same way, adaptation theory might be said to come into being when the approach to adaptations is no longer based on non-adaptive considerations. This is not to say that other theories cannot or should not be brought to bear on adaptations, but rather that, until adaptations have been theorized on their own terms, in relation to each other as adaptations, other theories from other disciplines, developed to treat other entities and subjects, can only offer delimited explanatory power. Just as scholars once fought to free film adaptation from literature to view it 'as film', 'in its own right', adaptation needs to be theorized as adaptation, on its own terms, in its own right. While this may be an oxymoron, since adaptation is always hybrid, there are nevertheless ways in which we can do this without essentialism.

Rhetoric has conceptual and theoretical substance, as countless critics have demonstrated. Most famously, Roman Jakobson posited that metaphor and metonymy *are* (rather than are analogies for) the paradigmatic and syntagmatic axes of language and communication, going further to theorize poetry and prose fiction, symbolism and realism, and various literary movements in terms of metaphoric and metonymic structures (Jakobson: 1956). Inspired by Jakobson, Jacques Lacan argued that metaphor and metonymy describe unconscious structures of substitution and displacement (Lacan: 1957). Christian Metz influentially carried semiotic and psychoanalytic figurations into film theory (Metz: 1977). Paul de Man extended them to rhetorical theory: "All language is … a conceptual, figural, metaphorical metalanguage" (De Man 1979: 152–3). Other scholars have addressed the politics of metaphor and metonymy, extending Roland Barthes's critique of the "metonymic fallacy" (Barthes 1974: 11) to identity politics (e.g., Alcoff and Potter 1993). The rhetoric of "adaptations as adaptations" is a similarly theoretical dynamic, working against the rhetoric of adaptations as theoretically bad. A self-reflexive simile that plays back and forth, it differentiates adaptation from itself through an exchange of adaptation as vehicle and adaptation as tenor. Studying adaptations as adaptations need never distil to tautology or essentialism; neither need it dissipate and disappear synecdochally into the other things that have sought to theorize it as other. Simile further resists the othering tendencies of theorization by making adaptation its own other as well as its own self.

It is neither vehicle nor tenor, nor even both together, that offers the most revolutionary potential for redressing the dysfunctional relationship between adaptation and theorization: it is the conjunction, 'as'. The 'as' forges the connection between vehicle and tenor; the 'as' sends 'theorizing' to the margins of the phrase 'theorizing adaptations as adaptations', displacing theoretical dominance over adaptation with a primary, reciprocal relationship of adaptation to adaptation. "As" is a democratising adverb "expressing a comparison of equivalence" (*OED*). Equivalence avoids essentialism and resists theories predicated on it. It does not consistently conform to any of the hierarchical, oppositional, or random modes of difference that dominate mainstream humanities theories, whether classical dualism, Marxist dialectics, poststructuralist *différance*, or postmodern diversity. Nor is it an adjective or noun that can be judged good or bad. Equivalence means "Equal in value, power, efficacy, import, meaning, or significance" (*OED*). It makes space for formal translation theories and democratizing political theories alike, both of which seek theoretical equivalences themselves.

Equivalence goes further to suggest how scholars might create a more mutual relationship between adaptation and theorization by establishing an equivalency between 'theorizing adaptation' and 'adapting theorization'. This is not mere rhetoric, but suggests a metatheoretical revolution in which our one-way, hierarchical, top-down preoccupation with theorizing adaptations is counterpointed by a two-way, democratizing discourse and practice of adapting theories to and through adaptation (Elliott 2013).

Even as recent theorists challenge media, disciplinary, social, and cultural hierarchies, the hierarchy of theory over practice remains unaccountably, paradoxically, and hypocritically unchallenged. 'Theo-rization' in the humanities carries the traces of theology, operating from a dogmatic point of view, levying moral, aesthetic, philosophical, political, and cultural judgments upon subject matter and denying the existence of theoretically resistant subjects such as adaptation, just as religious theologies deny the very existence of biological adaptation (Elliott 2003: 134). The tendency to view humanities theorization as the imposition of truth and correct ideologies upon wayward cultural materials requiring conformity to correct theoretical principles persists even among the proponents of radical and sceptical humanities theories, who hold left-wing political ideologies to be more ethical than right-wing views and postmodern indeterminacy and poststructuralist *différance* to be truer than modernism and structuralist difference.

More fundamentally, the undeconstructed binarism between theorization and adaptation is theoretically baffling, based as it is in a classical dualist metaphysics that has been deconstructed almost everywhere else. Such a one-sided emphasis on theorizing adaptations sits especially uneasily in the interdisciplinary, multi-disciplinary, intermedial, and multimedial field that is adaptation studies. The failure to create a reciprocal discourse does not arise from intellectual or theoretical difficulties: the binarism can be swiftly and readily deconstructed. Adaptation can be viewed as a process akin to as well as at odds with theorization: as a practice, it engages not only in criticism (Wagner 1975; Sinyard 1986) but also in theorization (Sanders 2005) and philosophy (Constable 2009). As Emig argues, "It is not only adaptation that must be positioned in theory, but adaptation is also always already a component of theory" (2012: 23). Concomitantly, theorization is itself a cultural practice: the moment that philosophy (or history or politics or psychology or theology) is applied to cultural materials, it becomes an academic practice. Conjoining these reciprocal inversions, scholars such as Jean Marsden have grouped critical and editorial writing with performed adaptations of literature (Marsden 1995).

The undeconstructed hierarchal binarism between theory and practice has furthermore obscured the relational dynamics of adaptation and theorization. Adaptation is not so much a theoretically bad object in the sense of *failing* theorization as it is in *resisting* it. It refuses to conform consistently to any theoretical tenet: its illogical, emotive, aesthetic, imaginative, unconscious, and abstract dynamics elude rational and empirical methodologies; its local, material, industrial, institutional aspects resist the pull of philosophical theories towards universality and abstraction. In its capacity to accommodate conflicting ideologies, adaptation rejects binaries and dialectics. In its theoretical promiscuity, it resists espousing any theoretical value or tenet universally. Adaptation always exceeds theorization; Hans-Bernhard Moeller and George Lellis are two among many who assess that "all creation can be considered adaptation" (2002: 6); Emig writes: "If culture is a continual process of adaptation, then adaptation is everything—and at the same time nothing that can be singled out and properly defined or theorized" (2012: 19).

Although critics such as Cartmell and Whelehan (2010: 21), Westbrook (2010: 41–2), and MacCabe (2011: 8) argue that the failure to theorize adaptations derives from the infinite numbers of variables in any given adaptation conjoined to the diversity and range of the field, this is not the only reason or, to my mind, even the main reason for the difficulties of theorizing adaptations in the humanities. There are infinite variables in all cultural materials, practices, or processes. The problem is not so much that the field is too large to theorize as that every infinitely variable aspect of adaptation potentially resists every kind of humanities theorization we apply to it: it does so precisely because it *is* adaptation and by definition refuses to be fixed, whether by the definition, taxonomization, and universal principles of traditional theorization or the ideologies and ethics of the post theories that reacted against traditional theorization.

What we need in adaptation studies is not new theory or more theories, but a major rethinking of the dysfunctional and oppositional relationship between adaptation and theorization. Amid longstanding discourses of adaptation's failure of humanities theorization, almost no discourse of humanities theorization's failure of adaptation has emerged. *Specific* theories have been critiqued for failing adaptations, but not theories or theorization *generally*. Typically, claims that theories have failed adaptations are immediately followed by recommendations of other theories to replace them (Elliott 2013). Emig astutely notes that theory "is itself an ongoing adaptation" (2012: 23); yet currently, humanities theorization adapts chiefly to other theories rather than to and through adaptations or any of the other materials that it theorizes: for example, postmodernism is modernism adapted; poststructuralism is structuralism adapted; New Historicism is old historicism adapted; posthumanism is humanism adapted, feminist psychoanalysis is patriarchal psychoanalysis adapted. In each instance, the new conditions to which theories adapt are theoretical ones. If theorization is to theorize adaptation effectively, it must adapt to and through adaptation rather than other theories alone. In a monograph, which I am now writing, I detail some ways in which this might be done; there are potentially limitless ways in which it could be done.

In place of universal theorization or piecemeal theorization or judgmental theorization or anything-goes theorization, I propose 'adaptive theorization', in which we no longer hold adaptations up to theories developed for things that are not adaptations and find them theoretically lacking, or ignore and castigate adaptations as bad because they violate our theoretical tenets—rather, I propose that we study adaptations to discover ways in which theorization is lacking, allowing adaptations to challenge and adapt even our most cherished theoretical beliefs. I am weary of reading humanities scholarship that tells us what we already know and believe, that uses cultural materials simply to affirm and contest theoretical principles and ideologies. I want to study adaptation to learn what I don't already know—to discover new ideas as well as to adapt old ones.

Note

1 Other books treating film and literature in the 1960s and 1970s devote only chapters to adaptation.

Works cited

Alcoff, L. and Potter, E. (1993), "Introduction: When feminisms intersect epistemology", in L. Alcoff and E. Potter (eds.), *Feminist Epistemologies*, London: Routledge, pp. 1–14.

Anon. (1856), "Review of *The Plymouth Collection of Hymns and Tunes*", *The New Englander* 14, February, pp. 92–114.

Anon. (1860), "To the reader", *The Players* 1, 2 January, p. 2.

Bal, M. (2007), "Working with concepts", in G. Pollock (ed.), *Conceptual Odysseys: Passages to Cultural Analysis*, New York: Palgrave, pp. 1–10.

Barthes, R. (1974), *S/Z: An Essay* (trans. R. Miller), New York: Farrar, Strauss & Giroux.

Blackstone, W. (1839), *The Rights of Persons, According to the Text of Blackstone*, ed. J. Stewart, London: Edmund Spettigue. *Commentaries of the Laws of England*. Originally published 1765–69.

Bluestone, G. (1956), "Word to image: The problem of the filmed novel", *The Quarterly of Film, Radio, and Television*, 11, pp. 171–80.

Bluestone, G. (1957), *Novels into Film*, Berkeley, CA: University of California Press, 1957.

Boggs, J. M. and Petrie, D. W. (2008), *The Art of Watching Films*, New York: McGraw-Hill.

Boyum, J. G. (1985), *Double Exposure: Fiction into Film*, New York: Plume, 1985.

Bullen, J. B. (1990), "Is Hardy a 'cinematic novelist'? The problem of adaptation", *Yearbook of English Studies*, 20, pp. 48–59.

Cardwell, S. (2002), *Adaptation Revisited: Television and the Classic Novel*, Manchester, UK: Manchester University Press.

Cartmell, D. (2010), "The Film industry and fiction", *The Encyclopedia of Twentieth-Century Fiction*, London: Blackwell, 2010, pp. 126–28.

Cartmell, D. and Whelehan, I. (2007), "Introduction: Literature on screen: A synoptic view", in D. Cartmell and I. Whelehan (eds.), *The Cambridge Companion to Literature on Screen*, Cambridge, UK: Cambridge University Press, pp. 1–12.

Cartmell, D. and Whelehan, I. (2010), *Screen Adaptation: Impure Cinema*, New York: Palgrave Macmillan.

Cattyrsse, P. (2014), *Descriptive Adaptation Studies: Epistemological and Methodological Issues*, Antwerp: Garant.

Coe, J. (2011), "Good book, great film", *The Guardian*, http://www.theguardian.com/books/2011/apr/01/book-adaptations-film-jonathan-coe. Accessed 15 July 2015.

Constable, C. (2009), *Adapting Philosophy: Jean Baudrillard and the Matrix Trilogy*, Manchester: Manchester University Press.

De Man, P. (1986), "The resistance to theory", in *The Resistance to Theory*, Minneapolis, MN: University of Minnesota Press, pp. 3–20.

De Man, P. (1979), *Allegories of Reading*, New Haven, CT: Yale University Press.

Del Villano, B. (2012), "Dramatic adaptation, authorship and cultural identity in the eighteenth century: The case of Samuel Foote", *Journal of Early Modern Studies*, 1 (1), pp. 175–91.

Dickens, C. the Younger (1892), "Introduction", *The Adventures of Oliver Twist by Charles Dickens, A Reprint of the First Edition, with the Illustrations and an Introduction, Biographical and Bibliographical, by Charles Dickens the Younger*, Chicago: Hooper, Clarke & Co., pp. xiii–xxxiii.

DiPaolo, M. (2007), *Emma Adapted: Jane Austen's Heroine from Book to Film*, Bern, Switzerland: Peter Lang.

Elliott, K. (2003), *Rethinking the Novel/Film Debate*, Cambridge, UK: Cambridge University Press.

Elliott, K. (2013), "Theorizing adaptations/adapting theories", in *Adaptation Studies: New Challenges, New Directions*, in J. Bruhn, A. Gjelsvik, and E. F. Hanssen (eds.), London: Bloomsbury, 2013, pp. 19–45.

Elliott, K. (2017), "Adaptation theory and adaptation scholarship", in T. Leitch, (ed.), *The Oxford Handbook of Adaptation Studies*, Oxford: Oxford University Press, pp. 679–97.

Emig, R. (2012), "Adaptation in theory", in *Adaptation and Cultural Appropriation: Literature, Film, and the Arts*, in P. Nicklas and O. Linder. (eds.), Berlin: Walter de Gruyter, pp. 14–24.

Friedmann, A. (2014), "The problem of adaptation", *Writing for Visual Media*, Waltham, MA: Focal Press, pp. 202–22.

Hutcheon, L. (2006), *A Theory of Adaptation*, London: Routledge.

Jakobson, R. (1956), "Two aspects of language and two types of aphasic disturbances", in *Fundamentals of Language*, The Hague, the Netherlands: Mouton.

Jameson, F. (2011), "Adaptation as a philosophical problem", in C. McCabe, K. Murray and R. Warner (eds.), *True to the Spirit Film Adaptation and the Question of Fidelity*, Oxford: Oxford University Press, pp, 215–234.

Jenkins, G. (1997), *Stanley Kubrick and the Art of Adaptation: Three Novels, Three Films*, Jefferson, NC: McFarland.

Kidnie, M. J. (2009), *Shakespeare and the Problem of Adaptation: Forms of Possibility*, London: Routledge.

Kraus, J. (2012), *The Laugh-a-Day Book of Bloopers, Quotes & Good Clean Jokes*, Grand Rapids, MI: Revell.

Lacan, J. (1957), "The agency of the letter in the unconscious or reason since Freud", in *Écrits: A Selection*, (trans. A. Sheridan), London: Tavistock, pp. 146–78.

M. G. (1922), "Adapted movies the bunk", letter to the editor, *Scenario Bulletin Digest* 8 (9), p. 28.

McCabe, C. (2011), "Introduction. Bazinian adaptation: *The Butcher Boy* as example", in C. McCabe, K. Murray, and R. Warner (eds.), *True to the Spirit: Film Adaptation and the Question of Fidelity*, Oxford: Oxford University Press, pp. 3–25.

McCaffrey, D. W. (1967), "Adaptation problems in two unique media: The novel and the film", *The Dickinson Review*, 1, pp. 11–17.

McKee, R. (2010) "The problem of adaptation", *Story: Style, Structure, Substance, and the Principles of Screenwriting*, New York: HarperCollins, pp. 364–70.

Manvell, R. (1979), *Theatre and Film: A Comparative Study of the Two Forms of Dramatic Art and of the Problems of Adaptation of Stage Plays into Films*, Rutherford, NJ: Fairleigh Dickinson University Press.

Marsden, J. (1995), *The Re-Imagined Text: Shakespeare, Adaptation, & Eighteenth-Century Literary Theory*, Lexington, KY: University of Kentucky Press.

Metz, C. (1977), *The Imaginary Signifier: Psychoanalysis and the Cinema* (trans. C. Britton), Bloomington, IN: University of Indiana Press.

Mitry, J. (1971), "Remarks on the problem of adaptation", *Midwest Modern Language Association Bulletin*, 4 (1), pp. 1–9.

Moeller, H. and Lellis, G. (2002), *Volker Schlöndorff's Cinema: Adaptation, Politics, and the 'Movie-Appropriate'*, Carbondale, IL: Southern Illinois University Press.

Murray, E. (1973), "*In Cold Blood*: The filmic novel and the problem of adaptation", *Literature/Film Quarterly*, 1, pp. 132–37.

Oxford English Dictionary (OED) (1989), 2nd ed., Oxford: Clarendon Press.

Pagello, F. (2011), "True to *The Spirit*? Film, comics, and the problem of adaptation", *Ol3Media*, 4 (10), http://host.uniroma3.it/riviste/Ol3Media/Pagello.html. Accessed 12 June 2012.

Phillips, G. D. (2006), *Beyond the Epic: The Life and Films of David Lean*, Lexington, KY: University Press of Kentucky.

Ponzanesi, S. (2014), *The Postcolonial Cultural Industry: Icons, Markets, Mythologies*, New York: Palgrave.

Sanders, J. (2005), *Adaptation and Appropriation*, London: Routledge.

Sinyard, N. (1986), *Filming Literature: The Art of Screen Adaptation*, New York: St. Martin's Press.

Sobchack, T. and Sobchack, V. C. (1987), "Film and literature: The problem of adaptation", in *An Introduction to Film*, Boston, MA: Little, Brown, pp. 312–24.

Stam, R. (2005), "Introduction: The theory and practice of adaptation", in R. Stam and A. Raengo (eds.), *Literature and Film: A Guide to the Theory and Practice of Film Adaptation*, London: Blackwell, pp. 1–52.

Verevis, C. (2014), "BADaptation. Is *Candy* faithful?", in C. Perkins and C. Verevis (eds.), *B is for Bad Cinema: Aesthetics, Politics, and Cultural Value*, Albany, NY: State University of New York Press, pp. 215–40.

Vincendeau, G. (2001), *Literature/Film/Heritage: A Sight and Sound Reader*, London: British Film Institute.

Wagner, G. (1975), "Three modes of adaptation", in *The Novel and the Cinema*, Rutherford, NJ: Fairleigh Dickinson University Press, pp. 219–31.

Westbrook, B. (2010), "Being adaptation: The resistance to theory", in C. Albrecht-Crane and D. R. Cutchins (eds.), *Adaptation Studies: New Approaches*, Cranberry, NJ: Associated University Press, pp. 25–45.

Zaiontz, K. (2009), "The art of repeating stories: An interview with Linda Hutcheon", in M. Macarthur, L. Wilkinson, and K. Zaiontz (eds.), *Performing Adaptations: Essays & Conversations on the Theory and Practice of Adaptation*, Newcastle upon Tyne, UK: Cambridge Scholars, pp. 1–10.

3

ADAPTATION AND THE CONCEPT OF THE ORIGINAL

Rainer Emig

Fidelity Criticism, a critical point of view that measures the success of an adaptation against the supposed value and meaning of the original, is as much a stalwart companion of adaptation studies as it is an embarrassment for it. Especially forms of adaptation studies that seem to assume that adaptation only comes into being with film, and that its traditional 'enemy' is literary studies, are often perplexed at the fact that critics are still tempted to search for truthful representations of original works or, worse, the intentions of their authors. Focused on the overt communal production of media such as film, such perspectives appear to them outdated and downright inappropriate.

The present chapter intends to show on the one hand that the question of originality is as old as the human concern with artistic expression, and on the other that it cannot simply be laid to rest by stressing the production modes of modern audio-visual media. Moreover, it tries to demonstrate that some central literary models also become testing grounds for contemporary adaptation and the theories that accompany them. Two famous linchpins of Fidelity Criticism in European culture, William Shakespeare and Johann Wolfgang Goethe, will appear as touchstones, but also as interrelated discursive reference points of the debate. This chapter's latter section will then shift from general cultural and literary questions to those of adaptation between media and its contemporary critics.

Perceiving a piece of art as a 'work' has an old as well as a relatively recent history. An early transition already occurred when the era of Classical Antiquity slowly mutated into the Middle Ages. The works of great Classical artists, such as those of the sculptor Praxiteles (c. 390–320 BC), whose works had spawned innumerable copies in the world of Antiquity, were then superseded by works whose origin was attributed to the one Christian creator-God, whose ideas were merely executed by human helpers. These executors could then justifiably remain anonymous.

Fidelity to an original therefore already means contradictory things at this early stage. In Classical thinking fidelity referred to the precise copying of models and their techniques, while in medieval thinking fidelity to an original meant conformity to an underlying theological idea (on Classical Antiquity, see Porter 2005; for the Middle Ages, see Tatarkiewicz 2010: 247).

This tension also characterises the Early Modern period, in which Michel Foucault locates the origin of modern authorship. While in Antiquity and in the Middle Ages so-called scientific works were authorised by the signatures of their supposed creators while artistic ones could remain anonymous, an artistic work now became an original by having an identifiable creator, while supposed science was acceptable without an identified discoverer (Foucault 1977: 125–127).

The authenticating creator (usually male) supposedly infused an original work with his ideas and skills. These could amount to genius. Since the combined power of creative genius and original work was believed to be strong, there could be no doubt as to its reception. Readings and interpretations were meant to pay tribute to artist and work of genius alike. They were thus usually both biographical and universal, even though this combination (like that of genius) carried within itself another contradiction. If the work was individual and inseparable from the life of the artist–genius, how could it at the same time carry timeless universal meaning on which all interpretations ought to agree unanimously?

The original author

The term 'genius' derives from the pagan Classical idea of inspiration as supernatural. In Greek and Roman art, genii were presented as external figures that attached themselves to humans and inspired them (Murray 1989: 9–31). With the Enlightenment, the idea of inspiration as external shifted into the individual human. This culminated in Romanticism, when it became *de rigeur* that a great artist and thinker had to be a genius. The first texts praising Shakespeare, later frequently held to be the greatest English author, as a genius appeared in the eighteenth century (see, for instance, Montague 1770). In his encyclopaedic study on Goethe, Shakespeare's German equivalent, Gero von Wilpert rightly states that by then the term had become so fashionable as to mean almost anything, but also that Goethe himself revised his definition of genius throughout his lifetime (von Wilpert 2007: 22).

When Goethe reviewed the poetry collection *Des Knaben Wunderhorn* by his colleagues Clemens Brentano and Achim von Arnim in 1806, he aired his own positions on genius:

> *Das wahre dichterische Genie, wo es auftritt, ist in sich vollendet, mag ihm Unvollkommenheit der Sprache, der äußeren Technik, oder was sonst will, entgegenstehen, es besitzt die höhere innere Form, der doch am Ende alles zu Gebote steht, und wirkt selbst im dunkeln und trüben Elemente oft herrlicher, als es später im klaren vermag. Das lebhafte poetische Anschauen eines beschränkten Zustandes erhebt ein Einzelnes zum zwar begrenzten doch unumschränkten All, so daß wir im kleinen Raume die ganze Welt zu sehen glauben.*
>
> *1806: 146–147*

[True poetic genius, wherever it manifests itself, is complete in itself, however strongly the incompleteness of language, of external technique or other issues may oppose it. It possesses the higher inner form that ultimately determines everything and sometimes even acts more splendidly in dark and murky spheres than later in clear ones. The lively poetic perception of a restricted state elevates a singular state to a bounded but unlimited universe, so that we believe to perceive the whole world in a small space.]

Translation Rainer Emig

Genius stands for completeness, even though the conditions inside which it manifests itself are incomplete and deficient. Goethe points at language as a transient and unstable code as much as at the changing fashions of genre and style. A plethora of other "dark or murky" aspects may hamper genius, and we might well assume that these comprise biographical as well as other contextual issues, i.e. ideological ones. Genius, however, and with it the truthfulness of correct reading, the ideal of Fidelity Criticism, outshines these problems. The effect is a paradoxical impression of universality within actual limitation.

It is noticeable that in this early eighteenth-century discussion of fidelity a central aspect of what would later become adaptation studies does not yet feature: the media. This is due to the fact that modern audio-visual media were not yet on the horizon and would only enter the scene in the second half of the nineteenth century (although Goethe was very interested in optics). Nonetheless, there already existed an awareness of adaptation, in the sense that popular literary material in prose and verse was frequently transferred to the stage. Goethe himself did not participate in this, but as a playwright, poet and novelist he not only understood the appeals of diverse genres, but also their mutual interplay. It is no coincidence that his famous Bildungsroman *Wilhelm Meisters Lehrjahre* [*Wilhelm Meister's Apprenticeship*] was first conceived as a novel about the protagonist's hopes of a theatrical career. In the case of Shakespeare, we do not only find adaptations within the dramatic medium, for example, John Dryden's adaptations of Shakespeare's *The Tempest*, the much freer reworking of Shakespeare's *Antony and Cleopatra* into *All for Love, or The World Well Lost* and the adaptation of *Troilus and Cressida*. In 1673, Thomas Shadwell had already turned *The Tempest* into an opera (see Fischlin and Fortier 2000). We also find prose adaptations of Shakespeare's plays, most famously those of the Romantic siblings Charles and Mary Lamb in *Tales from Shakespeare* of 1807.

Media change is crucial in modern views on adaptation (see, for instance, Cartmell and Whelehan 1999). The inevitable contradiction inherent in the simultaneous adherence to the idea of creation and interpretation (and transformation) of something already created is squeezed into a nutshell in Linda Hutcheon's shorthand for adaptation as "both process and product" in her *Theory of Adaptation* (2006: 31). A little earlier in the same study, she is even more explicit in describing adaptation as "always a double process of interpreting and then creating something new" (2006: 20).

Goethe, this much is evident, is not interested in the fact that creation is inevitably entangled in interpretation, something that later theorists would call intertextuality (Allan 2000). He is a traditionalist inasmuch as he assumes an inner integrity of artist and work that transcends external factors. A further conservative move is the claim of universality in the face of evident limitations. The combination of these effects apparently leads to an interpretation that can only be uniform and unified, an idea of reception that is ultimately authoritarian.

Yet the fact that Goethe also spells out these limitations together with the important disclaimers that we are talking about manifestations and their effects – and end up believing (rather than knowing) that we see a universe inside the limited sphere of art – also opens up the debate to more sceptical views on creativity and fidelity.

For Shakespeare, who lived and worked before the Enlightenment, though in an era that was acutely conscious of its Classical heritage, issues of universality inside a murky sphere of power, politics, religious strife and social mobility and its attendant destabilisations of nation, class, family and the individual, also assume a crucial role. Even more radically than Goethe, Shakespeare must have been aware of the instability of language as a code, since not even spelling was standardised at his time. As a prolific author of plays, he was also undoubtedly alive to the changing fashions that made certain conventions all the rage in one season and outmoded in the next.

Yet, much in contrast to Goethe, no claim to unity of an authorial genius that outshines external challenges and becomes one with the work would have made sense for Shakespeare. Drama, his main genre, possessed a high popular but low cultural status as one of several forms of cheap mass entertainment. An avid theatre-going audience demanded new plays every week. As a result, collaboration as well as plagiarism were rife – and untrammelled by modern copyright laws. This was not perceived as a problem for the works thus produced. Their status as uniform and unified works of one singular genius is in fact an erroneous attribution of later periods. The fact that Shakespeare never oversaw the publication of his works (and the resulting confusion of versions) is closely linked with the problem of attributing works to Shakespeare at all – and not his sources and/or collaborators (see Drábek, Kolinská and Nicholls 2008).

The original work of art

Ben Jonson was the first playwright to oversee the publication of his works in a complete edition. Jonson's first folio of 1616 proudly sports a frontispiece that places the title *The Workes of Beniamin Jonson* inside a Classical arch, flanked by allegorical figures in antique costumes (Brady and Herendeen 1991). The plural employed in Jonson's strategic publication, however, rather points at the notion of *oeuvre*, the collected products of a now proudly signatory artist, more so than to the integrity and privileged status of a unique work. Only the latter would be capable of demanding a treatment in accord with the demands of fidelity. It is no coincidence that it took a long time for the English language to employ 'work' for anything other than God's and nature's creations – or human activities in relation to God's plan. The *Oxford English Dictionary* lists John Skelton as the first proponent of such a use in 1523. Shakespeare himself is an early exponent of such a loaded use of 'work' to mean work of art. In *The Winter's Tale* of around 1611, the final reconciliation scene contains the lines "her mother's statue, / […] by that rare / Italian master, Julio Romano, who, had he himself / eternity and could put breath into his work, would / beguile Nature of her custom, so perfectly he is her / ape" (v.ii.97). Compared with their primary works, those of human artists must necessarily remain imitations and therefore deficient.

This changes in the course of later centuries until, once again in the Romantic era, the work of art acquires that which later critics call a nimbus or an aura that makes them resemble traditional theologically sustained creations and become primary works of genius themselves. This change was brought about by a detachment of nature from religion – or rather a projection of traditionally theological qualities into the newly discovered psychology of the now emancipated human being. Already the enlightened eighteenth century had started this trend by assuming that human rationality was in complete accord with the Divine rules of the universe. Rational association of ideas was therefore capable of constructing a world in agreement with both God and nature. External and internal nature were in correspondence in this thinking – and no longer a relation of distanced (and reverential) observation. Wordsworth's famous definition of poetry as the "spontaneous overflow of powerful feelings" from the Preface to his and Samuel Taylor Coleridge's *Lyrical Ballads* of 1800 applies to artistic creativity as much as to the reception of the art thus produced (Wordsworth 1800: xiii). This means that it encapsulates the art work in a safe position, where there can be no discrepancy between artistic intention and the audience's perception.

This, together with the already discussed genius concept, turned the artist into a now worldly creator-God, one who could unite hitherto separated art forms, as does Richard Wagner in his concept of the *Gesamtkunstwerk*, the total work of art. Since the work was apparently generated from nothing but the artist's genius, it was not bound by any rules or conventions. This opened the way for the radical experiments of Modernism and the avant-garde, yet also deprived art of use-value, something that the aesthetes of the *fin de siècle* proudly proposed in their slogan *l'art pour l'art* – art for art's sake (see Roberts 2011). It is interesting, though, that the independence of this new work of art by no means precludes adaptation. Wagner recycled the German national myth of the *Nibelungen* for many of his operas and used medieval legends for others. Oscar Wilde, the most prominent of the British aesthetes, turned a biblical tale into his play *Salome* of 1893.

The professed aim of the *Gesamtkunstwerk* and art for art's sake was the autonomy of the work of art that would make it independent of previous artistic works and the norms and rules institutionalised by them. It would also make it immune to the ideological issues of its time. Of course, this proved an empty promise. It was also a dangerous one, since it postulated an illusory freedom and a position of complete artistic irresponsibility. Theodor W. Adorno criticises the nimbus of the autonomous work of art as delusional and dangerous in his *Aesthetic Theory*. There he writes (in his usual dense style):

> Art works are things which tend to shed their thing-like quality. The authentic and the thing-like do not form distinct layers in a work of art; spirit is not superimposed on some supposedly solid objective basis. One of the key characteristics of works of art is, on the contrary, their ability to undo their own reified shapes in such a way that reification becomes the medium of its own negation.
>
> The two levels are mediated through each other. The spirit of art works evolves from their thing-likeness, and conversely their thing-likeness – i.e. their existence as works – springs from their spirit.
>
> *Adorno 1984: 389*

In other words: art works use the material conditions of their production as means of obscuring this very production. As works, now with the nimbus of original creation, they deny the labour that was put into them to make them what they are. This, of course, is exactly where adaptation produces a problem for the supposedly original work of art: it reminds the work of art of its object status – since adaptation relies on a source that it adapts. Fidelity Criticism tries to find an escape from this impasse by declaring adaptation false or at least inferior when it does not respect the supposed integrity and definiteness of the original. This original is miraculously exempt from a potential object status and remains rife for associations with nimbus and aura, with unquestioned greatness that exceeds the material, historical and biographical conditions of production.

With reference to Shakespeare, the materiality of the work is blatantly evident in the many versions that exist of diverse texts attributed to Shakespeare, including the so-called 'bad quartos' that nonetheless comprise the first editions of *Romeo and Juliet*, *Henry V*, *The Merry Wives of Windsor* and even *Hamlet* (Maguire 1996). But it also affects any production of any of his plays. If Shakespeare's apparent original is the model, then all actors would have to be male, since this was the convention of his time. If this convention is revived today, though, it comes across as a daring directorial decision.

Similar problems apply to costumes – which in Shakespeare's time were generally contemporaneous, i.e. Elizabethan and early Jacobean. This produces a conflict with the epochs and cultures in which Shakespeare's plays were set, from Classical Antiquity and the distant British past to the Danish Middle Ages to more or less contemporaneous Italy and the Mediterranean. If 'faithful' productions today put their actors into English Renaissance costumes, they do not escape this problem of anachronism. The issue could be expanded via stage technology and would make reconstructed Renaissance playhouses like the Globe on London's South Bank the only locations where truthful productions might take place.

A much more subtle, though nonetheless material, issue is the stability of language, of the linguistic codes used in adaptations and productions. Would the unstable Elizabethan spellings be the only truthful ones to which a faithful rendering and production must adhere? And what about pronunciation? Even Goethe, our much closer contemporary, is not treated to reissues of his works in their original spelling – or with their original mistakes.

Walter Benjamin was a modern thinker who made the fate of the work of art in the context of its mechanical reproduction a theme of his investigations. The perspective of his 1936 essay "The Work of Art in the Age of Mechanical Reproduction" is a twentieth-century one, yet his view is deeply coloured by the effects of the Industrial Revolution. One of its early manifestations was the introduction of the printing press, first manually operated and later steam-powered. For many scholars, the introduction of printing marks the shift from the European Middle Ages to the Early Modern period. Shakespeare's works were printed, as were Goethe's, and the printers introduced their own conventions and mistakes into the versions thus produced.

Benjamin's essay operates in a tellingly dialectic fashion. First it declares: "In principle a work of art has always been reproducible" (1969: 219), though later it seemingly reverts to art's originality when it states: "The uniqueness of a work of art is inseparable from its being imbedded in the fabric of tradition" (1969: 222). Yet in the very next sentence, it already qualifies this assertion again by insisting that "This tradition itself is thoroughly alive and extremely changeable" (1969: 223). There is an original for Benjamin, an authentic art work located in a specific time and space. But this location is far from stable. A similar ambivalence is visible in T.S. Eliot's essay "Tradition and the Individual Talent" (1975), first published in 1919. Tradition leads us to the final aspect in which adaptation enters a problematic relation with the idea of fidelity: interpretation.

Universally valid interpretations

When the German Romantics agreed that Shakespeare's character Hamlet was the prototypical German, they founded an interpretative community and a new tradition, one that has remained alive to the present. When Heiner Müller wrote *Die Hamletmaschine* [*Hamletmachine*] in 1977, he used a term in Hamlet's speech to his mother: "Thine evermore, most dear lady, whilst this machine is to him, Hamlet" (ii.ii.123–124) for his title and a fragmentary new translation of Shakespeare's text as the basis of his own. And yet, as critics have established, he also included ideas by Gilles Deleuze and Félix Guattari, especially that of the "desiring machine" from their *Anti-Oedipus* of 1972, a critique of capitalism through the psychoanalytic concept of schizophrenia. Lastly, Müller's text reflects on the role of the intellectual in the GDR and thus represents in a nutshell adaptation, reproduction and the dilemma that Benjamin had outlined for the contemporary art work (Fischlin and Fortier 2000: 208–214). Müller's works in general refrain from clear positions and messages and are thus prototypical postmodern works. But then again, this might be a contemporary prejudice, for did *Hamlet* convey unanimous messages to its original audience?

German readers not only identified the character Hamlet as one of their own; they also quickly claimed to be a better audience for Shakespeare than the English. Thus, Friedrich Theodor Vischers claims in his lectures on Shakespeare at the turn of the nineteenth into the twentieth century:

> *Die Deutschen sind nun also gewohnt, Shakespeare als einen der Unsern zu betrachten. [...]*
> *Ohne undankbar zu sein gegen England, das uns diesen größten aller Dichter geschenkt hat,*
> *dürfen wir es mit Stolz sagen: daß der deutsche Geist zuerst Shakespeares Wesen tiefer erkannte.*
> *1905: 2*

[Germans are used to considering Shakespeare as one of their own by now. [...] Without being ungrateful to England, who has bestowed on us this greatest of poets, we can proudly say that the German spirit was the first to recognise Shakespeare's essence more deeply.]

Translation Rainer Emig

Here we once again have an interpretative community, but now one that recognises itself as such and is capable of evaluating itself. In doing so, it postulates a unique essence in Shakespeare and tellingly no longer distinguishes between person and works, as is typical of genius approaches.

Modern literary and cultural theory knows, of course, that interpretation cannot rely on an essence of meaning that is hidden in texts and art works and merely requires expertise and authority for its correct identification. Reader-response criticism has pointed out the productive uncertainties in meaning that are the real triggers of interpretation (see Iser 1978). Poststructuralism and Deconstruction have driven this further to question whether an ultimate attainment and presence of meaning is even possible, or whether the desire for one is a mere reminder of the theological foundations of knowledge, something Deconstruction calls *logocentrism* (see Derrida 1976: 71).

A more moderate position is chosen by Stanley Fish, who nevertheless posits that interpretation actually precedes the work. In his famous essay "Is There A Text in this Class? The Authority of Interpretive Communities" (1980) he argues that

> communication occurs within situations and [...] to be in a situation is already to be in possession of (or to be possessed by) a structure of assumptions, of practices understood to be relevant in relation to purposes and goals that are already in place; and it is within the assumption of these purposes and goals that any utterance is *immediately* heard.
>
> *1980: 318*

What Goethe assumed to be the dark and murky conditions of artistic creation and reception by readers and audiences, and what Benjamin considered the inevitable but troublingly unstable context of tradition without which no work can be an original one, turns the table on the art work in Fish's argument. Interpretation decides what the work is – and what it means – before the work has any say in this. This might seem paradoxical, yet Fish makes a valid point when he continues:

> My students did not proceed from the noting of distinguishing features to the recognition that they were confronted by a poem; rather, it was the act of recognition that came first—they knew in advance that they were dealing with a poem—and the distinguishing features then followed.
>
> *1980: 326*

This spells the end of Fidelity Criticism – or perhaps its reversal: what is to be trusted is now the established meaning produced by an authorised community of readers, who become the new authors of meaning in the place of the original author. This corresponds to the conclusion of Roland Barthes' essay "The Death of the Author", which states: "the birth of the reader must be at the cost of the death of the author" (1977: 148).

Cultural studies have willingly taken up Barthes', Foucault's and also Fish's position, declaring the common ground of interpretative communities to lie in common political, economic and ideological conditions that enable a group identification under the banner of a correct reading. In a famous Shakespeare essay of 1773, the prominent German intellectual Johann Gottfried Herder tried to wrestle interpretative authority from the French and even the

English by promoting his own genius reading of Shakespeare that put him on a pedestal next to Sophocles (see Osinski 2007). Goethe eventually united his own self-declared genius with that of Shakespeare in a speech given on so-called 'Shakespeare Day' in Frankfurt am Main in 1777. It recuperated Shakespeare not only for Goethe's self-promotion, but also for the emerging programme of *Sturm und Drang*, Storm and Stress, a period also known as that of the cult of genius (Grange 2011: 260–261).

Modern Fidelity Criticism and its uncertain demise

Taking into consideration that the above-mentioned tenets of Fidelity Criticism all belong to a liberal Humanist paradigm that considers the link between authorial power and greatness, the indissoluble work of art and the unity and uniformity of interpretation solid and unbreakable, it is astounding that the first theoretically informed assault on it came from exactly that camp of liberal Humanists. The New Critics, an elitist group of writers and scholars of literature who worked in the 1920s–1940s, wished to rescue the artistic work (mainly poetry) from the onslaught of Marxism and psychoanalysis and revert it back to a notion of greatness that they saw lying within the text itself, in its structures. For their approach, they needed to exclude the focus on the author as the guarantor of meaning, and they did this with one of their famous fallacies (which means common mistakes in interpretation): the intentional fallacy (Wimsatt and Beardsley 1954). Recourse to explicit and supposed intentions of artists and writers were from now on considered erroneous projections. This, of course, also took the authority of the genial author away from any interpretation of a work and opened the path for adaptations that could be more than travesties.

The already mentioned Marxist and Psychoanalytic Criticism, early forms of alternative readings of literary and cultural texts that were later joined by Feminism, Gender Studies, Queer Theory and Postcolonial Criticism, insisted that the artistic work had to be seen in the force field of the political, economic and, in the widest sense, ideological conditions of its time – and that of its reception. A monolithic work to which one could remain true was declared a delusional fable, an expression of an unquestioned ideology rather than its artistic challenge. Coming from the opposite direction as the New Critics, these forms of ideological criticism achieved the same thing, and again made the positions of Fidelity Criticism appear outmoded and reactionary. Cultural studies, which relegated so-called 'high' art and literature to the rank of merely one expression of human culture, did not ask for the meaning of a work of art, but (again in the Marxist tradition) about what 'work' such a work did for a culture at a specific point in history (see Greenblatt 1995). What cultural studies has further done for adaptation studies is to declare the study of all manifestations of art – be they visual, aural, tactile or otherwise – textual studies in its insistence on an extended notion of the text that covers all human manifestations and refuses to differentiate between popular and supposedly high ones. For cultural studies, therefore, hierarchising originals and adaptations is, from the start, a mistaken endeavour, merely a symptom of cultural formations and their prejudices and not an answer to any questions that the hierarchised texts might pose.

The onslaught on the third column upholding Fidelity Criticism, that of a unified and uniform interpretation, came via Poststructuralism and Deconstruction. While Formalism (which emerged at the same time as New Criticism, although its origins in Russia made it forego a bourgeois praise of artistic greatness in favour of the efficiency of the textual machinery) and its successor Structuralism, though more focused on the mechanisms of texts and art works, did not per se abandon the notion of meaning in art and literature, their successors Deconstruction and Poststructuralism focused on the irreconcilable elements in artistic and literary texts and thus came to stress their openness (see Eco 1989), or what the Yale critics Paul de Man, J. Hillis Miller and Geoffrey Hartman called the "unreadability" of the text, by which they meant the impossibility of arriving at any final meaning (see de Man 1986: 21–26). Deconstruction, indeed, declared the very idea of the presence of a final meaning, *logocentrism*, a hangover from the theologically determined view of meaning.

Yet if this was meant to make the positions of Fidelity Criticism ultimately untenable by erasing at one stroke the authority of author, work and interpretation, it also had a paradoxical opposite effect: if all readings are misreadings, as some Poststructuralist and Deconstructionist works seem to imply, or – going even further with the likes of Jean Baudrillard – if simulacra are not the copies of originals, but have replaced originals, then the era of adaptations has dawned upon culture (see Baudrillard 1994). Among the many misreadings, there is then also space for those that postulate an attachment to a supposed spirit of the original, whether it resides in authors, works or interpretations. In the same way as the author has returned into criticism (see Burke 1992), for example through forms such as life writing that itself emerged out of ideological readings such as Marxist, Feminist and Postcolonial ones, Fidelity Criticism cannot justifiably be excluded from the possible realms of approaches to art and literature, even when its bases are exposed as ideological.

Already the early film critic André Bazin had tried to build a bridge between Fidelity Criticism and adaptation when he declared (in the context of adapting novels into films): "Fidelity meant respect for the spirit of the novel, but it also meant a search for necessary equivalents" (Bazin 1971: 141). As George Raitt (2010) points out in his provocatively entitled essay "Still lusting after fidelity?", this double-edged approach, which he sees pursued, among others, by the likes of Brian McFarlane, leads to the problem of differentiating between equivalence and difference, between the translation of material from one code and medium to another and the attendant deformations (McFarlane 1996: 14) – if this term is acceptable at all, since it implies loss rather than potential gain (see Raitt 2010: 50). Other recent contributions to adaptation studies even regard the structural turning away from idealised Fidelity Criticism as its very reinstatement through the back door. Thus, Nico Dicecco argues: "so long as we persist in thinking about the aesthetics of adaptation through the lens of formal repetition, we risk reinforcing the very problems of fidelity intertextual approaches purport to overturn" (2015: 163). Critiquing other recent repeat attacks on Fidelity Criticism, such as John Hodgkins' *The Drift: Affect, Adaptation, and New Perspectives on Fidelity* (2013), Casie Hermansson compares such "ritualistic" attitudes to the "Flogging" of an "(Un)Dead Horse" (Hermansson 2015: 147).

In his essay "Beyond Fidelity: The Dialogics of Adaptation", Robert Stam (2000) views the resilience of Fidelity Criticism as a way of retaining moralistic forms of criticism that have lost their hold in critical methods. Stam, like many adaptation studies scholars, takes the novels-into-film scenario as his starting point (see Stam 2000: 54), thus underestimating the long history of adaptation and its attendant criticism. Yet he is correct when he insists on an "automatic difference" that the change of medium introduces: "beyond such details of mise-en-scène, the very processes of filming – the fact that the shots have to be composed, lit, and edited in a certain way", in short, the difference between a "single-track" and a "multitrack" medium in addition to film's very different production processes (Stam 2000: 56). Yet Stam also acknowledges the inherent essentialism of Fidelity Criticism (see Stam 2000: 57), the insistence on one unique and indivisible meaning.

The alternative form of Fidelity Criticism is identified by Stam as an emphasis on truthfulness to "the medium of expression" (Stam 2000: 58). It assumes that some media are better at expressing some features than others. A common prejudice is, for instance, that film dazzles with the simultaneity of aesthetic features (picture, sound, movement, etc.), while the novel possesses greater depth through the abstraction inherent in language. Yet, as Stam also points out, both novels and films are already intertextual forms that contain within themselves adaptations of other texts and cultural artefacts (Stam 2000: 61–62). Stam eventually suggests a compromise that replaces the inadequate notion of fidelity with that of translation and finally dissolves it in a concept of dialogic intertextuality (Stam 2000: 62–68). Yet this is as convincing as it is facile. Intertextuality is always inherent in adaptation, but who is to say that all intertextuality is dialogic? Stam appears to make the mistake of replacing the intentionalism of Fidelity Criticism with that of a more modern but equally intentionalist adaptation scholar. Nevertheless, he precisely outlines the steps in this process that lead to a successful adaptation: "selection, amplification, concretization, actualization, critique, extrapolation, analogization, popularization, and reculturalization" (Stam 2000: 68). But he does not state if this "Grammar of Transformation" (Stam 2000: 68) is a prescriptive one for producing adaptations or for analyzing them. That the two ultimately fall into one proves that even Stam, with his demonstrated knowledge of the same shifts in literary and cultural theory that were outlined in the present chapter, cannot escape the attraction of a coherent monolithic meaning that Fidelity Criticism offers, even when it is presented as multiple, dialogic and translational.

But if detractors of Fidelity Criticism have a hard time escaping its clutches, its postmodern supporters also have problems sustaining its praise. Colin MacCabe, Kathleen Murray and Rick Warner's *True to the Spirit: Film Adaptation and the Question of Fidelity* (2011) pays much attention to the intertextual and dialogic nature of adaptation (almost as if it wanted to confirm Stam's contrary claims), but has very little that is new to say about fidelity. In his Introduction, MacCabe reminds us of the fact that literature also reacts to film, for instance, in the shape of modernist writing (MacCabe 2011: 5), but then jubilates prematurely when he discovers the term "ideal construct" in André Bazin's writings on film, a term that MacCabe apparently sees as a way out of the conundrum of fidelity (see MacCabe 2011: 6). For him, it combines an impersonal view of adaptation as a process with a residual insistence on value. But does it? Ultimately, as the conclusion of MacCabe's chapter proves, he simply shifts the ground of fidelity from author to director (MacCabe 2011: 21–22), even when he does so with the aid of Eliot and Barthes. Whether this new author and authority of adaptation is indeed a "super-author", as he claims (MacCabe 2011: 22), is questionable, especially when all the above rhetoric only ever leads to Raymond Williams' "structure of feeling" (Williams 1961: 41–71). Instead of individual genius to which our adaptations and their interpretations have to be true, we would then have communally shared expectations and perceptions, with the problem that these are equally unquestioned and unquestionable.

Works cited

Adorno, Theodor W. *Aesthetic Theory*. London: Routledge & Kegan Paul, 1984.

Allan, Graham. *Intertextuality (The New Critical Idiom)*. London and New York: Routledge, 2000.

Barthes, Roland. *Image – Music – Text*. New York: Hill & Wang, 1977.

Baudrillard, Jean. *Simulacra and Simulation*. Trans. Sheila Faria Glaser. Ann Arbor, MI: University of Michigan Press, 1994.

Bazin, André. *What Is Cinema?* Vol. II. Berkeley, CA: University of California Press, 1971.

Benjamin, Walter. "The work of art in the age of mechanical reproduction." *Illuminations*. Trans. Harry Zohn. New York: Schocken Books, 1969. 217–251.

Brady, Jennifer, and W.H. Herendeen (eds.). *Ben Jonson's 1616 Folio*. Newark, DE: University of Delaware Press, 1991.

Burke, Sean. *The Death and Return of the Author. Criticism and Subjectivity in Barthes, Foucault and Derrida*. Edinburgh: Edinburgh University Press, 1992.

Cartmell, Deborah, and Imelda Whelehan. *Adaptations. From Text to Screen, Screen to Text*. London: Routledge, 1999.

Deleuze, Gilles, and Félix Guattari. *Anti-Oedipus: Capitalism and Schizophrenia*. Athlone Contemporary European Thinkers. London: Athlone, 1972.

Derrida, Jacques. *Of Grammatology*. Trans. Gayatri Chakravorty Spivak. Baltimore, MD: Johns Hopkins University Press, 1976.

Dicecco, Nico. "State of the conversation. The obscene underside of fidelity." *Adaptation* 8.2 (2015): 161–175.

Drábek, Pavel, Klára Kolinská and Matthew Nicholls (eds.). *Shakespeare and His Collaborators over the Centuries*. Newcastle upon Tyne, UK: Cambridge Scholars Publishing, 2008.

Eco, Umberto. *The Open Work*. Trans. Anna Cancogni. Cambridge, MA: Harvard University Press, 1989.

Eliot, T.S. "Tradition and the individual talent." *Selected Prose*. Frank Kermode (ed.). London and Boston: Faber & Faber, 1975. 37–44.

Fischlin, Daniel, and Mark Fortier, Eds. *Adaptations of Shakespeare. An Anthology of Plays from the 17th Century to the Present*. London and New York: Routledge, 2000.

Fish, Stanley. *Is There A Text in this Class? The Authority of Interpretive Communities*. Cambridge, MA: Harvard University Press, 1980.

Foucault, Michel. "What is an author?" *Language, Counter-Memory, Practice*. Ithaca, NY: Cornell University Press, 1977. 113–138.

von Goethe, Johann Wolfgang. Review of *Des Knaben Wunderhorn. Alte deutsche Lieder*, by Clemens Brentano and Achim von Arnim. *Jenaische Allgemeine Literatur-Zeitung*. 18/19 (1806): 137–144 and 145–148.

Grange, William. *Historical Dictionary of German Literature to 1945*. Historical Dictionaries of Literature and the Arts 47. Lanham, MD: Scarecrow Press, 2011.

Greenblatt, Stephen. "Culture." *Critical Terms for Literary Study*. Frank Lentricchia and Thomas McLaughlin (eds.). 2nd ed. Chicago, IL : University of Chicago Press, 1995. 225–232.

Hermansson, Casie. "Flogging fidelity. In defense of the (un)dead horse." *Adaptation* 8.2 (2015): 147–160.

Hodgkins, Jack. *The Drift. Affect, Adaptation, and New Perspectives on Fidelity*. New York: Bloomsbury, 2013.

Hutcheon, Linda. *A Theory of Adaptation*. New York: Routledge, 2006.

Iser, Wolfgang. *The Act of Reading. A Theory of Aesthetic Response*. Baltimore, MD: Johns Hopkins University Press, 1978.

MacCabe, Colin, Kathleen Murray and Rick Warner (eds.). *True to the Spirit: Film Adaptation and the Question of Fidelity*. New York: Oxford University Press, 2011.

McFarlane, Brian. *Novel to Film. An Introduction to the Theory of Adaptation*. Oxford: Oxford University Press, 1996.

Maguire, Laurie E. *Shakespearean Suspect Texts. The "Bad" Quartos and Their Contexts*. Cambridge, UK: Cambridge University Press, 1996.

de Man, Paul. *The Resistance to Theory*. Minneapolis, MN: University of Minnesota Press, 1986.

Montague, Elizabeth. *An Essay on the Writings and Genius of Shakespeare*. 2nd ed. London: unknown publisher, 1770.

Müller, Heiner. *Die Hamletmaschine: Heiner Müllers Endspiel*. Cologne, Germany: Prometh, 1978.

Murray, Penelope (ed.). *Genius. History of an Idea*. Oxford: Wiley-Blackwell, 1989.

Osinski, Jutta. "Shakespeare als Sophokles' Bruder? Über Herders Shakespeare-Rezeption." In Roger Paulin (ed.) *Shakespeare im 18. Jahrhundert. Das achtzehnte Jahrhundert – Supplementa*. Göttingen, Germany: Wallstein, 2007. 167–180.

Porter, James I. "What is 'classical' about classical antiquity? Eight propositions." *Arion* 13.1 (2005): 27–61.

Raitt, George. "Still lusting after fidelity?" *Literature/Film Quarterly* 38.1 (2010): 47–59.

Roberts, David. *The Total Work of Art in European Modernism*. Ithaca, NY: Cornell University Press, 2011.

Stam, Robert. "Beyond fidelity: the dialogics of adaptation." In James Naramore (ed.) *Film Adaptation*. London: Athlone, 2000. 54–76.

Tatarkiewicz, Władysław. *A History of Six Ideas. An Essay in Aesthetics*. Trans. Christopher Kasparek. The Hague, the Netherlands: Martinus Nijhoff, 2010.

Vischers, Friedrich Theodor. *Shakespeare-Vorträge*. 2nd ed. Stuttgart and Berlin: J.G. Cotta, 1905.

Williams, Raymond. *The Long Revolution*. New York: Columbia University Press, 1961.

von Wilpert, Gero. *Goethe. Die 101 wichtigsten Fragen*. Munich: C.H. Beck, 2007.

Wimsatt Jr., W.K., and Monroe C. Beardsley. "The Intentional Fallacy." *The Verbal Icon. Studies in the Meaning of Poetry*. W.K. Wimsatt Jr. (ed.). Lexington, KY: University of Kentucky Press, 1954. 3–20.

Wordsworth, William. "Preface." *Lyrical Ballads: With Other Poems*. Second Ed. London: Longman & Rees, 1800. v–xlvi.

4

AN EVOLUTIONARY VIEW OF CULTURAL ADAPTATION

Some considerations

Patrick Cattrysse

Introduction

This chapter discusses what has been called the 'evolutionary view'[1] on cultural evolution and adaptation. In his study *Biological and Cultural Evolution: Similar but Different*, Mesoudi (2007: 119) points out that

> ever since *The Origin of Species*, but increasingly in recent years, parallels and analogies have been drawn between biological and cultural evolution, and methods, concepts, and theories that have been developed in evolutionary biology have been used to explain aspects of human cultural change.

While researchers agree that biological and cultural evolution show similarities and dissimilarities, they still debate on what these (dis)similarities are. Consequently, the following does not claim that cultural adaptation equals biological evolution, but rather assumes more generally that some specific features of an evolutionary approach might contribute to the traditional study of (film) adaptations. Hence, in what follows, the word 'adaptation' refers to *the process (or its result) of an item (or a set of items) changing or being changed to better fit their new surroundings*. This definition retains and discards at once some specific features that are common in the evolutionary paradigm. First off, it suggests looking into the distinction between the ipsative and the additive adaptation, i.e. the adaptation of the self and the adaptation of something into something else. While the former is more common in everyday language and the natural sciences, the latter is more common in film adaptation studies. Section 1 examines briefly if and how this distinction triggers new or at least different research questions in the field of cultural adaptation studies.

A second feature that is retained from the evolutionary paradigm concerns its target oriented perspective. The notion itself is not unproblematical, but it raises interesting questions with respect to traditional (film) adaptation studies. For example, on the backdrop of the never-ending debate for or against the concepts of 'fidelity' and 'equivalence', Section 2 discusses how an evolutionary view of adaptation offers ammunition to the opponents of the approach, and provides arguments to those adaptation critics who intend to step beyond the traditional fidelity-based discourse. Indeed, a target-oriented approach to adaptation studies deflates the explanatory power of concepts like 'fidelity' and 'equivalence' when adaptation complies with target (con)text requirements rather than with specific (dis)similarity relations between a source practice and a target practice. Moreover, an evolutionary view of cultural adaptation also includes adaptation processes that lead to end-products that no longer share features with the materials they adapted. In those cases, 'fidelity' and 'equivalence' seem to have left the adaptation scene for good. Not only have they become unhelpful to explain adaptations, in some cases they cannot even help to identify an item as an adaptation.

However, adaptations that do not share features with the adapted raise the question of how we can identify such phenomena as adaptations. For example, if an item P2 does not share a common feature with an item P1, we may think that P2 replaces rather than adapts P1. At first sight, the obvious solution seems to lie in a study of the making process of P2, assuming that a thing that results from an adaptation process may be called an 'adaptation'. Section 3 rekindles the old distinction between a study of 'adaptation as an end-product' and a study of 'adaptation as a process', and suggests that however obvious this solution may appear, it is less straightforward than one might think. Moreover, this shift of analytical focus leads us to the question of how to define adaptation as an evolutionary process. Section 3 suggests that even if transformational processes produce end-results that show no common features with their initial states, for them to be experienced as "adaptational" requires "invariance conditions" (Toury 1980: 24), irrespective of whether they are called 'fidelity', 'equivalence' or any other name. This suggests that if Section 2 seemed to allow adaptation scholars to throw 'fidelity' and 'equivalence' out of the discipline, Section 3 implies that they may just as soon have to sneak these concepts back in again.

Finally, an evolutionary or target (con)text-oriented approach to adaptation assumes that adaptation processes are goal-driven and aim at an optimal fit of the adaptation (as an end-product) in its target context. Looking at a specific set of *film noir* adaptations, Section 4 examines one possible conceptualization of what 'best fit' may mean in a specific historical context.

The working definition of 'adaptation' suggested above implies also that various features that are proper to evolutionary biology must be discarded here. For example, whereas biological evolution refers to a population-level process, cultural adaptation is considered both at the type and at the token levels of adaptation. I hereafter also leave aside the ongoing controversy about Darwinian and Lamarckian (epi)genetics and the intra- and intergenerational information transfer.

1 Ipsative versus additive adaptation

An evolutionary take on cultural adaptation suggests a first distinction I hereafter discuss in terms of 'ipsative' versus 'additive' adaptations. I borrow the terms from comparative and translation studies (see, e.g., Chesterman 1998: 18–19). Indeed, comparative studies distinguish between ipsative and self–other similarity relations: the former refer to a thing being identical to itself (e.g. A=A); the latter point to a thing being identical or similar to something else (e.g. A=B). For example, one five-euro bill is obviously identical to itself, but it may also be said to be identical to another five-euro bill. Based on this distinction, translation scholars have discussed the pros and cons of an ipsative or equative view of translation and an additive view of translation, and agreed to abandon the former in favour of the latter (Chesterman 1998: ibid.). When looking at adaptations however, it is understood that they comprise both ipsative and additive processes,[2] even though at first sight, it would seem that the ipsative adaptation is more commonly studied or considered in common parlance and in the natural sciences, while the additive adaptation figures more prominently in literature and film studies. Hence, from the point of view of (film) adaptation studies, it would seem that opening up the analytical focus to include also ipsative adaptations could enrich the research field. In 'To adapt or to adapt to? Consequences of approaching film adaptation intransitively', Leitch (2009) already suggests as much. Unaware of the previously existing terminology, the author distinguishes between what he calls the 'transitive' adaptation ('X adapts Y') and the 'intransitive' adaptation ('X adapts to Y'). Whereas the former corresponds with the additive adaptation, the latter matches the ipsative or equative adaptation. Leitch (2009: 95ff.) argues that a shift of focus from the additive to the equative adaptation raises some provocative questions, which involve a number of changes. Among them, the author mentions the changing of notions such as 'agency' and 'intention', the conceptualization of the process of adaptation as well as its motives, and the notions of 'teleology', 'substance' and 'identity'. I hereafter discuss the notions of 'identity' and the different conceptualization of the ipsative and additive process of adaptation, and I return to the concepts of teleology, agency and intention in Section 2.

I would argue that both ipsative and additive processes require the notions and the perception of identity, stasis or change and time. However, the equative or ipsative adaptation refers to a phenomenon that adapts itself, whereas the additive adaptation is said to adapt something into a separate entity. The distinction is oftentimes easy to make, and may then seem obvious. For example, when a species or an organism adapts itself, i.e. changes its proper features either during its lifetime or over the course of generations in order to better fit its new and changing environment, one would say that it evolves through an 'ipsative' or 'equative' process of adaptation. The same applies to the football team that must adapt its strategy against its opponent if it is going to win. Similarly, in a paper called "Cultural Adaptation and Translation", the British sociologist Ruth Cherrington explains how Chinese students, when coming to study in British universities, have to adapt to their new hosting environment (Cherrington 2012). What is common to these adaptation processes is that they all apply to the self, and since they occur in time, they install (dis)similarity relations between a prior version and a later version of the self. Note also that the end result of the ipsative adaptation of one item results in one item, and that, at least in these cases, the adaptation replaces the adapted and makes the latter disappear in the process. The relevance of this latter observation will become clear in Section 3. Finally, I point out that, as suggested above, these adaptation processes, as well as others that will be discussed hereafter, may be said to occur at the token level, not the type or species level, as is more common in evolutionary genetics.

By contrast, the additive adaptation adapts something into something else that becomes a separate entity. Unlike the ipsative adaptation, it does not delete the adapted in the process, in this paradigm, it actually leaves the adapted untouched, even though it may entail changes in its perception, and as the name suggests, it adds one or more items to the initial one, and thus installs self–other (dis)similarity relations. Translations are obvious examples, but so are film adaptations of literary texts, and are actually most commonly discussed intertextual practices in literary and film studies. The copy, the quote, the paraphrase, the parody, the pastiche, the compilation, the prequel, the sequel, the remake: all install self–other (dis)similarity relations and thus represent (the result of) additive processes.

Following Chesterman (1998: 19–20), one may formalize both processes as follows:

The sign '⇒' separates the situation before and after the operation. It also suggests the one-directional, and irreversible, line of time and ditto nature of the process.

- ipsative adaptation: P1 ⇒ P1'
- additive adaptation: P1 ⇒ P1+P1'+P1''+P1''' ...

Consequently, an evolutionary take on cultural adaptation raises the questions whether, besides the additive cultural adaptation, there are also ipsative cultural adaptations, and, if so, to what extent do they function in similar and different ways? Whereas the latter question requires further investigation, the former one is easily answered. Clear examples of ipsative cultural processes are the restoration of a damaged painting or the practice called 'bricolage'. Not everyone may label these processes adaptational, but, more to the point here, the process applies to the thing itself, and after the restoration the damaged painting no longer exists. Similarly, the word 'bricolage' refers to the creative use of whatever is around and available to make something new. The practice involves the displacement or re-use of one item, and results in one item, whereby when taken to its new use or environment, the item no longer exists in its initial use or environment. Hence, bricolage also installs (dis)similarity relations with a thing and itself. Furthermore, if ipsative adaptation means any kind of goal-driven self-referential evolution, any historical study of a cultural phenomenon may be studied as such. Consequently, a study of ipsative cultural adaptations may reveal adaptational features in processes that have been studied elsewhere as something else. Think, for example, of the evolution of a genre (e.g. how the Western adapted to its ever-changing surroundings, how literary or game genres evolved …), the evolution of a character in a feature film, a TV series, a feature franchise or a series of games (e.g. James Bond or Lara Croft), the way gender roles evolved or adapted in time and space (e.g. the 'Bond girl'), the way themes and ideologies gained and lost popularity, how narratives adapted to the specifics of various media, makers and audiences across cultures, etc. Note however that, unlike the ipsative adaptations mentioned above (cf. the species adapting to its new environment, the football team adapting its strategy to the opponent's, etc.), the ipsative adaptation process of genres, characters, gender roles and the like does not delete its initial status, nor does it erase its intermediary steps. On the contrary, unless other circumstances made them disappear, they remain as the manifest traces of the adaptation process, thus allowing the analyst to reconstruct how the initial item changed into its new state. Consequently, when considering Chesterman's scheme discussed above, these examples display features that match both the ipsative and the additive adaptation in that they install (dis)similarity relations with the self but do not erase the intermediate steps of the process. And once again, the relevance of this observation will become clear in Section 3.

To the extent that the distinction between ipsative and additive processes is clear, one may also study how additive adaptations have conditioned ipsative adaptations. For example, previous historical studies have shown how the 1930s gentleman-detective movie continued its success, i.e. adapted itself ($\approx$ ipsative adaptation) through the additive adaptation of literature, and how the 1940s *film noir* innovated the crime film partly in the same way, i.e. by adapting American *romans noirs* (see, e.g., Cattrysse 1990: 292ff.; 2014: 325ff.).

Recognizing adaptational characteristics in phenomena that are studied in other fields of research allows for literature-into-film studies to connect with these (sub)-disciplines, and to find its proper position and role in a larger field of cultural adaptation studies. It may also launch a comparative study of adaptational phenomena across disciplines and sciences, both natural and human. Adaptation studies may thus become a research field that examines how any paradigm, whether cultural or natural, deals with change in order to become or remain relevant.

Before moving on to the next paragraph, I add one caveat that concerns the conception of the additive adaptation. The questions of whether and how the additive adaptation process, like the translation or the film adaptation of a novel, actually impacts and modifies its source materials, like the ipsative adaptation does, remains a matter of contention. Some scholars state that an additive adaptation does not physically touch or change its sources, even though after being translated or adapted a text may be looked at differently (see, e.g., Toury 1980: 24; Cattrysse 1990: 36–37). Others have suggested redefining 'adaptation' as an ongoing and incessant two-way instead of one-way process, studying not only the changes of the content and form from novel to film, but also the changes being inferred on the originating text (see, e.g., Bruhn 2013). These commentators consequently object to the goal-orientedness of the adaptation process, and suggest that one

> de-hierarchize the relation between the primary and the secondary text, the source and the result, in order to make both texts results of each other.
>
> *Bruhn 2013: 83*

However, to conflate 'adaptation' with 'dialogics' and 'intertextuality' forsakes the possibility, which existed previously, of studying adaptations as one-directional, irreversible and goal-oriented processes, which represent the single turns that constitute the very makeup of wider, dynamic and more complex networks of polycentric and simultaneous or consecutive, multi-directional, inter-discursive forces. Hence, in this view, adaptations and particular influences offer the empirical input to describe and explain complex and dynamic interdependencies. The request to de-hierarchize the relation between pre-texts and post-texts, and to *a priori* reduce the former even to the status of a retroactive construct of the latter (see, e.g., Iampolski 1998: 9), echoes a debate that took place in the 1970s in literary studies. In an attempt to rid their discipline of the source-hunting reflex, commentators proposed to replace the concept of 'influence', which in their mind was too reverent towards the past, with the more present time-oriented term 'intertextuality'. Their aim was to replace at once the traditional vision of an author's passive submission to influences with either a more neutral or – why not? – a more positive view of writing as a creative interaction with previous texts (Orr 2008: 15, 83ff.; Juvan 2008: 54ff.). One could argue, however, that instead of claiming *a priori* that no hierarchical relations exist between pre-texts and post-texts, or that creation should be approached as 'active interpretation of' rather than 'passive submission to' previous influences, the analyst should keep an open mind and let the empirical data speak for themselves. To assume beforehand that all influences are equal reminds one of Orwell's *dicta*; we should know by now that some sources are more equal than others. Consequently, a proper study of power dynamics will be hindered rather than helped by *a priori* assumptions about hierarchical relations between pre-texts and post-texts. *Ad hoc* interdependencies may reveal themselves to be highly asymmetrical, and therefore require the analyst to look into one-directional or multi-directional vectorial forces. Following this, it may be safer for the analyst to assume *a priori* that the degree of passive submission versus active resistance in creation is not pre-given, and requires an *ad hoc* empirical investigation.

2 Evolution, fidelity and equivalence

To study adaptation in evolutionary terms also means to assume that the adaptation process is goal-driven. The idea to study adaptation as a teleological phenomenon is not new. Russian formalists like Viktor Žirmunski and Boris Eikhenbaum suggested already in the early 1920s that

> Even where an act of borrowing can be clearly established, the critic's prime concern ought to be not with the 'where from', but with the 'what for'; not with the source of the motif, but with the use to which it is put in the new 'system'.
>
> *Quoted in Erlich 1969: 268*

Half a century later, German and Israeli scholars picked up this teleological notion again in the field of translation studies (see, e.g., Even-Zohar 1978 [revised version published in 1990]; Toury 1980 on polysystem theory; and Vermeer 1978 on Skopos theory), and so did some film adaptation scholars the following decade (see, e.g., Andrew 1984; Orr 1984; Cattrysse 1990).

But what does it mean to claim that the adaptation process is teleological? The study of teleological behaviour is part of a larger field called theory of explanation.[3] Its tradition goes back to the ancient Greek (see, e.g., von Wright 2012). In a similar way, the evolutionary paradigm considers items that change in a changing environment. While most changes do not last, some of them do because they ended up providing the changed item with a better fit in its environment, at least temporarily, until the next environmental change. The first type of changes are called mutations, while the latter are called adaptations. A target-oriented study of cultural adaptations could thus in a similar way look for changes, which after the fact turn out to be adaptive, and describe and explain those adaptive changes.

It is also important to note that the current theories of teleological explanation agree that one should not reduce goal-directed behaviour to intentional, free-willed individual agency. Both non-human animate organisms (e.g. bacteria) and inanimate objects (e.g. a thermostatically controlled heat pump) may 'behave' or function in a teleological way, while not being assigned intentions (see, e.g., Salmon 2006: 26ff.). Matters change, however, when dealing with manmade artefacts, such as (film) adaptations or other art works. Then the aforementioned concerns about authorial agency and intention resurface. This question becomes all the more urgent in a Romantically biased setting where artistic creation is seen as a divine and immaculate process – i.e. unblemished by any influence – an active, free-willed and intentional process also, performed by an individual Genius Auteur. It is safe to say that for the last two centuries, the Romantic value system has prevailed among the elite in the Western art worlds.[4] And yet, the urge in culture and adaptation studies to consider other conditioners besides conscious individual agency is not new and seems to be increasing. For example, writing in 1927 on literary evolution, Jouri Tynjanov already pointed out that focusing on

> the structural function, that is, the interrelationship of elements within a work, changes the 'author's intention' into a catalyst, but does nothing more. 'Creative freedom' thus becomes an optimistic slogan which does not correspond to reality, but yields instead to the slogan 'creative necessity'.
>
> *Tynjanov 1971: 74*

Interestingly, more recent findings in multiple disciplines seem to confirm rather than refute this claim. Whereas intentions are generally assigned to individuals, sociologists of art have traditionally argued that 'all artistic work, like all human activity, involves the joint activity of a number, often a large number, of people' (Becker 2008: 1; see also Bourdieu 1998), and scholars have also studied institutional and collective agency in various forms, including 'collective intentions', herd behaviour and 'wisdom of the crowd' phenomena (see, e.g., Simonton 2003; Tolcffson 2004; Becker 2008, McIntyre 2008; Heath 2010; Swaab, 2016: 292ff.). However, more disturbingly, recent findings in the social neuro-sciences, social neuro-psychology and biology have raised doubts about the very existence of common notions like 'intention', 'free will' and 'decision-making'. While some experts have argued that they merely represent illusions,[5] others have suggested that we at least seriously reconceptualize these notions.[6]

Hence, when Leitch (2009: 95) writes that

> It is certainly true [sic] that individual authors and adapters intend goals and design strategic changes to meet them, it is much less clear that culture as such has goals and intentions.

the author reiterates a common belief. And when, in a later publication, Leitch (2013) suggests that, metaphorically speaking, texts such as screenplays, for example, could be said to 'want' something, his thoughts seem to go in a direction that is opposite to the one suggested by the studies mentioned above: instead of assigning metaphorical intentions to inanimate objects, these studies suggest that one should rather understand human intentions to be metaphorical.[7]

So far, adaptation critics have shown little interest in these kinds of ruminations, but some have called on the evolutionary, i.e. teleological nature of the adaptation process as a way to step beyond the traditional fidelity-based novel-into-film studies in order to highlight the originality of the adaptation rather than its faithfully adapting its source(s).[8] As such, it is interesting to note that, in the end, both the proponents and the opponents of the fidelity rule commonly serve (Romantically biased) judgmental rather than epistemic purposes. Put simply, the proponents of the fidelity rule discard adaptations as valuable cultural or artistic phenomena because of their derivative nature and lack of originality; at best, their value stems from the prestige of what they hope faithfully to adapt. Opponents of this view discard the relevance of fidelity, claiming that adaptations may be and have been creative and original in their own right.

The fidelity-based discourse is of course as old as (film) adaptation studies itself, and whereas opponents of this view may wish that it had disappeared today, wishing it does not make it so. There is still plenty of evidence to show that adaptation commentators, students, laymen and sometimes even adaptors continue to foster the notion that a film adaptation should represent a literary text in a respectful if not a faithful way.[9] The persistent prevalence of the fidelity or equivalence issue in film adaptation studies may also stem from the fact that, whatever definition was used, an adaptation, translation or any other intertextual practice always suggests represent-ing the adaptation or translation of something. The verbs 'to adapt', 'to translate', 'to paraphrase', etc. all represent transitive verbs, which grammatically suggest direct objects. Consequently, word users naturally expect there to be an umbilical cord between the adaptation and what it adapted, or the translation and what it translated, etc. This also explains why comparative source text–target text analysis has always been and remains a central part in adaptation studies, and why intertextual practices or relations in general have traditionally been classified according to source text–target text relations. Agreed, researchers have in the meantime complemented text analyses with studies of contexts, agency and wider media-oriented approaches, sometimes to the point of replacing the translation or adaptation as a primary object of study with a study of the transla-tor or the adaptor (see, e.g., Chesterman 2009; Cattrysse 2014: 193ff.). Be that as it may, it would be hard to conceive of adaptation studies without comparative text analysis, and here, perhaps, an evolutionary view of adaptation could offer support to the opponents of the fidelity discourse. I hereafter briefly discuss two arguments that could defend their cause: the first explains how, when comparing adaptations with their adapted materials, equivalence and fidelity lose their explanatory power. The second argument shows how an evolutionary view of adaptation may disable the concepts so far that it is impossible even to identify a phenomenon as an adaptation.

Fidelity and equivalence lose their explanatory value

If the adaptation process is understood to be teleological, i.e. to 'aim' at a better fit into the host-ing environment, a comparative source text–target text study may describe *what* shifts occurred during the passage from source to adaptation. But when it comes to explaining *why* these shifts occurred, the analytical focus will need to shift to the relationships of fit between the adaptation and its new target context. In other words, adaptations may or may not replace original items for reasons that lie outside the (dis)similarity relations that exist between the adapted and the adapting items. Take for example the second film adaptation of Dashiell Hammett's novel *The Maltese Falcon*, produced as *Satan Met a Lady* in 1936. Comparing book and film, one notices, among many other changes, that the movie replaces the cynical tough-guy detective Sam Spade with the comical gentleman-detective Ted Shane. Madame Barabbas takes Kasper Gutman's place, and the Maltese falcon becomes a horn. Comparing (dis)similarities between the exchanged items, however, does not help us understand why these features were replaced, or what the relevance of these changes could be. Not being a mind reader, I leave aside all assumptions about authorial intentions, but if nothing else, one thing is clear: the filmmakers did not make a tough-guy detective movie. Instead, they produced a combination of the gentleman-detective and the screwball comedy, two very popular film genres at the time. One may therefore assume that Ted Shane replaced Sam Spade not so much because he was similar or different, but rather because he fit into a long line of gentle-man-detectives played by William Warren. The character change extended the success of Warren's previous roles, including Perry Mason, Philo Vance and Michael Lanyard, a.k.a. *The Lone Wolf*. In other words, whatever the features of the selected source materials, they needed to be adapted, i.e. changed in order to better fit the target (con)text requirements.

Fidelity, equivalence and the identification of an adaptation

If one accepts an evolutionary definition of 'adaptation', the concepts of 'fidelity' and 'equivalence' may also lose their value as a distinctive feature to identify adaptations. Indeed, the aforementioned examples of ipsative adaptations show that to adapt is to change. If the organism or species needs to adapt in order to survive, they need to change, and the same applies to the football team that needs to adapt its current strategy. No change means no adaptation. Change is thus the *sine qua non* for adaptation, even though not all change is called 'adaptational'. As stated above, from this point of view, only those changes that imply what is felt by a beholder as an improvement of the entity – to make it fit better in its new hosting environment – qualify as 'adaptational'. From this, it follows that when it comes to identifying adaptation, any reference to sameness with respect to an initial state of affairs, for example, through the use of concepts like 'fidelity' or 'equivalence', is not only of secondary importance (see above), but points in the wrong direction; it refers to non-adaptation. On the other hand, change comes in degrees: it can be partial and it can be complete. The relevance of fidelity and equivalence lessens most when change is complete. If at the proper level of analysis change is partial, the adaptation process changed or adapted only part of its source(s) and left the rest un-changed or 'un-adapted'. In that case, a comparative study of source and target will still reveal common features. These common features may then be seen as leftovers of the 'fidelity' or 'equivalence' norm, and if they were, they may function as hypertextual markers (Broich 1985), suggesting hypotextual materials, and thereby trigger the identification of an item as a possible adaptation or hypertext. In that case, 'fidelity', 'equivalence' or similarity relations in general function more as 'smoking guns', so to speak. Conversely, if change is complete, fidelity no longer plays a role, and 'equivalence' is achieved in a rather counter-intuitive way, through difference only. In that case, neither concept can help to identify the end result as an adaptation.

Hence, in conclusion, one would think that the evolutionary view of adaptation brings about three (or at least two) cheers for the opponents of the fidelity rule and that, after all those years, it is finally 'Game Over' for fidelity.

3 Evolution and the identification of adaptation

The conclusion that evolutionary adaptation processes may lead to adaptations that do not share a common feature with the adapted may well be grist for the mill of the opponents of the fidelity discourse. But it raises at once the question, on what grounds can one identify such phenomena as adaptations? Indeed, if two items do not share a common feature, how is one to conclude or even guess that one is or could be the adaptation of the other? One way to find out may be to study or reconstruct the production process of the latter item. This shifts the analytical focus from the definition of an adaptation (as an end-product) to the study of the adaptation as a process.

Once again, a caveat may be in order for the beginning adaptation scholar. However obvious this solution may seem, the identification of an adaptation (as an end-product) through the reconstruction of its production process is not always that straightforward. Previous adaptation studies have shown that, in a somewhat counter-intuitive way, not all phenomena that are pre-sented and/or perceived as adaptations (as end-products) result from a process that is presented and/or perceived as adaptational, and vice versa, not all 'adaptation' processes lead to 'adaptations' as end-results. Here we enter the domain of what translation and adaptation scholars have called "illusory translations" (Levý 1969), "covert" translation (House 1977), "pseudo-original" (Cattrysse 1990) or "secret" or "hidden" adaptations (Grant 2002). More elaborate proposals to study the (film) adaptation process in terms of selection policies versus actual adaptation processes, and to study these processes also in relation with the function and position of their end-products in their hosting environment, appeared already in the early 1990s, if not sooner (see, e.g., Cattrysse 1990, 1992a, 1992b, see also Section 4).

So back to the study of adaptation as a process: various translation, adaptation and intertextuality scholars have suggested one consider the translation, adaptation and other types of text processing to consist of three or five logical operators producing 'shifts' between an initial and final state of affairs: the repetition, the addition, the deletion, the substitution and the permutation (see, e.g., Wienold 1972; Van Gorp 1978: 102–103; Cattrysse 1990: 203–205). The substitution may be seen as a combination of the deletion and the addition (e.g. A instead of B), and so may the permutation, which represents a change in order between the selected items (e.g. A-C-B instead of A-B-C). Interestingly, as I indicated elsewhere (see Cattrysse 2014: 268), none of these operators produces a change that 'feels' adaptive, except perhaps the permutation. The repetition lacks change, and therefore resembles the copy more than the adaptation. To add or delete an item is not perceived as an adaptation, and neither is the replacement of an item A with an item B. Imagine the following situation: John tries to drive a nail in the wall with his shoe. When he fails, he drops the shoe, takes a hammer and finishes the job. When describing what happened, we shall say that John substituted the shoe for the hammer, not that the shoe adapted into a hammer. Similarly, looking at *Satan Met a Lady* (1936), we would say that Ted Shane replaces Sam Spade and that Mme Barabbas takes Kasper Gutman's place, not that the one adapted into the other. And yet, film critics generally label *Satan Met a Lady* (1936) as an 'adaptation' of Hammett's *The Maltese Falcon*. How is it then that these five operators may produce a change that feels adaptational?

One explanation seems to lie in the interpretation of continuity versus discontinuity of identity. Since the five operators produce complete and abrupt changes, they operate on entire and discrete entities and thus trigger an interpretation of discontinued identity. By contrast, for change to be experienced as adaptational, it needs to be partial and gradual. Partial change leaves room for non-change or similarity relations, and thus allows for an interpretation of continued identity. The way this works is studied in *Gestalt* psychology and mereology, i.e. the study of the perception of parts and wholes. When considering the shoe and the hammer, or Ted Shane and Sam Spade, we experience the substitution of entire items, not the adaptation (in terms of partial change) of one into the other. However, if we adopt a wider perspective, where these items become the parts of a larger whole, our interpretation may change. I may say that John 'adapted' his working method to drive a nail in the wall, and the substitutions of Sam Spade, Kasper Gutman and the Maltese falcon may thus be seen as the parts of a larger whole – the narrative – which apart from some 'minor' changes remains 'the same'. In other words, when changes at the parts-level do not disturb the perception of continued identity at the whole-level, one may perceive these changes as constitutive of an adaptation process. Since it is not relevant here, I leave aside the additional condition that change serves improvement with respect to the hosting context.

But how then may a partial adaptational process lead to a target text that is entirely different from its initial state? Simple enough: through its second feature: 'gradualism'. Successive partial change may finally lead to complete change. This is what this image suggests[10]: if we look from left to right, we see Bush's face change gradually into a different face, i.e. Schwarzenegger's face.

Now if a partial and gradual change ends with an item P2 that shows no resemblance to its original state P1, where does this leave the notion of 'similarity' and the related concepts of 'fidelity' and 'equivalence'? Comparative studies call it 'transformational similarity' (Imai 1977). Indeed, comparatists apply various models to assess similarity, but two have been dominant in cognitive psychology: the feature or contrast model and the geometric or mental distance model (see, e.g., Chesterman 1998: 7; Minda 2015: 25ff.). In simplified terms, the former studies *comparanda* in terms of displaying features which do or do not overlap. The latter is based on the assumption that the similarity between two things is a function of the psychological distance between them, and that psychological distance is analogous to physical distance. However, one thing that neither model is able to account for is the kind of similarity that exists when two things are linked by non-featural aspects (Minda 2015: 31–32). This is the case with an adaptation P2 that adapts a P1 to the point of changing it completely. The transformational model is based on the notion that there is something similar about two things if one can be transformed into the other. Hence, according to the transformational model, water and steam are 'similar' to each other because one entity can be transformed into the other entity. This relationship is not strictly based on features that are shared, but rather on the number of steps it takes to turn water into steam. The similarity relation recalls Wittgenstein's (1953: 31) "family resemblance". The notion conveys the idea that the members of a category may share a series of a number of features, where no one feature is common to all. Consider, for example, the following table.

IF:	THEN:
A = 1,2,3,4	
B = 2,3,4,5	B is similar to A
C = 3,4,5,6	C is similar to B
D = 4,5,6,7	D is similar to C
E = 5,6,7,8	E is similar to D.

Note that, in the end, E does not share any features with A. While Wittgenstein's idea concerns the simultaneous existence of members in a category, one may also consider A, B, C, etc. as the steps in a partial and sequential changing process of one entity. Then it recalls Theseus' paradox and raises the question at what point in the successive and partial transformation process A ceases to exist and becomes E? At what point does Bush become Schwarzenegger or Theseus' ship become another ship? In other words, gradual partial change challenges the distinction between ipsative and self–other identity, and renders the distinction itself gradual. I venture to suggest that the question is unanswerable because of a human cognitive limitation. Whereas the identification of an entity requires the mental or conceptual fixation of it in the continuous flux of time, it cannot be conceived simultaneously with the identification of change. With reference to Heisenberg's uncertainty principle, physicists have made a similar observation when discovering that it is impossible to both determine the position and momentum of a particle.

The perception of continuity versus discontinuity of identity depends also on the representation of the changing process. Do we see the process occur at all, as in the Bush/Schwarzenegger image, or are we just left with the comparison of an initial and a final state of affairs? Hence, the representation of change in both the table and image suggests that the more detailed the representation of partial and gradual change, the stronger the feel of continued identity, and thus the stronger the adaptational effect of the process, irrespective of the ultimate difference or distance between the initial and the final state of affairs. If we could perceive the actual morphing process in the image, the effect of adaptation and continued identity might even be stronger than it already is. This is how family members, having lived together for decades, still recognize each other as 'the same' persons, even though their two-months-old version and their seventy-years-old version no longer share one common cell. Similarly, this explains how complex entities, like the Classical Hollywood Cinema for example, as described in Bordwell et al. (1985), may be conceived of as one entity, in spite of its numerous changes over the course of many decades.

Hence, when considering the additive (film) adaptation process where one item P2 does not share any feature with another item P1, analysts are confronted with two or more items, which may or may not be the result of an adaptational, translational or any other intertextual process operated on one or more previous items. Generally, they were not present when the process occurred, so if they want to know more about its nature, they need to reconstruct the process after the fact. Translation scholars have already developed a tradition in what Cordingley and Montini (2015) have coined 'genetic translation studies'. The authors hereby suggest another reference to the evolutionary view. The practice actually antedates the term by decades and examines the genesis of the creative process, be it a poem or a novel (see, e.g., Hay 1979; de Biasi 1988). Within translation studies, various sub-fields have emerged that study 'the little black box' (see, e.g., Toury 1995: 183ff.; Simeoni 1998: 3), i.e. the decision-making process of translators while translating, doing so-called TAP research (Think-Aloud Protocols) (see, e.g., Toury 1995: 232–238; Jääskeläinen 2002), or more recently tracking eye movements (see, e.g., O'Brien 2007), keystrokes, errors, corrections, hesitations and the like. All approaches attempt to study the mental process of translating. Needless to say, this type of research is less common in (film) adaptation studies. To scrutinize the mind of one translator is one thing; to examine the intentions of hundreds of people involved in the making of a (film) adaptation is quite another. However, this does not leave the adaptation scholar without any means to reconstruct the genesis of an adaptation. The usual procedures are still available: comparing screenplay drafts, analyzing production notes, interviewing people involved in the production process, etc. These obvious techniques certainly help, even though at times they may be unavailable (e.g. what about authors who died?) or problematical (e.g. what about authors who forget, invent intentions after the fact or lie deliberately?). Moreover, even though almost a hundred years old, the evolutionary view of adaptation still warns us against over-estimating the importance of individual intention and free will. As indicated above, suggestions made in the 1920s by Tynjanov, Eikhenbaum and others find corroboration in current neurosciences, claiming that free-willed individual intentionalism, as we commonly understand it today, may rely more on wishful thinking than on fact. Hence, if the adaptation process is not spread out in front of us, reconstructing a production process *ex post facto* in terms of sequential partial change leading to a similar but different item that better fits its new hosting environment may turn out to be more than challenging.

Conversely, and paradoxically, the reconstruction of the ipsative adaptation process may be less problematic when dealing with the self-referential evolution of cultural phenomena, such as the evolution or adaptation of genres, gender roles, characters and the like (see above), given that unlike the ipsative adaptation processes we discussed at the beginning of this chapter (e.g. the adaptation of a species or organism, football tactics, living on a British university campus), they do not erase the intermediate steps during the course of the adaptation process. For example, a scholar studying how the James Bond character evolved since the first time Sean Connery personified the character on the screen has free access to all the intermediate stages that led the evolutionary process to its last stage.

Finally, cognitive psychology teaches us that when dealing with knowledge and communication, we need to acknowledge the importance of perception. As Medin and Goldstone (1995) point out:

> Logically, similarity is a multi-placed predicate, not a two-place one (A is like B) nor even a three-placed one (A is like B in respect C); when we say that A is similar to B, what we

really mean is that '*A* is similar to *B* in respects *C* according to comparison process *D*, relative some standard *E* mapped onto judgments by some function *F* for some purpose *G*.

Quoted in Chesterman 1998: 12

From this it follows that the explicitation of the proper level of analysis is crucial to any study. Part 3 offered already an illustration of this when, instead of focusing on John's shoe and his hammer, we focused on the higher level of his working method to drive a nail in the wall. A similar parts-to-whole-level shift applies to the football team, exchanging one technique for another and so adapting its overall tactics in order to win. In other words, at what level we consider parts and wholes depends on the proper level of the analysis that is required for the *ad hoc* investigation. Looking back at the James Bond example: one critic may study the adaptation of an Ian Fleming novel as an additive adaptation while another may investigate the adaptation as one link in the longer chain of ipsative adaptation that progresses in the James Bond franchise. Looking back at the debate about intention and agency: proclaimed individual intentions may be superseded or inspired by infra- as well as trans-individual conditioners, which help the analyst to distinguish between more proximal and more distal causes.[11] As always, analytical relevance depends on the researcher's (inter-)subjective points of interest, the purpose of the investigation and how the world is.

4 'Best fit': the matter of quality

Finally, an evolutionary view of adaptation involves implications with respect to how adaptation users assign value to adaptations. Whereas in common language, a translation is often considered to be 'correct' or 'wrong', the words 'correct' and 'wrong' are rarely used in reference with adaptations, even though adaptations may be and often have been labelled more or less successful. If the aim of an adaptation process consists in changing a source element in order to make it fit better into its new hosting environment, then it follows that a 'good' or 'successful' adaptation refers to the process and its end-result that achieve just that. What this means in concrete terms becomes then a matter of empirical investigation of actually produced adaptations. Section 1 referred already to some features on the basis of which some specific *film noir* adaptations have been called 'successful' or not. For example, this study showed that in a hosting environment where certain conventions are stable and appreciated, a 'good' adaptation consists in assimilating to or morphing into the hosting background to the point of becoming 'invisible' as a foreign element. Exogenic features, i.e. features that are felt to be foreign, in adaptations are then generally depreciated. In this context, translation scholars sometimes use the term 'translationese' to refer to disturbing traces of foreignness in translations. Film adaptation critics may run into similar 'disturbing' exogenic features for example when the film adaptation of a theatre play limits its dramatic action to only a few settings, and thereby produces a 'theatrical', i.e. 'claustrophobic' feel, or when a screenwriter writes 'on the nose' dialogue, and tells instead of shows. Needless to say, these text features are audience related, i.e. related, among other things, to viewing competence and viewing conditions more in general. In other words, one needs to study them in an empirical way as text cues producing audience effects, whether intended or not. Also, when originality prevails in the target context(s), adaptations are forced to go into hiding and to appear as originals, hence 'pseudo-originals' (see above). In that case, exogenic features are once more eschewed, and 'good' adaptations are the ones that manage efficiently to cover their adaptational tracks. As indicated above, to identify such phenomena as adaptations may become difficult if not impossible. However, the goal of adaptations may also be the reverse: to import foreign features, and to 'loudly' call attention to them because they revitalize target conventions, or refer to the prestige of the imported items in order for that prestige to rub off on the adaptation and its environment. Well-known examples of the latter case are the film adaptations of prestigious literary authors like Shakespeare or Tolkien. Examples of the former case are found in the 1940s *film noir* adaptations of the tough guy detective variant, starting with *The Maltese Falcon* in 1941. Consequently, adaptation users may value exogenic or foreign features either in a positive or a negative way, depending on the historical specifics of the hosting situation and the role and position of the adaptation(s) in that context.

Conclusions

This chapter discussed some implications following an evolutionary definition of cultural adaptation. A first concerned the distinction between ipsative and additive adaptations. Whereas the clarity of the distinction depends on the perception of identity, the former type introduces 'new' phenomena into adaptation studies and allows one to study phenomena that were studied elsewhere as something else (e.g. genre evolution) in terms of adaptational features. Future research could examine if and how ipsative and additive adaptations, both as processes and as end-products, evolve and function in (dis)similar ways.

The above also argued that, on the one hand, the evolutionary view may be seen as the (target context-oriented) antipode of the (source text-oriented) fidelity-based discourse. Since the latter has dominated (film) adaptation discourse for decades, the former could be seen as a belated but still welcome change in adaptation studies. On the other hand, this essay argued that, to the extent that the word 'adaptation' suggests the adaptation of something, notions like 'continued identity', implying 'fidelity', 'equivalence' or at least 'invariance conditions', remain relevant. Consequently, the evolutionary view may complement rather than displace the concepts 'fidelity' or 'equivalence'. Put more simply, what-questions will continue to involve source text–target text comparisons, whereas why-questions are likely to lead to target context conditioners that lie outside the *comparanda*.

Moreover, in spite of its scientific antecedents, an evolutionary view of cultural or film adaptation does not prevent adaptation critics from applying it for judgmental, i.e. political rather than scientific purposes. In that case, adaptation criticism continues the paragone debate between the art forms, i.e. between literature and cinema. If the end-goal consists in using the fidelity rule versus the evolutionary view to argue why the book was better than the film or vice-versa, both paradigms may indeed be seen as each other's (source vs. target (con)text oriented) antipode, but they sit at the bi-polar ends of one and the same judgmental scale: that of the Romantic value system. Then advocates on either side of the scale continue to do battle, wielding the same Romantically biased arms called 'Originality', 'active free-willed creation' and 'individual Auteurist intention'. This leaves space for a descriptive-explanatory approach that disregards the (inter)-personal taste of the analyst. Section 4 presented one way that adaptation studies scholars could assess quality in a less judgmental manner.

Notes

1 See, e.g., Boyd (2009) and Boyd et al. (2010).
2 See, e.g., Gambier (1992).
3 Theorists have used the word 'explanation' in many ways. One common definition, however, refers to causal explanation, i.e. the detection or interpretation of cause–effect relations – as opposed to, say, statistical correlations – between events. Needless to say, whereas causes are supposed to occur *before* their effects, analysts can only identify causes *after* their having produced their effects. This may have

inspired the aforementioned post-modernist critics to conclude that causes occur after their effects, and that translations or adaptations actually 'make' their originals.

4 See, e.g., Becker (1982).
5 See, e.g., Harris (2012).
6 See, e.g., the work of Dan Ariely, Daniel Dennett, Daniel Kahneman, Alfred Mele, Steven Pinker, John Searle, Amos Tsversky.
7 See, for example, Daniel Dennett's Ted Talk called "Cute, Sweet, Sexy and Funny", where he discusses Darwin's strange inversion of reasoning: https://www.ted.com/talks/dan_dennett_cute_sexy_sweet_funny; visited on 14-04-2017. It raises the question whether we fear snakes because they look dangerous or whether they look dangerous because we fear them? And what about cars?
8 See, e.g., Bortolotti and Hutcheon (2007).
9 I refer to my film adaptation students who, year after year, assess adaptations on the basis of their faithfully adapting their beloved literary texts. See also, e.g., Bruhn et al. (2013: 2).
10 See image at: https://upload.wikimedia.org/wikipedia/it/0/06/Striscia_morphing.jpg. Accessed 27 -10-2017.
11 See Dan Dennett TED-talk mentioned above.

Works cited

Andrew, Dudley (1984) *Concepts in Film Theory*, Oxford: Oxford University Press.

Becker, Howard S. (1982) *Art Worlds*, Berkeley, CA: University of California Press.

Becker, Howard S. (2008) *Art Worlds*, 25th Anniversary Edition, Berkeley, CA: University of California Press.

de Biasi, Pierre-Marc (Ed.) (1988) *Carnets de travail*, Edition critique et génétique établie, Paris: Balland.

Bordwell, David, Staiger, Janet and Thompson, Kristin (1985) *The Classical Hollywood Cinema. Film Style and Mode of Production to 1960*, London: Routledge.

Bortolotti, Gary R. and Hutcheon, Linda (2007) "On the origin of adaptations: Rethinking fidelity discourse and 'success'—biologically". *New Literary History* 38: 443–458.

Bourdieu, Pierre (1998) *Les règles de l'art. Genèse et structure du champ littéraire*, Paris: Éditions du Seuil.

Boyd, Brian (2009) *On the Origin of Stories: Evolution, Cognition, and Fiction*, Cambridge, MA: Harvard University Press.

Boyd, Brian, Carroll, Joseph and Gottschall, Jonathan (Eds) (2010) *Evolution, Literature & Film. A Reader*, New York: Columbia University Press.

Broich, Ulrich (1985) "Formen der Markierung von Intertextualität", in *Intertextualität. Formen, Funktionen, Anglistische Fallstudien*, Tübingen, Germany: Niemeyer Verlag.

Bruhn, Jorgen (2013) "Dialogizing adaptation studies: From one-way transport to a dialogic two-way process", in J. Bruhn, A. Gjelsvik & E.F. Hanssen (Eds) *Adaptation Studies: New Challenges, New Directions*, New York: Bloomsbury Academic.

Bruhn, J., Gjelsvik, A., Hanssen, E. F. (Eds) (2013) *Adaptation Studies: New Challenges, New Directions*, New York: Bloomsbury Academic.

Cattrysse, Patrick (1990) "L'Adaptation filmique de textes littéraires. Le film noir américain", Katholieke Universiteit Leuven, Leuven. Available at: https://www.academia.edu/1098943/1990_Ladaptation_filmique_de_textes_littéraires_le_film_noir_américain [Accessed 27 July 2016].

Cattrysse, Patrick (1992) "Film (adaptation) as translation: Some methodological proposals", *Target. International Journal of Translation Studies* 4, 53–70.

Cattrysse, Patrick (1992) *Pour une théorie de l'adaptation filmique: Le film noir américain*, Bern, Switzerland: Peter Lang International Academic Publishers.

Cherrington, Ruth (2012) "Cultural adaptation and translation: Some thoughts about Chinese students studying in a British university", in *Translation, Adaptation, and Transformation*. London: Continuum.

Chesterman, Andrew (1998) *Contrastive Functional Analysis*. Amsterdam, the Netherlands: John Benjamins Publishing.

Chesterman, Andrew (2009) "The name and nature of translator studies", *Hermes—Journal of Language and Communication Studies* 42: 13–22.

Cordingley, Anthony and Montini, Chiara (2015) "Genetic translation studies: An emerging discipline", *Linguistica Antverpiensia, New Series–Themes in Translation Studies* 14: 1–18.

Dennett, Daniel (1991) *Consciousness Explained*. New York: Little, Brown and Company.

Erlich, Victor (1969) *Russian Formalism: History – Doctrine*, The Hague, the Netherlands: Mouton Publishers.

Even-Zohar, Itamar (1990) "The position of translated literature within the literary polysystem", in James S. Holmes, José Lambert, Raymon van den Broeck (Eds) *Literature and Translation: New Perspectives in Literary Studies,* Leuven, Belgium: Acco.

Gambier, Yves (1992) "Adaptation: une ambiguïté à interroger", *Meta: Journal des Traducteurs / Meta: Translators' Journal* 37: 421–425.

Grant, Catherine (2002) "Recognizing *Billy Budd* in *Beau Travail*: Epistemology and hermeneutics of an auteurist 'free' adaptation", *Screen* 43: 57–73.

Harris, Sam (2012) *Free Will*, New York: Free Press.

Hay, Louis (Ed.) (1979) *Essais de critique génétique*. Paris: Flammarion.

Heath, Joseph (2010) "Methodological individualism", [WWW Document] *Stanford Encyclopedia of Philosophy*. URL http://plato.stanford.edu/entries/methodological-individualism

House, Juliane (1977) "A model for assessing translation quality", *Meta: Journal des traducteurs / Meta: Translators' Journal* 22: 103–109.

Iampolski, Mikhail (1998) *The Memory of Tiresias. Intertextuality and Film*. Translated by Harsha Ram. Berkeley, CA: University of California Press.

Imai, Shiro (1977) "Pattern similarity and cognitive transformation", *Acta Psychologica* 41: 433–447.

Jääskeläinen, Riitta (2002) "Think-aloud protocol studies into translation", *Target. International Journal of Translation Studies* 14: 107–136.

Juvan, Marko (2008) *History and Poetics of Intertextuality*, West Lafayette, IN: Purdue University Press.

Leitch, Thomas (2009) "To adapt or to adapt to? Consequences of approaching film adaptation intransitively", *Studia Filmoznawcze* 30: 91–103.

Leitch, Thomas (2013) "What Movies Want", in J. Bruhn, A. Gjelsvik and E.F. Hanssen (Eds) *Adaptation Studies: New Challenges, New Directions*, New York: Bloomsbury Academic.

Levý, Jiří (1969) *Die Literarische Übersetzung: Theorie einer Kunstgattung*, Frankfurt, Germany: Athenäum.

Lorenz, Chris (1987) *De Constructie van het Verleden. Een Inleiding in the Theorie van de Geschiedenis*, Amsterdam: Uitgeverij Boom.

McIntyre, Phillip (2008) "The systems model of creativity: Analyzing the distribution of power in the studio", *JARP* 3: np.

Medin, D. L. & Goldstone, R. L. (1995), "The predicates of similarity", in C. Cacciari (Ed.) *Similarity*, Milan: Bompiani, pp. 83–110.

Mesoudi, Alex (2007) "Biological and cultural evolution: Similar but different", *Biological Theory* 2: 119–123.

Minda, John Paul (2015) *The Psychology of Thinking. Reasoning, Decision-Making & Problem-Solving*. London: Sage Publications.

O'Brien, Sharon (2007) "Eye-tracking and translation memory matches", *Perspectives: Studies in Translatology* 14: 185–205.

Orr, Christopher (1984) "The discourse on adaptation", *Wide Angle* 6: 72–76.

Orr, Mary (2008) *Intertextuality. Debates and Contexts*, Cambridge, UK: Polity Press.

Salmon, Wesley (2006) *Four Decades of Scientific Explanation*, Pittsburgh, PA: University of Pittsburgh Press.

Satan Met a Lady (1936) [film] Hollywood: William Dieterle.

Simeoni, Daniel (1998) "The pivotal status of the translator's habitus", *Target. International Journal of Translation Studies* 10: 1–39.

Simonton, Dean Keith (2003) "Creative cultures, nations and civilisations: Strategies and results", in *Group Creativity: Innovation Through Collaboration*. Oxford: Oxford University Press.

Swaab, Dick (2016) *Ons Creatieve Brein. Hoe Mens en Wereld Elkaar Maken*, Antwerp: Atlas Contact.

Toleffson, Deborah (2004) "Collective intentionality", *Internet Encyclopedia of Philosophy*.

Toury, Gideon (1995) *Descriptive Translation Studies and Beyond*, Amsterdam: Benjamins Translation Library. John Benjamins Pub. Co.

Toury, Gideon (1980) *In Search of the Theory of Translation*, Tel Aviv: Porter Institute for Poetics and Semiotics.

Tynjanov, Jurij (1971) "On literary evolution", in *Readings in Russian Poetics*, Cambridge, MA: MIT Press.

Van Gorp, Hendrik (1978) "La Traduction Littéraire parmi les autres Métatextes", in James S. Holmes, José Lambert and Raymond van den Broeck (Eds) *Literature and Translation: New Perspectives in Literary Studies,* Leuven, Belgium: Acco.

Vermeer, Hans (1978) "Ein Rahmen für eine allgemeine Translationstheorie", *Lebende Sprachen* 23: 99–102.

Wienold, Götz (1972) *Semiotik der Literatur*, Frankfurt.: Athenäum.

Wittgenstein, Ludwig (1953) *Philosophical Investigations*, Oxford: Blackwell Publishing.

von Wright, Georg Henrik (2012) *Explanation and Understanding*, London: Routledge & Kegan Paul.

PART II

Historiography

Katja Krebs

We swim in the past as fish do in water, and cannot escape from it.

Hobsbawm 1997: 31

History and historiography are, of course, established core areas of research within all academic fields that deal with adaptation, such as comparative literature, film studies, theatre and performance studies, cultural studies, media studies and so forth. Addressed especially widely have been issues related to the archive; ephemera's role within both historiographies and histories; the marginalisation of aspects of histories; and the politics embedded in histories and historiographies. Adaptation studies is no exception; history is at its core as it establishes as well as questions relationships between texts and their readers. Its concern with histories of adaptation practices and, importantly, its critique of temporal relationships between texts are important cornerstones of the field, cornerstones which are continually re-examined and reshaped in an effort to understand the phenomenon that is adaptation. The practice of adaptation is able to unearth and comment upon our own rootedness in the past. This section aims to investigate the many different ways in which adaptation engages with history, whether that is through an analysis of such rootedness or a critical engagement with adaptation studies' relationship to issues such as fidelity, to name a central, ever recurring trope.

Gregory Semenza charts adaptation studies' engagement with history, whether through the single case study, so prominent in the field, or, through what he terms "synchronic histories of adaptations" which concern themselves with particular moments in time. His assessment of the single case study, which compares adaptation to its source, finds that the tendency to privilege a so-called original can lead to a not necessarily critical embrace of fidelity. According to Semenza, the majority of these case studies fall into a so-called fidelity trap and even if they don't, they tend to lack the necessary historical rigour. Instead, he argues, adaptation studies needs to engage with diachronic histories as, "[w]hat's been missing for much of this time is a more telescopic view of adaptation".

In "Not just the facts: adaptation, illustration, history", Thomas Leitch takes as his starting point an investigation into the relationship between cinema and history, whereby he posits that "one of cinematic history's great contributions to the field is to shed new light on the constructed nature of all history". Exploring the relationship between the constructions of written histories, as is done by historians, and non-written ones, such as films, Leitch asserts that "historians assume that only words can tell stories, pictures can merely illustrate those stories". Offering examples of cinema's fictional and non-fictional accounts of the D-day landings in Normandy, in June 1944, it becomes apparent that the categories of fiction and non-fiction may have little use, as "they are not substantive but performative, both dependent on the ways they are framed by both producers and audiences". The consequence such a position may have for the role of adaptation studies is that it enables a more critical engagement with history as narrative rather than as fact.

Also concerned with the unpicking of conceptual positions vis-à-vis history and adaptation, Robert Geal's "Dialogism's radical texts, and the death of the vanguard critic" charts the way in which adaptation studies' own history has propelled it to the centre of textual studies. For Geal, the role Bakhtin's dialogism has played in the history of adaptation studies is central, and he argues that we have witnessed a dialogic turn, engendered partly by a critical engagement with the notion of fidelity. Charting a history of theoretical trends to which "adaptation studies was temporarily impervious to", Geal posits that adaptation studies' engagement with dialogism has allowed it to foreground the "shift from theory as radical to the uses of text as radical".

As this section on history and adaptation continues, Kyle Meikle, in "Adaptation and the media", takes a very intriguing instance of the slippery nature of qualifying a text as adaptation in order to make the case for "[e]nlarging the field of adaptation studies to encompass anchors, editors, and reporters" as that would reveal the "similarity between the business of adaptation and the business of journalism – on both an industrial and intertextual level". Turning stories into other stories is what adaptation and journalism have in common, and it is the untangling of their complex relationship which might allow adaptation studies as a field to pick up the gauntlet and grapple "with adaptation studies as media studies, as the study of media in all of its competing meanings".

Elaine Indrusiak and Ana Iris Ramgrab turn their attention to "Literary biopics: Adaptation as historiographic metafiction". Literary biopics are a contested genre: considered by many to be not as good as a biography, as the biopic transforms the legacy of individuals into "products of mass consumption", they are nevertheless exceptionally popular with a mass audience. Many a literary biopic further complicates the relationship between biographic accounts of individuals with their own position as authors: "literary biopics depict writers both as creations and creators, while the artists' works are rendered both as fact and fiction". Yet blurring "the lines between fact and fiction, high and low art, film and literature" could be the trait of adaptation which continues to make it so enduring both as object of consumption as well as object of study.

A different history altogether is what concerns Ricardo Fassone in "Notoriously bad: early film-to-video game adaptations (1982–1994)". As if to respond to Meikle's call to grapple "with adaptation studies as media studies", Fassone unpicks the relationship, both adaptive and commercial in nature, of early video games and their cinematic cousins. And such an unpicking relies on two histories: a "recapitulation of a specific trend within the video game industry" alongside "an analysis of larger cultural and technological shifts" within "different media systems". Taking us through the various stages of video game developments, from Atari through to Nintendo and LJN, Fassone elucidates how the industry developed from a "cluster of sub-par video games with a film license slapped on top" to a point where the relationship between cinema and video game adaptation has become fully integrated as part of "reshaping the paths of circulation and consumption of film".

Johan Callens' work on "Rosas: appropriation as afterlife" takes as its focus a renowned Belgian dance company, Rosas, and traces the central position intertextuality, interdisciplinarity and intermediality holds in Anne Teresa De Keersmaeker's work. Furthermore, it problematises what some call re-appropriation and others identify as plagiarism by Beyoncé, and "probes the question where to draw the thin line between, on the one hand, originality and its attendant rights and rewards, and on the other hand, the ready availability for consumption of cultural materials circulating in a globalized landscape" – a landscape which by no means offers equal access or rights to such materials.

No section on history and adaptation would be complete without at least one reference to the nineteenth century. Not only has that period seen a prolific production of adaptations itself, but adaptations of texts from that era have delighted and continue to delight audiences of all media, be it television, theatre, cinema, radio, graphic novels, fanzines, blogs etc. Lysette Lopez Szwydky completes this section with "Adaptations, culture–texts, and the literary canon: on the making of nineteenth-century 'classics'". Taking a historical perspective to explain the phenomena of the contemporary adaptation's apparent obsession with a nineteenth-century canon, Szwydky looks back "to the nineteenth-century to see the beginnings of the commercial adaptation industry that dominates the present". She argues for a central position for adaptation, not so much as part of the process that is canon preservation, but rather as part of the process of canon formation: "Historically, adaptation has served as the catalyst for literary canonization, both then and now." Basing her discussion on arguably the most prolific writers in terms of a nineteenth-century canon, that is, Charles Dickens and Mary Shelley, Szwydky concludes that such a historic approach to adaptation studies can uncover "the symbiotic relationship between popular adaptation and literary canonization".

5

TOWARDS A HISTORICAL TURN?

Adaptation studies and the challenges of history

Gregory Semenza

Let us acknowledge at the start that adaptation scholarship, generally speaking, is sophisticated and thriving. *Someone* needs to say it. For decades now—since Dudley Andrew's characterization of adaptation discourse as "frequently the most narrow and provincial area of film theory" (1984: 96)—many scholars have seemed almost guilty about their professional interest in adaptations. As I have argued elsewhere (2015: 6), authorial laments about the field's backwardness, in both article and book introductions, have hardened into tropes almost as predictable and ritualistic as the anti-fidelity declaration still proffered by so many authors. Andrew was right about the field thirty years ago, and even today, it is easy to understand the sorts of neuroses an institutional identity crisis such as ours (we're not quite film studies, and we're not quite literary studies) might cause. I think even the most formidable critics of adaptation studies would be hard pressed, however, to argue convincingly that the field is *in toto* more provincial or narrow than other ones. Moreover, in the past decade, the influences of poststructuralism, postcolonialism, feminism, cultural studies, and even evolutionary biology have helped to redefine adaptations as sites of a fascinating and limitlessly complex cultural and intertextual dialogism that has profoundly impacted the historical development of both literature and film.[1]

I begin with this positive revaluation of adaptation studies because I believe the developments sketched above followed specifically Andrew's call for adaptation studies to take "a sociological turn" (1984: 104), a somewhat vague prescription which, although almost never acknowledged, spoke to the crucial significance of *history* in the field's evolution. Consider the questions and facts that Andrew's "sociological turn" intended to highlight:

> How does adaptation serve the cinema? What conditions exist in film style and film culture to warrant or demand the use of literary prototypes? Although adaptation may be calculated as a relatively constant volume in the history of cinema, its particular function in any moment is far from constant. The choices of the mode of adaptation and of prototypes suggest a great deal about the cinema's sense of its role and aspirations from decade to decade. Moreover, the stylistic strategies developed to achieve the proportional equivalences necessary to construct matching stories not only are symptomatic of a period's style but may crucially alter that style.
>
> *1984: 104*

One of the more powerful emphases of Andrew's key paragraph has to do with the constantly changing functions, modes, and structures of adaptation "from decade to decade." Although George Elliott Howard famously described the intersections of history and sociology through the claim that "History is the past Sociology and Sociology is the present History" (1904: vii), I'm not sure I understand entirely why Andrew chooses the term "sociology" rather than "history." Presumably, he wishes to suggest the usefulness of studying adaptations as trans-temporal phenomena while keeping an emphasis on how adaptations function in the present and might function in the future. In spite of the emergence of various historical turns at precisely the moment he composed his famous essay—new historicism, cultural materialism, new cultural history, and so forth—he understandably opts for a term capable of transforming a traditionally provincial and narrow field into one whose future relevance could be more ambitiously imagined. One wonders, though, whether a logical emphasis on the present and future has caused us to downplay the importance of the past, and of the *longue durée*, and what these neglected subjects might teach us moving into the future.

One particular limitation of even the most cutting-edge theoretical approaches to adaptation studies is their striking transhistoricity. Too often, scholars have insisted on defining the terms according to which adaptations should be evaluated without equal insistence on the contingent and constantly changing nature of historical events, let alone of literary, dramatic, musical, and filmmaking practices and priorities.[2] Even when scholars acknowledge in some way the importance of historical change and contingency, they just as often seem comfortable culling their definitions of adaptation from the smaller contexts on which they almost inevitably choose to focus. In Simone Murray's *The Adaptation Industry* (2012), for example—to my mind, the most effective example of a game-changing sociological study of contemporary adaptations—she admits that "[a]ny mapping of the *contemporary* adaptation industry … needs to take account of aspects of the book trade which have become pervasive only in recent decades" (2012: 20)—choosing to italicize 'contemporary' as a way of signaling her awareness of the historical specificity of such cultural phenomena. In spite of her reasonable decision to limit her subject matter to the cultural economy of adaptation since 1980, she nevertheless feels confident arguing for the applicability of her approach even to "studies of 'classic' text adaptions occurring in earlier eras of the book, radio, film and television industries" (Murray 2012: 21). Conclusions drawn from a synchronic approach to adaptation, then, are in this case assumed to apply to all periods in the history of film and even various other media. The scholarly quest for a grand theory or methodology may trump sustained attention to the stubborn realities of historical specificity.

Indeed, a relentless drive towards a grand theory of adaptation seems one of the most problematic consequences of the field's self-conscious insistence on its own inadequacy. In the quest to discover something like a unified field theory—in the hope, that is, of fixing the thing we've always insisted is broken—scholars often project a greater homogeneity onto the material at the expense of the incredible "range of topics invited by the multi-dimensionality of the discipline itself" (Cartmell and Whelehan 2007: 5).[3] When Linda Hutcheon (2006) poses *A Theory of Adaptation*, for example, as "a bold rethinking of *how adaptation works across all media and genres*" (2006: back cover) she risks reducing the complexity and diversity of the amazingly protean and constantly evolving art form of adaptation to a transhistorical, and more or less inert, formal structure.

How might we better recognize the importance of historical change and contingency in our study of adaptations? What would be the benefits of more long-term histories of adaptation? What obstacles and challenges would we face in undertaking such histories? In the remainder of this chapter, I wish to explore in greater depth the value of a *historical turn* in adaptation studies. For the sake of economy, I will remain focused on the specific case of *film* adaptation, though I am confident that this discussion will be relevant to adaptations of history, music, and so forth. What follows is organized into three distinct but overlapping sections: the first two can be said to reflect on the most prevalent methodological approaches to studying adaptations—the so-called case-study and synchronic historiography; the third focuses on the largely neglected diachronic historical approach, or what we might refer to as big histories.

Case studies

The single-text analysis (e.g., "*Shakespeare After Columbine: Teen Violence in Tim Blake Nelson's 'O'*" [2005]) remains the dominant methodology for analyzing adaptations of all sorts, as you will see from the table of contents of any adaptation journal you randomly pluck off your shelf. I will be referring to such essays as case studies, because they focus on an in-depth examination of a single so-called case, and because they often seek to apply their conclusions about the case to a wider phenomenon. In the example above, conclusions drawn about *O*'s treatment of teen violence are used to analyze how Shakespeare films work after the tragic 1999 shootings in Colorado. In other words, analysis of a single text authorizes generalizations of various types.

The adaptation case study has obvious strengths and limitations. On the positive side, case studies promote the close reading of texts, yielding valuable insights about source material and hypertexts that can be useful to scholars interested in those particular texts. Depending on how seriously they value historical/sociological contextualization, they can teach us much about the distinctive aesthetic, ideological, and cultural priorities of different industries, audiences, and even time periods. From an institutional/professional perspective, the case study offers the benefit of accessibility to less experienced scholars of adaptation. In a semester-long seminar on adaptations, I can only teach my students so much of film history, say, or even the general history of a specific time period; students can write strong seminar papers, however, perhaps even article drafts, based on their 14-week study of a particular text and its critical heritage. Robert B. Ray has gone further in theorizing links between the case-study approach and academic professions by tying the ascendancy and endurance of the approach to the demands of the American tenure system:

> Restricted in scope, demanding neither sustained research into nor *historical study* about the two media [my italics], the typical adaptation study had things in common with that undergraduate staple, the comparison-contrast paper—it was easy to turn out, it was easy to satisfy the requirements, and it could be done over and over again.
>
> *Ray 2000: 47*

Though Ray generously acknowledges that the case study's popularity is logical or understandable under certain institutional circumstances, his critical judgment of the approach is clear enough.

Indeed, on the negative side, case studies often rely on comparison as their primary mode of analysis, a practice which feeds tendencies to privilege a so-called original text, and thus lends itself to considerations of the hypertext's degree of fidelity to that original. At best, even when case studies avoid the fidelity trap, they too often fail to move beyond simple comparisons that "overwhelmingly give rise to the frankly unilluminating finding that there are similarities between the two mediums, but also differences, before moving on to the next book–film pairing to repeat the exercise" (Murray 2008: 4). From a professional perspective that considers the wider discourse community of adaptation scholars, case studies tend only to be useful to individuals working on particular texts, Shakespeareans who work on *Othello*, say, or film scholars working on teenpics; such a fact reinforces the field's traditional problem of being more extra-disciplinary than interdisciplinary, and promoting over-specialization that limits productive discussion and intellectual solidarity between adaptation scholars, as well as literature scholars and film scholars, who should be more engaged with one another.

More important, I am deeply suspicious of whether, in our field, conclusions drawn from single cases are actually generalizable at all. Whereas in the sciences strict falsifiability testing often results in more and more refined theses, or outright refutations of them, conclusions in adaptation studies are very often based on non-random and exceedingly small samples, and the majority are rarely, or never, tested by subsequent studies. What seems true to me, though, is that the key factor determining the reliability of case studies may be their degree of historical rigor—meaning both the reliability of their historical methods and the depth of their historical knowledge, as I explain below. Again, though, even once we recognize these facts, adaptation studies will continue to present unique challenges to scholars working on single or limited cases. Because of the field's bizarre institutional position, even the luckiest aspiring adaptation scholars will have few opportunities to take more than a single course on adaptation at the college or post-graduate levels, and such courses may or may not delve into research-related (let alone pedagogical) methodological issues. Immersion in literature and/or film studies classes will, of course, provide scholars with many of the tools they will need to be successful, but neither of these curricula are likely to concern themselves adequately with adaptation history specifically. So what are we to do? Settling for less rigor should not be an option, especially if the study of specific adaptations is to keep pace with all those positive recent developments in adaptation theory. In lieu of curricular solutions, then, some simple consciousness raising about the methodological alternatives open to us may be a good starting point.

Synchronic histories of adaptations

Most historical work on adaptations is *synchronically* focused, meaning that it concerns itself with a particular moment in time. Studies of silent one-reel adaptations of literature, studies of Laurence Olivier's Shakespeare films directed between 1944 and 1955, studies of the 1990s craze for teenpic adaptations of canonical literature—all these might be said to constitute synchronic histories, though the degree to which they prioritize historical questions and contexts will differ drastically from study to study. For example, is the hypothetical Olivier study interested in situating the actor–director's films in relation to the Second World War, the post-war moment, or even British wartime cinema, or is it merely performing comparative analyses of the film's relationships to source texts, and maybe to each other? Whereas the former study might yield conclusions interesting to war and post-war historians, film scholars, and literature scholars—suggesting the important role played by certain adaptations in a crucial moment in world and cinematic history—the alternative study will likely only interest certain teachers and scholars of Shakespeare. Another way of putting this, whereas the former can serve as something like its own argument for the relevance of adaptations to more established cognate fields, the latter tends to reinforce the idea that adaptation studies is merely provincial and narrow.

Let us continue just a bit longer with this example by considering some of the most basic questions we might expect a scholar or student, tackling the subject of Olivier's Shakespeare films, to ask him- or herself: How many other films do I need to watch featuring Olivier as either director or actor? Is my biographical knowledge of Olivier adequate? How many other contemporary Two Cities and London Film Productions should I study? Previous *Hamlet* films? Influential stage productions preceding his films? How familiar am I with 1940s British social history? How familiar am I with the 1940s Hollywood cinema that so profoundly impacted contemporary British cinema and TV? How do the economic and legal realities of film production and distribution in the 1940s and 1950s impact each of these films? And what about the critical, general, and financial reception of these films, both at the time of their release and over the decades since?

Though hardly exhaustive, it is a daunting list of questions to be sure, and a clear indication of why ahistorical case studies continue to proliferate in spite of their limitations and dead ends—not to mention the numerous calls by adaptation scholars such as Andrew and Ray for an end to their reign. We should be clear, though, that the challenges posed by such projects are in no way unique to adaptation studies. In fact, the work that must be undertaken by literary scholars to become experts in, say, Medieval Studies, or by film scholars to profess on silent film, are *at least* equal to the sorts of challenges our Olivier scholar faces. No, the main difference is that adaptation scholars have so few opportunities to receive the sort of focused, systematic training from which a Ph.D. in Medieval Studies will benefit. Furthermore, the common turn to theory as a solution to the field's woes has often served to obscure the fundamental methodological questions that adaptation scholars should be asking. As I suggested above, a greater balance between theory and history—as well as a more historically informed (or non-formalist) practice of theory—would be salutary. Simone Murray has, like Andrew before her, insisted on the importance of a sociological or a materialist turn, suggesting we need to "rethink adaptation, not as an exercise in comparative textual analysis of individual books and their screen versions, but as a *material* phenomenon produced by a system of institutional interests and actors" (2008: 10). Murray's methodology fuels the superb scholarship she produces, the best contributions of which demonstrate not just which adaptations are produced, but *how* they are produced, distributed, and oftentimes, received.

Other useful book-length examples of synchronic adaptation history include Guerric DeBona's *Film Adaptation in the Hollywood Studio Era* (2010), Thomas Leitch's *Film Adaptation and its Discontents* (2007), and Judith Buchanan's *Shakespeare on Silent Film* (2009), to mention only a few. Essays in Part I of Deborah Cartmell's edited collection *A Companion to Literature, Film and Adaptation* (2012) do excellent work situating adaptations within specific periods of film history. A model essay-length piece I have started showing all of my students is R. Barton Palmer's "The Sociological Turn of Adaptation Studies: The Example of *Film Noir*" (2004: 259–77), which demonstrates clearly how "adaptation provides the cinema not only with new texts, but with new norms and models [...] productive influences from 'outside'" (2004: 275). Each of these scholarly studies reveals a clear solution to the dead-end case-study approach precisely by shifting analysis away from the texts themselves and onto the historical, sociological, or intermedial phenomena the multiple, selected texts serve best to elucidate. In other words, these are works that concern themselves with textual analysis not as an end in itself but, rather, as part of a process of seeking answers to larger questions—though in somewhat circumscribed spaces. In this methodological commitment, the studies are precursors and companions to this very *Routledge Companion*, whose emphasis from the start has been on *adaptive attitudes, processes, and histories* rather than single case studies.

Diachronic histories of adaptation

> The farther back you can look, the farther forward you are likely to see.
>
> *Winston Churchill*

Diachronic histories, or studies of changes across larger periods of time, have been extraordinarily rare in adaptation studies. I have already suggested, more or less directly, three interrelated reasons why this has been the case, but it may be useful to reiterate them here: first, institutional and professional realities have impeded adaptation studies from developing its own rigorous standard curriculum or set of ideal practices, which would allow younger scholars to receive adequate preparation in, for example, film theory and history as well as literary studies. Second, the term sociology was logically assimilated by the field in such a way that Andrew's (1984) larger emphasis on change and evolution was overshadowed by his focus on the present and the future; history was overshadowed by sociology, which was hardly surprising considering the dominance of the case study approach to adaptations that Andrew was critiquing in the early 1980s. Finally, major theorists of adaptation—who, in the long run, revolutionized the field— tended to grasp at singular solutions or Grand Theories that could solve the apparent problem of a much-denigrated area of study; under such circumstances, the vicissitudes of time—the frus- tratingly contingent nature of historical events and structures—offered only more uncertainty and professional fragmentation.

Certainly the cultural turn of the 1980s facilitated the emphasis on sociology over history. According to George C. Iggers, a "key reason for the decline of macrohistorical conceptions" (2012: 102) of history such as those favored by the Annales School,[4] not to mention the rise of extreme methodological alternatives such as microhistory, "was to be found in the loss of faith in […] [an] optimistic view of the beneficial social and political fruits of technological progress" (Iggers 2012: 102). Certainly one of the weaknesses of many macrohistories, which included social science approaches popular since the 1960s, was an over-confident conception of mod- ernization as a positive force or a more general investment in other teleological narratives. In contrast, an increasingly narrow focus on more circumscribed subjects of research, *thickly described*, could disrupt the sort of optimistic teleology that often characterized the macrohis- tories of the previous decades. Indeed, Carlo Ginzburg and Carlo Poni, arguably the founders of microhistory, argued that the new methodology should "take as its objective a series of case studies" and then "choose from the mass of available data those cases that are relevant and sig- nificant" (1979: 7).

Whereas the *problem* in adaptation studies in the 1960s and 1970s, then, had been a too- exclusive focus on case studies (though without any methodical sorting to determine "relevant and significant" ones) at the expense of deep diachronic and synchronic contextualizing work, other fields such as history and literary studies were turning deliberately towards case stud- ies—though with a serious commitment to deep contextualization always being implied. And these fields would eventually experience precisely the same problems adaptation scholars did in attempting to manage case studies. The New Historicism, for example, at its best, inspired many article-length studies that productively and deeply contextualized literature, films, and other artworks and cultural artifacts within specific historical contexts. At its worst, however, it led to an endlessly proliferating number of case studies that were often totally disconnected from one another and devoid of any interest in—or worse, ignorant of—the larger patterns of historical development and change.[5]

No humanities scholar even remotely aware of our profession's constantly shifting critical fads and priorities will be surprised to hear that interest in the *longue durée* has recently returned to departments of history. As David Armitage explains the situation in an article anticipating *The History Manifesto* (2014), his book-length defense (with Jo Guldi) of long-term historiography,

> [a]cross the historical profession, the telescope rather than the microscope is increas- ingly the preferred instrument of examination; the long-shot not the close-up is becoming an ever-more prevalent picture of the past. A tight focus has hardly been abandoned, as the continuing popularity of biography and the utility of microhistory

both amply show. However, it is being supplemented by broad panoramas of both space and time displayed under various names: "world history", "deep history", and "big history."

Armitage 2012: 493

Note that Armitage's call for broader panoramas does not replicate the common error of scholars in previous generations of demanding an end to other historiographic methods. There is no real telos implied in the return of such histories, just approval of a new sense of balance between synchronic histories, diachronic histories including microhistories, and historically informed case studies.

This is precisely the type of balance needed in adaptation studies, and it is my main reason for suggesting that the term 'history' may prove more useful moving ahead than the term 'sociology.' A historical turn, as more encompassing than a sociological one, would place more emphasis on the dialectical relationship between long-term historical analyses, more circumscribed ones, and micro-analyses. Our field continues to provide numerous forums for the publication of case studies, and sophisticated synchronic histories of adaptation have emerged in recent years, as we have seen. What has been missing for much of this time is a more telescopic view of adaptation, though this situation may be changing.

Timothy Corrigan's *Film and Literature: An Introduction and Reader* (1999), whose first section is titled "Film and Literature in the Crosscurrents of History," laid crucial foundations for thinking about the longer history of film adaptation. Composed of eight chapters, his long essay focuses on many of the highlights of cinematic adaptation in the twentieth century. Bloomsbury's new book series *The History of World Literatures on Film* prioritizes longer-term histories of film adaptation, focusing on histories of national and regional literatures on film, so as to allow analytical depth while promoting a goal of general comprehensiveness. Greg Semenza's and Bob Hasenfratz's *The History of British Literature on Film: 1895–2015*, the inaugural volume in the series, was published in 2015, and several other volumes written by leaders in their respective fields are currently under contract: Thomas Leitch on *The History of American Literature on Film* (forthcoming); Christiane Schönfeld on *The History of German Literature on Film* (forthcoming); Andrew Watts and Kate Griffiths on *The History of French Literature on Film* (forthcoming); and David Gillespie on *The History of Russian and Soviet Literature on Film* (forthcoming). The details for volumes of Japanese, Italian, and sub-Saharan African literatures on film, to mention only a few, are currently being negotiated. By the time the series is complete, our understanding of how film adaptations are conceived, produced, marketed, and received in different locations and time periods, will be much greater.

These are certainly large-scale histories in comparison with what has come before them, and yet it is easy to envision much larger ones that would take into account changing conceptions of adaptation and appropriation since the earliest days of artistic production. Regardless of how big they decide to go, diachronic histories of adaptation are crucial to illustrating how contingent on specific historical and cultural factors—and therefore how very fragile and mutable—even our most basic assumptions happen to be. Even the most cherished clichés of the field will experience extreme tension when analyzed against the larger currents of history. To take the most obvious example, fidelity has managed to remain a central preoccupation of cinematic adaptation criticism and audience reception from as early as the 1910s, when the fidelity debate and discourse truly began to emerge, mainly from within the film industry itself. In spite of the frequent production of adaptations in the first decade or so following the invention of cinema, fidelity was a strikingly irrelevant concept to most filmmakers, as well as their audiences and critics, during that initial phase of film history (see Semenza and Hasenfratz 2015: 70, 94–103).

This observation suggests the degree to which fidelity needs to be studied as a discursive and analytical category as subject to the influences of time and place as any other. Understanding that the fidelity discourse actually emerged in the 1910s—rather than springing fully formed into the universe somewhere near the beginning of time and remaining more or less the same until our day—prompts us to consider the functionality of this discourse for certain time periods, as well as its evolution in recent decades: why did fidelity emerge as a particular concern of filmmakers, critics, and popular audiences at this specific moment in time? What different meanings has the concept of fidelity borne in vastly different contexts? What functions has fidelity served at different historical moments and in different locations—and for whom? Rather than positioning fidelity as a transhistorical concept and then viewing its demise as the natural end of a certain teleological trajectory, we must be vigilant about subjecting the concept to historical scrutiny. Although fidelity has worn out its welcome under the pressure of more complex modes of analysis, it very likely was necessary to the development and evolution of various cross-media and intertextual theories and methodologies over the course of the twentieth and early twenty-first centuries. We adaptation scholars might find the fidelity concept today frustrating and intolerable for precisely the same reasons that many thin-skinned members of *Homo sapiens* find intolerable the idea of their derivation from non-human primates.

On this note, it may be worth concluding with some acknowledgment of Gary R. Bortolotti and Linda Hutcheon's (2007) provocative pursuit of the homology between biological and cultural adaptation. Inspired by Richard Dawkins's (1976: 32) concept of the meme, their interdisciplinary work represents the cutting edge of adaptation theory but may, I fear, remain unpursued by the wider scholarly community because the methods it requires are so decidedly non-formalist and dependent on long-term historical research. The only way that mutations can truly be observed, after all, is diachronically. Still, this fact merely suggests another argument in favor of a collective turn to diachronic historiography and the *longue durée*. The sociological turn has served us extremely well over the past three decades and now, with years of excellent scholarship serving as our foundation, the time seems ripe for an even broader and more ambitious approach to adaptation studies, one systematically tuned into the importance of the past as a way into the future.

Notes

1　For a useful overview of the major theoretical developments in our field since the 1950s, but especially since about 2005, see Murray (2012: 7–12).
2　For more on this critical tendency, see Semenza and Hasenfratz (2015: 6–11).
3　The quote is from Deborah Cartmell and Imelda Whelehan (2007: 5), who argue in favor of considering the full range and diversity of adaptation studies approaches rather than lamenting the field's impoverishment or myopia.
4　Derived from the journal *Annales d'historie économique et sociale*, the term describes a group of historians, active since the late 1920s, committed to long-term historiographic studies.
5　This problematic tendency in New Historicism was noted almost immediately by scholars; see especially Vincent P. Pecora's "The Limits of Local Knowledge" (1989: 243–76).

Works cited

Andrew, J. D. (1984) *Concepts in Film Theory*, Oxford: Oxford University Press.
Armitage, D. (2012) "What's the big idea? Intellectual history and the *longue durée*," *History of European Ideas* 38 (4): 493–507.
Armitage, D. and J. Guldi (2014), *The History Manifesto*, Cambridge: Cambridge University Press.
Bortolotti, G. and L. Hutcheon (2007) "On the origin of adaptations: Rethinking fidelity discourse and 'success'—biologically," *New Literary History* 38 (3): 443–58.

Buchanan, J. (2009) *Shakespeare on Silent Film: An Excellent Dumb Discourse*, Cambridge, UK: Cambridge University Press.

Cartmell, D. (2012) *A Companion to Literature, Film and Adaptation*, Hoboken, NJ: Wiley-Blackwell.

Cartmell, D. and I. Whelehan (eds.) (2007) *Cambridge Companion to Literature on Screen*, Cambridge, UK: Cambridge University Press.

Corrigan, T. (1999) "Film and literature in the crosscurrents of history." In Timothy Corrigan (ed.) *Film and Literature: An Introduction and Reader*, 2nd edition, London and New York: Routledge, 5–51.

Dawkins, Richard (1976) *The Selfish Gene*, Oxford: Oxford University Press.

DeBona, G. (2010) *Film Adaptation in the Hollywood Studio Era*, Urbana, Chicago, and Springfield, IL: University of Illinois Press.

Ginzburg C. and C. Poni (1979) "The name and the game: Unequal exchange and the historiographic marketplace." In Edward Muir and Guido Ruggiero (eds.) *Microhistory and the Lost Peoples of Europe*, Baltimore, MD: Johns Hopkins University Press, 1991, 1–10.

Howard, G. E. (1904) *A History of Matrimonial Institutions*, Chicago, IL: University of Chicago Press.

Hutcheon, L. (2006) *A Theory of Adaptation*, New York: Routledge.

Iggers, G. C. (2012) *Historiography in the Twentieth Century*, Middletown, CT: Wesleyan University Press.

Leitch, T. (2007) *Film Adaptation and its Discontents: From "Gone with the Wind" to "The Passion of the Christ*, Baltimore, MD: Johns Hopkins University Press.

Murray, S. (2008) "Materializing adaptation theory: The adaptation industry," *Literature/Film Quarterly* 36 (1): 4–20.

Murray, S. (2012) *The Adaptation Industry: The Cultural Economy of Contemporary Literary Adaptation*, New York: Routledge.

Palmer, R. B. (2004) "The sociological turn of adaptation studies: The example of *film noir*." In Robert Stam and Alessandra Raengo (eds.) *A Companion to Literature and Film*, Malden, MA: Blackwell, 258–77.

Pecora, V. P. (1989) "The limits of local knowledge." In H. Aram Veeser (ed.) *The New Historicism*, New York and London: Routledge, 243–76.

Ray, R. B. (2000) "The field of literature and film." In James Naremore (ed.) *Film Adaptation*, New Brunswick, NJ: Rutgers University Press, 38–53.

Semenza, G. (2005) "Shakespeare after Columbine: Teen violence in Tim Blake Nelson's 'O,'" *College Literature* 32 (4): 99–124.

Semenza, G. and R. Hasenfratz (2015) *The History of British Literature on Film: 1895–2015*, New York and London: Bloomsbury.

6

NOT JUST THE FACTS

Adaptation, illustration, and history

Thomas Leitch

Writing as a lowly adaptation scholar, I presume in this chapter to offer advice and assistance to historians who clearly do not believe they need either one. In fact, their published descriptions of cinematic history make it unlikely that they would be inclined to give a serious hearing to a scholar of either adaptation or cinema. One of the great ironies of history is that just as historians were becoming more self-conscious and tentative about their own claims to be telling the truth, cinema arrived on the scene to provide a convenient scapegoat, an increasingly extensive corpus of works that could never, in the view of a surprising number of historians, tell the truth. The determination of what William Guynn has called the "scientific" (2006: 16) history that arose toward the end of the nineteenth century to distinguish itself from earlier history, "an unregulated form of discourse" produced by "amateurs" (2006: 16), was followed a century later by a crop of historians who

> acknowledged the role of subjectivity in the production of scientific discourse. […] Indeed, it is only in taking their critical distance from the facts stored in historical archives of all sorts that historians are able to winnow the grain of history from the chaff of myth and ideology and pass judgment on the errors of their predecessors.
>
> *2006: 26*

Historians who dismiss cinematic history as abnegating its responsibility to tell the truth, the whole truth, and nothing but the truth often act as if their own work is devoted to presenting just the facts, even though history is defined by its critical distance from those facts.

Ever since the turn of the twentieth century, historiographers have been increasingly sceptical of positivistic claims that history is progressive or that the ideal historian is in a position to tell the truth about the past. But even though "cinema had not yet been born when history assumed its habits, perfected its method, and ceased to narrate in favour of explaining" (Ferro 1988: 23), historians have no reservations about condemning the truth-telling standards of cinematic histories, even though "they consistently fail to raise the question of their own discursive practice. Work on the rhetoric of historical representation is not part of the literature of the discipline" (Guynn 2006: 18).

The pioneer in liberating cinematic history from the strictures of professional historians is Robert A. Rosenstone, who argued as early as 1988 that since "[h]istory does not exist until it is created" (1988: 1185), it is just as true of written history, "especially narrative history" (1988: 1180) as of cinematic history that it is "shaped by conventions of genre and language" (1988: 1180) and therefore urges us "to think about history on film not simply in comparison with written history but in terms of its own" (1988: 1180). In his more recent work, Rosenstone acknowledges that

> filmmakers—even the best and most serious—have defined history in a different way from historians. How to put it simply? Filmmakers are less concerned with empirical truth. They are more willing to use the past for some special ends. Or at least more willing to overtly admit they are doing so than any academic historian can or will admit to doing.
>
> *1995: 242*

Hence, one of cinematic history's great contributions to the field is to shed new light on the constructed nature of all history:

> Film is so obviously constructed of bits and pieces that to view any film is to face the issue of the production of meaning. One might say that it is written history's ability to escape this question that makes so much history uninteresting to read—a collection of details that are of interest only if one is already interested in those details.
>
> *Rosenstone 1995: 245–46*

Despite Rosenstone's considerable influence on recent accounts of cinematic history, even writers who are reflective and even-handed about historical films often continue to apply a double standard to written history and filmed history. In *Reel History: In Defense of Hollywood*, Robert Brent Toplin (2002: 204) aims to create a closer accord between "tradition-minded enthusiasts of history who are eager to dismiss Hollywood productions as only commercial entertainment" and "cinema specialists who look askance at the historians' interest in finding important elements of historical understanding in Hollywood dramas" because they are "[p]reoccupied with claims about the relativity of all historical interpretations." Urging a generic conception of cinematic history, Toplin lists nine leading features of the genre:

> Cinematic history simplifies historical evidence and excludes many details
> Cinematic history appears in three acts featuring exposition, complication, and resolution
> Cinematic history offers partisan views of the past, clearly identifying heroes and villains
> Cinematic history portrays morally uplifting stories about struggles between Davids and Goliaths
> Cinematic history simplifies plots by featuring only a few representative characters
> Cinematic history speaks to the present
> Cinematic history frequently injects romance into its stories, even when amorous affairs are not central to the historical events
> Cinematic history communicates a feeling for the past through attention to details of an earlier age
> Cinematic history often communicates as powerfully in images and sounds as in words
>
> *Toplin 2002: 17–53*

Much as I admire both Toplin's general project and this list in particular, I wonder if he realizes how many of its items, most of which carry an implicitly negative charge, apply to a much broader range of historical projects than Hollywood has produced. All historians who deal with massively documented events like the reign of Louis XIV or the two World Wars are painfully aware of the need to select and simplify from a much wider range of material lest it overwhelm them. Histories are not all shoehorned into three-act structures, but they are all governed by the conventions of coherence that by definition impose narrative form on inchoate data that cannot speak for itself. The reason military histories or biographies of statesmen are not routinely castigated for offering partisan views of the past is that their partisanship amounts to a generic imperative. Toplin himself recalls the debates that broke out among the "blue-ribbon panel of scholarly experts" he had assembled to advise the makers of *Denmark Vesey's Rebellion*, a 1982 film about the Charleston slave trade on which he served as consultant, "[o]n issues as vital as the personal motivation of the conspirators, their fundamental goals, and their chances of success" (2002: 159). Not all histories portray morally uplifting stories, or add questionable romantic elements, but all histories speak to unstable and contingent problems in the present; if they did not, as Benedetto Croce observes, "there would no longer be any reason for looking back and the very capacity to understand would disappear" (1941: 103). As for the charge that cinematic history communicates a feeling for the past through (presumably disproportionate) attention to details of an earlier age, the most judicious conclusion is that the standards for the presumably proportionate attention to the details of an earlier age are set by what Rosenstone calls the "ruling ideology" (1995: 228) of written history, not by the facts it surveys.

Because Rosenstone and later theorists have demonstrated so effectively the extent to which historians in general have adopted the generic norms Toplin has reserved for cinematic history, I wish to emphasize the very last of these norms, which is clearly applicable to filmed history alone: "Cinematic history often communicates as powerfully in images and sounds as in words" (Toplin 2002: 53). This premise, which is incontestably true of filmed history in a way it is not true of written history, amounts to a bedrock prejudice even a sympathetic media historian like Toplin has against the movies, which add sounds and images to what ought to be histories composed in words alone. In reviewing the first edition of Rosenstone's *History on Film/Film on History*, Alun Munslow notes that in the face of post-structuralists' sustained attacks on the very possibility of historical truth, historians have concluded that "print just about holds the line because of the belief that nothing other than print can possibly defend epistemology" (2007: 567) in the face of apocalyptic threats from epistemological nihilists:

> The end result of epistemological skepticism for many historians is apostasy and deviant professional practices. Indeed, it might mean (as Keith Jenkins has so doggedly and forensically demonstrated) the end of history as conventionally understood. This (assumed to be) terrifying prospect can only be kept at bay according to conventional thinking (which construes the notion as terrifying rather than think it through) by means of a heavily policed printed history. Only in print can history correspond to the past.
>
> *Munslow 2007: 568*

If it is true that "[w]hen we ask whether film can do history, in most cases it seems to me that we are *actually* unconsciously asking whether film can do *written* History, which is basically asking whether one medium is able to ape another effectively" (Elliott 2011: 13), then adaptation scholars should be uniquely qualified to bring a useful new perspective to a debate in which individual filmed histories, and cinematic history generally, continue to be framed as a series of adaptations of written history. This highly tendentious assumption is built into the distinction between "'historio*photy*' (the representation of history and our thought about it in visual images and filmic discourse" and "'historio*graphy*' (the representation of history in verbal images and written discourse)" first formulated by Hayden White (1988: 1193), who makes his agenda clear in the sentence immediately following these definitions: "Here the issue is whether it is possible to 'translate' a given written account of history into a visual–auditory equivalent without significant loss of content" (1988: 1193). In contrast to White's measure of historiophoty's adequacy as history by its ability to serve as an adequate adaptation of historiography, adaptation theorists like Linda Hutcheon, who defines adaptation as "a derivation that is not derivative—a work that is second without being secondary" (2013: 9), suggest a more even-handed way of thinking about cinematic history by looking more closely at the nature of historical illustration.

Andrew B.R. Elliott has argued that Satish K. Bajaj's restatement of White's definitions—"History […] has begun to be represented and thought about in visual images and filmic discourse. As distinguished from historiography which is a representation of the past in verbal and written discourse, the former is described as historiophoty" (Bajaj 1998: 69)—places the latter term "in alignment not with History, but with historiography" (Elliott 2011: 225). But White's own focus on the question of historiophoty's ability to translate or adapt a historiography whose discursive primacy is never questioned implies a closer identification between historiography and history than historiophoty can ever hope to achieve. White's historiography gets to set standards for historiophoty from which it is itself exempt because of its undefined and undefended intimacy with history itself. White's definitions of historiophoty and historiography in terms of their "representation of history" ought to allow historiophoty to cover both presentations *of* history and reflections *on* history, just as historiography, which is used more often in the second sense than the first, has long been applied to both activities. In practice, however, historians and historiographers like Toplin limit their assessments of historiophoty to its ability to present history, glossing over its aspiration to serve as what White, in the title of his best-known book, has called "metahistory." In Rosenstone's terms, written history is assumed to have the option of incorporating reflections on the nature and adequacy of history, no matter how seldom it takes advantage of this option, but cinematic history's reflections on the nature and adequacy of history, whether written or cinematic, are routinely ignored by historians who "believe the past belongs to them" (Rosenstone 2012: xviii).

Adaptation scholars are likely to find this argument familiar because they have been through another version of it themselves during the period when they struggled to make a case for particular adaptations, and adaptation in general, as something different, greater, and more interesting than a more or less accurate copy of a set of adapted texts. The historians' version of this argument, which assigns primary authority to written histories that subordinate cinematic histories can only adapt, faithfully replicates the comparative evaluations earlier generations of commentators assigned to novels and movies.

More specifically, Toplin's observation that "[c]inematic history often communicates as powerfully in images and sounds as in words" (2002: 50) reduces the images and sounds of cinema to the status of illustrations of the propositional content that are historical films' true contribution to history. In *Schindler's List,* for example, "[Amon] Goeth's poor posture and protruding stomach" indicate "that the movie's villain is a self-indulgent and effete snob who is given to excess" (2002: 51), and "[Steven] Spielberg and [Janusz] Kaminski employ a documentary filmmaking style to suggest that their accounting of the Holocaust aims to be realistic" (2002: 52). Although "Spielberg accentuates the tragedy of a ghetto's destruction by providing the audience with a frightening three-dimensional auditory experience" (2002: 52–53) that encourages them to feel shocked, helpless, and overwhelmed, Toplin implies that that film's historical contribution is directly proportional to the verbal propositions about the Holocaust that its images and sounds illustrate.

Considering its importance to any discussion of history, it is surprising to note how seldom historians have reflected on the nature of illustration as such. In his discussion of *Schindler's List,* Toplin unwittingly conflates two distinct meanings of illustration: the secondary visual material— traditionally paintings and photographs—that accompanies visual texts whose importance is assumed to be primary, and the examples in printed or spoken discourse or any other medium offered in support of more general arguments. The first of these meanings subordinates visual illustrations to their verbal texts they supplement; the second leaves open the question of which is primary, illustrations or the arguments they support and inform. The conflation of the two meanings rarely leads to misunderstanding or confusion, even in history textbooks that make liberal use of both kinds of illustrations. But when historians like the sixty contributors to Mark C. Carnes's collection *Past Imperfect: History According to the Movies* turn to cinema, they readily accept Carnes's distinction between "[p]rofessional historians" who "pluck from the muck of the historical record the most solid bits of evidence, mold them into meanings, and usually serve them up as books that […] can be held and cherished, pondered and disputed," and "Hollywood history," which "fills irritating gaps in the historical record and polishes dulling ambiguities and complexities. […] Hollywood history sparkles because it is so morally unambiguous, so devoid of tedious complexity, so *perfect*" (Carnes 1995: 9). The volume further confuses the distinction between the two senses of *illustration* by adding to all but one of its essays pairs of pictorial illustrations, the first labelled "HISTORY," the second "HOLLYWOOD." The illustrations in the first group include photographs of Michelangelo's statue of Moses, Hans Holbein the Younger's portrait of Thomas More, John Trumbull's painting of the presentation of the Declaration of Independence to John Hancock, a series of photographs beginning with a daguerreotype of Abraham Lincoln, and a painting of Christopher Columbus attributed to Pedro Berruguete described as "agree[ing] extraordinarily well with written descriptions of Columbus" (Phillips and Phillips 1995: 61). Just as the authority of Berruguete's likeness of Columbus is secured by its consistency with written histories, the authority of the photographic stills taken from the films is implicitly rooted in their similarities to or departures from visual representations characterized as "HISTORY," even if they were painted rather than photographed or indeed created centuries after the fact. Movies are assumed to be more or less accurate illustrations of earlier paintings or sculptures or photographs, which are both illustrations of history and conflated with history.

The resulting confusion makes it easier to assume that cinematic history is always and necessarily illustrative in both the senses I have described because they lose sight of the distinction between them. Histories, according to the line of reasoning these confusions foster, extract general propositions about history (the fall of the Roman Empire was preceded by a systemic decline, the rise of the English middle class was fostered by advances in technology and literacy, the assassination of the Archduke Franz Ferdinand precipitated the start of the First World War a month later) by plucking the most solid bits of evidence from the historical record. When histories are written, however, the record is molded into meanings, the general propositions that mark each history's distinctive contribution to the discipline—complex, often ambiguous meanings that Hollywood history takes it upon itself to polish in the interest of mass entertainment. As Rosenstone summarizes the process of writing history:

> We come to understand the past in the stories we tell about it, stories based on the sort of data we call fact, but stories which include other elements that are not directly in the data but arise from the process of story telling. Through the work of recent theorists, we have come to know that this narrative of the past is itself a device—our narratives select some of those traces, and in doing so, 'constitute' them, that is make them into the 'facts' that we then link together to show and explain and interpret what happened—to, in short, produce meaning. By now we also know enough about narrative history to suggest that a great deal of this 'meaning' often precedes the 'facts' and is part of the process that helps to constitute them.
>
> *Rosenstone 2012: 176*

In short, the facts that are assumed to serve as the basis for historians' propositions are sought and selected specifically to illustrate "a reality we historians would prefer to consider to be 'found' in the events themselves or, if not there, at least in the 'facts' that have been established by historians' investigations of the past" (White 1988: 1195). A foundational question of history is therefore the question of what is illustrating what. When they consider cinematic histories, Carnes and his contributors do not pause to consider this question because they assume that movies, which they consider primarily visual texts, always illustrate written discourse, never the other way around. Greg Dening, for example, contends: "Writing the history of a cultural memory created by a film can be difficult because films as cultural artifacts have a disembodied feel. Their function is to be seen, not read," and seen by audiences in communities rather than solitary individual analysts. "Fortunately for historians," Dening adds, "even these shared experiences leave behind some texts" (Dening 1995: 101)—that is, the kinds of written texts more amenable to dispassionate historical analysis. Hence White's observation that "[w]e are inclined to use pictures primarily as 'illustrations' of the predications made in our verbally written discourse" (White 1988: 1194) is echoed nearly twenty years later by Anna Pegler-Gordon, who argues:

> Most history textbooks and many academic histories use images to illustrate the history that they tell. However, when we assign these books in our classes, we rarely spend time exploring the images in them, focusing our attention instead on the written content. We do so because of the way that images are presented in many historical texts and also because of the way that historians are trained to view images—as illustrations of written history rather than sources of history themselves.
>
> *Pegler-Gordon 2006*

The historians whom White and Pegler-Gordon describe continue to call books that include such illustrations written histories rather than verbal/visual histories, because the latter description would undermine historians' sense of their own primacy by placing verbal and visual signifiers on an equal footing. Those same historians commonly characterize cinematic history as visual or audio-visual history, despite Kamilla Elliott's persuasive argument that "the hybrid verbal/visual nature of illustrated novels and worded films" (2003: 2) indicates the impossibility of maintaining categorical distinctions between verbal and visual representational systems in discussions of either novels or films, or indeed assumptions that visual illustrations have always been, and always should be, considered subordinate to the verbal texts they accompany.

Even Rosenstone accepts this verbal/visual duality, a duality fundamentally informed (or misinformed) by the conflation of the two distinct meanings of "illustration" and freely transgressed in practice, though not in theory, by both historiography and historiophoty, when he focuses on "what seems a most basic question: whether our written discourse can be turned into a visual one" (1988: 1176)—a question that assumes that history is always intrinsically written unless it is "turned into" some other medium. In his response to Rosenstone, White puts the relationship between the two differently:

> The historical evidence produced in our epoch is often as much visual as it is oral and written in nature. Also, the communicative conventions of the human sciences are increasingly as much pictorial as verbal in their predominant modes of representation.
>
> *1988: 1173*

But White does not pursue the implications for historians of liberating visual representations from their subordination to verbal representations.

The frequent use of voiceover commentary in historical documentaries and the addition of verbal captions to the illustrations in history books continue to suggest that in both cases the visuals are illustrating the verbal propositions that are identified with history as such. Instead of assuming that every picture tells a story, historians assume that only words can tell stories; pictures can merely illustrate those stories, even though police departments' videotaping of confessions, for example, indicates that for at least some authorities an audio-visual record is more reliable and authoritative evidence than a verbal record—revealingly labelled, when it is produced in court, a transcript—would be. Finally, they assume that movies overstep the boundaries of history whenever they encourage emotional reactions from an audience that written histories would encourage to react more analytically.

I propose to test these assumptions by looking at four movies about D-Day, two of them clearly fictional, two of them avowedly non-fictional, but all of them mixing fictional and non-fictional tropes in different combinations and to very different effects. In support of my argument that historians have much to learn from adaptation scholars, I will treat all four of these films as adaptations. This may seem a wilful misreading of their status, for although two of the films present themselves as adaptations—one of a widely recognized work of written history, the other of a forgotten novel—the other two do not, claiming instead to present history directly, without the intervention of any particular written text they are adapting. In treating them as adaptations, I assume that what adaptations adapt is not texts themselves but what Lawrence Venuti (2013: 181), speaking of translations, has called "interpretants": ideas about texts, readings of texts, interpretations of texts, summaries of texts, memories of texts. A great deal of evidence could be marshalled in support of this position. Alfred Hitchcock told François Truffaut, "What I do is to read the story only once, and if I like the basic idea, I just forget about the book and start to create cinema" (1984: 71). John Huston acknowledged that he focused on adapting not Rudyard Kipling's story *The Man Who Would Be King* but "my reading of the story at age twelve or fifteen, and my impressions of it. [...] Most of my pictures begin with this kind of inbred idea, something that lives in me from long ago" (Huston cited in Sarris 1967: 255). Story departments covered potential properties throughout the Hollywood studio era by churning out summaries and treatments that producers and filmmakers could read in lieu of the properties themselves. But the most economical and elegant support is provided by Jack Boozer: "It is the screenplay, not the source text, that is the most direct foundation and fulcrum for any adapted film" (2008: 4). Every film that is based on a screenplay—and that includes the vast majority of feature-length films that have ever been publicly displayed—is adapting that screenplay much more directly than any source text the screenplay in turn has adapted. Since the filming of a screenplay always involves a transition from verbal to visual or audio-visual signifiers, historians might well argue in this connection that films illustrate screenplays rather than adapting them. But I suspect that both historians and adaptation scholars would resist this argument, because thinking about adaptation as illustration would make explicit the conflation between the two senses of illustration historians are unwilling to acknowledge, and adaptation theorists would see it as threatening to return adaptation to the subordinate role to which it seemed condemned until twenty years ago.

My aim in treating all four of these films as adaptations is not to show that some of these films are more fictional than others—although it is easy enough to rank them from the least to the most fictional—but to examine the way their presentations of history are differently framed by different textual and contextual or paratextual tropes that conflate and confuse but ultimately illuminate questions about what is illustrating what. In response to historians who regard cinematic histories as adaptations or illustrations of written histories, I will focus particularly on the question of exactly how the films illustrate their respective histories, and which kinds of information are used to illustrate which propositions in accord with the suspicion that "a great deal of this 'meaning' [historians purport to derive from history] often precedes the 'facts' and is part of the process that helps to constitute them" (Rosenstone 2012: 176).

The obvious film to begin with is 20th Century Fox's *The Longest Day*, not only because it is one of the most successful, honoured, and influential films about the Second World War, but because it inspired the highly characteristic reaction of Stephen E. Ambrose, compounded of admiration for the film's "logistic achievements," nostalgia for "the sense of pride I felt in the autumn of 1962 when I saw *The Longest Day* for the first time," and criticism of the filmmakers for their inability "to make use of [the] individual stories [presented by Cornelius Ryan's best-selling historical account] within a coherent, integrated story" (1995: 236). Ambrose ascribes the film's major failures to its distortions of the historical record in the interests of satisfying the large-scale action sequences the film's producer, Darryl F. Zanuck, demanded. Ambrose is particularly critical of the film's handling of the climactic American landing at Omaha Beach, calling it "a wonderful scene" even as he sternly notes that "nothing remotely like it ever happened" (Ambrose 1995: 238).

Ambrose emphasizes in a sidebar the film's use of "famous stars":

> Moviegoers in 1962 were overwhelmed by a cast that included John Wayne, Robert Mitchum, Henry Fonda, Robert Ryan, Rod Steiger, Robert Wagner, Mel Ferrer, Jeffrey Hunter, Sal Mineo, Roddy McDowall, Eddie Albert, Edmund [sic] O'Brien, Red Buttons, Richard Burton, Kenneth More, Peter Lawford, Sean Connery, Curt Jurgens, Paul Anka, and Fabian.
>
> *1995: 240*

He suggests that Zanuck's showmanship—"Every time a door opened, in would come a famous star" (Ambrose 1995: 240)—diverts the audience's attention from the combatants to the stars playing them in ways that inevitably heighten the film's status as grand-scale Hollywood fiction. Ambrose prudently declines to criticize the many scenes of invented dialogue the film uses to illustrate its portrayal of D-Day, a criticism historians often level at cinematic histories. Instead he notes approvingly that Ryan "interviewed nearly one thousand survivors" (Ambrose 1995: 236) in researching his book without criticizing Ryan's reliance on the accuracy of his subjects' memories or remarking the many brief dialogue scenes sprinkled throughout his book, which would have made an irresistible target for many historians if they had been presented in a movie. The reason that scenes of invented dialogue, or dialogue reported second hand or recalled over a gap of many years, are permissible in written histories but not in cinematic histories must be that in written histories, they are illustrating a history that is somehow independent of them, whereas in cinematic histories, they are taken to constitute a history that is itself an illustration.

Both Zanuck's determination to make *The Longest Day* a landmark film and Ambrose's fondness for it in spite of himself can be more sharply defined by considering one element of the film's context: the 1956 release of Henry Koster's *D-Day, the Sixth of June*, also for 20th Century Fox. Koster's film, based on a 1955 novel by Lionel Shapiro, a Canadian journalist who had landed in Normandy with the 3rd Canadian Division on D-Day, uses the event as a historical frame through which to examine the shifting relations among the American Capt. Brad Parker (Robert Taylor), the English Lt. Col. John Wynter (Richard Todd), and Valerie Russell (Dana Wynter), the military nurse they both love. The narrative structure of the film reflects this focus. Though it begins and climaxes on D-Day, most of it is a series of flashbacks dramatizing the romantic, as opposed to the military, problems its three leads face. The film can be read either as a weepie that focuses on its fictional characters or as an allegory using the characters as tropes to probe questions about wartime relations between the United Kingdom and the

United States—a topic explored in some detail in the film's Internet Movie Database message board "Yanks in Britain," which quickly spirals away from the particulars of the film and into a general, highly tendentious but inarguably non-fictional debate about who won the war, the Yanks or the Brits. But the primary emphasis the film places on its romantic triangle makes it clear that, despite its ability to provoke debates like this one and despite its own title, *D-Day, The Sixth of June* uses the war as an illustrative background for its fictional plot and characters, not, as Ambrose accurately perceived of *The Longest Day,* the other way around.

The primary status it reserves for its version of history, which is illustrated by numerous invented dialogues, scenes, and characters, makes *The Longest Day* less fictional than *D-Day, the Sixth of June.* But it is clearly more fictional than BBC/Discovery's *D-Day: Reflections of Courage* (2004), a program televised in connection with the eponymous book by Dan Parry (2004) to commemorate the sixtieth anniversary of the Normandy landings. Given the simultaneous launch of the book and the television program, it would be invidious to say that either one intends to illustrate the other in the sense of providing historical examples that support the other one's argument. Nor can either one be described as independent and non-illustrative; Parry's book, for example, includes many photographic illustrations that provide both supplementary visual material and evidence for its larger analytical claims. More to the point, the television program adds several obviously fictional elements the book does not include, elements that can be traced directly to its cinematic, or more precisely its televisual, status. Its voiceover narration uses the historical present, with occasional excursions into the future. It freely intercuts black-and-white archival footage of Dwight D. Eisenhower, Erwin Rommel, Robert Capa, Sir John Montgomery, and the double agent code-named Garbo with colour footage of actors playing them, as when the actor playing Eisenhower, sitting in the back of his car, complains about reporters to the actress playing Kay Summersby. The film invites the audience to interpret this invented scene in light of their historical knowledge that Summersby was Eisenhower's lover, a relationship the scene illustrates but never makes explicit. In fact, it does not even identify Summersby by name until the closing credits. These fictionalizations, clearly intended to provide illustrative detail for each of these public figures, raise several questions. What sorts of detail (still photographs? interviews? remembered anecdotes?) would not be illustrative? Is it possible to write history without any illustrative material? If not, what kinds of illustrations are permissible and what kinds are not?

As far as I can tell, no historians took the opportunity provided by the release of *D-Day: Reflections of Courage* to consider these questions. But my claim that historians generally dismiss cinema's ability to present history must be modified in the face of *D-Day: Code Name Overlord* (1998). The surest sign that historians of D-Day endorse this film is that over a dozen military historians participated in its making, providing on-camera interviews that supplement the film's interviews of dozens of participants in the invasion. Although it is the most purely non-fictional of the four films I am considering, its purity is relative. In fact, the persistence of illustrative strategies and tropes that are frankly fictional in even this three-part, six-hour history lesson makes it in many ways the most interesting of the four.

The second part of *D-Day: Code Name Overlord,* the part focusing on the actual invasion, is dominated by talking-head interviews with participants and black-and-white footage intended to illustrate their recollections. Many of the speakers have donned their military uniforms, caps, or medals for the occasion, establishing iconic identifications between present and past that confound the contemporaneous time frames of the interviews. Dr. Harold Baumgarten, of the 29th Infantry Division, takes a fictional narrative as his point of departure: "Lt. Harold Donaldson of Texas, he was machine-gunned right away—never got out of the boat. And a few of the people around him were machine-gunned. Like they showed in *Saving Private Ryan.*" In Part 3, Les

Anderson, of the 83rd Infantry Division, agrees: "If you want to know what it was like, uh, if you saw the motion picture *Private Ryan,* uh, it was a lot like that." Sometimes the participants appear on camera as they are speaking; sometimes their recollections are used as voiceovers to accompany period photographs they could be said to illustrate. It is equally clear, however, that the footage, chosen to synchronize with their recollections, are illustrating the recollections.

The ambiguity about whether the spoken testimony is illustrating the photographic images that accompany them or vice-versa is unimportant as long as the speakers are testifying to their own experience, since both the words and the images can be construed as illustrating that experience. Problems arise, however, when the speakers are historians rather than participants. Throughout the second part of the program, personal interviews with participants in the invasion are supplemented by on-camera analyses by seven military historians evidently meant to provide context to oral testimonies that are presented as primary sources, even when they have been filtered through such later fictional frames as *Saving Private Ryan.* In the first and third parts, however, this balance shifts, with many more talking-head shots of historians such as Gary Rhay, John Bruning, Jack Radey, Dr. Christopher Gabel, Dr. Robert Bauman, Charles Sharp, Dana Lombardy, Hugh Ambrose, Ben Roberts, David Glantz, Frank Chadwick, Walter Boyne, David Isby, and Marty Morgan. The participants in Part 3 include "Dr. Arthur Schlesinger," identified as an "OSS veteran" whose remarks on the demoralizing effects of the renewed V-1 and V-2 attacks neatly straddle the line between the participant's and the historian's perspective. In two of the program's three parts, it is clear that the participants' accounts are illustrating the historians' generalizations, not the other way around.

Even as it presents much of the narration by participants and historians over period footage, *D-Day: Code Name Overlord* conscientiously identifies every speaker except for Dale Reed, who provides a voiceover narrative that remains anonymous and unattributed until the end credits of each of the program's three parts. The effect is to restrict the importance of knowing who is speaking to the segments narrated by individual participants and historians, and reducing the footage accompanying their remarks to visual illustrations of their oral narrative. *D-Day: Code Name Overlord* raises the question of which examples are used to illustrate which propositions more urgently than any of the other three D-Day films. Practically every individual chapter of the program is introduced by the anonymous narrator, then supplemented by other voices whose remarks are illustrated by period footage before the anonymous narrator summarizes the material at the end of the chapter by *explaining the meaning of events* whose participants can testify only to their atomistic experiences. Describing the American landing at Omaha Beach, for example, historian Marty Martin announces:

> Nowhere was it worse that in the Dog Green sector of Omaha Beach, in front of a little beach community called Vierville-sur-Mer. [...] When A Company [of the 29th Infantry Division] landed, they had approximately 165 men; within five minutes of combat, 91 were killed in action. In addition to the 91 killed, they suffered about 65 wounded, so that literally in five minutes' time, an entire company of men, a reinforced rifle company, had been reduced to a handful of men who were capable of fighting.

Historian Steve Zaloga adds, "The first waves took awful casualties," before the narrator summarizes: "After the first few waves, the Omaha landing was in utter chaos. Hundreds were already dead, and no real movement inland had been achieved. Omaha Beach was in jeopardy."

D-Day: Code Name Overlord is so assiduous in identifying all of its speakers, historians as well as participants, that it is easy to overlook the way it not only renders the participants' recollections illustrations of the historians' glosses on their significance, but also turns the historians' glosses into illustrations of the unidentified voiceover narrative, earning the historians' imprimatur by subordinating illustrations to written generalization, the only mode of communication historians recognize as capable of telling the truth about the past. The still and moving images synchronized with the historians' narration clearly illustrate it, rather than vice-versa. It is unlikely that many of the sequences played back in slow motion or the shots showing recognizable military and political figures, from Eisenhower to Hitler, were filmed any time near D-Day. Since this footage is never identified as to time and place, it is best consumed as illustrations in a representative, metonymic sense rather than in a selective, microcosmic sense. So this production frames itself as less fictional, more severely historical, than *D-Day: Reflections of Courage* by multiplying its sources of information but subordinating them more systematically to the summarizing written narrative characteristic of history as a genre.

The third part of *D-Day: Code Name Overlord* ends with the Allied drive to the fall of Paris, including a summary of some criticisms of Eisenhower's commitment to a broad-front advance; some contemporary shots of the landing sites and Normandy cemeteries as a few participants take a more retrospective stance; and, finally, a freeze-frame on smiling GIs facing the camera as one participant says in voiceover: "Was I a hero? I wasn't a hero—but there were heroes all around me"—a sentiment presumably offered as the takeaway from this segment and from the entire program. The end credits, like those of *The Longest Day,* emphasize the program's fictionality even as they establish its historical credentials by identifying three historians—John Bruning, Gary Rhay, and Jack Radey—as its writers, emphasizing the selecting and shaping power of their viewpoints above those of their collaborators, whose narration is thereby revealed as more strictly illustrative than the narration of the three screenwriters. The single most obvious example of this power of selection and shaping is the determination to include in *D-Day: Code Name Overlord* a two-hour prologue and a two-hour epilogue that frame the participant-dominated story of the invasion by calling on a dozen historians to explain exactly what was at stake in the invasion, what all those individual experiences that ultimately illustrate the historians' story meant.

Thirty years ago, I argued that the difference between fiction and non-fiction is that "the same [film] sequence can be fictional or documentary, depending on its context" (Leitch 1986: 193). Footage of Henry Fonda as Norman Thayer is fictional when it appears as part of *On Golden Pond,* but the same footage is non-fictional when it appears as a clip accompanying Fonda's nomination for an Academy Award in the 1982 Oscar telecast, which uses it to illustrate the achievement of Fonda's performance, and non-fictional in still a different way in a television announcement of Fonda's death, which uses it to illustrate a chapter of his life. Now I would add that showing *On Golden Pond* in a classroom, for example, imposes different expository functions on its fictional elements, depending on whether the film is used to illustrate the career of Henry Fonda or Jane Fonda or Katharine Hepburn or Mark Rydell or domestic dramas or the history of Universal Studios or Hollywood film practices in the 1980s. Adaptation scholars, who have been increasingly inclined to define adaptation as an activity not marked by distinctive textual features but framed by distinctive intertextual practices and expectations, would do well to urge a more serious consideration of these distinctions on historians whose consideration of historical films is marked by the assumption, both self-aggrandizing and self-limiting, that historical truth is congruent with propositional content. As historians from White to Rosenstone have increasingly acknowledged, history can only aspire to propositional truth; at best, it presents arguable truths. The most urgent lesson adaptation scholars have to share with historians is that even though history depends on harnessing texts that may seem more fictional than the propositions they illustrate, neither propositions nor illustrations are purely fictional or nonfictional. More generally, fictionality and non-fictionality cannot be categorically distinguished because they are not substantive but performative, both dependent on the ways they are framed by both producers and audiences.

My analysis leads inevitably to four questions. Revealing the importance of two of them—what is the relation between illustration and fictionality, and who and what determines which elements in a given discourse function as illustrations of which other elements?—has been the primary aim of this essay. A third—what frames some accounts as history and others as fiction?—can be reframed itself in a way that might help us attack it from a new perspective: who gets to decide whether utterances are framed as fictional or non-fictional, producers or audiences, and what are the consequences of deciding that either producers or audiences are responsible for that decision? My final question is more pointedly directed to historians: what would happen if we questioned more radically "the criteria of truth and accuracy presumed to govern the professional practice of historiography" (White 1988: 1193)? Is there any single genre, medium, or mode of presentation that has a privileged ability to tell the truth? Historians who define "history as the non-authorial reporting of discovered meaning" (Munslow 2007: 571) too easily ignore the stiff competition historians face from other reporters whose medium is written language—journalists, constitutional lawyers, philosophers, and especially experimental scientists, who have the significant advantage of dealing with texts whose truth-claims are in principle disconfirmable in ways that those of historians are not. The propositional content so often taken as the gold standard of historical writing always takes the form of arguable propositions about the facts, not statements of the facts themselves. As White acknowledges, "[e]vents happen or occur; facts are constituted by the subsumption of events under a description, which is to say, by acts of presentation." (1988: 1196) However strongly it may be tinged with fictional and generic and representational tropes, however determined it remains to disavow those tropes, history is never just the facts.

Works cited

Ambrose, S.E. (1995) *D-Day: June 6, 1944: The Climactic Battle of World War II*, New York: Simon & Schuster.

Bajaj, S.K. (1998) *Recent Trends in Historiography*, New Delhi: Anmol.

Boozer, J. (2008) "Introduction: The screenplay and authorship in adaptation." *Authorship in Film Adaptation*, ed. J. Boozer, Austin, TX: University of Texas Press. 1–30.

Carnes, M.C., ed. (1995) *Past Imperfect: History According to the Movies*, New York: Henry Holt.

Croce, B. (1941) *History as the Story of Liberty*, trans. Sylvia Sprigge, London: Allen and Unwin.

D-Day, Code Name Overlord (1998) dir. Lanny Lee. Marathon. DVD.

D-Day, the Sixth of June (1956) dir. Henry Koster. 20th Century Fox. DVD.

D-Day: Reflections of Courage (2004) dir. Richard Dale. BBC Video. DVD.

Dening, G. (1995) *Mutiny on the Bounty*. Carnes (1995) 98–103.

Elliott, A.B.R. (2011) *Remaking the Middle Ages*, Jefferson, NC: McFarland.

Elliott, K. (2003) *Rethinking the Novel/Film Debate*, Cambridge, UK: Cambridge University Press.

Ferro, M. (1988) *Cinema and History*, trans. by Naomi Greene, Detroit, MI: Wayne State University Press.

Guynn, W. (2006) *Writing History in Film*, New York: Routledge.

Hutcheon, L., with O'Flynn, S. (2013) *A Theory of Adaptation*, 2nd ed., New York: Routledge.

Leitch, T.M. (1986) *What Stories Are: Narrative Theory and Interpretation*, University Park, PA: Pennsylvania State University Press.

The Longest Day (1962) dir. Ken Annakin, Andrew Marton, Gerhard Wicki. 20th Century Fox. DVD.

Munslow, A. (2007) "Film and History: Robert A. Rosenstone and history on film/film on history." *Rethinking History* 11(4): 565–75.

On Golden Pond (1981) dir. Mark Rydell. Universal. DVD.

Parry, D. (2004) *D-Day: Reflections of Courage,* London: BBC Books.

Pegler-Gordon, A. (2006) "Seeing images in history." *Perspectives on History*, Feb. 2006. American Historical Association. 31 Mar. 2017.

Phillips, C.R., and Phillips, W.D., Jr. (1995) "Christopher Columbus: Two films." Carnes (1995) 60–65.

Rosenstone, R.A. (1988) "History in images/history in words: Reflections on the possibility of really putting history onto film." *American Historical Review* 93(5): 1173–85.

Rosenstone, R.A. (1995) *Visions of the Past: The Challenge of Film to Our Idea of History*, Cambridge, MA: Harvard University Press.

Rosenstone, R.A. (2012) *History on Film/Film on History*, 2nd ed., Harlow, UK: Pearson.

Ryan, C. (1959) *The Longest Day: June 6, 1944*, New York: Simon and Schuster.

Sarris, A., ed. (1967) *Interviews with Film Directors*, New York: Avon.

Saving Private Ryan (1998) dir. Steven Spielberg. Dreamworks/Paramount. DVD.

Schindler's List (1993) dir. Steven Spielberg. Universal. DVD.

Shapiro, L. (1955) *The Sixth of June*, Garden City, NY: Doubleday.

Toplin, R.B. (2002) *Reel History: In Defense of Hollywood*, Lawrence, KS: University Press of Kansas.

Truffaut, F., with H.G. Scott. (1984) *Hitchcock*, 2nd ed., New York: Simon and Schuster.

Venuti, L. (2013) *Translation Changes Everything: Theory and Practice*, London: Routledge.

White, H. (1988) "Historiography and historiophoty." *American Historical Review* 93(5): 1193–99.

"Yanks in Britain." Message board, *D-Day, the Sixth of June*, Internet Movie Database. Web. 3 December 2015. URL http://www.imdb.com/title/tt0049117/reviews?ref_=tt_ov_rt

7

DIALOGISM'S RADICAL TEXTS, AND THE DEATH OF THE RADICAL VANGUARD CRITIC

Robert Geal

It is not only artworks that can be grouped into historical and cultural contexts such as Sumerian or Anglo-Saxon epic poetry, Yoruba or Olmec statuary, Baroque *chiaroscuro* or Mughal miniature painting, Ming or Attic vases, the French *nouvelle vague* or Brazilian *cinema novo*. Academic interpretation of art, too, is located within historically specific networks of thought in which any one form of understanding art interacts in complex ways with prior and overlapping forms of understanding. This is true of all arts disciplines, but it has been particularly instrumental in relation to the study of adaptation. In part, this is no great revelation for the field. Because of certain consequences of its historical development, relating to the comparative valorisation of its intersecting media, adaptation studies has a strong record of historical self-analysis. The long domination of a fidelity-based model, which attempted to account for how filmmakers might 'faithfully' negotiate what Jack Jorgens calls the "expressive possibilities of shifting relations between words and images" (1977: 17), eventually ushered in a new theory which critiqued fidelity analysis by locating it within a historically specific context. But the precise form of the model which displaced fidelity analysis is not subject to the same historical explanation. That is not to say that its intellectual roots are not thought of historically. The new model is usually called dialogism because of the way that it principally draws on early twentieth-century Russian philosopher Mikhail Bakhtin's (1981) idea that all works of art are constantly informed by and informing other works of art, so that adaptations are just more acute examples of this dialogue between texts. There are specific historical reasons, however, why it should be dialogism, rather than some other methodology which might make a similar criticism of fidelity analysis, that replaced fidelity as adaptation studies' new orthodoxy at around the turn of this century.

The reasons why it would be dialogism that critiqued and usurped fidelity analysis are located within the intellectual histories of the field's intersecting parental disciplines: film studies and literary studies. Film studies, from its outset in the early twentieth century, defensively strove to define its subject matter as legitimately artistic in relation to the older, more established arts. Attempts were made to define *Film as Art*, as Rudolph Arnheim's (1957) influential panegyric would have it, as a specific, unique and, for Sergei Eisenstein "unprecedented art" (1949: 233), in which the film artist must "consciously stress the peculiarities of his medium" (Arnheim 1957: 35). Early film scholars were loath to compare their medium with another lest it be found wanting. Meanwhile, throughout most of the twentieth century, literary scholars who looked at adaptations were yet to be convinced that film as of itself was a worthy subject of study, and their analyses of filmic adaptations wavered between Robert B. Ray's argument that, under the dominance of undertheorised mid-twentieth-century New Criticism, "scholars could only persist in asking about individual movies the same unproductive layman's question (How does the film compare to the book?), getting the same unproductive answer (The book is better)" (Ray 2000: 44), to George Bluestone's claim that a successful adaptation can "accomplish the dual purpose of accommodating both its [the original's] meaning and its structure in filmic terms" (Bluestone 1957: xi). This undertheorised literary bias, then, favoured a perceived fixed authorial meaning in the valorised 'original', against which an adaptation might or might not be successfully judged.

This partly explains why theoretical inquiry into the study of adaptation was delayed, so that when, in the final third of the twentieth century, its disciplinary parents waged epistemological warfare over the likes of Marx, Freud, Nietzsche, Saussure, Barthes, Foucault, Lacan, Derrida and Althusser, the atavistic field of adaptation studies, as Colin MacCabe has eloquently (and dialogically) put it, "rather like Don Quixote, continue[d] to fight the day before yesterday's battles" (2011: 7). This still does not explain, however, why it was specifically dialogism that emerged from the field's undertheorised past.

In part, adaptation studies' dialogic turn was informed by the specific criticism of undertheorised fidelity analysis which preceded it. There were two main elements to this criticism, both of which point somewhat towards a dialogic solution. First, fidelity was thought of as being located within a Kantian understanding of aesthetics, in which "both the making and the appreciation of art were conceived as specialized, autonomous, and transcendent activities having chiefly to do with media-specific form" (Naremore 2000: 2). In problematising the notion of different media's essential properties, this criticism facilitated a dialogic focus on the complex interrelationships between and across various media and texts so that, as Robert Stam puts it, "[a]daptations in a sense make manifest what is true of all works of art – that they are all on some level 'derivative'" (2005a: 45). The second broad criticism of fidelity analysis was its valorisation of the text within one ('original') medium at the expense of another. James Naremore (2000) links fidelity analysis' propensity to think along these lines with the hierarchical cultural tradition exemplified in Matthew Arnold's *Culture and Anarchy* (1869), which valued high over mass culture, and tradition over innovation. This criticism, too, partly points to a dialogic solution, so that instead of Arnold's textual binary of worthy 'original' and vulgar 'copy', Thomas Leitch argues that adaptation studies should think of art as existing within a more symbiotic and less hierarchical context which encourages us to understand "all texts as intertexts, all reading as rereading, all writing as rewriting" (Leitch 2005: 239). Adaptation's grounding within "the infinite and open-ended possibilities generated by all the discursive practices of a culture" (Stam 2000a: 64) means that Arnoldian claims to moral, political or aesthetic hierarchies are open to question.

There is nothing deterministic, however, about these dialogic remedies to the problems inherent in fidelity analysis' biases. They are, rather, dependent on specific historical and cultural conditions. It is true that there is a logic to the way that adaptation studies responded to the field's fallacies, as Leitch (2003) has put it, by replacing media specificity with a focus on what all art forms share, and by replacing Arnoldian moralistic and aesthetic binaries with an understanding of texts as complex and conflicting cultural artefacts. There are, however, other possible responses to the problems of fidelity analysis which share the criticism, but which might formulate different remedies. It is telling, in this context, that Naremore notes that fidelity analysis is "constitutive of a series of binary oppositions that poststructuralist theory has taught us to deconstruct" (2000: 2), because dialogism has a very different understanding of the relationships between texts and readers/spectators than the poststructuralism which Naremore sees as the 'teacher' of this critique of fidelity. There are a number of specific historic reasons why poststructuralism has been thought of as part of the diagnosis of, but not part of the remedy to, fidelity analysis' limitations.

The chronological delay of adaptation studies' shift from fidelity analysis is central to this. The poststructuralism which Naremore sees as a critique of fidelity's binary biases was enormously influential in both film studies and literary studies in the 1970s and into the 1980s. Never a single unified methodology, it was composed of various strands of politically informed semiotics, Marxism, psychoanalysis and feminism. Although David Bordwell's characterisation of these methodologies within film studies as "subject–position theory" (Bordwell 1996: 8) is reductive, it serves as a useful categorisation here because it emphasises the understanding that "cinema constructs subject positions as defined by ideology" (Bordwell 1996: 8). Subject–position theory was concerned less with the vagaries and subtleties of texts, and more with the unconscious interactions between text, cinematic apparatus and subject–spectator. Both text and apparatus were conceived as acting to deceive and constitute the subject–spectator. Only politically engaged theory and theoretically informed avant-garde filmmaking could expose the deception of realist cinema. Variations of this theory were influential both in literary studies, so that, for Frank Lentricchia "it is the task of the oppositional critic to re-read culture so as to amplify and strategically position the marginalised voices of the ruled, exploited, oppressed, and excluded" (1983: 15), and in film studies, in which Claire Johnston argued that a non-hegemonic form of cinema necessitated "a revolutionary strategy which can only be based on an analysis of how film operates as a medium within a specific cultural system" (1973: 4).

But, as has already been discussed, because of the particular historical reasons why adaptation studies laboured under Kantian and Arnoldian approaches to its respective media while its disciplinary parents engaged with theory which dispensed with these approaches, no substantial poststructuralist account of adaptation was ever articulated. Such an approach would have repeated, albeit with a very different political intention, one of Leitch's *Twelve Fallacies in Contemporary Adaptation Theory* (2003), fidelity's fallacy that "[d]ifferences between literary and cinematic texts are rooted in essential properties of their respective media" (2003: 150), since the focus of a poststructuralist film critic like Johnston is "an analysis of how film operates *as a medium* within a specific cultural system" (1973: 4, my emphasis). But adaptation studies was temporarily impervious to theoretical trends which were sweeping across the rest of the humanities. By the later 1980s and 1990s, as adaptation studies was drawing closer to a substantial criticism of its fidelity biases, the primacy of poststructuralism was questioned and then displaced. Bordwell has written extensively about the reasons for poststructuralism's decline, which he partly puts down to the efficacy of philosophical criticisms of its *a priori* philosophical premises (1996: 8–9), and partly puts down to the failure of the radical left to achieve meaningful revolution in the world at large, and the failure of leftist philosophy like poststructuralism to offer a solution to this failure. Thus, he gives the example that poststructuralism encouraged feminists

> to adopt the sexists Freud and Lacan strategically, as analysts of patriarchy. This theory, articulated in the wake of lost battles of the 1960s, was more diagnostic than prescriptive. It arose at a period when explaining why revolutions fail had a higher priority than showing how successful rebellion might occur.
>
> *Bordwell 1996: 11*

As a result of these broad theoretical transformations, poststructuralism was no longer understood as the only radical solution to the inevitably ideological text. Those who still had hope for resistance to hegemony had to look beyond the confines of a narrow coterie of radical academics and esoteric avant-garde practitioners. This reorientation called for an understanding of the subversive potential of both texts and audiences. The turn, therefore, was away from textual deception and towards textual pluralism. Robert Lapsley and Michael Westlake, for example, could argue that

> [i]n a climate where the notion of an elite vanguard group of intellectuals seeking mastery came to be seen as impossible and undesirable, Theory's authority could only decline. What was needed was not direction from and legislation by an elite supposedly in the know, but *radical democracy in which every voice could be heard in difference.*
>
> *Lapsley and Westlake 2006: xii, my emphasis*

Dialogism's multiplicity of voices, in which there is no single authoritative authorial articulation, therefore stems from a wider epistemological turn towards pluralism. If subject–position theory had a tendency to characterise audience responses as monolithic, new studies attempted to investigate pluralistic audience responses. In film studies, analysing audience reception was one way to do this, and this allowed scholars to construct what Tony Bennett calls reading formations, each of which is "a set of intersecting discourses that productively activate a given body of texts and the relations between them in a specific way" (1983: 5). Importantly, these reading formations could resist a text's ideological potential. Jacqueline Bobo's study of black female responses to Steven Spielberg's *The Color Purple* (1985), for example, led her to conclude that a film which "Tony Brown, a syndicated columnist and the host of the television program *Tony Brown's Journal* has called […] 'the most racist depiction of Black men since *The Birth of a Nation* and the most anti-Black family film of the modern film era'" (1988: 90) could be used to "examine the way in which a specific audience creates meaning from a mainstream text and uses the reconstructed meaning to empower themselves and their social group" (1988: 93). A potentially regressive, conservative text could thereby by reclaimed by (certain) audiences in a progressive sense, without needing the intervention of the poststructuralist vanguard critic to facilitate that reclamation.

Dialogism shares this optimistic approach, but shifts the focus from the interpretation(s) of the audience *outside* the text to the act of interpretation(s) from the 'original' *within* the adapted text. Just as the study of the spectator–subject turned to the diversity of multiple and conflicting forms of audience reception, so too the study of the text–object turned to the diversity of multiple and conflicting forms of textual hybridity. Reception theory had borrowed its methodology from the broad subject of cultural studies, and specifically Stuart Hall's (1980) notion of encoding/decoding, which held that although a text might encode ideological meaning, the audience's decoding of that text might involve negotiated and/or oppositional readings of it, as well as ideology's preferred reading. But cultural studies also suggested a way in which interpretations *within* the text–object might be thought of as oppositional in the way that reception theory understands an audience–subject response as potentially oppositional. John Fiske's 1989 account of popular culture was a critique of what Max Horkheimer and Theodor W. Adorno (1972) called the culture industry, which, much like the poststructuralist approach to art that was then coming under the kind of assault described above, saw ideology as a constraining force which could only be overcome by a vigilant academic and avant-garde vanguard. Dwight Macdonald, a proponent of Horkheimer and Adorno's Frankfurt School approach, characterised mass culture within this context as being "imposed from above. It is fabricated by technicians hired by businessmen; its audiences are passive consumers, their participation limited to the choice between buying and not buying" (Macdonald 1957: 60). For Fiske, this account fails to recognise how that which is imposed from above can be appropriated in counter-hegemonic contexts. Giving the example of how young people from various non-hegemonic subcultures customise and individualise an industrial product, such as a pair of jeans, he argues that "[t]he creativity of popular culture lies not in the production of commodities so much as in the productive use of industrial commodities. [...] The culture of everyday life lies in the creative, discriminating use of the resources that capitalism provides" (Fiske 1989: 27–8). Fiske's distinction between production and productive use of commodities, which he called excorporation, could be thought of as a prelude to dialogism's focus on adaptation's productive uses of source texts, which one might call excanonation. Even if culture consists largely of an attempted ideological imposition from above, the product or text does not necessarily impose either Arnold's patronisation or the hegemony identified by the Frankfurt School. Just as Fiske thought that capitalism's false choice between Levi and Wrangler jeans could be transcended by a personalising customisation of those jeans, so too dialogism thinks that canonical culture's false choice between, for example, Shakespeare and Dickens can be transcended by adaptation's dialogic customisation of those texts. There is a historically and culturally determined similarity, then, between Fiske's analysis of how "popular forces transform the cultural commodity into a cultural resource, pluralize the meanings and pleasures it offers, evade or resist its disciplinary efforts, fracture its homogeneity and coherence" (1989: 28) and Stam's claim that

> [m]any revisionist adaptations of Victorian novels [...] 'de-repress them' in sexual and political terms; a feminist and sexual liberationist dynamic releases the sublimated libidinousness and the latent feminist spirit of the novels and of the characters, or even of the author, in a kind of anachronistic therapy or adaptational rescue operation. Postcolonial adaptations of colonialist novels [...] retroactively liberate the oppressed colonial characters of the original.
>
> *Stam 2005a: 42*

This broad historical and cultural sensibility helps to explain both dialogism's historical emergence and its subsequent successes. Due to the long dominance of the Arnoldian and Kantian approaches to adaptation, by the time that adaptation studies understood these earlier methodologies as "constitutive of a series of binary oppositions that poststructuralist theory has taught us to deconstruct" (Naremore 2000: 2), it did not adopt the approaches of poststructuralism *tout court* because they had already been discredited within the then more epistemologically current disciplines of film studies, literary studies and cultural studies. Instead, adaptation studies both engaged with and foregrounded the era's shift from theory as radical to the uses of texts as radical. If it was no longer possible to be optimistic about the interventions of the radical vanguard critic, then it was important to instead be optimistic about the possibilities of more widespread textual and interpretative radicalism.

Dialogism could then think of texts as replacing the radical function that theory had until then claimed only for itself. So, in the examples given by Stam (2005a: 42) quoted a moment ago, it is not the feministic critic who "de-represse[s]" the Victorian novel, or the postcolonial critic who "retroactively liberate[s] the oppressed colonial characters" of the colonial novel. Rather, it is the adapted text which is itself either feminist or postcolonial, and which makes this intervention *within* its text. Stam (2005a: 46) argues, therefore, that "[w]e can still speak of successful or unsuccessful adaptations, but this time oriented not by inchoate notions of 'fidelity' but rather by attention to [...] 'readings' and 'critiques' and 'interpretations' and 'rewritings'". Replacing the obsolete vanguard critic, "[a]daptations, then, can take an activist stance toward their source[s]" (Stam 2000a: 64). Dialogism, then, understands epistemology and text in a strikingly similar pluralistic emancipatory manner. Stam (2005b: 15) calls his simultaneous deployment of "literary theory, media theory, and (multi)cultural studies [...] a kind of methodological cubism" and likewise notes that "cinema can literally include painting, poetry, and music or it can metaphorically evoke them by imitating their procedures; it can show a Picasso painting, or emulate cubist techniques" (2005a: 24). The same pluralist, modernist, emancipatory art movement, cubism, is applicable to both adaptation and adaptation studies. Moreover, Stam also links this back to the broader turn which I have identified as facilitating the shift from radical criticism to radical texts, writing in the introduction to a film theory reader from the dawn of adaptation studies' dialogic era, "*Film and Theory* offers a kind of cubist collage of theoretical grids" (2000b: xv).

This optimistic spirit is a defining feature of dialogism and, indeed, spreads beyond an optimistic account of texts' emancipatory potential into an optimistic understanding of how the until recently much maligned field of adaptation studies can move out from what Timothy Corrigan (2007: 30) calls the forlorn "gap" between literary studies and film studies. Both adapted text and adaptation studies are understood within this optimistic context. Deborah Cartmell and Imelda Whelehan can thereby write, in their recent survey of the field, both about "the excitement of encountering in every site of adaptation an entirely new set of relations which allows us to draw promiscuously on theoretical tendencies in film and literary studies and to observe how, in that process of adaptation, something unique is produced" (2010: 22) and about their "increasing confidence in the space we [adaptation studies] occupy across the disciplines of literary, film and TV studies, and beyond" (2010: 9). Adaptation studies, then, emerged from its fidelity 'gap' at a historical and cultural moment which both facilitated its optimistic approach to textual hybridity and which legitimated that approach to textual hybridity to each of its parent disciplines. It could thereby go, in a short period of time, from "being stuck in the backwaters of the academy" (Leitch 2008: 63) to its rightful place "at the very center of intertextual – that is, of textual – studies (2008: 168).

Works cited

Arnheim, R. (1957) *Film as Art*, Berkeley, CA: University of California Press.

Arnold, M. (1869) *Culture and Anarchy*, London: Smith, Elder and Co.

Bakhtin, M.M. (1981) *The Dialogic Imagination: Four Essays*, Austin, TX: University of Texas Press.

Bennett, T. (1983) "Texts, readers, reading formations," *Midwest Modern Languages Association*, 16(1): 3–17.

Bluestone, G. (1957) *Novels into Film: The Metamorphosis of Fiction into Cinema*, Berkeley, CA: University of California Press.

Bobo, J. (1988) "*The Color Purple*: Black women as cultural readers," In E. Deidre Pribram (ed.) *Female Spectators: Looking at Film and Television*, London: Verso, 90–109.

Bordwell, D. (1996) "Contemporary film studies and the vicissitudes of grand theory," In David Bordwell and Noël Carroll (eds.) *Post-theory: Reconstructing Film Studies*, Madison, WI: University of Wisconsin Press, 3–36.

Cartmell, D. and Whelehan, I. (2010) *Screen Adaptation: Impure Cinema*, Houndmills, UK: Palgrave Macmillan.

Corrigan, T. (2007) "Literature on screen, a history: In the gap," In Deborah Cartmell and Imelda Whelehan (eds.) *The Cambridge Companion to Literature on Screen*, Cambridge: Cambridge University Press, 29–43.

Eisenstein, S. (1949) "Dickens, Griffith, and film today," In Sergei Eisenstein (ed.) *Film Form: Essays in Film Theory*, New York: Harcourt, Brace and World, 195–255.

Fiske, J. (1989) *Understanding Popular Culture*, London: Routledge.

Hall, S. (1980) "Encoding/Decoding," In Stuart Hall, Dorothy Hobson, Andrew Lowe and Paul Willis (eds.) *Culture, Media, Language: Working Papers in Cultural Studies, 1972–79*, London: Hutchison, 128–138.

Horkheimer, M. and Adorno, T.W. (1972) *Dialectic of Enlightenment*, New York: Herder and Herder.

Johnston, C. (1973) *Notes on Women's Cinema*, London: Society for Education in Film and Television.

Jorgens, J.J. (1977) *Shakespeare on Film*, Bloomington, IN: Indiana University Press.

Lapsley, R. and Westlake, R. (2006) *Film Theory: An Introduction*, 2nd ed., Manchester, UK: Manchester University Press.

Leitch, T. (2003) "Twelve fallacies in contemporary adaptation theory," *Criticism*, 45(2): 149–171.

Leitch, T. (2005) "Everything you always wanted to know about adaptation *especially if you're looking forwards rather than back," *Literature/Film Quarterly*, 33(3): 231–245.

Leitch, T. (2008) "Adaptation studies at the crossroads," *Adaptation: The Journal of Literature on Screen Studies*, 1(1): 63–77.

Lentricchia, F. (1983) *Criticism and Social Change*, Chicago, IL: University of Chicago Press.

MacCabe, C. (2011) "Introduction: Bazinian adaptation: *The Butcher Boy* as example," In Colin MacCabe, Kathleen Murray and Rick Warner (eds.) *True to the Spirit: Film Adaptation and the Question of Fidelity*, Oxford: Oxford University Press, 3–26.

Macdonald, D. (1957) "A theory of mass culture," In Bernard Rosenberg and David Manning White (eds.) *Mass Culture*, New York: Free Press, 59–73.

Naremore, J. (2000) "Introduction: Film and the reign of adaptation," In James Naremore (ed.) *Film Adaptation*, London: Athlone, 1–16.

Ray, R.B. (2000) "The field of 'literature and film,'" In James Naremore (ed.) *Film Adaptation*, London: Athlone, 38–53.

Stam, R. (2000a) "Beyond fidelity: The dialogics of adaptation," In James Naremore (ed.) *Film Adaptation*, London: Athlone, 54–76.

Stam, R. (2000b) "Introduction," In Robert Stam and Toby Miller (eds.) *Film and Theory: An Anthology*, Oxford: Blackwell, xiv–xviii.

Stam, R. (2005a) "Introduction: The theory and practice of adaptation," In Robert Stam and Alessandra Raengo (eds.) *Literature and Film: A Guide to the Theory and Practice of Film Adaptation*, Oxford: Blackwell, 1–52.

Stam, R. (2005b) *Literature through Film: Realism, Magic, and the Art of Adaptation*, Oxford: Blackwell.

The Color Purple (1985), dir. Steven Spielberg. Warner Bros. DVD.

8

ADAPTATIONS AND THE MEDIA

Kyle Meikle

In early November 2004, "The Men Who Stare at Goats," the first of journalist Jon Ronson's three-part documentary series *Crazy Rulers of the World*, aired on Britain's Channel 4 with the disclaimer, "The things we reveal in this film have remained, until now, US military intelligence secrets." The same month, Picador published Ronson's *The Men Who Stare at Goats*, a companion to the series that begins with the sentence, "This is a true story" (2004: 1). Half a decade later, in November 2009, BBC Films released Grant Heslov's *The Men Who Stare at Goats*, a Hollywood movie "Inspired by the book by Jon Ronson," whose opening credits include the intertitle, "More of this is true than you would believe." To be sure, much of the history that "The Men Who Stare at Goats" (2004)—and *The Men Who Stare at Goats* (2004) and *The Men Who Stare at Goats* (2009)—recounts is unbelievable: the heretofore "secret history" (as Ronson says in the opening moments of the documentary) of the US military's forays into psychological operations in the 1980s, including, but not limited to, a sergeant who attempted to stop a goat's heart with his mind. Despite the connections between these texts—Ronson's book is implicitly inspired by his documentary, while Heslov's film is explicitly inspired by Ronson's book—they qualify their believability in different ways. Indeed, Ronson does not pause to qualify the "truth" of his documentary at all, letting his use of news footage, talking heads, and voiceover do that work for him; instead, he focuses on the relative and revelatory newness of the truths he is about to present. Ronson's book is similarly bound by its back cover, which categorizes its contents as "HISTORY/SOCIAL SCIENCE" (a categorization reflected on Amazon); the first sentence serves only to underscore the believability of the unbelievable truths that follow. Heslov's movie, meanwhile, lacks the generic hallmarks of documentary or nonfiction literature, anticipating its viewers' skepticism by playfully suggesting that at least some of what they are about to see is based in truth. While the documentary emphasizes the exposure of "long-held secrets," the feature film follows the book's lead by emphasizing that the truth is sometimes stranger than fiction—two typical claims, as Thomas Leitch argues, of adaptations "based on a true story" (2007: 286–88).

If these texts qualify their believability in different ways, so too do they qualify themselves differently *as adaptations*. Only one of the three even announces itself as such: Heslov's movie, which is not "based on a true story" but "inspired by" Ronson's book, whose first sentence affirms that it *is* "a true story," not based on one—and certainly not based on Ronson's documentary. The movement between documentary and nonfiction book conceals adaptation, while the movement between book and film reveals it. *The Men Who Stare at Goats'* pivot from documentary to feature film suggests that what Leitch calls the "slippery slope from adaptation to allusion" (2007: 126) often intersects with the slippery slope from fiction to nonfiction. And if, as Leitch says, adaptation scholars have tended to huddle at the adaptation end of the former slope, so too have they tended to focus on fictionalized texts over and above the true stories that inspire them—on one pair in the Ronson triad (book/movie) over the other (documentary/book). Adaptation as a nonfiction context—or adaptation as nonfiction text—proves particularly slippery.

As Ronson's story moves through the contexts of documentary, nonfiction literature, and feature film, adjusting its claims to believability along the way, it also reveals a surprising blind spot in the study of adaptation: the role of the media itself in the adaptive process. *The Men Who Stare at Goats* stresses the fact that adaptation scholars, who speak of adaptation as an intra-, inter-, or transmedial phenomenon, tend to overlook that other meaning of media: media as the so-called Fourth Estate, an intra-, inter-, and transmedial network of newspapers, news channels, and the journalists and reporters who work for them. Yet the media's reach within what Simone Murray (2012) has called the adaptation industry is clear enough with even a brief turn to the 2016 Oscars, where the awards for Best Picture and Best Original Screenplay went to *Spotlight* (a film about the *Boston Globe's* investigation of abuses in the Catholic Church) and the award for Best Adapted Screenplay went to *The Big Short* (based on financial journalist Michael Lewis's nonfiction book about the housing bubble of the early 2000s). *Spotlight* and *The Big Short's* respective trajectories from the *Boston Globe* and the *New York Times Magazine* to the Dolby Theatre in downtown Los Angeles mirror Ronson's own path from Channel 4 to Hollywood— all arcs of successive adaptation(s).

This chapter, then, is an attempt to follow some of those trajectories, to take measure of the distance between the Fourth Estate and the house of adaptation. Enlarging the field of adaptation studies to encompass anchors, editors, and reporters reveals a striking similarity between the business of adaptation and the business of journalism—on both an industrial and intertextual level. Janet Staiger observes in her essay on *The Return of Martin Guerre* (a 1982 movie based on a true sixteenth-century story) that our sense of "the real" is always "mediated through specific sociohistorical discourse"; for Staiger, intertextuality is the most useful notion for approaching any film that claims to be based on a true story, because "other texts suggest what the real is to us, and every kind of text is in some sense a representation of the real" (2000: 193). The specific discourses of adaptation and journalism converge on "the real"—on what Staiger labels "the original [that] is never there but always absent" in representation (2000: 193). If the process of signification is "a compulsive attempt to point to and to fix the real" (2000: 193), then adapters and journalists are particularly compulsive, or particularly significant. As the return of *The Men Who Stare at Goats*—first as book, then as feature film—shows, adaptation and journalism each qualify the other *as adaptation* and *as journalism*, while at the same time they qualify our sense of "the real." Adaptation and journalism are not only intertwined institutions but intertwined processes of mediation and remediation—a point at which we will arrive if we shed our own spotlight on the nonfictional actors who play different roles within the adaptation industry.

To be sure, adaptation scholars have already taken some tentative steps in such an investigation, attending to other nonfictional frames pillared, at least in part, by the Fourth Estate. In *Adaptation, Intermediality and the British Celebrity Biopic*, for instance, editors Márta Minier and Maddalena Pennacchia argue for the biopic as an "adaptation *par excellence*" (2014: 7) whose perception as such is "inextricably linked to the problematic issues of 'truth-value' and the fiction-versus-fact debate" (2014: 11). That is, the Jane Austen Society of North America (JASNA) post on "*Becoming Jane*: Sorting Fact from Fiction"—an online response to Julian Jarrold's 2007 biopic—is hardly removed from the recent JASNA review of "the number of important ways" that *Pride and Prejudice and Zombies* (2016) "deviates" from the 2009 Seth Grahame-Smith novel upon which it is based, "and from Jane Austen's story." Even ostensibly fictional adaptations may come burdened with the politics of believability—not of *a* true story but of *the* true story; will the real Jane Austen please stand up? Biopics foreground this fact.

While Minier and Pennacchia (2014) focus on the specific genre of biography, Laurence Raw and Defne Ersin Tutan broaden their focus to *The Adaptation of History* itself in a 2012 edited collection. They open by proposing that "all historical documents should be treated as *adaptations*" (2012: 10), since "'history' consists of a series of conflicting stories that reveal how individuals have adapted to a particular event, or series of events" (2013: 12). Raw and Tutan seek to move beyond "binary oppositions (between 'facts' and 'interpretation,' or 'accurate' and 'inaccurate' history)" to make the point that "adapting history is not just the preserve of professional historians, but something undertaken by everyone, regardless of age, race, gender or class, as a means of coming to terms with their particular worlds" (2013: 21). Journalism, a more democratic practice in principle than the rarified kind of historicizing to which Raw and Tutan allude, could prove a useful complement to their work. Any study of adaptation and history should begin with a study of adaptation and the first rough draft of that history.

Other scholars have come closer still to the specific intersection of adaptation and reportage. David Johnson, for instance, has asked "what adaptation in the field of documentary might mean" (2009: 3), looking at how Stacy Peralta's *Dogtown and the Z-Boys* (2001) adapts a series of photographic essays that Craig Stecyk composed for *Skateboarder* magazine in the 1970s. But in repeating Walter Metz's claim that documentary is rarely discussed alongside adaptation because "this has not historically been the way documentary films get produced" (cited in Johnson 2009: 2), he overlooks those cases in which it has: in documentaries like Kevin Macdonald's *Touching the Void* (2003), about a near-fatal climb in the Peruvian Andes, based on Joe Simpson's 1988 nonfiction book; Alex Gibney's *Enron: The Smartest Guys in the Room* (2005), adapted from *Fortune* reporters Bethany McLean and Peter Elkind's 2003 book; and James Marsh's *Project Nim* (2011), based on Elizabeth Hess's book *Nim Chimpsky: The Chimp Who Would Be Human* (2008). In 2015, Gibney adapted Lawrence Wright's *Going Clear: Scientology, Hollywood, and the Prison of Belief* (2013); a sticker that proclaimed that the book was "Now an HBO Documentary" replaced the "National Book Award Finalist" label that graced its cover before. Nonfictional adaptations are, in fact, frequently inspired by nonfictional sources, even if, as in the case of *The Men Who Stare at Goats,* adaptation between nonfictional contexts often goes unmarked.

The sticker adorning *Going Clear*'s (2013) cover not only marks that particular form of adaptation—it also points to the inextricability of the adaptation and news industries themselves. As Simone Murray has attuned us to "the various institutional, commercial and legal frameworks surrounding adaptations [that] profoundly influence the number and character of adaptations in cultural circulation" (2012: 4), we would do well to remember that a number of adaptations circulate through the institutional, commercial, and legal frameworks of the news media. The news industry runs parallel and often perpendicular to the world of book agents, book festivals, authors, and screenwriters who—as Murray (2012) thoroughly illustrates—give rise to literary adaptations as such. Consider an October 2015 *Deadline* report that Leonardo DiCaprio's production company Appian Way acquired a four-page proposal for journalist Jack Ewing's as-yet-unwritten book on the Volkswagen scandal, a book that Ewing himself will adapt from his series of articles on the subject for the *New York Times* (Fleming Jr. 2015). The *Hollywood Reporter* story on the option describes Ewing's book as "the *Too Big To Fail* of the auto industry" (Ford 2015), drawing a direct comparison to yet another source text: *New York Times* reporter Andrew Ross Sokin's 2009 book on the 2008 financial crisis, adapted into a feature film for HBO in 2011.

In both cases, the institution of the news media directly connects to the commercial framework of the adaptation industry. A number of adaptations characterized as true stories (or based on them) are founded in the daily news.

While this practice is hardly new—1976's *All the President's Men* (another Oscar-winner for Best Adapted Screenplay) followed a similar arc from reportage to book to film—the business of adapting news has accelerated in recent years. The *Times* may only be naïve in not selling both stories to Appian Way itself, since, as Michael Cieply writes in a March 2015 article in the newspaper, "some of the most aggressive contemporary purveyors of information, journalistic and otherwise, are seeking future growth from what has not seemed novel since Edison's day: the feature-length motion picture." Companies like CNN, Condé Nast, and Newsweek (among the Fourth Estate's oldest guards) "have all built units or alliances aimed in part at creating long-form narrative or documentary films that will be seen in theaters" (Cieply 2015)—narrative and documentary films drawn from their very archives. Newsweek, for instance, hired Apex Entertainment's Mark Ciardi to "mine [the magazine's] coverage from the moment of conception, as potential grist for both television shows and studio-level films," and Condé Nast has "spent years hunting movie material from the company's magazines, including Vanity Fair and The New Yorker, after watching others mine their output to produce films like 'Argo,' 'Brokeback Mountain' and 'A Beautiful Mind'" (Cieply 2015). Two of Condé Nast's roughly eighteen projects include *Army of One*, a movie based on Chris Heath's 2010 *GQ* piece about a "one-man quest to apprehend Osama Bin Laden" starring Nicolas Cage and helmed by *Borat's* Larry Charles; and a film for Fox Searchlight based on Carl Carmer's 1936 *New Yorker* piece "Voices Through the Trumpet," about a haunted house in upstate New York (Cieply 2015). Cieply paraphrases Condé Nast Entertainment's president Dawn Ostroff as saying that "[i]f film development is painfully slow … the alternative – letting others profit from the Condé Nast properties"—roughly 75,000 "stories and articles"—"is much worse." Spurred, perhaps, by the success of properties based on true stories, like 2014's *Serial* (the true crime podcast, set to be a documentary TV series produced by directors Chris Lord and Phil Miller) and 2015's *The Jinx: The Life and Deaths of Robert Durst* (Andrew Jarecki's 2015 documentary series, which followed his earlier, fictionalized movie about Durst, 2010's *All Good Things*), Condé Nast and company are, in effect, streamlining the process of adaptation that turned *The Men Who Stare at Goats* from a documentary into a book into a feature film. The Fourth Estate is increasingly indistinguishable from the adaptation industry's major players.

While studios that produce adaptations have long held ties to multinational corporations that may also include news media networks (Fox being the biggest and best example), this streamlining drastically foreshortens the distance between adapting and reporting. The stakes of such streamlining become clear if we turn to yet another trio of texts centered on Ronson: an article, a book, and a film all about the journalist's late 1980s stint as the keyboardist for the Frank Sidebottom Oh Blimey Big Band, a novelty act notable for its nasal, northern frontman—really the English comedian Chris Sievey—who wore a cartoonish papier-mâché head on stage. Ronson recalled his time in the band in a May 2006 *Guardian* article under the headline, "Oh blimey!" Later, he adapted that article into a feature-length film (2014's *Frank*) and a short book (2014's *Frank: The True Story That Inspired the Movie*). The journalist describes the genesis of the film therein, as he sits poolside in Puerto Rico while Heslov shoots *The Men Who Stare at Goats*. He wonders aloud to the movie's screenwriter, Peter Straughan, if "the story [he]'d written for the *Guardian* [could] be adapted into a film" (Ronson 2014: ebook loc 363–369). It could, he decides, if he and Straughan "fictionalized the whole thing" (loc 390). Ronson realizes that "[i]t could be a fable instead of a biopic" (loc 396). (His language reflects A.O. Scott's disclaimer in a 2015 review of Danny Boyle and Aaron Sorkin's *Steve Jobs* that "[t]he accuracy of this portrait is not my concern. Cinematic biographies of the famous are not documentaries. They are allegories.") Ronson describes the process of adapting *Frank* as "the opposite of journalism." "In journalism," he says, "you write what's unfolding in front of you. Journalism is a game with rules. In journalism what's acceptable is what happened, and what's not acceptable is what didn't happen. But with fiction comes a daunting infinity" (loc 403).

The opposition that Ronson establishes here between the daunting infinity of adaptation and the more orderly finitude of journalism raises the question of how adaptation may change the more—and the more directly—it becomes the domain of journalism. Cieply's mention of Mark Ciardi mining *Newsweek*'s coverage "from the moment of conception" is particularly germane in this respect. As news media merges with the adaptation industry, adaptation may never have to leave the house; adaptation may change the Fourth Estate's house rules. The *New York Times*: All the news that's fit to adapt. Reportage could be conceived as adaptational from the first, as yielding stories that will become other stories that will become other stories—or, at least, yielding stories that transcend the text and context of the *Times* to settle in other adaptational contexts (e.g. book publishing) and become other adaptational texts (e.g. feature films). This perpetual motion echoes William Randall's sense that when news stories break, they break like waves or dams: "It's a metaphor for the restless, ever-changing nature of *all* stories, which never stand still, which have to *go* somewhere by definition" (2015: 136). It is a metaphor for adaptation, too.

Such a similarity is not lost on Leitch, who argues that reportage has always been adaptational, both in the way that it "depends on earlier sources and agendas," and in the way that "news stories that draw on primary sources – observations, interviews, discoveries of new evidence – are interpretations of earlier texts" (2015: 11); today's headlines, updated information, emendations, and corrections represent nothing less than adaptations of yesterday's news. The convergence of the news and adaptation industries, then, foregrounds the "rules"—as Ronson calls them—by which certain stories are recognized as adaptations and certain stories are recognized as journalism. Even Ronson himself arrives at the point where "fiction and journalism meet" when he realizes, in adapting *Frank*, that his "decisions are no longer haphazard, but informed by the things [he's] already written" (2014: ebook loc 409). The point of origin—the source text— shifts from "the real" (or his first impression of "the real") to what he's written. He continues, "You have a mass of material and you start to whittle it down, like a sculptor chipping away at a slab … And at the end, whether it's fiction or nonfiction, you have a story" (loc 409)—story being, of course, the privileged term of news journalism, as in "'Is this really a story?' or 'Where's the story in this?'" (Dell 1991. 147), and adaptation studies being a field in which, as Johnson says in his *Dogtown* essay, adaptation "almost always connotes narrative" (2009: 2).

Both adaptation and journalism represent breaking points at which stories become stories and are then able to become other stories—all with varying claims to believability and with varying levels of profitability. If Randall defines a "story" as "event(s) *plus*—or *times*—interpretation" (2015: 136), then this is what CNN, Condé Nast, and Newsweek are selling: not events themselves (the most public of public domains) but particular interpretations thereof. And the producers of adaptations are buying those particular interpretations for any number of reasons: their specific agendas, their specific authority, their specific character, their specific currency, their specific comprehensiveness, their specific salability. To recall an example as infamous as *All the President's Men*, Richard Brooks's 1967 film adaptation of Truman Capote's 1966 true crime novel *In Cold Blood* is not (or not only) an adaptation of the Clutter family murders and their fallout but an adaptation of Capote's particular version thereof—or an adaptation designed to capitalize on Capote's reportage. DiCaprio may want the *New York Times*'s version of the VW scandal, or he may want Jack Ewing's version of the VW scandal, or he may want *Jack Ewing's book* of the VW scandal, which would likely expand the film's knowing (and paying) audiences—or he may want all three. But his film will only represent yet one more adaptation, yet one more interpretation, yet one more report, of an earlier adaptation, an earlier interpretation, or an earlier report. Journalism is not just the first rough draft of history; journalism is drafty. Adaptations raise the question of why certain reports are fixed—to use Staiger's term—as somewhat final drafts at particular moments in time, and why certain reports are named as sources over others (according to *Argo*'s IMDB credits, the film is based both on a "selection" from Tony Mendez's autobiography *The Master of Disguise: My Secret Life in the CIA* (1999) and Joshuah Bearman's 2007 *Wired* article "How the CIA Used a Fake Sci-Fi Flick to Rescue Americans from Tehran"). Today's news is tomorrow's fish wrap—except those stories that are picked up by actors, agents, authors, and award juries, leading to new headlines, new lifelines, new stories the day after tomorrow.

Both adaptation and journalism break stories like incessant waves against the shore, reshaping the coastlines from which audiences must constantly adjust their horizons of expectation. Staiger's view of intertextuality as a "fundamental and unceasing spectatorial activity" applies here, since "none of the discourses invoked nor the real is ever fixed, halted, and cured but only referred to compulsively and repetitiously" (2000: 196). Every new interpretation serves only to confirm the instability of the one that came before it, and the one that came before that. As Randall says, "Getting back to the original, pure event is impossible" (2015: 137)—as impossible, we might say, as getting back to an original, pure source text. Likewise, Randall's description of reportage could just as well apply to the production and reception of adaptations:

> Two of us witness the same event and maybe even read the same account of it later on, yet what story is each of us experiencing? It's a quicksand of relativity. For this reason, newscasters seldom acknowledge that there might well be several *sides* to the same story, or, if they do, that the sides that they're espousing might possibly be less informed than those the other networks are.
>
> *Randall 2015: 138*

The pleasure of consuming adaptations, like the pleasure of consuming the news, comes from the ways in which both seem to delimit the daunting infinity of reality into discrete products (stories), even while they advert to that infinity's infinitude in their processes (condensing, correcting, expanding, retracting, revising, updating, etc.). Just as there are several sides to one news story, there are several sides to one source text; or, rather, there is no one source text—no Jane Austen's *Pride and Prejudice*, only their Jane Austen's *Pride and Prejudice*, her Jane Austen's *Pride and Prejudice*, your Jane Austen's *Pride and Prejudice*, my Jane Austen's *Pride and Prejudice*, the first time I read Jane Austen's *Pride and Prejudice*, the last time I read Jane Austen's *Pride and Prejudice*. Randall writes that news is a many-sided hydra because "in the final analysis, [it's] big business"; networks must maintain viewership by distinguishing their stories as "more 'hard-hitting' and 'up-to-the-minute'" than others," so that "we develop preferences among such differences and lean toward one network's rendering of the world rather than another's" (2015: 138)—just as, I would hasten to add, the big business of adaptation is often predicated on selling a given adaptation as grittier, more contemporary, more faithful, or more heretical than the last (your new favorite *Pride and Prejudice*). Adaptation and journalism are both paradoxical forms that claim authority and primacy for texts that are always, in some ways, already secondary. This is/not a/the true story.

Curiously, a number of recent fictional texts dwell in this "quicksand of relativity," compounding interpretations with no beginning, no end, no event in sight. While 2015 led to the coronation of *Spotlight* and *The Big Short* (as well as the more typical true histories of

The Revenant and *Bridge of Spies*), it also saw the critical and commercial success of texts like E.L. James's bestselling *Grey* (2015), which revisits the first installment of the *Fifty Shades* trilogy from the eponymous Dominant's perspective; Lauren Groff's National Book Award long-listed *Fates and Furies* (2015), whose first half ("Fates") tells the story of a marriage from the husband's perspective, the second ("Furies") from the wife's; and the return of Showtime's critically acclaimed *The Affair*, which presents its titular transgression—and a hazily related murder—from the dueling, often contradictory perspectives of its participants (indeed, the second season doubled those perspectives to include the adulterers' husband and wife). These texts' perspectival vertigo affords their audiences decidedly different, but glancingly related, pleasures. *Grey* allows readers to revisit Ana and Christian's relationship from a more knowing position, hyper-aware of the childhood trauma that drives the wealthy billionaire's sadomasochistic tendencies—a trauma that remains a mystery through much of the original trilogy. Groff uses the break in her book to make a more poetic point about the unspoken distances of marriage, which includes the wife's imbalanced account, if not of adultery, then of an undisclosed sexual past and sterilization, informed in large part by her own traumatic childhood. Groff concludes the book with the wife revisiting her and her husband's origin story—their meet-cute—which he tells differently than she remembers. And *The Affair* sets different accounts against each other both to highlight the moral uncertainties and ambiguities of its central act and, on a more fundamental level, to drive its murder mystery.

All of these texts echo the relativism of Akira Kurosawa's *Rashomon* (1950), a film itself based on multiple stories (Ryūnosuke Akutagawa's "In a Grove" and "Rashomon"), featuring multiple stories (a bandit's, a wife's, a samurai's, a woodcutter's) about a single event (an alleged rape and murder). Such texts leave us not with the question of whose story is true, but whether any 'true story' can ever be true at all, can ever be more than a story, given the daunting infinity of interpretations; journalists even call this "quicksand of relativity" the "Rashomon effect" after the movie. Some stories, of course, can be truer than others, depending on how and by whom they are authorized. *Rashomon*'s own conflicting stories take the form of testimonies, and the second season of *The Affair* moved closer to the courtroom, an arena in which the veracity of competing stories is challenged, tested, and verified by impartial third parties. We need look no further than *The Men Who Stare at Goats* to see how nonfiction and adaptation are not only subject to the same scrutiny, but instrumental in the authorization of each other. As Videogum's Gabe Delahaye complains in a blog post about the trailer for Heslov's film, "*The Men Who Stare at Goats* Should Have Been a Documentary." He writes, "On the one hand, I'm glad that they turned Jon Ronson's book into a movie, because Jon Ronson's book was great," but "[on] the other hand, they should have just made this a documentary" (Delahaye 2009). A commenter below the post explains that the book already was a documentary, but Delahaye's confusion emphasizes the role that adaptation plays in authorizing certain stories as nonfiction and the role that nonfiction plays in authorizing certain stories as adaptations. Just as journalism authorizes some adaptations as adaptations, some adaptations authorize journalism as journalism.

The Affair, *Fates and Furies*, and *Grey* may be fictional texts (or they may be shelved as such) but they point to one final context in which the media's imbrications in adaptation become clear. Alongside the top-down model of media offered at the beginning of this chapter—the new adaptation industry of the news adaptation industry—exists another model that moves towards the first from the bottom up. Indeed, the main subject of Cieply's piece is *not* CNN, Condé Nast, or Newsweek, but BuzzFeed, the "viral content start-up" that's "building the breadth of a large media company"—including BuzzFeed Motion Pictures—largely atop of its successes in social media, with "150 million average monthly viewers" and counting (Isaac 2014). And BuzzFeed is not alone in its attempts to bridge social and traditional media. At the beginning of October 2015, Twitter introduced its new "Moments" feature, "a stream of text, photos and videos"—sorted according to categories like "Entertainment" and "Sports"—"that adds up to 'What's News'" on the social network, as *The New Yorker's* Om Malik writes. He's quick to point out, however, that each stream "is not algorithmic but has been put together by editors" (Malik 2015). As such, Twitter has begun to resemble the Fourth Estate of old media, a resemblance confirmed by tech analyst Ben Thompson's sense that "Twitter just reinvented the newspaper … not just any newspaper though – it has the potential to be the best newspaper in the world" (cited in Malik 2015). Malik writes that this is Twitter CEO Jack Dorsey's hope—for users to see Twitter as an "'information news network,' an idea originally floated by the company's co-founder, Biz Stone, who once described it as the new CNN" (2015).

The Twitter app's initial welcome page for Moments included cartoon images of a stadium, an Oscar statuette, and a rocket launch, as well as an invitation to "Discover the biggest things happening on Twitter." This parade of things was followed by the promise that users could "Get the whole story at a glance" and "Easily find the best pics, clips and conversations" on the site—accompanied by a series of illustrations expanding upon the rocket-launch cartoon: a control panel, a robot on Mars, a control room. Each set of images emphasized the multiplicity of perspectives (different scales, different locations) from which Twitter offers users purchase, a promise borne out by the service itself. Each stream in Twitter's Moments includes a mix of GIFs, photos, and tweets from both verified (i.e. authorized) and unverified (i.e. unauthorized) accounts, so that users see the attempt to make the world's largest peace sign in Central Park on 6 October 2015, in honor of John Lennon's 75th birthday, in real time from the accounts (and literal perspectives) of CBC News, Congressman Joe Crowley, the NYPD, and those who are forming the peace sign itself.

These streams are reminiscent of the recent rash of oral histories that have proliferated on the web, covering everything from film to television to theater to baseball. *The Hollywood Reporter's* "'Taxi Driver' Oral History" (Kilday 2016), *Esquire's* "Oral History of Ghostbusters" (Matloff 2014), and *Vanity Fair's* "Definitive Oral History" of *Clueless* (Chaney 2015) take those iconic movies of the seventies, eighties, and nineties (respectively) as their subjects. The Smithsonian's "Oral History of 'Star Trek'" (Altman and Gross 2016) covers the original incarnation of the classic television series, while the AV Club offers a two-part oral history of cable network Comedy Central (Seabaugh 2016). Vulture offered an oral history of *Rent* (Milzoff) in 2016 to mark the musical's twentieth anniversary, and *Slate* celebrated *Angels in America's* twenty-fifth anniversary with a "Complete Oral History" (Butler and Kois) in 2016. While many of these subjects' productions spanned years—even decades—the subject of Yahoo! Sports' "oral history of Kirk Gibson's 1988 World Series home run" (Brown 2016) spanned only a few seconds. Like Twitter's Moments, these oral histories upend notions of scale, scope, and authority, presenting their subjects through a multitude of sometimes contradictory voices.

Both the oral histories and the Twitter Moments pretend to be "Definitive" (or "Complete"), even while they gesture towards their incompleteness (a hallmark, as we have seen, of both adaptation and journalism). Indeed, Twitter explains, "When the Moment ends, so do the Tweets." That is, the story only develops so long as the multiplicity of perspectives sustains it on the social network. These streams, even more so than the oral histories or *The Affair* or *Fates and Furies*, keep competing interpretations in view alongside one another, at once gesturing to a Moment (or event) and denying its very possibility. And in these Moments, the news media represents only one such interpretation, one perspective, alongside myriad other institutions (the legal, the political) and other participants (the personal). These Moments reveal nothing less than the real-time processes of adaptation itself. They also suggest that adaptation scholars will similarly have to bridge social and traditional media—that in expanding their own perspectives to better account for adaptation and the Fourth Estate, they may find themselves grappling with the Fifth. They may find themselves grappling, finally, with adaptation studies as media studies, as the study of media in all of its competing meanings. Stay tuned as this story develops …

Works cited

The Affair (2014–present) Showtime.

All Good Things (2010), dir. Andrew Jarecki. Magnolia Pictures. DVD.

All the President's Men (1976), dir. Alan J. Pakula. Warner Bros. DVD.

Altman, M. and E. Gross (2016) "An oral history of 'Star Trek,'" *Smithsonian* May: n. pag. Web. [20 April 2017].

"Argo" (2012), *IMDB*. Web [20 April 2017].

Army of One (2016), dir. Larry Charles. Anchor Bay. DVD.

Bearman, J. (2007) "How the CIA Used a fake sci-fi flick to rescue Americans from Tehran," *Wired* 24 April: n. pag. Web [20 April 2017].

Becoming Jane (2007), dir. Julian Jarrold. Buena Vista and Miramax.

Bell, A. (1991) *The Language of News Media*, Oxford: Blackwell.

The Big Short (2015), dir. Adam McKay. Paramount. DVD.

Bridge of Spies (2015), dir. Steven Spielberg. Disney. DVD.

Brown, T. (2016), "An oral history of Kirk Gibson's 1988 World Series home run," *Yahoo! Sports* 16 Oct.: n. pag. Web [20 April 2017]

Butler, I. and D. Kois (2016) "*Angels in America*: The complete oral history," *Slate* 28 June: n. pag. Web [20 April 2017].

Carmer, C. (1936) "Voices through the trumpet," *The New Yorker* 16 May: 27–. Web [18 April 2017]

Chaney, J. (2015) "The definitive oral history of how *Clueless* became an iconic 90s classic," *Vanity Fair* 9 June: n. pag. Web [20 April 2017]

Cieply, M. (2015) "News companies see movies as opportunity for growth," *New York Times* 29 March: n. pag. Web [30 March 2015].

Delahaye, G. (2009) "*The Men Who Stare at Goats* should have been a documentary," *Videogum* 28 Aug.: n. pag. Web [16 Oct. 2015].

Elkind, P. and B. McLean (2003) *The Smartest Guys in the Room: The Amazing Rise and Scandalous Fall of Enron*, New York: Penguin.

Enron: The Smartest Guys in the Room (2005), dir. Alex Gibney. Magnolia Pictures. DVDd.

Fleming Jr., M. (2015) "Volkswagen's clean diesel scandal set for movie by Paramount, Appian Way," *Deadline* 12 Oct.: n. pag. Web [30 Oct. 2015].

Ford, R. (2015) "Volkswagen scandal movie in the works from Paramount, Leonardo DiCaprio," *The Hollywood Reporter* 12 Oct.: n. pag. Web [16 Oct. 2015].

Frank (2014), dir. Lenny Abrahamson. Magnolia Pictures. DVD.

Going Clear: Scientology and the Prison of Belief (2015), dir. Alex Gibney. HBO. DVD.

Groff, L. (2015) *Fates and Furies*, London: William Heinemann.

Heath, C. (2010) "The true story of Gary Faulkner, the man who hunted Osama bin Laden and inspired Nic Cage's *Army of One*," *GQ* 8 Sept.: n. pag. Web [18 April 2017].

Hess, E. (2008) *Nim Chimpsky: The Chimp Who Would Be Human*, New York: Bantam.

In Cold Blood (1967), dir. Richard Brooks. Sony. DVD.

Isaac, M. (2014) "50 Million new reasons BuzzFeed wants to take its content far beyond lists," *New York Times* 10 Aug.: n. pag. Web [16 Oct. 2015].

James, E.L. (2015) *Grey*, New York: Vintage.

The Jinx: The Life and Deaths of Robert Durst (2015), dir. Andrew Jarecki. HBO. DVD.

Johnson, D. (2009) "Playgrounds of unlimited potential: Adaptation, documentary, and *Dogtown and Z-Boys*," *Adaptation* 2(1): 1–16.

Kilday, G. (2016) "'Taxi Driver' oral history: De Niro, Scorsese, Foster, Schrader spill all on 40th anniversary," *The Hollywood Reporter* 7 April: n. pag. Web [20 April 2017].

Leitch, T. (2007) *Film Adaptation and Its Discontents: From* Gone with the Wind *to* The Passion of the Christ, Baltimore, MD: Johns Hopkins University Press.

Leitch, T. (2015) "History as adaptation," in Dan Hassler-Forest and Pascal Nicklas (eds.) *The Politics of Adaptation: Media Convergence and Ideology*, London: Palgrave Macmillan, 7–20.

Malik, O. (2015) "Jack in a box: Can Twitter be saved?" *The New Yorker*, 16 Oct.: n. pag. Web [16 Oct. 2015].

Mattloff, J. (2014) "An oral history of *Ghostbusters*" *Esquire* 24 Feb.: n. pag. Web [20 April 2017].

"The Men Who Stare at Goats" (2004), dir. Jon Ronson. Universal Eye. YouTube.

The Men Who Stare at Goats (2009), dir. Grant Heslov. Anchor Bay. DVD.

Mendez, A.J., with M. McConnell (1999) *The Master of Disguise: My Secret Life in the CIA*, New York: William Morrow.

Milzoff, R. (2016) "*Rent*: The oral history," *Vulture* 1 May: n. pag. Web [20 April 2017].

Minier, M. and M. Pennacchia, eds. (2014) *Adaptation, Intermediality and the British Celebrity Biopic*, London: Routledge.

"Movie review of 'Pride and Prejudice and Zombies'" (2016) *JASNA Oregon & SW Washington Region*, 7 Feb.: n. pag. Web [3 June 2016].

Murray, S. (2012) *The Adaptation Industry: The Cultural Economy of Contemporary Literary Adaptation*, New York: Routledge.

Project Nim (2011), dir. James Marsh. Lionsgate. DVD.

Randall, W. (2015) *The Narrative Complexity of Ordinary Life: Tales from the Coffee Shop*, New York: Oxford University Press.

Rashomon (1950), dir Akira Kurosawa. Criterion. DVD.

Raw, L. and D. Tutan, eds. (2013) *The Adaptation of History: Essays on Ways of Telling the Past*, Jefferson, NC: McFarland.

The Revenant (2015), dir. Alejandro G. Iñárritu. 20th Century Fox. DVD.

Ronson, J. (2014) *Frank: The True Story That Inspired the Movie*, New York: Riverhead.

Ronson, J. (2004) *The Men Who Stare at Goats*, New York: Simon & Schuster.

Ronson, J. (2006) "Oh blimey!" *The Guardian* 30 May: n. pag. Web [10 Oct. 2015].

Scott, A.O. (2015) "Review: 'Steve Jobs,' Apple's visionary C.E.O. dissected," *New York Times* 8 Oct.: n. pag. Web [8 Oct. 2015].

Seabaugh, J. (2016) "*Night After Night* to *@midnight*: An oral history of Comedy Central (Part 1)." *The A.V. Club* April 4: n. pag. Web [20 April 2017].

Serial (2014–present). WBEZ.

Simpson, J. (1988) *Touching the Void: The True Story of One Man's Miraculous Survival*, New York: Harper Collins.

Sorkin, A.R. (2009) *Too Big to Fail*, New York: Viking.

Spotlight (2015), dir. Tom McCarthy. Universal. DVD.

Staiger, J. (2000) "Securing the fictional narrative as a tale of the historical real: *The Return of Martin Guerre*," in *Perverse Spectators: The Practices of Film Reception*, New York: New York University Press, 191–209.

Too Big to Fail (2011), dir. Curtis Hanson. HBO. DVD.

Touching the Void (2003), dir. Kevin Macdonald. MGM. DVD.

Wright, L. (2013) *Going Clear: Scientology, Hollywood, and the Prison of Belief*, New York: Knopf.

9

LITERARY BIOPICS

Adaptation as historiographic metafiction

Elaine Indrusiak and Ana Iris Ramgrab

The definition of genre remains one of the most slippery concepts in film theory, a fact that should be celebrated as a clear sign of cinema's maturing into a complex and highly diverse art form. Despite the ongoing debate on the whys and wherefores of film categorization, biographical films have been thoroughly examined and labeled an autonomous genre by a number of fairly recent studies (i.e. Anderson 1988; Custen 1992/2000; Bingham 2010; Brown and Vidal 2013; Minier and Pennacchia 2014). This effort has helped attenuate the "toxicity and pejorative odor" of the term *biopic* (Bingham 2010: 13), turning it into a pet name of sorts, an endearing initiative that almost compensates for decades of little scholarly and critical attention paid to this prolific yet somewhat downplayed film category or, as Dennis Bingham puts it, a "respectable genre of very low repute" (2010: 3). Needless to say, this renewed academic interest followed a surge in popularity of such films, said to have reached its peak in 2004, "[t]he Year of the Biopic" (Anderson and Lupo 2008: 50).

Like biographies, biopics constitute "[a]n amalgam of literature, history, psychology, and sociology" (Anderson 1988: 331) and, as such, feed the never-ending debate over facticity – a debate that brings to the fore the role and limitations of literary and filmic narratives in addressing historical facts. Adding fuel to the fire of this debate is the postmodernist assumption that history's performance is equally flawed and partial in its attempt to account for past events. Unlike biographies, though, biopics face additional bias for projecting on the screen representations of historical characters, therefore transforming the legacy and memories of actual individuals into products of mass consumption, and, of course, for doing so by means of performers, actors and actresses often deemed unfit for the job and unworthy of the historical characters' names and glory. In this respect, biopics are subject to the same illogical demands that prey on any other films said or believed to be "based on a true story", a phrase that, according to Leitch:

> begins with an ambiguous verb – just what does it mean to be "based" on a true story? what sort of fidelity to the historical record is offered? – and ends with the implication that even before the film was made, a story was circulating that was not only about actual events but was a true account of them, as if extracting a story from actual events or imposing a story on them was unproblematic.
>
> *Leitch 2007: 283*

This "true story" behind a biopic may be from a single source, usually a previous written biography, or from a number of texts and documents brought together so as to compose a mosaic portrait of the character and his/her times. Though one may tend to question the former and praise the latter for its apparent commitment to historical accuracy, it is obvious that choosing the pieces that will make up the mosaic is as subjective as the decision to stick to one single source. In addition to that, as Minier and Pennacchia point out, the claim to having a "single source" is further complicated by the inescapable intertextual nature of any given text, fictional or historical, which inevitably rules out "a mere one-to-one correspondence between source and adaptation or straightforward one-way traffic from A to B" (Minier and Pennacchia 2014: 9). In any case, the resulting film is an adaptation of one or more previous texts; therefore, it partakes in all the complications typical of adaptations, such as expectations of fidelity as well as hierarchical relations based on seniority and prestige of arts and media, as thoroughly analyzed by Hutcheon (2006). But if there is one type of biographical film that takes this craving for complications a step further, that is the literary biopic.

The literary biopic is a subcategory of biographical films in which lives of writers are told. The name may suggest an obvious definition, but, as with most film genres, when one searches for examples to support the category, hitches arise. How much of a writer's life must be covered by the narrative for the label to apply? If any amount will do, do *The Hours* (2002) and *Finding Neverland* (2004) qualify as biopics? What if the focus slips from the writer onto the writing, such as in *Capote* (2005)? What if the story questions authorship and biographical accounts readers have so long taken for granted, as does *Anonymous* (2011)? And, more importantly, what if the film replaces historical facts or fills in a faulty biography with fictional events taken from the writer's own literary works? Though this may be exactly the point at which most historians and biographers shudder and draw the line, this is precisely what some of the most influential and genre-defining biopics have been known to do, blurring lines between fact and fiction not only in the biographical (filmic) texts, but also in the literary oeuvres that preceded them. In such blurred scenarios, "Will" Shakespeare falls in love with Viola (*Shakespeare in Love*, 1998), Jane Austen might be mistaken for a Bennet sister (*Becoming Jane*, 2007), Yukio Mishima's highly performative career is less verisimilar than his fictional creations (*Mishima: A Life in Four Chapters*, 1985) and Goethe contemplates suicide over forbidden love (*Goethe!*, 2010).

Drawing from Paul Murray Kendall's categorization of biographies, Carolyn Anderson (1988: 331) states that "[t]he majority of bio-pics would probably fall into the fictionalized or interpretive biography categories." Fictionalization, therefore, seems to be a prerequisite of the genre, one that tends to be a highly marked feature in the literary subgenre, since a "writer might seem unpromising subject matter for a film. A life of reflection, observation, composition and self-abstracting *literariness* does not self-evidently offer the sort of cinematic dynamism and narrative pulse usually considered the staple fare of the movies." (Buchanan 2013: 3). In attempting to account cinematically for the great deeds of writers – the unfathomable creation of literary masterpieces – filmmakers have often replaced knowledge with conjecture. As a result, the nature, forms and triggers of literary creation have been represented "through imaginative projections of various sorts" and "become the site of both earnest and impish speculative on-screen enquiry." (Buchanan 2013: 3–4) Evidently, no source other than the literary work itself offers more or better meat for speculations concerning its own creation.

Despite all the efforts of literary theory and criticism of the past hundred years to detach writers' works from their biographies, most literary biopics blatantly reinforce romantic genius. Resisting Roland Barthes' philosophical proposal of the death of the author, these seemingly anachronistic productions imply that "The *explanation* of a work is always sought in the man or woman who produced it, as if it were always in the end, through the more or less transparent allegory of the fiction, the voice of a single person, the *author* 'confiding' in us." (Barthes 1977: 143). According to Belén Vidal,

> The contemporary biopic is culturally rooted in star-driven genre formulas cast in the mould of the conventions of classical storytelling, and industrially situated in the traditions of quality cinema and television. Central to cinema's incursions into the lives of artists is the mystery of artistic genius. The biopic attempts to portray the uniqueness of the artist's talent more or less counterweighted by human fallibility – a trajectory which, as Griselda Pollock [1980] has noted, more often than not entails the commodification of art through the conventions of the romanticized biography.
>
> *Vidal 2007: 7*

On the other hand, literary biopics perform "important extra-cinematic work, that of bringing the 'original' artwork to a new kind of attention" (Andrews 2013: 369), oftentimes renewing its appeal and readership, reinvigorating or even relocating it from a peripheral to a central position in literary systems, a role played by different forms of adaptations (Indrusiak 2013). In addition to that,

> Screened writers do not conform neatly to philosophical paradigms or theoretical schools. Few films subscribe to a single theory of authorship; most support, contest, undermine, and parody various theories simultaneously. [...] Far too much cultural information would be suppressed, distorted, ignored, and lost were philosophy to be the dominant discourse of this chapter; moreover, such an approach allows screened writers to challenge and demonstrate the limitations of philosophy in accounting for cultural practice.
>
> *Elliot 2012: 180*

Deaf to the complications brought about by authorism and biographical approaches to literature, literary biopics depict writers both as creations and creators, while the artists' works are rendered both as fact and fiction. Films that build biographic accounts along the lines of, or interwoven with, plots from fictional works created by the biographees themselves do double duty. Not only do they adapt historical accounts and fictional oeuvres to the screen, but they raise questions about the nature of those texts which are adapted and, by extension, about their own nature as both historical *and* fictional works. Therefore, literary biopics are a perfect example of *historiographic metafiction*: works "which are both intensely self-reflexive and yet paradoxically also lay claim to historical events and personages" (Hutcheon 1988: 5). They are also intrinsically postmodernist, since

> [h]istory's referents are presumed to be real; fiction's are not. But [...] what postmodern novels teach is that, in both cases, they actually refer at the first level to other texts: we know the past (which really did exist) only through its textualized remains. Historiographic metafiction problematizes the activity of reference by refusing to either bracket the referent (as surfiction might) or to revel in it (as non-fiction novels might). This is not an emptying of the meaning of language, as Gerald Graff seems to think (1973: 397)[1]. The text still communicates – in fact, it does so very didactically.

> There is not so much a 'loss of belief in a significant external reality' 403) as there is a loss of faith in our ability to (unproblematically) know that reality, and therefore to be able to represent it in language. Fiction and historiography are not different in this regard.
>
> *Hutcheon 1988: 119*

Blurring boundaries seems to be one of the characteristic traits of postmodern art, and literary biopics score high in this requisite not only for mixing up historiography, literary criticism and fiction, but also for reconciling high and mass culture. Though biopics are as old as cinema itself, Aleid Fokkema considers the author the "postmodern stock character", whose story "is all about representation, querying its (im-)possibilities, its relation to knowledge, language, and power" (1999: 49). Films like *Shakespeare in Love* (1998) and *Bright Star* (2009) bring the canon into the multiplex, adapting some of the most celebrated and influential literary texts to meet the conventions of commercial film genres, such as melodrama and romantic comedies. These productions bring poets, playwrights and novelists of high esteem and status down from the ivory tower by showing their human and flawed nature; paradoxically, this seems to add more than it takes from the fame and glory of such historical celebrities.

As previously explored (Ramgrab 2013), in *Shakespeare In Love* the intertextuality between the film's story, Shakespeare's work and the gaps in the author's biography is topped with "the literary topos of the romantic poet in love" (Fokkema 1999: 47). According to most biopics, as common as literary luminaries appear to be, they still have something that differentiates them from us mortals: genius. In such films, true literary artists may go through dry spells, writers' blocks, unrequited love, depression, bipolar disease, editorial failure, anxiety of influence and all other forms of hardship of their craft, but they persevere and eventually purge all these traumas and demons through an outpouring of their soul and intellect onto pages, one that, if not immediately, will be duly acclaimed by future generations and will turn these tormented, sensitive individuals into immortal characters. Therefore, by reinforcing this view of literary creation and creators, biopics do not usually allow viewers to

> take a peep behind the scenes, at the elaborate and vacillating crudities of thought – at the true purposes seized only at the last moment – at the innumerable glimpses of idea that arrived not at the maturity of full view – at the fully-matured fancies discarded in despair as unmanageable – at the cautious selections and rejections – at the painful erasures and interpolations – in a word, at the wheels and pinions – the tackle for scene-shifting – the step-ladders, and demon-traps – the cock's feathers, the red paint and the black patches, which, in ninety-nine cases out of a hundred, constitute the properties of the literary *histrio*.
>
> *Poe 2004: 676*

In so doing, biopics undermine efforts made by scholars and critics to promote an understanding of the literary text as intrinsically dialogic and intertextual, and to make a distinction between originality and primacy. When depicting poetry writing, filmmakers might go as far as casting actresses to body forth ethereal muses; when narratives are the target, they will merge writers and narrators, subverting the narratological principle of separation of story worlds from the real one. Evidently, biopics are not expected to be bound to literary theory, so they often suggest that literature is basically the result of a sensitive person's re-elaboration of personal experiences. Moreover, they do not necessarily uphold literary values or notions of prestige, frequently resorting to popular performers and resources of commercial cinema to represent the lives of canonical authors, blurring lines between highbrow and lowbrow culture products.

Yet, these films are loved for the opportunities they bring us to materialize onto moving and talking images those beloved authors Michel Foucault had us believe were merely a function in our most treasured texts. As childish as this statement may read, one should neither dismiss nor downplay the mesmerizing power of images in materializing cherished memories and projections. According to Alison Landsberg, film scenes and images allow viewers to experience facts and events they did not take part in, yet, film language and technology imprint these scenes and images onto viewers' brains as if they were real memories. These

> prosthetic memories [...] are those not strictly derived from a person's lived experience. Prosthetic memories circulate publicly, and although they are not organically based, they are nevertheless experienced with a person's body as a result of an engagement with a wide range of cultural technologies. Prosthetic memories thus become part of one's personal archive of experience, informing one's subjectivity as well as one's relationship to the present and future tenses. Made possible by advanced capitalism and an emergent commodified mass culture capable of widely disseminating images and narratives about the past, these memories are not "natural" or "authentic" and yet they organize and energize the bodies and subjectivities that take them on.
>
> *Landsberg 2004: 26*

The power of filmic prosthetic memories is such that one may not only incorporate them to account for events not experienced, but may even replace actual visual memories with fictional ones. In the case of literary biopics about recent writers whose photographs or maybe even videos are available, this phenomenon is particularly noticeable: how many viewers and readers remember Truman Capote's physical traits from his role in *Murder by Death* (1976) or from his many TV interviews, most available today on YouTube? How many simply associate his name to the image of a well-groomed, slimmer and effeminate Philip Seymour Hoffman? Likewise, given the wide circulation of commercial film images, even people who have not seen Attenborough's *Gandhi* (1982) may have replaced the visual memory of the Indian separatist leader with that of a skinny Sir Ben Kingsley. If that is the case with historical characters whose images have been captured by the assumed reliable objectivity of photography and video, imagine the potential films have ultimately to determine the looks and manners of long-gone writers whose public image has relied on painted portraits, marble busts or other representations.

In this respect, Jane Austen's contemporary representations are a particularly appropriate case study. The perception that we have of Jane Austen on both personal and artistic levels has always been molded by the surviving letters from her lifetime, and because these records are scarce and chronologically incomplete, the reading of these documents can be fallible. A prolific correspondent, Jane Austen must have written much more than the documented 160 letters attributed to her. Her sister Cassandra was responsible for censoring and destroying much of the correspondence, with the surviving ones distributed among family members as mementos. In a way, Cassandra Austen helped shape future generations' perception of Jane Austen: on the one hand, if we think in terms of genetic criticism, we mourn the loss of invaluable material and information. On the other hand, this interference may have created gaps that feed Austen's fans' imagination and add mystery and appeal to her work. The selection of the surviving material made by Cassandra Austen after her sister's death left many blank spaces in Jane Austen's history. Because of that, biographers, scholars and even readers in general would have to rely on speculation to be able to convey a likely narrative. Many of these gaps ended up being filled by the content of her fiction, and the resulting texts have become a form of continuation or complementation of her work. Therefore, the derivative body of fictional work created about Jane Austen is not to be taken as attempts at biographical truth, but as creative outlets for readers to engage with Austen's life and work.

Produced by Miramax, the same studio responsible for *Shakespeare in Love*, Julian Jarrold's *Becoming Jane* (2007) is a film that dramatizes historical aspects of Jane Austen's life, as presented in the biography *Becoming Jane Austen* (2003), written by Jon Spence – who also acted as consultant to the film production – mixing them with episodes lifted from her novels, especially *Pride & Prejudice*. Focusing on an episode mentioned in one of the earliest surviving letters written by Austen, the script by Sarah Williams and Kevin Wood explores a supposed relationship between a young Jane Austen at the beginning of her writing career and an Irish Law student, Tom Lefroy. It also speculates that the trauma of a failed relationship with Lefroy was the inspiration for Austen's mature novels. Fictional biographical reinventions such as these fall within the category of historiographic metafiction for the appropriation they make of historical events and characters that are inserted into a fictional universe that may or may not resemble that which it reflects. Even more,

> [a]s a genre, the author as character differs from other historical fiction in that the modern author is engaged in a dialogue with a more or less illustrious predecessor. Almost inevitably, therefore, the genesis of a literary work becomes one of its main themes and some degree of self-projection on the part of the modern author seems inevitable.
>
> *Franseen and Hoenselaars 1999: 20*

Certainly, if we take Franseen and Hoenselaars's conjecture to be true for writers, there is no reason to believe that screenwriters and directors would be less inclined to self-projecting or obsessing over creative processes simply because the language and generic conventions of their texts differ from those of their literary predecessors.

The story being told in *Becoming Jane* resembles that of the novel *Pride & Prejudice*, inasmuch as the affair between Jane and Tom mirrors that of Elizabeth and Darcy, and becomes the genesis of Jane's manuscript of *First Impressions* (as the early draft of *Pride & Prejudice* was called). The film opens as Jane struggles to write a simple text as a gift for her sister's engagement. As the relationship with Lefroy evolves, she becomes more focused on her writing, to the point where she becomes completely immersed in it when the relationship inevitably collapses. Therefore, the Jane Austen we see in *Becoming Jane* resembles the Shakespeare we meet in *Shakespeare in Love*: a poet in love. Such reading would not be possible had the restraints of historical accuracy been held up for both biopics. As such, the exercise in possibilities opens avenues of discussion on the breech between fact and fiction, and to what degree those concepts really apply.

Historiographic metafiction may be employed in varying degrees of both historicization and fictionalization. For example, when Jon Spence wrote *Becoming Jane Austen*, he moved from historical records to conjecture and back again, and did so very carefully, as would be expected of a scholar writing a biography. Later, Spence became a consultant to the production team of *Becoming Jane*, a biopic. Not bound by the same criteria of fidelity to historical events, the traffic between fiction and historical documentation was much less restricted and, therefore, much more malleable (Ramgrab 2013).

But *Becoming Jane* establishes conversations with texts other than Spence's biography and Austen's novel: mostly, it dialogues with Joe Wright's 2005 film adaptation of *Pride & Prejudice*. Telling Jane and Tom's story as a parallel to Elizabeth and Darcy's, readers/viewers can attribute to the character Jane Austen the characteristics that are most attractive in Elizabeth Bennet: her intelligence, her wit, her style and her views on family and marriage – and vice versa. In *Becoming Jane*, the choice of Tom Lefroy as the male representative of Austen's romantic entanglements can be attributed to more than just the fact that Spence's biography focuses on this relationship. The underlying assumption here is that, in having been the first documented romantic interest in the author's life, Lefroy can be taken, in a way, as part of her formation as a writer. The visual choices made by cinematographers Roman Osin and Eigil Bryld are similar, which reinforces the dialogue between the story being told in *Becoming Jane* to that of Wright's *Pride and Prejudice*. Such visual links are made even clearer when we pay attention to the similar casting choices in the two films, and Joe Wright's following feature, *Atonement* (2007): Keira Knightley plays Elizabeth Bennet in *Pride and Prejudice* (Wright 2005); Anne Hathaway plays Jane Austen in *Becoming Jane* (Jarrold 2007), where the writer falls in love with Lefroy (James McAvoy) and creates Elizabeth Bennet as her alter ego; Knightley's character falls in love with McAvoy's in *Atonement* (Wright 2007). Getting the record straight, Jane Austen, Elizabeth Bennet, Anne Hathaway and Keira Knightley are one and the same … and they all fall in love with James McAvoy. Though *Atonement* is in no way related to Jane Austen's work or legacy, the textual nature of the performers' personae is used by the cultural industry to reinforce associations beyond the fictional world.

Such complicated relationship between actors and characters, in what are essentially three works of adaptation, brings to the fore another aspect of movies that has heightened meaning in biographical films: namely the fact that the images of the performers are texts themselves (Mast 1982: 284) and, as such, allow for intertextual relations. At surface level, this is what feeds the maze of references fans revel in. However, as amusing and lighthearted as this crisscrossing of actors and roles may be, at a deeper and more complex level it highlights the textual nature of characters, both fictional and historical. The mere presence of a well-known face performing in a film supposed to be historical and/or biographical foregrounds the textual, therefore subjective and flawed, nature of film representation – a didactic, unpretentious and effective way of deconstructing master narratives, history included. Hence, by appropriating and adapting literary works along with visual and historical accounts of their creative processes, biopics work as "a formal manifestation of both a desire to close the gap between past and present of the reader and a desire to rewrite the past in a new context" (Hutcheon 1988: 118).

This new context Hutcheon refers to is one in which literature and film coexist as both highbrow art and mass media; history and fiction are perceived as narratives, open to a number of interpretations and adaptations; and though creative genius is celebrated, it does not grant the artist more authority over a work of art than readers and viewers are entitled to have. In short, this context is a postmodern one, and though much criticism has targeted the self-legitimizing trait of the postmodern debate, where critics endorse art that meets the criteria appointed as postmodern by the very same critics, the fact that literary biopics have experienced a remarkable increase in popularity in recent decades seems to indicate that they deliver what contemporary audiences want to watch: "the very postmodernism that proclaimed the death of the author and the demise of character delights in resurrecting historical authors as characters" (Franssen & Hoenselaars 1999: 11). And if film audiences want to see writers at work, the benefits of popularity may easily outweigh the contradictions and controversies literary biopics are prone to bring about with their blurred lines between fact and fiction, high and low art, film and literature.

Note

1 Graff, G. (1973) "The Myth of the Postmodernist Breakthrough," *TriQuarterly* 26: 383–417.

Works cited

Anderson, C. (1988) "Biographical film," In Wes Gehring (ed.) *Handbook of American Film Genres*, Westport, CT: Greenwood Press, 331–351.

Anderson, C. & Lupo, J. (2008) "Introduction to the special issue," *Journal of Popular Film and Television* 36(2): 50–51.

Andrews, H. (2013) "Recitation, quotation, interpretation: Adapting the oeuvre in poet biopics," *Adaptation* 6(3): 365–383.

Anonymous (2011), dir. Roland Emmerich. Columbia Pictures. DVD.

Atonement (2007), dir. Joe Wright. Working Title. DVD.

Barthes, R. (1977) "The death of the author," *Image, Music, Text*, New York: Hill and Wang, 142–148.

Becoming Jane (2007), dir. Julian Jarrold. Miramax. DVD.

Bingham, D. (2010) *Whose Lives Are They Anyway?: The Biopic as Contemporary Film Genre*, New Brunswick, NJ: Rutgers University Press .

Bright Star (2009), dir. Jane Campion. Pathé Renn Productions, Screen Australia. DVD.

Brown, T. & Vidal, B. (eds.) (2013) *The Biopic in Contemporary Film Culture*, New York: Routledge.

Buchanan, J. (2013) "Image, story, desire: the writer on film," *The Writer on Film: Screening Literary Authorship*, Basingstoke, UK: Palgrave Macmillan, 3–34.

Capote (2005), dir. Bennett Miller. Sony Pictures Classics. DVD.

Custen. G.F. (1992) *Bio/Pics: how Hollywood constructed public history*, New Brunswick, NJ: Rutgers University Press .

Custen, G.F. (2000) "The mechanical life in the age of human reproduction: American biopics, 1961–1980," *Biography* 23(1): 127– 59.

Elliot, K. (2012) "Screened writers," In Deborah Cartmell (ed.) *A Companion to Literature, Film and Adaptation*, Chichester, UK: Wiley-Blackwell, 179–197.

Finding Neverland (2004), dir. Marc Forster. Miramax. DVD.

Fokkeima, A. (1999) "The author: Postmodernism's stock character," In Paul Franssen and Ton Hoenselaars (eds.) *The Author as Character: Representing Historical Writers in Western Literature*, Madison, Fairleigh Dickinson University Press, 39–51.

Franssen, P. & Hoenselaars, T. (eds.) (1999) *The Author as Character: Representing Historical Writers in Western Literature*, Madison, NJ: Fairleigh Dickinson University Press.

Gandhi (1982), dir. Richard Attenborough. International Film Investors, National Film Development Corporation of India (NFDC), Goldcrest Films International. DVD.

Goethe! (2010), dir. Philipp Stölzl. Senator Film Produktion. DVD.

Hutcheon, L. (1988) *A Poetics of Postmodernism*, New York: Routledge .

Hutcheon, L. (2006) *A Theory of Adaptation*, New York: Routledge .

Indrusiak, E. (2013) "Adaptation and recycling in convergent cultural polysystems: A case study," *Ekphrasis* 10(2): 96–109.

Landsberg, A. (2004) *Prosthetic Memory: The Transformation of American Remembrance in the Age of Mass Culture*, New York: Columbia University Press .

Leitch, T. (2007) "Based on a true story," *Film Adaptation and Its Discontents: From Gone with the Wind to The Passion of the Christ*, Baltimore, MD: Johns Hopkins University Press, 280–303.

Mast, G. (1982) "Literature and film," In JeanPierre Barricelli and Joseph Gibaldi (eds.) *Interrelations of Literature*, New York: MLA, 278–306.

Minier, M. & Pennacchia, M. (2014) "Interdisciplinary perspectives on the biopic: An introduction," *Adaptation, Intermediality and the British Celebrity Biopic*, Farnham, UK: Ashgate, 1–31

Mishima: A Life in Four Chapters (1985), dir. Paul Schrader. Zoetrope Studios, Filmlink International. DVD.

Murder by Death (1976), dir. Robert Moore. Sony Pictures Home Entertainment. DVD.

Poe, E. A. (2004) "The philosophy of composition," In G. R. Thomson (ed.) *The Selected Writings of Edgar Allan Poe*, New York: W. W. Norton & Co., 675–684.

Pollock, G. (1980) "Artists, mythologies and media – Genius, madness and art history," *Screen* 21(3): 57–96.

Pride and Prejudice (2005), dir. Joe Wright. Universal. DVD.

Ramgrab, A. I. M. (2013) *Meet Jane Austen: The Author as a Character in Contemporary Derivative Works.* (Master's Thesis)

Shakespeare in Love (1998), dir. John Madden. Miramax. DVD.

Spence, J. (2003) *Becoming Jane Austen*, New York: Hambledon Continuum.

The Hours (2012), dir. Stephen Daldry. Paramount Pictures, Miramax. DVD.

Vidal, B. (2007) "Feminist histographies and the woman artist's biopic," *Screen* 48(1): 70–90.

10

NOTORIOUSLY BAD

Early film-to-video game adaptations (1982–1994)[1]

Riccardo Fassone

In defense of bad games

In October 2015 I spent three weeks in Rochester, New York, perusing the archives of The Strong National Museum of Play. I was granted access to an extensive library, a host of archival material, and several collections of design documents, letters, and contracts donated by toy manufacturers, game designers, collectors, and enthusiasts. I also visited the International Center for the History of Electronic Games, a series of neon-lit rooms that looked like a consumerist fantasy of my fourteen-year-old self: shelves full of consoles, old video games, arcane and rare controllers, remnants of both the grand history of video games and of short-lived, largely forgotten gaming fads. I quickly became the researcher playing bad games. Archivists would walk by my CRT TV[2] and inquire about a sub-par platformer starring a digital rendition of Macaulay Culkin, or a take on *Terminator 2: Judgement Day* (James Cameron, 1991) that looked nothing like the film and relied on an ominous 16-bit version of a T-800 to convey a sense of aesthetic consistency. My research project, devoted to the reconstruction of a history of film-to-game adaptations, had tapped into a largely unexplored corpus of unaccomplished, dysfunctional, unbalanced, and often just plain bad video game design.

This chapter is both a historical account of film-to-game adaptations produced between 1982 and 1994, and a plea in favour of what Guins describes as "critical historical studies of video games" (2014: 22), a historical excavation that eschews both grand histories and minute, often vacuum-sealed, vernacular descriptions of specific phenomena. In other words, it is a history of the medium that accounts for technological, social, and political transformations within the wider media ecosystem, and follows their trajectories as they intersect with video games. It is also, and more importantly, a retelling of video game history that does not need to converge with the canon of video game criticism, and that re-frames rhetorics of obsolescence and disruption by considering supposedly lower intensity or marginal processes within the context of the wider historical dynamics of the medium. A history of video games that deals with good games as well as bad games. Finally, this chapter analyses a specific practice within the video game industry: that of licensing[3] and adaptation. In this sense, the chapter aims to describe the ways in which video games of the 1980s and 1990s worked on pre-existing cultural material in order to adapt it into playable artefacts.

A periodisation

This chapter analyses film–to–game adaptations produced and distributed between 1982 and 1994, with a specific focus on games produced for home console platforms. While these dates may seem non-canonical within the context of video game historiography – they fall slightly shy of major events such as the 1982–1983 market crash and the release of Nintendo 64 in 1996 – they are distinctly relevant to the history of film adaptations. Most histories of video games (Kent 2001; Donovan 2010) tend to associate the American market crash of 1982–1983 with a complete halt in the production and distribution of video game cartridges, consoles, controllers, and other related artefacts. While multiple causes were at the origin of the crash, the game *E. T. the Extra-Terrestrial* (Atari, 1982) is often used as a synecdoche for the collapse of what had been one of the booming sectors in the entertainment industry in the late 1970s. Developed in less than two months in the summer of 1982 in order to meet the holiday season release deadline, the game was an attempt to capitalise on the success of Spielberg's film, but failed to generate the expected revenue and left Atari with heaps of returned or unsold cartridges. While the title of "Worst video game of all time" (Townsend 2006) may be undeserved, *E. T. the Extra-Terrestrial* has come to signify a clearly recognisable rupture in video game history. Pre-crash and post-crash video games seem to be two different breeds of media objects, and the fact that an adaptation such as *E. T. the Extra-Terrestrial* stands on the brink of this historical rupture may lead historians to think that this should be especially true for licensed games. The periodisation proposed in this chapter, while acknowledging the relevance of the market crash in shaping future design, development, and distribution strategies in the game industry, proposes to treat the crash as one of the many accidents that punctuate the history of film-to-game adaptations. This interpretation of the 1983 crash as 'soft rupture' derives from a series of considerations that seem to emerge when analysing the corpus of film-to-game adaptations. First, a wave of adaptations produced for the home console market surfaced in the years around the crash, making the historical rupture all the more artificial. In the case of Atari 2600, for example, a number of adapted games were released before: *Towering Inferno* (US Games, 1982), *Star Wars: The Empire Strikes Back* (Parker Brothers, 1982), *Raiders of the Lost Ark* (Atari, 1982) among others; in the midst of: *Halloween* (Wizard Video, 1983), *The Texas Chainsaw Massacre* (Wizard Video, 1983), *Krull* (Atari, 1983); and after the crash: *Gremlins* (Atari, 1984), *Star Wars: The Arcade Game* (Parker Brothers, 1984), *Ghostbusters* (Activision, 1985). Furthermore, while certainly relevant to the whole industry, the market crash had the most significant impact on the console market. Other sectors, such as the hobbyist computer market, were arguably less impacted by the crash, due to the diffusion of personal computer technologies outside of the United States, and the peculiar industrial and commercial practices informing the sector.[4] Adaptations released for various micro-computers during the crash, such as *Bruce Lee* (Datasoft, 1983), *Alien* (Concept Software, 1984), *James Bond 007: A View to a Kill* (Domark, 1985), and *Rambo: First Blood Part II* (Platinum Productions, 1985), seem to suggest that, in the history of film-to-game adaptations, a certain continuity beyond the crash may be observed. 1982, then, despite being on the brink of a historical collapse, seems to be the year in which the practices of licensing and adapting blockbuster films became widespread in the game industry. Certainly, there had been precursors to the trend: Atari's 1979 version of *Superman*, despite not being explicitly sourced from the film, was a clear attempt at cashing in on Warner's popular intellectual property. An almost inevitable convergence, since Warner had owned Atari since 1976. In arcade rooms of the late 1970s, games derived from films were not an exceptional sight: in 1975 Bally had produced a pinball machine called *Wizard,* themed after the film *Tommy* (Ken Russell, 1975), and in the same year Atari had released *Shark Jaws,* an unlicensed video game version of *Jaws* (Steven Spielberg, 1975), whose cabinet art clearly pointed at Spielberg's film by having the word 'Jaws' displayed in a considerably larger font.

Nevertheless, 1982 saw a veritable explosion of adaptations. The release of the film *Tron* (Steven Lisberger, 1982), in which a video game programmer saves the world from the domination of computerised machines, spawned at least three adaptations for the popular home consoles of the time: *Adventures of Tron* (Mattel, 1982) for the Atari 2600, *Tron: Deadly Discs* (Mattel, 1982) for the Atari 2600 and Intellivision, and *Tron Maze-A-Tron* (Mattel, 1982) for the Intellivision, alongside with an arcade game produced by Bally Midway simply titled *Tron*. Moreover, games such as *Star Wars: The Empire Strikes Back* and *Raiders of the Lost Ark* contributed to reinforcing the widespread sentiment that video games were on the verge of becoming a valuable tie-in market for block-buster films and, more generally, intellectual properties designed for younger audiences. In the wake of such a notable convergence, most video games magazines ran stories on what *Atari Age* dubbed "movie games", and the cover of the first issue of *Electronic Fun with Computers and Games* (November 1982) depicted a series of digitised versions of popular film characters – Indiana Jones, Princess Leia, Mr. Spock – accompanied by a title claiming "Hollywood Video Game Explosion: Raiders! ET! Star Wars! Star Trek!"

As for the other end of the spectrum, 1994 is an arbitrary date in the context of a fuzzy area of transition from bi-dimensional to tri-dimensional representation in video games. As argued by Arsenault et al. (2013), in the early 1990s 3D technologies became a prominent theme within the medium; the convergence of the industry's technological teleology, the distinct tech-nophilia found in the discourses on video games, and a series of emergent trends in game design brought tri-dimensional representation to the fore. The success of games such as *DOOM* (id Software, 1993), the release of Sony PlayStation in late 1994, and the production of the first 3D graphics processing units for home PCs in 1995 grounded in technology what magazines and sales representatives had predicted: 2D was out, 3D was in. This literal shift in perspective ushered in an era in which video games explored the possibilities of tri-dimensional space exploration as a narrative tool (Jenkins 2004; Nitsche 2008) and, at the same time, as a way to remediate older and arguably better established media forms such as cinema (Bolter and Grusin 2000). While earlier film-to-video game adaptations had to work their way around the abstract-ness of bi-dimensional representation (Wolf 2003), 3D allowed designers to build believable and explorable worlds and appropriate some of the aesthetic markers of cinema – camera movement, editing, depth of field, et cetera – for dramatic purposes. 3D allowed video games to "borrow heavily from cinematic styles and effects" (Bryce and Rutter 2002: 71), ultimately reconfiguring the relations between cinema and video games and influencing the booming genre of film-to-game adaptations. This decade-long shift seemed to peak in 1994 with the release of the ultimate 3D machine, Sony PlayStation. As the industry and the market were changed forever by the advent of 3D technology, video game criticism and the emerging field of game studies soon followed suit. Researches on the exchanges and adaptations between cinema and video games invariably focus on games produced around or after the shift towards tri-dimensional representation (King 2002) or on the very issue of cinematic aesthetics in – mostly 3D – games (Brooker 2009; Fassone et al. 2015). With the exception of the occasional contribution on the matter (Aldred 2012), the researcher willing to reconstruct the history and aesthetics of early film-to-video game adaptations is left with a lot of work to do and a pile of bad games.

A history of marginality

A history of adaptations should always consist of at least two histories. On the one hand, the his-tory of the adapted text, its modes and times of production, its reception and use; on the other hand, the history of the text into which it is adapted. In this sense, as noted by Murray, adapted texts "enter the contemporary adaptation economy already freighted with critical approbation and/or notoriety" (2012: 21). In the case of film-to-video game adaptations produced in the 1980s and early 1990s, the complexity, vastness, and relative obscurity of the corpus makes the task of telling two asynchronous histories almost unattainable. While working on specific case studies is certainly possible and even fruitful, the variety of the strategies and methods adopted by game designers when translating a film into a video game prevents this sort of analysis from capturing larger phenomena and concurrent processes. Nevertheless, this chapter will rely on a form of historiographical "double vision" (Hutcheon 2006: 15) that tries to correlate the extent and characteristics of a phenomenon – that of intellectual property licensing between two media industries – with synchronic and diachronic processes at work within the wider ecosystem of western audio-visual media during the time-period taken into account. In other words, the chapter will present two histories that differ in the level of granularity. One is the recapitulation of a specific trend within the video game industry, the other an analysis of larger cultural and technological shifts involving several different media systems, video games being one of them. Their superimposition will point at a process of convergence that involved two media industries well before "media convergence" (Jenkins 2006) became a ubiquitous buzzword.

The first history is that of the hundreds of games derived from films of variable success between 1982 and 1994. This is in most cases a history of marginality. In *A Theory of Adaptation*, Hutcheon claims that "in both academic criticism and journalistic reviewing, contemporary popular adaptations are most often put down as secondary, derivative […]" (2006: 2). While this seems to be true of most adaptations in which a 'younger' medium is often considered a degrading agent to the content of a nobler, 'older' ancestor, during the decade in question video game adaptations came to be considered almost universally as a necessary evil for both cinema and games. Writing about pre-crash adaptations, Aldred claims that "movie-licensed games have come to be viewed as a blight on both the film and video game industries at the time" (2012: 91), mere cash-grabbers that misrepresented beloved films and, at the same time, smears in the catalogues of respectable game manufacturers. This was not always the case; as we have seen, the adaptations produced for the home console market in 1982 sparked the enthusiasm of the specialised press. So-called movie games seemed to be a viable way of marketing successful cartridges by relying on popular content; on the other hand, the forerunners of media conglomeration such as Warner saw the video game market as an ideal source of revenues from tie-in products. While the market crash may have instilled a dose of cynicism in both the press and the industry, a general optimism towards film-to-game adaptations may be observed well beyond the traditional cut-off date of 1983. The Strong's collection of trade documents, contracts and memoranda from the arcade division of Atari shows the company's genuine interest in liaising with Lucasfilm for the production of two arcade games, *Star Wars* (Atari, 1983) and *Return of the Jedi* (Atari, 1984). While the latter was developed in conjunction with Lucasfilm from the very beginning, the former was based on a game tentatively titled *Warp Speed*, whose design started in 1979 but was completed only in light of the deal with the producers of the *Star Wars* saga. Both games achieved considerable success, to the point that *Star Wars* was immediately licensed for conversion on several home consoles and micro-computers. While the arcade game industry thrived on successful film adaptations, the languishing home console market was about to find in Nintendo, a Japanese toymaker, its saviour. In 1985, Nintendo launched the Nintendo Entertainment System, a version of the Famicom, Nintendo's flagship Japanese 'family computer', designed to appeal to the American consumer electronic market. Between 1985 and 1986, Nintendo's aggressive industrial strategy and the system's astounding technological features were enough to revive the interest in video game consoles and open new possibilities for game developers and manufacturers in the United States. By the end of the 1980s Nintendo had become the most significant force in the American toy market. According to Sheff:

> Nintendo ate up a larger and larger share of the toy business, becoming far bigger than Hasbro and Mattel combined. Of the estimated $11.4 billion spent on toys in 1989, 23 percent was spent on Nintendo products. Of the thirty top-selling toys in America, twenty-five were Nintendo or Nintendo-related.
>
> *Sheff 1999: 203*

Nintendo's hegemony forced several established toymakers to join forces with the Japanese giant. Among these, New York's LJN distinguished itself as the most prominent producer of film-to-game adaptations. LJN started releasing games for the NES in 1987, but the company, founded in 1970, had a history of licensed products. A domestic price list from 1983 features a *Magnum P.I.* action figure, a *Chips* gift set, and, most notably, a series of *E.T.*-related vinyl figurines. All through the 1980s, LJN's interest in licensed products had manifested itself in a vast production of traditional toys based on popular TV shows or blockbuster films. The company's 1988 trade catalogue shows an apparently seamless transition into the production of NES cartridges: LJN's new lines of games is introduced emphatically as the avant-garde of video game renaissance:

> Video games are back and LJN has the hottest new games in this action-packed category! And we've teamed up with Nintendo to create an unbeatable combination! Our new line of Enteractive Video Games includes strong American licenses with enormous staying power.
>
> *LJN 1988: 20*

While the reference to video games 'being back' underlines Nintendo's role in reviving a market considered dead, LJN's trade catalogue clearly displays the company's confidence in the economic viability of licensed video games. Between 1987 and the mid-1990s LJN's various subsidiaries – Enteractive Video Games being the most notable – produced over a dozen video games derived from film properties. More specifically, between 1987 and 1990 – when LJN was acquired by video game developer Acclaim – Enteractive Video Games released several adaptations for the NES whose game design, graphics, packaging, and artworks showed significant aesthetic consistency, thus creating a recognisable sub-series (Fassone 2014) within LJN's video game production. Games such as *Jaws* (1987), *Friday the 13th* (1989), *Back to the Future* (1989), and *A Nightmare on Elm Street* (1990) were not merely serialised through aesthetic markers, but shared a common method of production: they were developed by sub-contracted companies, such as Beam in Australia or Rare Ltd. in the UK, and then released by LJN via Enteractive Video Games. These adaptations were generally received rather poorly by the press, that started associating LJN's cinematic adaptations with sub-par tie-ins, engendering a critical discourse that would eventually come to encompass most film-to-game adaptations.[5] A review of *A Nightmare on Elm Street* from 1990 reads:

> The king of movie-licensed video games, LJN, kicks of 1990 with the vid version of everybody's favorite horror character (excuse me Jason), Freddy Krueger! […] This game capitalizes on a movie name and does not deliver a decent game to back it up.
>
> *Anonymous 1990: 18*

Other companies, such as Ocean in the UK and Sunsoft in Japan, would adopt similar licensing strategies when developing games for the NES, relying on sub-contracted developers or internal teams whose task was to produce adaptations of acquired intellectual properties. By 1990, players of NES games had seen a variety of film adaptations – everything from hit films, such as *Batman* (Tim Burton, 1989), to unlikely oldies, such as *The Three Stooges* – and had learned to steer clear of them. With the advent of the 16-bit era of consoles in the beginning of the 1990s and the advancement in graphics and sound that helped Nintendo and SEGA sell their new-generation consoles, film adaptations could rely on a relatively higher degree of visual fidelity. Players could finally control characters that looked like their real-life counterparts and explore recognisable environments, although much of the contextualisation was still provided by the game's paratextual apparatus. *Terminator 2: Judgement Day* (LJN, 1991) for Nintendo's SNES features a playable character that resembles the droid played by Arnold Schwarzenegger in Cameron's film and is set in a series of locations (a bar, an office, a foundry) that distinctly follow the film's plot. Nevertheless, it is what Gerard Genette would describe as "the publisher's peritext" (1997: 169), that is the visual and textual references found in the game's packaging and manual, that cements the connection between the film and the game. The game's box depicts a series of screenshots and captions them with iconic punchlines from the film, such as "I need your motorcycle" and "Hasta la vista, baby!" Like games from the previous generations, whose elaborate box arts promised an unattainable degree of visual fidelity, adaptations of the 16-bit era had to rely on paratextual materials to crystalise their consistency with the adapted film, while they usually employed rote or overly generic game mechanics when it came to actual gameplay. Despite the relative popularity of some adaptations and the amount of films produced in the early 1990s that, just like *Tron* in 1982, seemed to be prime choices for a video game conversion,[6] the attitude of both the press and the players towards adaptations remained generally negative. In an article published in 1995 by the British magazine *Edge*, 'old' adaptations, that is, those that came before the era of 3D representation in computer games, are characterised as notoriously unredeemable:

> In the bad old days of C64s and Spectrums, games based on films were notoriously bad. Of course, back then the programmers were crippled by tiny memory and pathetic colour palettes, so their games were bound to be pale imitations of the celluloid masterpieces. How could you possibly do justice to the magnificent morphing of *T2* on a C64? You couldn't – you just had to create a badly animated T1000 and hope the kids got sucked in by the hype.
>
> *Anonymous 1995: 70*

If it is true that, even in modern adaptations, "games play second fiddle to films" by representing "yet another opportunity to buy into the film franchise" (Moore 2010: 186), between 1982 and 1994 video game adaptations seemed to represent little more than a cluster of sub-par video games with a film license slapped on top.

A history of domestication

This first history is that of a breed of games of marginal interest to video game historians and critics, whose existence seems to relate to the medium's "cinema envy" (Zimmerman 2002: 125) and the volatility of a market based on low-quality licensed products. It is only in the light of a second, larger-scale history that the relevance of early film-to-game adaptations in reconstructing the vicissitudes of the medium and its positioning in the wider media ecosystem becomes evident. Media historian Erkki Huhtamo (2012) claims that gaming consoles of the 1970s and 1980s promoted a process of domestication of video games; while arcade games were located in public spaces that functioned as hubs for the social patterns of youth culture, consoles were often presented as playthings for the whole family. A useful companion to Huhtamo's theory is Murphy's claim that "the widespread public acceptance and use of home video game systems by a broader audience indicates that consumers were rethinking television's role as a home technology in the mid-1970s" (2009: 202). Designed to be played in the living room as an ideal companion to a TV set – think of the wooden finish of the Atari 2600 – video game consoles were meant to become part of the American "electronic hearth" (Tichi, 1992) as both a domesticated form of electronic entertainment and a refashioning agent for the TV set. While Huhtamo's archaeology of home video games manages to describe a distinct transition in the history of the medium, it should be noted that several other domestications were happening at the same time. The video game console stood as one of the symbols of an era in which a series of media technologies found their way into the domestic space, often in conjunction with the TV set, influencing the times and practices of consumption of audio-visual content. Another notable household appliance that started gaining traction in the mid- and late-1970s was the VCR. Conceived as a tool for the preservation and reproduction of both private and acquired videos (Moran 2002), the VCR contributed to "increase the 'channel capacity' of the average home" (Wasser 2001: 80) by allowing users to access a variety of media contents in their living rooms. Just as television had reconfigured the domestic space and the habits of western families (Meyrowitz 1985), domestic technologies such as the game console and the VCR were changing television, reconfiguring its identity as a machine for broadcasting, and turning it into the companion of a number of new devices whose function was to allow users to access audio-visual products in a personalised fashion, using the TV screen as a canvas on which to project their own content.

While the new devices were battling for supremacy in the living room and, at the same time, cooperating in rewriting the uses, discourses, and ultimately the protocols (Gitelman 2006) associated with television, American cinema was entering "the era of blockbuster movies" (Sutton and Wogan 2009: 117). The success of *Jaws* in 1975 had convinced Hollywood producers that high-budget, spectacular summer blockbusters aimed at a young audience were financially viable. In the following years, films such as the *Star Wars* saga (various directors, 1977–2015), *E. T. The Extra-Terrestrial* (Steven Spielberg, 1982), and the *Indiana Jones* series (Steven Spielberg, 1981–2008) established a stylistic canon that retrieved the narrative transparency of classical Hollywood cinema and reworked it in the context of what David Bordwell describes as "intensified continuity" (2006: 54). Post-classical Hollywood blockbusters had relatively simple plotlines, memorable set-pieces, and iconic, immediately recognisable characters and locations. These "spectacular narratives" (King 2000), often built in a modular fashion with bits of story connecting quasi-autonomous spectacular sequences, were a perfect fit for adaptations and licensing. LJN's aforementioned lines of *E. T.* toys and *Rambo* branded toy guns tapped into the iconic potential of 1980s Hollywood cinema and, more significantly, constituted an integral part of larger process of domestication of a medium that had been historically characterised by its public nature. Film historian Francesco Casetti describes the ongoing process of de-institutionalisation of cinema as "relocation", a complex dynamic in which cinema as a medium is increasingly experienced outside of traditional places of consumption, in "new spatial systems and [...] new viewing conditions" (2009: 62). While Casetti refers to the portability of contemporary digital cinema and its decoupling from analogue supports such as film, it could be argued that in the late 1970s and 1980s, Hollywood blockbusters were relocated to the domestic space. While toys reified salient icons from the films, VHS tapes allowed viewers to own their favourite films and adapt screening times and spaces to their personal schedule. Furthermore, viewers could segment their viewing experience, or decide to revisit a dramatic set-piece rather than watch the whole film. Film-to-game adaptations seem to constitute a peculiar *trait d'union* between licensed toys and home video products. As we have seen, gaming consoles shared the living room with home video devices and, more importantly, were conceived as an extension of the TV set, thus negotiating their screen presence with the VCR and with regular broadcast. On the other hand, video games – both adaptations and original properties – were often sold in toy stores *and* rented in video rental chains at the same time, a condition that reinforced their ambiguity as both toys and audio-visual media products. Finally, film-to-game adaptations of the 1980s and early 1990s established a relation with the adapted material that merged the iconisation of the toy industry and the potential for segmentation and modularity offered by the VHS tape and later home video technologies. Few games exemplify this tension as clearly as *Krull*, an adaptation of the 1983 film released by Atari in September of the same year. On one hand, *Krull* employs a set of standard practices often seen in video game adaptations of the time. The film is referenced aesthetically mostly through the representation of iconic props; just like the whip in *Raiders of the Lost Ark* had an indexical function in identifying an anthropomorphic aggregate of pixels as Indiana Jones, the glaive, an iconic star-shaped weapon used by Colwyn (Ken Marshall) in the film, constitutes the most recognisable object in the game. On the other hand, *Krull* participates in a double process of discretisation and abstraction of the narrative of the original film. Hutcheon (2006: 13) suggests that video game adaptations often allow the player to 'play out' the second act of a film, that is, the segment of the narrative in which, typically, the conflict established in the first act is confronted and, eventually, resolved. In the case of *Krull*, the game seems to operate differently from the source material; instead of representing only the main conflict of the film – that of Colwyn against the evil creature who stole his wife, a narratological cliché if ever there was one – the game instils a series of conflicts within several sequences taken from the film. In order to progress through the game, the player is required to clear four distinct stages, representing salient sequences from the film: the marriage, the journey through a hostile land, the encounter with a giant spider and, finally, the confrontation with the monstrous kidnapper. This segmentation of the original material is aided by the film's distinct modularity: as many fantasy–action epics of the time, *Krull* is essentially a linear juxtaposition of set-pieces that can be easily exported to other formats as relatively independent narrative modules. Nevertheless, the game's attempt at following the plot of the film is counterbalanced by a strict observance of some of the design conventions of the time: instead of ending with the death of the evil foe by the hand of the player's avatar, the game loops and prompts the player to re-play at a higher level of difficulty. This oscillation between the abstracted, albeit recognisable, format of the film and the emergence of toy-like qualities may be seen as one of the defining features of early film-to-game adaptations, a form of media production that stands at the heart of an inter-media negotiation that involves cinema, television, home video, and the toy industry. The video game *Krull* is an actor within the complex process of domestication of the film *Krull*, whose narrative and iconographic affordances made it an ideal candidate for dissemination in various formats: a series of toys (the film spawned a set of playing cards, replica weapons, frisbees and a tabletop game), a video game adaptation, and a home video version, thus demonstrating the adaptability of film-based properties to a wide spectrum of formats and artifacts.

Conclusions

Tracing the histories of film-to-game licensing and analysing *Krull* as a part of a complex relation among media rather than as an exercise in inter-semiotic adaptation means recognising that a history of adaptations is always also a history of media. In reconstructing the histories of film-to-game adaptations, domestic appliances such as TV sets and VCRs are relevant, unavoidable actors, and the large-scale shifts observed in the history of American cinema in the 1980s may be able to reveal more about the object of this study than any single film or game could. Studying the history of the apparently marginal production of video game adaptations may provide media historians with useful insights on the process of domestication that cinema underwent with the advent of technologies aimed at reshaping the paths of circulation and consumption of films. Video game consoles, VCRs, licensed toys and other early transmedia products contributed to the ongoing relocation of the larger experience of cinema, through which "this previously physically remote and transitory medium has thus attained the solidity and semi-permanent status of a household object, intimately and infinitely subject to manipulation in the private sphere" (Klinger 2006: 57–58). A process of which video game adaptations – bad as they may be – stand as relevant historical traces.

Notes

1 This research was made possible by a fellowship offered by the Strong Museum of Play in October 2015. I am sincerely grateful to The Strong for this opportunity. I also wish to thank the staff of the Library, the Archives, and the International Center for the History of Electronic Games, whose invaluable help greatly contributed to the advancement of my research project and the writing of this chapter. My friend and colleague Luca Barra offered an insightful bibliographic suggestion, and I thank him for it.
2 A CRT TV is a cathode ray tube TV, a technological standard that was eventually superseded by digital TV sets.
3 The term licensing refers to the practice of acquiring the license for an intellectual property (e.g. *E.T. The Extra-Terrestrial*) in order to derive various types of adaptations and other forms of bi-products.
4 Micro-computers such as the Commodore 64 were very popular in Europe, especially in computer clubs and other forms of hobbyist communities, and relied on an economic model which allowed for grass-roots practices, home-made software, and various forms of piracy to emerge. For this and other reasons, this sector was less impacted by the crash.
5 LucasArts, a sub-division of LucasFilm, is a notable exception. Their games, often based on proprietary intellectual properties such as Indiana Jones, were generally well received and were considered among the best examples of point-and-click adventure games available on personal computers at the time.
6 Following literary movements such as cyberpunk, the late 1980s and early 1990s saw a significant increase in the production of films whose plots offered a dystopic take on the evolution of computer technologies and robotics. Among these were *RoboCop* (Paul Verhoeven, 1987), *Total Recall* (Paul Verhoeven, 1990), *The Lawnmower Man* (Brett Leonard, 1992), and *Johnny Mnemonic* (Robert Longo, 1995). All of these films were adapted into video games.

Works cited

Aldred, J. (2012) "A question of character. Transmediation, abstraction, and identification in early games licensed from movies," in Mark J.P. Wolf (ed.) *Before the Crash. Early Video Game History*, Detroit, MI: Wayne State University Press, 90–104.

Anonymous (1990) "Nightmare on Elm," *Electronic Gaming Monthly* 11: 18.

Anonymous (1995) "Johnny Mnemonic," *Edge* 23: 70–71.

Arsenault, D., Côté, P., Larochelle, A., Lebel, S. (2013) "Graphical technologies, innovation and aesthetics in the video game industry: A case study of the shift from 2d to 3d graphics in the 1990s," *Game. The Italian Journal of Game Studies* 2(1).

Atari (1982), *E.T. – The Extra-terrestrial*, Atari 2006 console, Atari Inc.

Atari (1983), *Star Wars*, Atari 2006 console, Atari Inc.

Atari (1984), *Return of the Jedi*, arcade game, Atari Inc.

Batman (1989) Directed by Tim Burton [Film]. USA: Warner Bros.

Bolter, J., Grusin, R. (2000) *Remediation. Understanding new media*, Cambridge, MA: MIT Press.

Bordwell, D. (2006) *The Way Hollywood Tells It. Story and Style in Modern Movies*, Berkeley, CA: University of California Press.

Brooker, W. (2009) "Camera-eye, CG-eye: Video games and the 'cinematic'," *Cinema Journal* 28(3): 122–128.

Bryce, J., Rutter, J. (2002) "Spectacle of the deathmatch: Character and narrative in first-person shooters," in Geoff King, Tanya Krzywinska (eds.) *ScreenPlay. Cinema/Videogames/Interfaces*, London: Wallflower, 66–80.

Casetti, F. (2009) "Filmic experience," *Screen*, 50(1): 56–66.

Domark (1985), *A View to a Kill*, Domark.

Donovan, T. (2010) *Replay. The History of Video Games*, Lewes, UK: Yellow Ant.

E. T. the Extra-Terrestrial (1982) Directed by Steven Spielberg [Film]. USA: Universal Pictures.

Fassone, R. (2014) "Sometimes they come back (some of them). Serialization, remaking, nostalgia. The case of Xbox Live Arcade," in Federico Giordano, Bernard Perron (eds.) *The Archives. Post-Cinema and Video Game between Memory and the Image of the Present*, Milano: Mimesis International, 125–138.

Fassone, R., Giordano, F., Girina, I. (2015) "Re-framing video games in the light of cinema," *Game. The Italian Journal of Game Studies* 4(1).

Genette, G. (1997) *Paratexts: Thresholds of Interpretation*, Cambridge, UK: Cambridge University Press.

Gitelman, L. (2006) *Always Already New. Media, History, and the Data of Culture*, Cambridge, MA: MIT Press.

Guins, R. (2014) *Game After: A Cultural Study of Video Game Afterlife*, Cambridge, MA: MIT Press.

Huhtamo, E. (2012) "What's Victoria got to do with it? Toward an archaeology of domestic video gaming," Mark J.P. Wolf (ed.) *Before the Crash. Early Video Game History*, Detroit, MI: Wayne State University Press, 30–52.

Hutcheon, L. (2006) *A Theory of Adaptation*, New York: Routledge.

Indiana Jones franchise (1981–2008) Directed by Steven Spielberg [Film]. USA: Paramount Pictures.

Jaws (1975) Directed by Steven Spielberg [Film]. USA: Universal Pictures.

Jenkins, H. (2004) "Game design as narrative architecture," in Noah Wardrip-Fruin, Pat Harrigan (eds.) *First Person. New Media as Story, Performance, and Game*, Cambridge, MA: MIT Press, 117–130.

Jenkins, H. (2006) *Convergence Culture. Where Old and New Media Collide*, New York: NYU Press.

Kent, S. (2001) *The Ultimate History of Video Games. From Pong to Pokémon and Beyond – The Story behind the Craze That Touched our Lives and Changed the World*, New York: Three Rivers Press.

King, G. (2000) *Spectacular Narratives. Hollywood in the Age of the Blockbuster*, London: I.B. Tauris.

King, G. (2002) "Die hard/try harder: Narrative, spectacle and beyond, from Hollywood to videogame," in Geoff King, Tanya Krzywinska (eds.) *ScreenPlay. Cinema/Videogames/Interfaces*, London: Wallflower, 50–65.

Klinger, B. (2006) *Beyond the Multiplex. Cinema, New Technologies, and the Home*, Berkeley, CA: University of California Press.

Krull (1983) Directed by Peter Yates [Film]. USA: Columbia Pictures.

LJN (1988) *LJN and Entertech Trade Catalogue.*

Mattel (1982), *Adventures of Tron*, Atari 2006 console, Mattel.

Meyrowitz, J. (1985) *No Sense of Place. The Impact of Electronic Media on Social Behavior*, Oxford: Oxford University Press.

Moore, M.R. (2010) "Adaptation and new media," *Adaptation* 3(2): 179–192.

Moran, J.M. (2002) *There's No Place Like Home Video*, Minneapolis, MN: University of Minnesota Press.

Murphy, S.C. (2009) "'This is intelligent television'. Early video games and television in the emergence of the personal computer," in Bernard Perron, Mark J.P. Wolf (eds.) *The Video Game Theory Reader 2*, New York: Routledge, 197–212.

Murray, S. (2012) *The Adaptation Industry: The Cultural Economy of Contemporary Literary Adaptation*, New York: Routledge.

Nitsche, M. (2008) *Video Game Spaces. Image, Play, and Structure in 3D Worlds*, Cambridge, MA: MIT Press.

Robo Cop (1987) Directed by Paul Verhoeven [Film]. USA: Orion Pictures.

Sheff, D. (1999) *Game Over. How Nintendo Conquered the World*, Wilton, CT: GamePress.

Star Wars franchise (1977–2015) Various directors [Film]. USA: Various distributors

Sutton, D., Wogan, P. (2009) *Hollywood Blockbusters. The Anthropology of Popular Movies*, Oxford: Berg.

Terminator 2: Judgement Day (1991) Directed by James Cameron [Film]. USA: TriStar Pictures.

The Lawnmower Man (1992) Directed by Brett Leonard [Film]. USA: New Line Cinema.

Tichi, C. (1992) *Electronic Hearth. Creating an American Television Culture*, Oxford: Oxford University Press.

Tommy (1975) Directed by Ken Russell [Film]. UK: Hemdale.

Total Recall (1990) Directed by Paul Verhoeven [Film]. USA: TriStar Pictures.

Townsend, E. (2006), "The 10 worst games of all time," *PC World 23* Oct.; Web [15 Oct. 2015].

Tron (1982) Directed by Steven Lisberger [Film]. USA: Buena Vista Distribution.

Wasser, F. (2001) *Veni, Vidi, Video. The Hollywood Empire and the VCR*, Austin, TX: University of Texas Press.

Wolf, M.J.P. (2003) "Abstraction in the Video Game," in Bernard Perron, Mark J.P. Wolf (eds.) *The Video Game Theory Reader*, New York: Routledge, 195–221.

Zimmerman, E. (2002) "Do independent games exist?" in Lucien King (ed.) *The History and Culture of Videogames*, London: Laurence King Publishing, 120–129.

11

ROSAS

Appropriation as afterlife[1]

Johan Callens

In 1990, Rosas, the Belgian dance company led by Anne Teresa De Keersmaeker, created *Stella: A Woman's Piece*, a dance theatre piece which initiated a sequence of closely related works, *Achterland* (1990) and *Erts* (1992), as well as drew on the earlier *Ottone, Ottone* (1988), based on Monteverdi's *L'incoronazione di Poppea*. Typical of *Stella*—performed by Fumiyo Ikeda, Johanne Saunier, Carlotta Sagna, Nathalie Million and Marion Levy—is that it muddles the discreteness of text, genre and media in crossings characteristic of adaptations. It may be argued that more mixed or hybridized forms like this act as a bid to undo the power that comes with exclusiveness, whether that of authorial origins or gender categories and identities. The following discussion, focusing more on dramaturgical conception than on choreographical execution, inflects this notion of exclusiveness, by exploring the interpretative ramifications of the recycled source material and by treating the debate over Beyoncé's appropriation of *Achterland* as well as Rosas's inaugural production, *Rosas danst Rosas* (1983). Such appropriation extends a creative artistic practice in which the adaptation of foreign artistic sources and media images is always already supplemented by the reappropriation of one's own work and gender image as a female artist. *Stella*'s disintegration of the essentialized, unified female subject problematizes authorship as much as its inverse, plagiarism, and its later appropriation put De Keersmaeker in an awkward position. Beyond De Keersmaeker's paradoxical frustration over seeing her experimental dramaturgy mainstreamed without being credited or compensated, there was also the devious issue of retrospectively claiming a feminist artistic stance, which the Flemish choreographer at the time had resisted, as well as reclaiming the kind of authorial control against which her own recursive and intertextual working method had consistently rebelled.

Reappropriating oneself and the other

De Keersmaeker's three dance pieces are, indeed, deeply recursive and intertextual. Each of the pieces relies in their inception on Goethe's eighteenth-century play *Stella*; Akira Kurosawa's 1950 screen adaptation of two early twentieth-century stories by Ryunosuke Akutagawa, "Rashomon" and "In a Grove" (1917); as well as on Elia Kazan's 1951 movie version of Tennessee Williams's *A Streetcar Named Desire* (1947). Moreover, the meanings of *Achterland*'s title (hinterland) may be said retroactively to include the backstage setting of *Stella*. Just so, *Erts* (ore) figures the intertextual and interdisciplinary layering of all three productions.

Music usually is the point of departure for Rosas dance productions, even more theatrical ones like *Stella*. Typically, De Keersmaeker's transhistorical and multilingual source material would have been combined in a single dance production, in this case starting with Ligeti's *Études pour piano*, his *Symphonic Poem for 100 Metronomes* and three sonatas for solo violin by the Belgian virtuoso, Eugène Ysaÿe. The difference is that this time De Keersmaeker insisted on using live music after experimenting with both live and recorded music written by Reich and Bartok (Hughes 1991: 18). That decision gradually led to a plan to have a kind of diptych. *Stella* would be danced by five women only, with a second piece performed by male dancers (Van Kerkhoven 1990–1991: 52). This idea was soon abandoned, however. Ultimately, the material ended up being developed in at least two supplementary choreographies with gender-mixed casts. Each treatment, nevertheless, acquired sufficient autonomy to be viewed independently.

The original intention seems to have been to deliver the text fragments in the three original languages, though only French and Japanese seem to have been used in the Haarlem premiere of *Stella* on 9 March 1990. For theatre critic Jac Heijer (1990), the ostensible meanings of the original texts had been pushed aside to make room for the bodily movements which these texts generated and the inner emotions given access to. Thus, the former context of the women's speeches was largely avoided, and where the text had been retained, it was turned into mono-logues by excising their interlocutor's contribution. What mattered to De Keersmaeker, in short, were the vicissitudes of the human heart rather than a faithful rendition of Williams's, Goethe's and Kurosawa's stories and characters. At the same time, as she made clear to interviewers at the time, the radical appropriation of these male authors buttressed the female choreographer's critique of the authority attendant upon a masculine logocentrism and the western urge to understand and rationalize. It does not seem to matter to De Keersmaeker that the complex structurings of the dances may be seen as another manifestation of this rationalism, or that she somehow tells the stories without really telling them, as she put it for *Ottone, Ottone*. *Stella* as a composition also deconstructs the Aristotelian unities, plot and character and the underly-ing sense of selfhood and identity, as do the multiple interpretations she allows of her work (Kottman 1991; Hughes 1991: 18–19).

The structural relation between *Stella* and *Achterland* was not settled from the start, as becomes most conspicuous in the promotional text for the 1990–1991 season's packet of deSingel, Antwerp (52), where *Achterland* ran from 9–12 January 1991. This anonymous text presumably was written by *Stella*'s dramaturg, the late Marianne Van Kerkhoven (1946–2013), shortly after the March 1990 Haarlem premiere, since this season's brochures seems to have gone to the printer in late spring. The many options for understanding these relationships (embedding, recursion, sequentiality, organic fusion or parallelism) were left open in Van Kerkhoven's evocation, for the simple reason that at that point *Achterland* had not yet been made. Strangely enough, this variety of relationships still characterizes the final string of choreographies (now including *Erts*), in which the assembled intertextual components accrue a welter of new meanings by 'reacting' with each other in their new environments. Thus, Rosas's eponymous Stella would seem to refer first and foremost to the lover in Goethe's play, but by combining it with Williams's *Streetcar*, the lover automatically also becomes wife and sister, thereby multiplying the female character's functional identities and invit-ing, to the dramaturg, consideration of the piece from the perspective of Lévi-Strauss's structuralist theorizing of the family less as a natural unit than a patriarchal institution regulating the exchange of women and building of the community (1956: 261–285). Van Kerkhoven also compared the relationships of *Stella*'s various figures to the interactions of electrons, in keeping with her desire to align the performing arts with the new sciences: "Sometimes it teems and collides. Still, there is structure in it, somewhat as in the world of electrons" (1990–1991: 52, my translation). The desire to bridge these two cultures is expressed in "The Weight of Time," a companion text written by the dramaturg at the time *Stella* was created and explicitly referencing the French structuralist (Van Kerkhoven June 1990: 6).

These complex interrelationships are important because De Keersmaeker's work has always been a continuous work-in-progress. In discussing the interrelationships between her choreographies, Philippe Guisgand speaks of "un entrelacs," the tracery or open pattern of interlacing ribs in a running decoration (as on walls, plates, etc.) (2007: 13). This is perhaps an unfortunate comparison, considering the implied superficiality, since the connections between De Keersmaeker's works run much deeper. Pieter T'Jonck, on the other hand, connected this interweaving to the choreographer's desire to build a repertory in which the revival of older material not only exposes the development of the artist's work but also generates new pieces (2006: n.p.). This recursive method is particularly relevant for the string of productions under discussion, since it sets up a dialogue between Williams, Kazan, Kurosawa, Akutagawa and Goethe, whose 1809 novel *The Elective Affinities* applies the idea of chemical reactions to human passions. The same idea can be extended to the intertextuality, interdisciplinarity and intermediality of Rosas productions, and in turn to the kind of fragmented, constructed and distributed identity they materialize in a protean manner. This may explain De Keersmaeker's recycling of *Rashomon*, whose double crime is refracted by several contradictory testimonies which never permit the viewer to settle on a single absolute truth. Accordingly, in *Streetcar* the conflict between Blanche's magic and Stanley's realism can be redefined as that between different perceptions of reality (Harris 1993: 85), even if the more radical and unsettling point *Rashomon* spells out may well be that we are all liars and the truth is ultimately unknowable. This also helps explain dancer Fumiyo Ikeda's cross-dressing when re-enacting bits from the movie, since cross-dressing destabilizes conventional gender binaries and unsettles fixed categories of meaning and identity.

To De Keersmaeker, "nothing ever is totally unequivocal [...] Even science has long ago discovered that linear thinking isn't sanctifying. There's doubt and chance is taken into account, chaos, and the erratic" (qtd in Kottman 1991). Elsewhere, she noted, "Closed systems aren't really very credible anymore, in mathematics and biology and the other sciences" (qtd in Hughes 1991: 19). Hence, *Stella*'s lack of closure and prolongation in *Achterland* and *Erts,* whose image of stratification conveys the assembled productions' complex layerings. These obvious layerings are meant to jolt habitual perceptions. Hence also the quotation from Adorno's *Minima Moralia* (1951), "Das Ganze ist das Unwahre" (The whole is the untrue), recycled as a caption underneath the still from *Stella* adorning the cover of the issue of the performing arts journal *Etcetera,* which contains two reviews of the production and the late Marianne Van Kerkhoven's dramaturgical companion piece, "The Weight of Time" in which she pleads for a rapprochement of the arts and sciences without forgoing feminine-gendered qualities like intuition, imagination and emotionality.

The set, designed by Herman Sorgeloos, as well as the costumes by Annette de Wilde and Rosas, contributed toward this programmatic constitutional and interpretative multiplicity. *Stella* takes place in what looks like a backstage area or dressing room with clothes racks, a conflation of the Rosas rehearsal space in the Werkhuizenkaai, Brussels (down to the architecture, volume and paint colour) and the stage space at the Toneelschuur Haarlem, to which the box set for *Bartok/Aantekeningen* was added when the production went on tour (Van Kerkhoven 1992; Hughes 1991). The black school dress costumes, too, harked back to *Bartok* and made for dance sequences in which the women seemed liberated from self-awareness, utterly engrossed and transported as they were by the joy of pure movement: what De Keersmaeker calls the child's pleasure at losing itself in dance (Heijer 1990: 4). Still, the resemblance of these dresses to boarding-school outfits, typical for "girls of proper families" (*filles de bonne famille*) (Kottman 1991: n.p.), insinuates a repressed sexuality as much as girlish freedom.

Throughout most of the production of *Stella*, the female dancers certainly displayed a conspicuous preoccupation with how they "(are forced to) appear in the eyes of man" and "how that gaze affects their appearance and especially their being" (T'Jonck 2001: n.p.). The attendant tension between voyeurism and exhibitionism is evident in *Streetcar* from Blanche's constant worries about her looks and her behavior during the first poker game, when she catches Mitch's eye (1: 289, 296). It is also evident in the furtive glances between Goethe's Fernando and Stella during their first encounter before their eyes meet, both text bits retained in the Rosas production. This existential, social and theatrical dynamic between showing and looking, between being on one's own terms rather than being defined by others, automatically fragments identity. This effect is compounded by mediatization and foregrounded by Rosas's explicit use of the screen adaptations of Akutagawa and Williams's texts rather than the primary material. T'Jonck largely dismisses the mediatization of Kurosawa and Kazan's screen adaptations by insisting on the dancers' "naked presence," which is enhanced by the fragmentary nature of the borrowed filmic and textual material (1991: n.p.). He insists that the performers also refused to subordinate themselves to full-fledged characters inscribed within continuous narratives. This is true despite the fact that dancers Nathalie Million and Carlotta Sagna did recite lines from *Streetcar*, Johanne Saunier from Goethe's *Stella* and Fumiyo Ikeda from *Rashomon*, leaving only Marion Levy speechless. For T'Jonck, this "naked presence" is intensified to discomforting "exhibitionist proportions" as the five different women persistently try to "realize" themselves (*waar maken*) on stage (1991: n.p.).

By comparison, Heijer perhaps better gauges the importance of Rosas's reliance on the movies. He recognizes that Brando's "stellar" performance, in which he acts up the "richly feathered male bird among hens" (1: 265), was crucial to the canonization of *Streetcar* but also unhinged the more balanced play Williams had written (1990: 4). Brando's Stanley outdoes Blanche's coquettish behavior, as Split Britches and Bloolips realized only too well when starting on their cross-gender *Belle Reprieve* (1991), shortly after Rosas's *Stella* (1990). This is why to Heijer the outcome of Kazan's 1951 movie may be rigged after all. Brando's Stanley arguably is so irresistible that, despite her repeated battering, as well as her sister's rape, Stella may have to return to him. "She can't do otherwise, because he has set her on fire, Brando being such a hot piece" (Heijer 1980: n.p.) It is this premise which Ivo van Hove rejected in his stagings of *Streetcar*, first with Globe (1995) and then with the New York Theatre Workshop (1999). To van Hove, Brando's equivocal interpretation of Stanley behaving toward women as an "undeniably sympathetic" "male pig" has long ago become impossible (Callens 2014: 306).

T'Jonck's Dutch phrase (*waar maken*) in this regard invokes the "truthfulness" (*waarachtigheid*) of the female performers' authenticity in Rosas's *Stella*, but also suggests the degree of self-assertion necessary, either to rupture the projections imposed on them as women or, conversely, to persuade the men that they are getting what they want. For Harris, too, Blanche could have avoided the rape had she not complied with Stanley's whorish image of her, an image that inescapably leads to death and madness (1993: 96). And Kazan's movie adaptation only enhances this sense of seeing Blanche from Stanley's, i.e. Brando's, macho perspective on women—yearning to be seduced and loved, untrustworthy yet idolized as well as infantilized and needing to be pacified when displeased or ignored. This is Stanley's/Brando's refracted image of femininity with which Rosas's female dancers confront the theatre spectators. The playscript already makes this mediation explicit in Blanche's rejoinder to Mitch's "I was fool enough to believe you was straight" (1: 385): "Who told you I wasn't—'straight'? My loving brother-in-law. And you believed him" (1: 385). Ironically, the choreography has Blanche's movie rejoinder: "What is straight? A line can be straight, or a street, but the human heart, oh, no, it's curved like a road through mountains" (Heijer 1990: 4). For a choreographer, of course, all kinds of trajectories are precious, as means of moving through space, dividing it into volumes, sizing up its dimensions, embodying and inhabiting emotions (Hughes 1991: 17).

On the other hand, Fumiyo Ikeda's dizzying opening turn, in which she condenses the different versions of Masago, as they emerge from the men's testimonies, right away foregrounds the choreography's gendered perspective. As Gilbert and Gubar have demonstrated in *The Madwoman in the Attic* with regard to feminist identity politics, women still need to define themselves in opposition to the patriarchal misconceptions and appropriations, causing them to suffer, not from Harold Bloom's agonistic and psycho-historical male "anxiety of influence" but from a female "anxiety of authorship." This distinctly female anxiety led nineteenth-century women to seize upon male mimicry and male impersonation as an intermediate stage of assuming authorship and creatively inscribing themselves into a female tradition (Gilbert and Gubar 1984: 46–49). It is true that poststructuralist critics like Luce Irigaray have rejected the underlying exclusivist and proprietary assumptions of patriarchal canon-building, and with it the essentialist and individualist identity politics buttressing traditional authorship (Randall 1989: 267; Bennett 2005: 86–87). But since the 1980s, De Keersmaeker's recursive repertory has embodied the kind of female authorship Gilbert and Gubar define.

One choreographic strategy in *Stella* that tends to undermine essentialist gender thinking was the carefully studied yet always inevitably differential repetition of movement sequences. These sequences were radically differential, even contrapuntal, compared to the forceful unison and subtle shifts of De Keersmaeker's earliest minimal choreographies like *Fase* (1982), though they were no less systematic in their confrontation of the material explored (Kottman 1991: n.p.). Thus Nathalie Million from the edge of the stage desperately tried to win over the spectators and, failing to do so, teetered back on her high heels to the upstage wall, viciously kicked it, then returned downstage to try once more. Conversely, Carlotta Sagna in her white ball gown, her eyes directed upwards, initiated a sequence in which her lower body slowly gyrated as she raised herself on the tip of her toes and stretched her neck as if she wanted to leave behind this crass world and reach the heavens for a union with God. And failing to do so, several times started all over again, each time contemptuously turning away from the audience. Then, after changing into one of the black school dresses familiar from *Bartok/Aantekeningen*, she repeated her sequence one final time, now looking straight at the spectators, pulling faces and ridiculing her own ecstatic longing. Still, the effect is more tragicomic than plainly ludicrous, as with Johanne Saunier's protracted, irrepressible giggling, which Heijer speculates might well be triggered by Stella's realization that Fernando, the man she loves so impetuously, is actually married to her best friend (1990: 4). For Hughes, too, the women in *Stella* seemed "to cry out for contact and companionship in a world where their texts and trajectories and repetitive actions lock them into islands" (1991: 17).

Another expression of the differential repetition in *Stella* is the extent to which the women's parts were strongly shaped by the dancers themselves (Van Kerkhoven: *Theaterschrift* 1991) yet always virtual, too, insofar as they depend on men's consumerist, mediatized desires and projections. While such dependence on others makes for a distributed self, a notion enhanced in the information age, it is also adumbrated in *Streetcar* where Stanley's superiority partly derives from the operationality of his network of acquaintances, as opposed to the virtuality of Blanche's, consisting as it does of imaginary beaux, lovers and long-deceased poets. However, insofar as either character embodies evolutionary ideas—whether Stanley's technology or Blanche's humanist culture strenuously instilled in generations of students—both characters figure as way stations toward Rosas's multilingual choreographic subject "becoming beside itself, plural, transalphabetic, derived from and spread over multiple sites of agency, a self going parallel: a para-self" (Rotman 2008: 9). This is why Van Kerkhoven and De Keersmaeker insisted on referencing several Stellas—Williams's as well as Goethe's—and why Fumiyo Ikeda's performance encompassed the wife, her husband and the bandit from Kurosawa's screen adaptation of Akutagawa's stories. Ultimately this feminine, distributed para-self or non-identity is not just a combination of Blanche, Stella and Masago, as reflected by the men in their lives, or the five female dancers displaying themselves to the theatre audiences, but all of them together, i.e. "woman" in her most diverse manifestations, as evidenced by the subtitle of Rosas's production, "A Women's Piece."

Because many of these agencies remain unacknowledged and invisible, they revalorize Williams's wonted spectral dramaturgy as well as the ghostly presence of Takehiro, the murdered husband, whose story in *Rashomon* is told through a medium. Re-appropriation—whether that of 'media' relaying the spirits of the dead, actors upstaging characters, writers and dramaturgs adapting extant material or artists creatively recycling their own repertory—can be considered one of these agencies, as the quite successful distributed, recursively constituted choreography, here under consideration, illustrates with a vengeance. An-Marie Lambrechts in this regard views *Stella's* choreography as writing in the tradition of Luce Irigaray's *écriture féminine* and as practiced and developed in *Ce sexe qui n'en est pas un* (1977). But this anti-patriarchal practice can be defined as an appropriatory poetics or 'poétique du plagiat' (Randall, 1989: 275–277) because it mimics and parodies masculine models in order to subvert them. In this same manner, De Keersmaeker exposes how women are perceived in the works she adapts in the hope of changing that perception (Hughes 1991: 19).

Beyoncé's re-appropriation

Since its initial performances, *Achterland* has gained worldwide popularity. It won the London Dance and Performance Award in 1991, and performances of the piece recorded at the Monnaie Dance Studio in August 1993 were released as a video work in 1994. This video production was directed by De Keersmaeker with the technical assistance of Herman Van Eyken, and received the 1994 Lyon Dance Screen Award and the Award for Best Adaptation at the 1995 Festival International du Film sur l'Art in Montreal. The video was re-edited in 2006 for a slightly shorter work released on DVD. Since then the choreography has become accessible via YouTube, where it still proves extremely popular. This popularity may be at least partially ascribed to Beyoncé Knowles's video clip of *Countdown* (2011, codirected by Adria Petty), which appropriated choreographic material from *Achterland* that was, by then, readily available on video and DVD, as well as from Thierry De Mey's 1997 film adaptation of the earlier *Rosas danst Rosas* (1983).

Knowles copied diverse elements from the video release of *Achterland,* including the split-screen technique, costumes, haircuts, light design and set (originally Henry Van de Velde's iconic, now overhauled technical school in Leuven, Belgium). And since Rosas tended to showcase *Rosas Danst Rosas* together with *Achterland*, as in Japan on the occasion of the fifth anniversary of Tokyo's Bunkamura Arts Center (March 4–11, 1994), even the specific combination of works could be considered part of Beyoncé's appropriation. This forced the Belgian choreographer to take legal action, but also invited comparison by critics like Alastair Macaulay with the choreographer's own treatment of some of her inspirational sources. Beyoncé's appropriation and De Keersmaeker's reaction, then, both exemplify the vicissitudes of creating in an age given to recycling. Put differently, the case here presented probes the question of where to draw the thin line between, on the one hand, originality and its attendant rights and rewards, and on the other hand, the ready availability for consumption of cultural materials circulating in a globalized mediascape.

In *Rosas danst Rosas* four female dancers perform sequences of "repetitive moves, fiercely exaggerated versions of ordinary gestures like fiddling with their hair, adjusting their clothes, slouching, slumping and twitching" (Mackrell 2013: n.p.). The ordinariness of the gestures is crucial to the rights issue, since it is hard enough to claim ownership over a single dance move, much less a woman's frequently repeated everyday moves like adjusting the shoulder of a blouse or tucking a strand of hair behind the ear. Which is why the plea to copyright dance by Culturebot founder, Andy Horwitz (2011: n.p.), may be hard systematically to implement, and why *New York Times* critic Alastair Macaulay cannot "get exercised about the subject" because the moves "are scarcely of striking originality" (2011: n.p.).

Beyoncé's copyright infringement, however, was no simple matter. The deeper target of Macaulay's similarly lacklustre reactions to some of Duchamp's and Warhol's appropriations of commonplace objects and images is the work's very artisticity. Macaulay seems to question the art value of De Keersmaeker's appropriation of everyday actions, as well as Knowles's reappropriation of those same actions. What matters, though, from the legal point of view, is the unicity of the moves' sequence and its length, besides the ability to demonstrate that the sequence was deliberately created for a work of art and has been substantially reappropriated without permission. In 1992, shortly after *Achterland* had been created, the Belgian artist Kobe Matthys established his agency which in publications, installations and interactive talks with artists and legal experts explores controversial borderline cases between nature and culture as they pertain to intellectual property rights (patents, trademarks, etc.). The concept of intellectual property indeed revolves around the, at times, fluid distinction between what exists naturally and what has been created. Viewed thus, consumer products like DVDs—including Beyoncé's popular music videos as well as Rosas's more experimental dances—tend to assume a self-evidence that naturalizes them and invites their appropriation despite the fact that they remain, at the same time, legally protected. To make matters more complicated, Rosas's gendered and discursive take on the female characters in various textual and filmic sources challenges a unified essentialist subject, simultaneously undermining notions of intellectual authorship and plagiarism (Bennett 5, 96–97). Structuralism's complex functional social understanding of woman, invited by Van Kerkhoven's reference to Lévi-Strauss's essay on the family in "The Weight of Time," and the distributed notion of self it anticipates equally affect authorship and plagiarism issues, much as Foucault's and Barthes's declarations of the death of the author shifted artistic responsibility to social and linguistic structures.

The Rosas case was not Beyoncé's first. In 2007 she was accused and acquitted of plagiarizing the lyrics of her single "Baby Boy," and earlier in 2011 she and her co-director Kenzo Digital admitted to using Lorella Cuccarini's choreography and visual effects for Beyoncé's performance of *Run the World (Girls)* at the Billboard Music Awards (Trueman 2011: n.p.; McKinley 2011: n.p.). Given the renown of De Keersmaeker, the more recent 2011 case was extensively covered in the press. Many cultural commentators defended De Keersmaeker's intellectual and creative rights, but some, like Luke Jennings in *The Guardian* (11 Oct. 2011) and the already mentioned Macaulay in the *New York Times* (21 Nov. 2011), also ventured that artists reference and borrow often enough. When it comes to recycling, modern choreographers like De Keersmaeker may not be that different from classical ones like Balanchine, whose debts can be traced back, via Lev Ivanov and Marius Petipa, to Anna Pavlova. After all, adds Jennings, "In her ballet *D'un soir un jour* [De Keersmaeker] includes part of the Vaslav Nijinsky choreography of *L'Après Midi d'un Faune* and a snatch of the Antonioni film *Blow-up*" (2011: n.p.).

The comparison seems warranted by the appropriatory strategy of "auto-legitimization" through "intertextualization", which Rosas initially followed regarding its use of styles, forms and norms. The company did this to position itself within the international dance scene, since there was no institutionalized Flemish contemporary dance scene to speak of during the 1980s. Rudy Laermans and Pascal Gielen cite *Rosas danst Rosas* and *Fase* as cases in point for being, quite paradoxically, "a highly original appropriation of American minimal dance as developed by, for instance, Lucinda Childs" (2000: 15–16). The originality, deriving in this case from the rejection of Flemish ballet as well as the excessive abstraction of analytical American postmodern dance, was indeed crucial to avoid accusations of plagiarism. It could of course be argued that Yvonne Rainer in the early 1960s had already introduced everyday gestures and actions into contemporary dance at the Judson Memorial Church. *Ordinary Dance* (1962) was programmatic in this regard (Burt; 2006: 19, 88–115). Rosas's novelty, however, consisted in the mix of repetition, minimalism and the everyday, emotionally enriching the rigorous patterning. All the same, the attention to her dancers' individuality, the reliance on text material or the combination of film and dance in *Stella* and *Erts* could still be considered a return to Rainer.

Macaulay's and Jennings's comparison of De Keersmaeker's practice with that of classical ballet is also warranted by the already mentioned appropriatory poetics of Rosas's choreographic *écriture féminine*, which involves mimicry and plagiarism to distance the work performed from predecessors or the dominant regional context, where modern ballet still held sway. A case in point is *Erase-E(X)* where Johanne Saunier briefly joked about Maurice Béjart, artistic director of the Ballet of the XXth Century, founder and artistic director of the international Mudra dance school, where De Keersmaeker and Fumiyo Ikeda were trained, and former choreographer-in-residence of the Royal Monnaie Opera House (Callens 2008). So the effect of the appropriation can be double: to install continuities (with international contemporary dance) as well as discontinuities (with the dominant Flemish ballet tradition and with the all too abstract American minimal dance). Incidentally, in 1998 Béjart lost a court case against Frédéric Flamand, for copying from the latter's Plan K. production of *The Fall of Icarus* its idea of a practically naked winged man with TV monitors attached to his feet. Béjart thereby very much seemed to prove Macaulay's and Jennings' assessment of classical ballet's careless attitude toward appropriation. His defence argued that ideas cannot be copyrighted, but this argument was overruled by the resemblance of execution in Béjart's *Le presbytère* (Gabriels & Mostmans 2011), an argument that may well have been used in De Keersmaeker's legal case about which no details have been made public.

Still, the comparison of De Keersmaeker's practice with classical ballet's appropriations, including Béjart's, is also flawed, insofar as the choreographic and filmic excerpts in her *D'un soir un jour* are acknowledged in the production's program and on the company's website, acknowledgments that legitimize the quotations. Jennings and Macaulay doubt whether De Keersmaeker suffered any financial damages as a result of Beyoncé's plagiarism, but a share from the proceeds of *Countdown*'s sales may have been a fair gain in return for the creative rights and intellectual alienation. As to the symbolic capital at stake, it is not entirely clear who gained most, despite Marilyn Randall's conviction that this benefit devolves more to the accuser of plagiarism than the plagiarist (2001: 159–188). It is true that in the 1980s Rosas needed symbolic capital to validate its work, capital extracted from referencing already artistically validated international models of contemporary dance or being associated with them by organizations like Kaaitheater and Klapstuk or performing arts journals like *Etcetera* (Laermans & Gielen 2000). But three decades later, a pop artist like Beyoncé may actually have been accruing artistic prestige by borrowing from Rosas. Equally relevant to the discussion about appropriation are the involved artists' aesthetics, similar though they may seem, though not as similar as Macaulay and Jennings insist. To the former, *Rosas danst Rosas* is concerned "not with original movements but with recontextualizing ordinary ones [...] Assemblage, not invention," i.e. a core legal distinction but also a dominant feature of the contemporary subject (Jennings 2011: n.p.). By comparison, the aesthetics of *Countdown* maybe that of "a very slick, very new-retro piece of film-making," but "borrowed imagery is absolutely the point of it" (Jennings 2011: n.p.).

This is very much the line Beyoncé took when publicly venturing in her own defence that the *Countdown* video is a "tribute" (Merritt 2014: n.p.) inspired by Audrey Hepburn in *Funny Face* (1957), Stanley Donen's film adaptation, co-starring Hepburn and Fred Astaire, of George and Ira Gershwin's 1927 Broadway musical. In addition, *Countdown* was said to rely on the 'looks' of 1960s and 1970s cultural icons like Andy Warhol, Twiggy, Diana Ross (her Supremes incarnation) and Brigitte Bardot (*Telegraph* 13 Oct. 2011; BBC Newsbeat 14 Oct 2011). For Warhol, though, the term 'looks' may be an understatement, as he refused to distinguish between the copy and the original in the consumerist age. Accordingly, he may or may not have minded that Pontus Hulten produced his own Warhol Brillo Boxes, just as Rauschenberg did not mind that his friends Cy Twombly and Brice Marden gave his all *White Paintings* (1951) a new coating of commercial Latex paint. The latter paintings, harking back to Malevich's *White on White* paintings from 1918, are possibly alluded to in Ligeti's "White on White" from Book 3 of the Piano Studies De Keersmaeker's *Stella* and *Erts* incorporated, even if Ligeti's title strictly speaking means that only the white piano keys are to be used (Ross 2001: n.p.). After all, Ligeti attended three of John Cage's lectures at the 1958 Darmstadt Summer Courses and must have been familiar with Cage and Rauschenberg's collaborations, like the 1952 ground-breaking *Untitled Event* at Black Mountain College, in which the *White Paintings* were used (Steinitz 2003: 118–119). Several years after the *Stella* trilogy, De Keersmaeker would also collaborate on *Erase-E(X)* (2004-2007), whose title references, among others, Rauschenberg's *Erased de Kooning Drawing,* and foregrounds once again a recursive creative practice building on the reprocessing of extant material, whether made by oneself or by others (Callens 2008).

Re: Rosas

De Keersmaeker granted that her work would never have reached the larger public which pop music is capable of mobilizing without the publicity over the plagiarized video clip, a clip downloaded more than two million times. This is why Carolyn Merritt, in her reflection on the case, criticized Philadelphia's FringeArts for its "unabashed use of the scandal to promote the show" when *Rosas danst Rosas* was revived in September 2014. The whole affair also made the choreographer wonder in her public reaction why it took three decades for her experimental performances to be assimilated by a now globalized mainstream culture. Yet she regretted the blunting of the emotional and critical edge of *Rosas danst Rosas*. De Keersmaeker deemed its reenactment by some Flemish schoolchildren to the music of Madonna's *Like a Virgin* more "touching" than *Countdown*. In the 1980s, the original production of *Rosas danst Rosas* was also seen "as a statement of girl power [...] assuming a feminine stance on sexual expression," whereas *Countdown* is only "seductive in an entertaining consumerist way" (qtd in Van Frankenhuyzen 8 Oct. 2011). Another vital difference, as mentioned, is that the inspiration of Hepburn is acknowledged while that of De Keersmaeker is not, both artists, ironically, being born in Brussels. Beyoncé, however, also styles herself a feminist (Ellison 2013). She actually started her career as lead singer in the girl band Destiny's Child, and with her third album created the alter ego of Sasha Fierce. The Transatlantic quarrel, then, was a sophisticated sparring game revolving around artistic image and identity.

As if to conclude her brief public reaction with a conciliatory gesture, De Keersmaeker remarked upon the "funny coincidence" of Beyoncé's dancing the *Countdown* clip while being pregnant, just as De Keersmaeker was in 1996 when working with Thierry De Mey. Presumably the Belgian dancer and director at that point was not aware of the irony that Beyoncé was pregnant with the daughter whose name, Blue Ivy, she tried yet failed to copyright (Ellison). Beyoncé's quixotic attempt to preserve the name's unicity recalls the debate over whether the everyday gestures De Keersmaeker choreographed in *Rosas danst Rosas* could actually be copyrighted. With her conciliatory closing remark, however, De Keersmaeker seems to link Beyoncé's plagiarism to a (feminine?) (re)creative power, which they both successfully have mined. As mentioned, feminists like Irigaray levelled plagiarism as a (potentially counterproductive) subversive force against phallogocentrism, Derrida's fusion of phallocentrism and logocentrism (Randall 1989: 270, 273, 277), which gives the radical reductions of the male-authored source material in *Stella* and *Erts* a further critical edge. When realizing that her choreographies were leading an afterlife unbeknownst to her, De Keersmaeker even issued an official invitation for adults and children all over the world to reenact a section of *Rosas danst Rosas* and upload a video of the remake to the company's website (Mackrell 2013: n.p.). This online invitation is accompanied by a sound clip of the original music, some background info and a tutorial in which De Keersmaeker (who performed in the premiere production) and Samantha Wissen (who performed in the revivals of the piece) explain its structure and go through a simplified version of the choreography's different moves. So far "Re: Rosas!, The fABULOUS Rosas Remix Project" has produced more than ten hours of footage, and the Open Access repository, with a fascinating trailer, keeps growing. It thereby attests to the importance of the internet for cultural dissemination and participation, with all the copyright issues this entails. It also keeps alive the memory of the very production through which Rosas, by its own admission on the Remix website, "put itself on the map."

Note

1 The research conducted for this article is part of the "Interuniversity Attraction Poles" programme financed by the Belgian government (BELSPO IAP7/01).

Works cited

Adorno, Theodor W. *Minima Moralia: Reflections on a Damaged Life*. Trans. E.F.N.Jephcott. London: Verso, 2005 [1951].

Bennett, Andrew. *The Author*. London: Routledge, 2005.

Burt, Ramsay. *Judson Dance Theater: Performative Traces*. London: Routledge, 2006.

Callens, Johan. "The double recursiveness of postmodern dance." *PAJ: A Journal of Performance and Art* 30.3 (2008): 70–80.

Callens, Johan. "Tennessee Williams and Ivo van Hove at Home Abroad." *Tennessee Williams and Europe: Intercultural Encounters, Transatlantic Exchanges*. Ed. John S. Bak. Amsterdam, the Netherlands: Rodopi, 2014. 301–319.

Ellison, Jo. "Mrs.Carter uncut." *Vogue* (UK) May 2013. http://www.vogue.co.uk/news/2013/04/04/beyonce-interview-may-vogue.

Gabriels, Katleen, and Lien Mostmans. "Beyoncé danst Rosas danst Rosas." *Rekto Verso* 49 Nov.–Dec. 2011: n.p.

Gilbert, Sandra, and Susan Gubar. *The Madwoman in the Attic: The Woman Writer and the Nineteenth-Century Literary Imagination*. New Haven, CT: Yale University Press, 1979, 1984.

Guisgand, Philippe. *Les fils d'un entrelacs sans fin: La danse dans l'oeuvre d'Anne Teresa De Keersmaeker*. Villeneuve d'Ascq, France: Presses Universitaires Septentrion, 2007.

Harris, Laurilyn J. "Perceptual conflict and the perversion of creativity in *A Streetcar Named Desire*." *Confronting Tennessee Williams's A Streetcar Named Desire: Essays in Cultural Pluralism*. Ed. Philip C. Kolin. Westport, CT: Greenwood, 1993. 83–103.

Heijer, Jac. "Een ego-document (Dat moet dan maar)." [An ego-document (So be it).] *Dramatisch Akkoord* 13 (1980). Rpt. in *Jac Heijer: Een keuze uit zijn artikelen*. Ed. Judith Herzberg, Gerrit Korthals Altes, Michael Matthews and Jan Ritsema. A'dam: IFTB, 1994. 335–337.

Heijer, Jac. "Ze stelt zich aan, ze kan niet anders: Verliefde dansen van Anne Teresa De Keersmaeker." [She is showing off, she cannot help it: Enamored dances by Anne Teresa De Keersmaeker] *NRC Handelsblad* 23 Mar. 1990: 4. Rpt. in *Jac Heijer: Een keuze uit zijn artikelen*. Ed. Judith Herzberg, Gerrit Korthals Altes, Michael Matthews and Jan Ritsema. A'dam: IFTB, 1994. 750–753.

Horwitz, Andy. "Anne Teresa De Keersmaeker vs Beyonce" 24 Oct. 2011. http://www.culturebot. org/2011/10/11496/anne-teresa-de-keersmaeker-vs-beyonce/. Accessed January 18, 2018.

Hughes, David. "Stop making sense." *Dance Theatre Journal* (Summer 1991): 16–19.

Jennings, Luke. "Beyoncé v De Keersmaeker: Can you copyright a dance move?" *The Guardian* 11 Oct. 2011.

Kottman, Pieter. "Anne Teresa De Keersmaeker over abstractie, loodzware stoelen en uitbundigheid; Onze voeten zoeken naar woorden." [Anne Teresa De Keersmaeker on abstraction, chairs heavy as lead and exuberance; our feet search for words] NRC 26 Apr. 1991: n.p.

Kurosawa, Akira, and Shinobu Hashimoto. *Rashomon*. The Criterion Collection, 2002 [1950]. DVD RAS040.

Kurosawa, Akira, and Shinobu Hashimoto. *Rashomon*. New York: Grove Press, 1969.

Laermans, Rudy, and Pascal Gielen. "Flanders. Constructing identities: The case of 'the Flemish dance wave.'" *Europe Dancing: Perspectives on Theatre, Dance, and Cultural identity*. Ed. and introd. Andrée Grau and Stephanie Jordan. London: Routledge, 2000. 12–27.

Lambrechts, An-Marie. "Wie wat vindt heeft slecht gezocht." [Whoever finds something looked badly.] *Etcetera* 30.6 (June 1990): 8–10.

Lévi-Strauss, Claude. "The family." *Man, Culture and Society*. Ed. Harry L. Shapiro. New York: Oxford Univeristy Press. 1956. 261–285.

Macaulay, Alastair. "In dance, borrowing is a tradition." *New York Times* 22 Nov. 2011: C1.

McKinley, James C. Jr. "Beyoncé accused of plagiarism over video." *New York Times* 10 Oct. 2011.

Mackrell, Judith. "Beyoncé, De Keersmaeker – and a dance reinvented by everyone." *The Guardian* 9 Oct. 2013.

Merritt, Carolyn. "Re: Rosas." *thINKingDANCE: Upping the Ante on Dance Coverage and Conversation*, 24 Oct. 2014. http://thinkingdance.net/articles/2014/10/24/Re-Rosas.

Randall, Marilyn. "L'écriture féministe: Une poétique du plagiat?" *Queen's Quarterly* 96.2 (Summer 1989): 263–278.

Randall, Marilyn. *Pragmatic Plagiarism: Authorship, Profit, and Power*. Toronto, ON: Toronto Univeristy Press, 2001.

Ross, Alex. "Ligeti Split." *The New Yorker* 28 May 2001: n.p.

Rotman, Brian. *Becoming Beside Ourselves: The Alphabet, Ghosts, and Distributed Human Beings*. Foreword Timothy Lenoir. Durham, NC: Duke Univeristy Press, 2008.

Steinitz, Richard. *György Ligeti: Music of the Imagination*. Boston, MA: Northeastern University Press, 2003.

T'Jonck, Pieter. "Op de koord tussen yin en yang: Dansgezelschap Rosas schetst eigen evolutie met een drieluik in de Munt." [Balancing between yin and yang: Dance company Rosas sketches own evolution with a trilogy in the Monnaie] *De Morgen* 8 Nov. 2006.

T'Jonck, Pieter. "Vragen bij een verjaardag." [Questions on the occasion of a birthday] *De Tijd* 22 Sept. 2001: n.p.

T'Jonck, Pieter. "Tweemaal hevige verlangens." [Twice strong emotions.] *De Standaard* 23 Feb. 1991.

Trueman, Matt. "Beyoncé accused of 'stealing' dance moves in new video." *The Guardian* 10 Oct. 2011.

Van Frankenhuyzen, Ingrid. "Beyoncé in de fout met videoclip." [Beyoncé fouls with videoclip]. *Cultuurpers* 8 Oct. 2011.

Van Kerkhoven, Marianne. "The Weight of Time." *Ballet International* 14.1 (January 1990).

Van Kerkhoven, Marianne. "Het gewicht van de tijd." *Etcera* 30.6 (June 1990): 2–7.

Van Kerkhoven, Marianne. Program notes to *Achterland*. DeSingel Season's Pocket 1990–1991. 52.

Van Kerkhoven, Marianne. "Niet meer spelen, maar gespeeld worden." [No longer to play, but to be played.] *Theaterschrift* Oct. 1991.

Van Kerkhoven, Marianne. "Tussen hemel en aarde: Een gesprek met Anne Teresa De Keersmaeker." "Between heaven and earth: A conversation with Anne Teresa De Keersmaeker." *Theaterschrift* 2 (1992): 169–197.

12

ADAPTATIONS, CULTURE-TEXTS AND THE LITERARY CANON

On the making of nineteenth-century 'classics'

Lissette Lopez Szwydky

When Sir Walter Scott died in 1832, he was the most famous novelist of the day and continued to be popular as long as his works were dramatized, appropriated, and inscribed into cultural memory (Rigney 2012). By comparison, Jane Austen achieved high critical acclaim but limited popular celebrity in the nineteenth century; to date she is the most famous nineteenth-century novelist lacking a record of theatrical adaptations staged during her lifetime (Bolton 2000). The picture is the complete opposite today, where Austen's cultural visibility has eclipsed Scott's in mass culture. Each author's cultural influence hinges on the proliferation of adaptations in a particular historical moment. Scott dominated in the Romantic period, when "the 'Author of Waverley' sold more novels than all the other novelists of the time put together" (St. Clair 2004: 221). Austen perseveres today, where her heroines have regularly inspired new iterations for television, film, and multimedia since the turn of the twentieth century—most recently exemplified in the *Pride and Prejudice and Zombies* series that includes a mash-up novel (2009), a graphic novel (2010), and an action film (2016) (Looser: 2017).

What conclusions may be drawn from the fluctuating popularity of Scott and Austen alongside their respective adaptation histories? What larger conclusions might be drawn when comparing these case studies to other nineteenth-century literary classics? Adaptations are the common experience shared by celebrity authors, as well as lesser-known writers known as one-hit-wonders. Pick up any nineteenth-century classic novel and you will likely find several film adaptations, perhaps also graphic novels or abridged, illustrated versions for children. For characters that seem to have lives of their own in contemporary culture (think Frankenstein, Quasimodo, Ebenezer Scrooge, Alice, Dracula), you will find more adaptations than you can track. One can be overwhelmed by how much the characters of nineteenth-century fiction continue to inhabit contemporary society. Across film, television, stage, graphic novels, board games, or digital media, the proliferation of adaptations around today's most widely read nineteenth-century novels might seem like evidence of a literary work's so-called greatness, ripe for commercial exploitation in today's media-saturated environment. However, this essay challenges that assumption, arguing that we must look back to the nineteenth century to see the beginnings of the commercial adaptation industry that dominates the present.

The ubiquity of adaptations in contemporary culture has given rise to the interdisciplinary field of adaptation studies, which has developed into a robust area of scholarly inquiry. (For useful overviews of the history of adaptation studies, see the introductions to Murray

2012 and Slethaug 2014.) The last decade in particular has seen a boom in exciting scholarship spearheaded by Robert Stam (2005), Linda Hutcheon (2006), and Thomas Leitch (2007), whose collective projects—alongside the work of others—have developed theoretical frameworks for understanding the cultural and educational power of adaptations in our day.

Scholars who specialize in adaptation studies no longer see adaptations as merely the derivative, parasitic offspring of canonical literature. Nevertheless, adaptation scholars often invoke these metaphors in order to develop more productive frameworks for understanding both the prevalence and importance of adaptation in both cultural production and literary history. Consider, for example, Hutcheon's claim that "an adaptation is not vampiric, it does not draw the life-blood from its source ... nor is it paler than the adapted work. It may, on the contrary, keep that prior work alive, giving it an afterlife it would never have had otherwise" (2006: 176). Leitch amplifies the adaptation-as-vampire metaphor but takes a decidedly different approach by highlighting the relationship between "vampire adaptations" and the "immortality" of literature:

> [T]exts do not have a natural history, a series of straightforward relationships with authors and readers and texts earlier and later that are challenged by such marginal or outlawed intertextual relationships as vampire adaptations. If adaptations were not in the picture, simplifying or abridging or flattening or transforming or otherwise distorting their originals, the history of textual production and reception would doubtless be very different, but it would be no more natural because texts and textuality are by their very nature subject to cultural—that is, humanly contrived—forces. ... The word "natural" is rarely invoked to describe textual relationships purged of vampire interlopers like unauthorized or inexpert adaptations. But even if it were constantly invoked, it would be only one more rhetorical weapon in an endless power struggle among different versions, different texts and different modes of textuality and intertextuality, and reception and consumption.
>
> *Leitch 2011: 7–8*

Despite their disagreement on the desirability of vampires and the application of vampiric metaphors to the study of adaptations, Hutcheon and Leitch agree on the role that adaptations play in the cultural staying power of texts, especially canonical ones. In this regard, they echo another of the often-invoked "afterlife" metaphors in studies of adaptations (see Kucich and Sadoff 2000; Brewer 2005; Carroll 2009; John 2010). All of these recuperative models position adaptation as a catalyst for widespread recognition and long-term cultural viability, regardless of whether or not they articulate a direct cause and effect relationship between adaptation and canonicity.

Although discussions of longevity and canonicity may seem better suited to the study of literary classics rather than popular adaptations, the idea deserves more consideration. Adaptations are central to the development of cultural literacy, argues Thomas Leitch in the Introduction to *Film Adaptation and Its Discontents* (2007). Even when packaged as animated cartoons, which Leitch calls "entry-level classics," adaptations "teach [audiences] the value of literary classics, even if what it teaches them is remote from whatever Dickens, or the cultural custodians of Dickens's reputation, may have in mind" (Leitch 2007: 71). A substantial body of scholarship champions the educational potential of literary adaptations, including several volumes dedicated to classroom use and pedagogical possibilities (see Cutchins et al. 2010; Cartmell and Whelehan 2014). Using film to illuminate texts is common in the classroom. Deborah Cartmell's claim that "Shakespeare on screen is now firmly placed with the literary canon" should come as no surprise to faculty who teach literature (1999: 29). Film adaptations are largely thought to make canonical literature more interesting or accessible to contemporary students whose literacies are increasingly visual, or who may otherwise find the nuances of historical contexts difficult to understand.

But is this really a recent phenomenon? Or does it just appear so? Despite contributions from a range of disciplines, adaptation studies remain largely focused on film and television, with the most historically oriented work beginning with the silent film era. Exciting work in multimedia and transmedia storytelling also focuses on contemporary culture. The picture presented suggests that the cultural ubiquity of adaptation—whether driven by artistic or commercial interests—appears to be a modern or postmodern phenomenon. This perception is inaccurate.

Cultural histories suggest a different story—one where adaptation is an equal partner in making so-called great literature, not a derivative byproduct of mass consumer culture. Adaptation was a standard commercial practice in London's theaters and printing houses in the eighteenth and nineteenth centuries. During this time, many texts were adapted for the popular stage and later reissued in various print media such as inexpensive abridgements, luxury gift books, and illustrated editions for children. The adaptations inspired paintings and other products. Sometimes, these adaptations and appropriations were a significant reason the novels found critical acclaim or commercial success in the first place.

This essay reconsiders the role of adaptation in the making of literary classics from a historical perspective. I follow Yvonne Griggs's work on the interrelationship between canonical literature, literary adaptation, mass entertainment, and cultural consumers. She writes:

> We are now more likely to view canonical texts not as works of individual genius but as cultural artefacts that are reliant for their construction and consumption on more than the writer's imaginative outpourings. We may ask what processes are involved in the construction of such a product; how and why it may attain canonical status; and how its canonical status may influence our relationship (and *its* relationship) with other related texts.
>
> *2016: 10*

Viewing the literary canon as a cultural construct influenced by a range of factors is hardly new. The idea of a literary "canon" itself has been rightly challenged for decades, especially by scholars working primarily from frameworks grounded in poststructuralism, feminism, critical race studies, postcolonialism, and multiethnic literatures since at least the 1970s. (For useful overviews of these debates and major works addressing this conversation, see Smith and Gless 1991; Banks 1993; Bona and Maini 2006). What remains underexplored, however, is the extent to which adaptation and reimaginings function to support, critique, or revise the very notion of canonicity itself.

Instead of approaching adaptations as popular forms of canon extension or canon preservation, I argue for adaptation's importance at a much earlier stage in canon formation. This essay repositions adaptation not as the consequence of a cause and effect relationship, but as an equal player in the process of canonization. Examining historical case studies through the lens of contemporary adaptation studies shows that early stage adaptations largely determined which nineteenth-century novels were transformed into "culture-texts" (Davis 1990; defined in the next section of this essay). Continued adaptation and appropriation into a range of popular forms eventually ensured their place in the literary canon. The historical perspective is necessary because it demonstrates that commercial successes in the past have influenced academic interests in specific texts in the present. Historically, adaptation has served as the catalyst for literary canonization, both then and now.

Literary classics, or culture-texts

What do Romeo and Juliet, Robinson Crusoe, Oliver Twist, Jane Eyre, and Dracula have in common? They are culture-texts. Culture-texts exist beyond the scope of their respective "original." In some cases, the original source is itself contested, forgotten, or otherwise ignored. Although each of the aforementioned names corresponds to titles of published literary texts, they owe their widespread recognition and cultural visibility to regular adaptation, appropriation, and allusion.

In *The Lives and Times of Ebenezer Scrooge,* Paul Davis coins the term "culture-text" to distinguish between Dickens's novella *A Christmas Carol* and what he calls "the Carol" (no italics), the version that exists in the cultural imagination more broadly. He writes:

> The *Carol* has inverted the usual folk process. Rather than beginning as an oral story that was later written down, the *Carol* was written to be retold. Dickens was its creator, but it is also the product of its re-creators who have retold, adapted, and revised it over the years. … We remember the *Carol* as a cluster of phrases, images, and ideas. … The text of *A Christmas Carol* is fixed in Dickens' words, but the culture-text, the Carol as it has been re-created in the century and a half since it first appeared, changes as the reasons for its retelling change. We are still creating the culture-text of the Carol.
>
> *1990: 3–4*

Throughout his book, Davis switches back and forth between italicized forms—referring to the text of Dickens's published novella (and subsequent authorized editions) as the *Carol* and everything else (adaptations, spin-offs, merchandise) as the Carol. The Carol consists of all retellings and adaptations across media and time. Davis's terminology is valuable for its commitment to break down (or open up) the ubiquitous text: adaptation binary, a typical tension expressed in studies of adaptations that include ideas of original versus derivative works and high versus mass culture (Stam 2005; Corrigan 2007).

More so, Davis's "culture-text" definition provides a productive model for studying the cultural history of canonical novels, which necessitate an approach that can accommodate adaptations over a long period of time. Davis's term complements more recent adaptation models, such as Regina Schober's understanding of "adaptations as intricate webs of connections" among media that work alongside the adaptation's relationship to its source (2013). Schober is right to consider influences outside of an adaptation's direct relationship to a named source text. The culture-text model allows for such intermedial considerations but also accounts for the particular difficulties of tracing multiple versions of the same story, not only across media but also across centuries.

For example, Brian Rose sees Davis's culture-text model as the product of a "body of adaptations extended over time that […] has the potential of becoming a larger, reflexive body of narratological, performative, and cultural elements" (1996: 2).[1] This open, comprehensive approach is especially important when looking at the adaptation histories of canonical novels, since there are typically dozens, and sometimes hundreds, of individual additions to a culture-text over decades, centuries, and even millennia (when one thinks of how the legends and myths of antiquity exist in contemporary popular culture). The model can be extended and expanded for most culture-texts with similar widespread visibility across centuries. In fact, such a historically flexible model is necessary in order to accurately understand how culture-texts are made in the first place.

For example, consider the famous juggernaut of nineteenth-century gothic fiction—Mary Wollstonecraft Shelley's *Frankenstein; or, The Modern Prometheus* (1818). Paul O'Flinn has famously summed up the phenomenon in a sentence frequently cited by *Frankenstein's* cultural historians: "There is no such thing as *Frankenstein*, there are only *Frankensteins*, as the text is ceaselessly rewritten, reproduced, refilmed, and redesigned" (1983: 194). Steven Forry's *Hideous Progenies: Dramatizations of* Frankenstein *from Mary Shelley to the Present* (1990), the most complete catalog of *Frankenstein* on stage and screen, lists 96 theatrical adaptations of Shelley's novel through 1986— not counting film versions, which Forry counts at just over 100. More recent work expands the scope of adaptations to children's books, comics and graphic novels, and later films, but do not include full listings in the style of Forry's book, nor do they provide a thorough survey of contemporary theatre (Hitchcock 2007; Fisch 2009). A search for "Frankenstein" filtered by title turns up 200 results on the Internet Movie Database (including television series, individual titles in anthology-style shows, short films, and two video games).[2] Collectively, scholarship and internet sources show that adaptation is the primary mode of circulation for the Frankenstein culture-text.

The number of people who will read Shelley's novel at some point in their lives is relatively small compared to how many people will encounter dozens of "Frankensteins" on film, television, radio, or in various print and digital media (Cutchins and Perry 2018). Walk into a preschool classroom and ask them about Frankenstein; you are sure to be amused by four-year-olds giving their best impressions of a heavy-footed, shuffling giant. What is their "original" Frankenstein? They have first encountered him in animated films from *Alvin and the Chipmunks Meet Frankenstein* (1999) to *Frankenweenie* (2012). There's Samantha Berger's illustrated book *Crankenstein* (2013) for early childhood and a variety of young adult fiction titles, such as *This Dark Endeavor* (2012), the first book in *The Apprenticeship of Victor Frankenstein* series by Kenneth Oppel—all targeted to ages and reading levels not yet prepared for Shelley's novel (Coats and Norris Sands 2016). Frankenstein's monster also has a notable presence in comic books and graphic novels (Murray 2016). One way or another, most people recognize Frankenstein regardless of whether they've ever read *Frankenstein*—or even know who wrote the novel. Monster and creator (sometimes together, sometimes separately) are part of today's collective cultural memory, much as Scott existed throughout the nineteenth century (Fisch 2009; Rigney 2012). Each new iteration adds to the Frankenstein culture-text.

Outside of specialized circles, such as Romantic-period literary and cultural historians, what is less known about *Frankenstein* the culture-text is that, much like today, Shelley's novel was also primarily recognizable through adaptations in the 1820s. Forry (1990) identifies at least 15 distinct dramatizations of the story between 1823 and 1826, most notably Richard Brinsley Peake's (1823a) gothic melodrama *Presumption; or, The Fate of Frankenstein*. Peake's play established most of the visual iconography that we associate with the Frankenstein story in film and other popular media, including the elimination of the creature's ability to speak and the introduction of the comic lab assistant Fritz (renamed Igor in later films). By focusing so much on the creature, early stage productions are also at least partly responsible for the conflation of the scientist's name with his monster—an almost ubiquitous confusion in contemporary culture, where the name "Frankenstein" typically refers to the monster more so than the scientist—in contrast with Shelley's book where Frankenstein's creation remains nameless (Fisch 2009).

Early adaptations clearly affected reception and printings of *Frankenstein*. As St. Clair (2004) shows, when the novel was published anonymously in 1818, only 500 copies of the text were printed. No other printings occurred until the stage premiere of *Presumption*, at which time William Godwin (Mary Shelley's father) arranged for a new printing of 1,500 copies (this time with the author's name revealed). A new edition of the novel was published as part of the *Bentley's Standard Novels* series in 1831. Between 1823 and 1831, only 2,000 copies of the novel were in print, approximately the same number of seats in many of London's theaters. St. Clair puts these numbers into perspective:

The English Opera House, where *Presumption* opened in 1823, was able to hold about 1,500 persons. The Coburg Theatre, where *Frankenstein of the Demon of Switzerland* opened shortly afterwards, held over 3,800. […] Every single night when one of the *Frankenstein* plays was performed brought a version of the story of the manmade monster to more men and women than the book did in ten or twenty years.

2004: 369

St. Clair's figures require some context in order to better understand the scope of adaptation's reach in this case. During the 1820s, more than 15 distinct adaptations were produced with varied success in major and minor London theatres (Forry 1990). There were also productions in Paris. No matter the venue, on the night of any given performance, more people might encounter the monster than the total number copies of the novel available in print. Without Peake's *Presumption* and other early *Frankenstein* adaptations that immediately followed in its footsteps, such as *The Man and the Monster* (1826) and *The Monster and the Magicien* (1826), the now-iconic monster may have been overlooked or forgotten due to limited availability of the printed novel (Szwydky 2018). In this light, we clearly see how and why *Frankenstein* is an exemplary model of a culture-text. Indeed, adaptations, parodies, and other forms of imitation (not the novel) kept the creature alive in the popular imagination in the nineteenth century; today the monster is a global icon. New media continues to bring new adaptations not only for theater, film, and television, but also radio, comics, web series, novelizations, and video games.

The Frankenstein culture-text is also a notable historical case study because we see clear evidence that adaptations inherited more from the adaptations before them than they took from the novel. Some of the earliest adapters created multiple stage versions of the Frankenstein story. Peake wrote *Another Piece of Presumption* (1823b), a farce based on his popular melodrama. Another adaptation, *Frankenstein; or, The Man and the Monster* (1826) by Henry Milner was more successful than the dramatist's first attempt because it drew on Peake's play, while adding its own original flourishes.[3] In fact, Milner's 1826 play marked the first time the Creature was animated on stage, and it enjoyed both a considerable run and importance in developing the iconography associated with the Creature's animation (Forry 1990). Since Milner's play, every notable Frankenstein dramatization on stage or screen features a creation scene. Victorian versions riffed off the plays of the 1820s, and the earliest films more closely resemble their nineteenth-century stage brethren than they do Shelley's novel (see Cutchins and Perry 2018; Szwydky 2018).

The cumulative effects of *Frankenstein's* early stage history stretch beyond the realm of popular culture. *Presumption's* early commercial success also launched Shelley's writing career. None of her later novels came close to the recognition of her debut novel (notably, no adaptations of the later works have been identified). Nevertheless, Shelley continued to publish and indeed supported herself and her son through her writing and editorial work. The commercial success of the early plays served as a useful negotiation and marketing tool for future book sales, just as Godwin predicted when he learned that his daughter's novel was to be adapted for the stage in 1823. In a letter to Mary less than one week from *Presumption's* premiere, Godwin wrote:

It is a curious circumstance that a play is just announced, to be performed at the English Opera House in the Strand next Monday, entitled, *Presumption, or the Fate of Frankenstein*. I know not whether it will succeed. If it does, it will be some sort of feather in the cap of the author of the novel, a recommendation in your future negociacions [sic] with booksellers.

Forry 1990: 3; Robbins 2017

Indeed, Shelley made more money on the next edition of *Frankenstein* (Robbins 2017). The frontispieces to Shelley's subsequent novels all refer to her as "the author of *Frankenstein*" despite the fact that *Frankenstein* never became a best-selling novel in Shelley's lifetime, at least not compared to the works by Scott or Austen (St. Clair 2004).

Similarly, stage adaptations greatly influenced the careers of many nineteenth-century writers, most notably Charles Dickens. By the late 1830s, serial publication was on the rise, and Dickens is widely understood as one of the first authors made famous by this new storytelling format. His novels were adapted in London theaters while they were still in serialization, complicating the idea that the 'greatness' of these texts served as the primary reason behind decisions to adapt them in the first place. William L. Rede's *The Peregrinations of Pickwick; or, Boz-i-a-na The Pickwick Papers* was first staged at the Adelphi Theatre on April 2, 1837 while Dickens's first novel was only one-third completed (Koger 2012). Dickens's second novel, *Oliver Twist*, began its theatrical 'afterlife' more than six months before the first serialization was complete, that is, before it was fully born. Sue Zemka suggests that "Dickens might have been inspired to kill Nancy by the lackluster performances of *Twist*, as there were two produced in London playhouses before November 1838, and both failed for want of a stirring climax" (2010: 29).

The serial publication model created some unique challenges for the early novels, giving Dickens little or no authority in directly shaping the earliest adaptations of his first few novels. His lack of control over these adaptations concerned him primarily from a financial perspective, given that throughout the nineteenth century theaters did not pay royalties to novelists for adapting their works. Copyright did not cover dramatic adaptations until the late Victorian period, making the eighteenth- and nineteenth-century adaptation industries a commercial free-for-all. As an emerging writer who supported himself entirely through his publications, Dickens was concerned that these adaptations would deter people from buying the printed works. Luckily for him, they only fueled his popularity.

Early stage adaptations influenced both sales of Dickens's works as well his writing. Dickens grew increasingly frustrated when adaptations of his third novel began appearing in November 1838, approximately half-way through the novel's serialization.[4] He took the opportunity to voice his disapproval. Chapter 48 of *Nicholas Nickleby* (first published in May 1839) includes Dickens's response through the critical portrait of a "literary gentleman who had dramatized in his time two hundred and forty-seven novels as fast as they had come out—some of them faster than they had come out—and who *was* a literary gentleman in consequence" (Dickens 1839: 632). Dickens leaves no doubt as to his feelings about the practice, allowing a heated confrontation between the "literary gentleman" and Nicholas, who likens the adapter's work to "picking a man's pocket in the street" (633). Dickens understood adaptations primarily as a commercial business practice, not only an artistic endeavor. (John 2010: 54–5). The sheer number of adaptations suggested as much. By 1840, London had seen 60 distinct plays based on Dickens's first three novels (Bolton 2011; 1987).

The adaptation industry that formed around Dickens's early novels complicates the idea that adaptation naturally follows a so-called great work. Instead, it shows us how adaptation can fuel a text's or a novelist's popularity, helping to canonize them or make their works classics—not the other way around. Gordon Slethaug articulates the need for a poststructuralist model of adaptation studies that "is open, dialogically negotiated, and distributed across the system […] so that adaptation, derivation, translation, and copy have the same cultural currency as the original" (2014: 25–6). The early reception history of *Nicholas Nickleby* demonstrates an early model of feedback with Dickens watching and responding to reception of his work in multiple mediums and versions. This early adaptation and reception positions originals and adaptations and subsequent editions as equally important in the circulation of his novels and the culture-texts they became. In this case, adaptations are not simply derivative works but also natural extensions of the writing process, sometimes contributing to a cycle of engagement between writers, readers, adapters, and audiences.

For Dickens, the speed of adaptation was a double-edged sword. Dramatists were creating scenes and climatic endings that had little to do with authorial intent on the part of Dickens, but those dramatists were also driving interest and sales in the serial fiction. Being a lifetime patron of the theatre no doubt also contributed to his respect for the theatre's power to increase visibility, drive sales, and secure recognition (Glavin 1999). As his career developed, Dickens was able to exert more influence on how his works were adapted and even challenge adaptations. He eventually joined forces with fellow writers, such as Wilkie Collins and Charles Reade, in the 1860s to change theatrical practices and ensure that authors maintained some ownership of their intellectual property, but this changed after his death in 1870, as theatres clamored to celebrate his already-established legacy (Bolton 2011). Later, film and television kept the author's immortal characters alive in popular culture as well as in print (Glavin 2003; 2012). As John makes clear: "Dickens would not have attained his posthumous position as an international icon whose fame rivals that of Shakespeare were it not for the screen" (2010: 188).

Dickens produced so many culture-texts that he himself became a cultural icon during his lifetime. Successful adaptations and record-breaking book sales worked together to transform Dickens into a literary celebrity.[5] Dickens's work eventually became synonymous with the period, and he now serves as a synecdoche for Victorian literature, culture, and social issues (John 2010). Today, Dickens is undoubtedly the most recognized nineteenth-century novelist alongside Austen. Nevertheless, it is still difficult to discern which came first—critical success, commercial success, or visibility via adaptations. As Juliet John explains:

> Central to Dickens's cultural survival and pervasiveness has been the evolution of the so-called "Dickens industry", engineered in the Victorian period by Dickens himself. In his lifetime, his reading tours, public speaking engagements, journals, travels, and acting projects, made Dickens a celebrity, the most visible author of the nineteenth century. This visibility meant the duplication of his image in newspapers, advertisements, and on commodities, and the ubiquitousness of the idea of Dickens in Victorian mass culture.
>
> *2010: 15*

The simultaneity of Dickens's success across media, class, and national boundaries makes it impossible to tease out a direct cause-and-effect narrative for Dickens's writing career, literary merit, cultural afterlife, and literary canonization. The implications for both adaptation studies and historically focused literary and cultural studies are significant and provide an opportunity for productive conversations that extend beyond the separate academic circles where these discourses typically circulate, in this case contemporary adaptation studies and nineteenth-century cultural studies.[6]

A historical model of adaptation studies and canon formation

Shelley's and Dickens's places in the English canon are hardly contested today, but that was not true in the first half of the twentieth century. *Frankenstein* was not a regular staple of the literature classroom until the 1970s, though today, according to the Open Syllabus Project, it is one of the most frequently taught and studied English novels in college curricula.[7] What prompted the change? The proliferation of adaptations of works by these two authors between the nineteenth century and today begins to answer that question.

In the case of *Frankenstein*, some scholars have suggested that Shelley's novel has reached canonical status because it is a scholarship magnet. Fred Botting, for instance, has provocatively suggested that *Frankenstein* is a "product of criticism, not a work of literature" (1995: 1). Complicating this claim, Diane Long Hoeveler (2004) responds by situating *Frankenstein* as one of the most important novels, both as a work of artistry and criticism, while admitting its centrality to the development of feminist literary criticism as a whole (rivaled perhaps only by *Jane Eyre*). I would argue that the basic structure and premise of Botting's and Hoeveler's arguments can be applied similarly when considering adaptation's role in creating literary classics. Adaptations kept Shelley's story alive in the popular imagination between the novel's nineteenth-century publication history and feminist recovery in the 1970s. One needs to look no further than adaptations from the melodramas of the 1820s to the Universal films of the 1930s and 1940s to the Hammer films of the 1950s and 1960s to identify the missing link between *Frankenstein*'s early (and relatively lukewarm) critical reception and its later success in academic circles and classrooms. *Frankenstein* is as much a product of adaptation as it is a work of literature and criticism, showing us how adaptation can fuel a novel's cultural visibility, keeping it visible in culture so that it can eventually be visible to scholars and eventually become canonical. The example of Dickens's literary career extends this model beyond the single-text case study to an author known for a range of culture-texts.

Indeed, there are many examples of culture-texts that show a strong correlation (if not a causal relationship) between early commercial adaptations and later literary canonization. The vast majority of nineteenth-century novels that have attained culture-text status have rich adaptation histories dating back to their original publication. The phenomenon has inspired many case studies on individual novels (Forry 1990; Meer 2005), authors (Rigney 2012; Glavin 1999; 2003; 2012), and even genres (Hoeveler 2010; Saggini 2015). However, by focusing on a single text or author or genre, these in-depth studies typically (and unintentionally) promote the idea that specific works are extraordinary in their ability to draw in new audiences when adapted to new mediums. The more recent trend in adaptation studies to complicate the traditional single case study approach is especially helpful here in recovering a more robust picture of the cultural history of adaptation, as well as its role and impact in other areas of literary and cultural production, impacting so-called 'high' and 'low' art forms alike.

The collective examples mentioned above pose larger questions about the role of adaptation in the formation of literary classics and canons. I understand canon formation to be open to continuous critique and constant change through feminist and other counter-discursive, critical models, similarly to Griggs, who writes:

> re-visionist adaptations become part of the ongoing debate surrounding the canonical texts that engender their creation but they can also attain their own place within that canon: they are neither consumed by nor solely defined by it but rather present us with other manifestations of the cultural anxieties that circulate around the initiating canonical text *and* its various adaptations. The canonization of texts is, like adaptation itself, and ongoing process that reacts to and interacts with the cultural and critical preoccupations of its time of production.
>
> *2016: 8*

My invocation of canonicity in this essay similarly rebukes the upholding of a traditional canon of so-called great books that are determined by vague notions of universality and timelessness. The historical approach I propose disrupts a top-down model of canon formation characterized by aesthetic and cultural elitism. We have historical evidence that places popular culture and entertainment on equal ground in the making of culture-texts, or literary classics. For this reason, the study of popular adaptations should be seen as a natural extension of the study of literature and the history of the novel.

"Today canonical status is not only assigned to a work by a single literary critical guru ... or by the number of citations it receives long after the death of the author, but is often bestowed in recognition of the number of films it has generated," write Cartmell and Whelehan (2014: 3). History supports this claim, and it suggests that today's adaptation industry will likely have equal (or perhaps stronger) influence on the making of future classics and canons, given the closely linked literary publishing and popular entertainment industries of the present (Murray 2012). Canonization requires availability. Canonical works are rarely out of print, and adaptation is a major driver of contemporary books sales. St. Clair's work on the emergence of a reading public in the Romantic period shows us that this was also the case in the nineteenth century. The popular series *Bentley's Standard Novels* (which ran from 1831–56) reproduced 126 titles. Each title sold for approximately six shillings, a fraction of the previous average cost of 30 shillings (Wallins 1991: 43). St. Clair situates the importance of Bentley's publishing endeavor: "Bentley did for romantic-period novels what the editors of the initial old-canon series had done at the end of the previous century. Novels ... which failed to be selected, disappeared from public attention until the twentieth century" (2004: 361).

Instead of situating academic institutions as the primary arbiters of aesthetic taste and cultural value, the cultural history of commercial adaptations suggests that popular appropriations play an equally important (and perhaps even stronger) role in determining which texts survive both in mass culture and in the classroom. As Sarah Cardwell notes in *Adaptation Revisited: Television and the Classic Novel*:

> The authors who are most favored by the makers of [television] adaptations—Austen, the Brontes, Dickens, Eliot, Hardy, and so on—tend also to dominate the academic construction and study of English literature (particularly at the secondary-school level), and likewise predominate in the catalogue lists of publishers of classic novels. [...] There are no simple chains of cause and effect here; the reasons for this convergence are multiple and complex. Books that are adapted for television will sell more copies; books on school syllabuses and those that are most widely read are more likely to be adapted; and so on. There is a circular affirmation of a certain range of books commonly perceived as classics. Consistency across various areas of the public sphere means that certain texts (or [...] certain authors) are held by the reading and viewing public, programme-makers and educationalists to be classics; in this way the identity of some novels as 'classic novels' is accepted and perpetuated in common parlance.
>
> *2002: 2–3*

Cardwell sketches an ongoing cycle in the production of adaptations, the printing of novels, and the teaching of particular texts. Artistic creativity and academic inquiry work alongside commercial entertainment and popular culture; they are not mutually exclusive forms of cultural production. Together, they determine which narratives get transformed into culture-texts and, as a result, remain part of the critical conversation.

The benefits of incorporating similar historical and cyclical models in adaptation studies are substantial. Combining the history of popular entertainment, commercial publishing, and the academic study of literature follows Schober's understanding of adaptations as a "complex assemblage of cross-influences rather than a seemingly unidirectional procedure between two media" (2013: 92). Several scholars support a similarly dynamic approach, including

Jørgen Bruhn, who proposes a dialogical model for adaptation studies where "Adaptation … ought to be regarded as a two-way process instead of a form of one-way transport" (2013: 73). To Bruhn's dialogical understanding of adaptation as a two-way process between media, I add the case for understanding canon formation as a dialogical process that includes the interrelated-ness of texts and their adaptations, the interplay between 'high' and 'low' art, and the connec-tion between past and present. The historical connection opens new avenues of investigation in line with Timothy Corrigan's call for delineating a history of film adaptation that includes the pre-cinema period. He writes, "Between the high cultural ground being articulated in universi-ties and museums, and the low cultural fairgrounds from which the movies would spring, the nineteenth century predicted a long line of anxieties and claims about the relationship of the movies to the other arts" (2007: 38). Studying the history of commercially driven adaptations in the nineteenth century uncovers blueprints for early film versions and shows us how much contemporaneous cultural anxieties continue to drive interpretation and cultural production (Rose 1996).

In addition to enhancing our understanding of the history of cinema and other modern entertainment forms, a historical approach to adaptation studies also illuminates the place of older texts in contemporary adaptation studies. The back and forth interplay of a dialogical model that includes past and present is particularly relevant when studying the long-term effects of adaptation on older works of literature. Such a model, as John Glavin posits, is a way of

> refocusing the relation between the page and the stage, between reading, adapting, and performing. It's a way that positions reading as a close cousin to adapting. A way that updates both of them as versions of performance. And that specifies such performance as modeling, fundamentally, what it means to find yourself coming *after* an original.
>
> *1999: 2*

Glavin's understanding of Dickens's *oeuvre* foregrounds historical distance as an important con-sideration in reading the canon and other older works of literature. For contemporary readers, there is no 'outside' of adaptation because the culture-text we have inherited is composed not only of Dickens's novels, but also of all of their respective adaptations, as well as of all that we now know about the author and the historical milieu that produced the writer and the works. Similar claims can be made for most famous authors, culture-texts, or popular genres. Can we realistically see Shakespeare without our favorite film versions affecting our reading of his plays? Is it possible for today's students to approach nineteenth-century gothic fiction free from our cultural inheritance of classic Hollywood monsters or their contemporary CGI renditions? It's not only impossible to ignore the value of these adaptations, it's also historically inaccurate to suggest there was a time when texts and adaptations did not work together in the formation of cultural literacy and popular entertainment.

The sheer number of adaptations on the nineteenth-century stage deserves consideration in its own right. Pick almost any British author of the period, and you are sure to find adaptations of their works staged soon after publication (several French and American authors might also be included here—especially Victor Hugo and Harriet Beecher Stowe). In the 1820s, Sir Walter Scott proved to be one of the safest and most profitable investments for theaters, as his historical romances broke sales records in bookstores. Theater managers commissioned adaptations from different writers, and they scrambled to compete for ticket sales by advertising "innovative" and "original" takes on Scott's novels, especially the following three novels in the *Waverly* series: *Waverly* (1814), *Rob Roy* (1817), and *Ivanhoe* (1820). Multiple versions of James Fennimore Cooper's *The Pilot* (1823) and *The Red Rover* (1827) competed with several adaptations of

Scott's *The Pirate* (1822) during the mid-1820s (Burwick and Powell 2015). In the 1830s, at least two different versions of Hugo's *Notre Dame de Paris* (1831) were staged, with another major adaptation in the 1870s (Szwydky 2010). Stowe's *Uncle Tom's Cabin* (1852) was an immediate transatlantic hit, with at least a dozen distinct adaptations staged during the 1852–3 London theatrical season alone (Meer 2005). At least eight versions of *Jane Eyre* (1847) saw the English stage between 1848 and the turn of the twentieth century (Stoneman 2007). *The Strange Case of Dr. Jekyll and Mr. Hyde* (1886) was staged in New York within seven months of its publication, making it to the London stage shortly thereafter (Rose 1996; Danahay and Chisholm 2004). Famous authors even added their own versions to the mix, as with Hugo's own *La Esmeralda* (1836) for the Paris stage and Collins's contribution to *The Woman in White* (1860) in 1871 and again for *The Moonstone* (1868) in 1877 (Szwydky 2010; Pedlar 2012). Adaptations were everywhere in nineteenth-century England. Commercial adaptation is not a phenomenon of the modern culture industry or of postmodern artistic production; it has been at the center of entertainment since the emergence of commercial mass culture.

Despite this rich history, the critical conversation around adaptation remains largely engaged in the present. We have yet to uncover a comprehensive history of adaptation's functions in the past or a complete understanding of early adaptations' continued impact on today's entertainment industries and academic trends. Contemporary adaptation studies illuminate the extent to which the current literary canon is influenced by a history of adaptation in the nineteenth century. As a result, canonization might also be seen as one of the long-term effects of continuous adaptation. Together, the examples of *Frankenstein* and the literary career of Charles Dickens demonstrate, in part, the extent to which early adaptations and appropriations created widespread visibility and readership in the years immediately following their initial publication. Their contemporary relevance as staples of both literary curricula and popular entertainment alike suggests a clear tie to their historical reception via stage adaptations and other cultural appropriations.

A historical approach to adaptation studies uncovers a symbiotic relationship between popular adaptation and literary canonization, making adaptations central (not tangential or marginal) to what is regularly taught in classrooms and published in academic circles. History shows us that adaptation is rarely a simple commercial byproduct of cultural production. Instead, adaptation should be understood as an active partner in cultural literacy, narrative evolution, and book sales. Adaptations largely determine which texts experience a long cultural life and which ones eventually fade from cultural memory.

Notes

1 Rose works through another great example of nineteenth-century fiction that has eclipsed its author—Robert Louis Stevenson's 1886 novella *The Strange Case of Dr. Jekyll and Mr. Hyde*, which has more than 100 dramatizations or appropriations to date. Geduld (1983) catalogs 136 film and television titles (including direct adaptation and loose retellings). The Robert Louis Stevenson Archive lists at least 40 additional adaptations, spin-offs, appropriations, or character cameos in film and television since the publication of Geduld's guidebook.

2 Internet Movie Database can be found at www.imdb.com.

3 Milner also wrote a now-lost drama *Frankenstein; or, The Demon of Switzerland* (1823), which turned out to be a financial flop (see Forry 1990).

4 The first stage adaptation was Edward Stirling's farce *Nicholas Nickleby; or, Doings at Do-the-Boys* Hall, which premiered at the Adelphi Theatre on November 19, 1838 (see Adelphi Theatre Project Online).

5 Similar adaptation- and appropriation-fueled histories can be traced for other nineteenth-century authors, such as Scott (Rigney 2012). In fact, Scott is often hailed as Dickens's closest competitor for nineteenth-century British celebrity (John 2010).

6 Here, I echo Kamilla Elliot's observation that adaptation scholarship too often happens in isolation or closed conversations, which in effect limit the development of the field.
7 The Open Syllabus Project tracks and collates citations across college syllabi submitted to its database. *Frankenstein* ranks #5 in its list of most frequently taught texts. http://explorer.opensyllabusproject.org

Works cited

Austen, J. and Grahame-Smith, S. (2009) *Pride and Prejudice and Zombies*. Philadelphia, PA: Quirk Books.

Austen, J., Grahame-Smith, S., Lee, T., and Richards, C. (2010) *Pride and Prejudice and Zombies: The Graphic Novel*. New York: Ballantine.

Banks, J. A. (1993) "The canon debate, knowledge construction, and multicultural education," *Educational Researcher* 22(5): 4–14.

Berger, S. (author) and Santat, D. (illus.) (2013) *Crankenstein*. New York: Little, Brown, and Company Hachette Book Group.

Bolton, H. P. (1987) *Dickens Dramatized*. Boston, MA: G. K. Hall.

Bolton, H. P. (2000) *Women Writers Dramatized: A Calendar of Performances from Narrative Works Published in English to 1900*. New York: Mansell.

Bolton, H. P. (2011) "Dramatizations and dramatizers of Dickens's works." In P. Schlike, (ed.) *Oxford Reader's Companion to Dickens*. Oxford University Press. n. pag. Web [19 Apr. 2017].

Bona, M. J. and Maini, I. (2006) *Multiethnic Literature and Canon Debates*. Albany, NY: State University of New York Press.

Botting, F. (1995) *Frankenstein/Mary Shelley*. New York: St. Martin's.

Brewer, D. (2005) *The Afterlife of Character 1726–1825*. Philadelphia, PA: University of Pennsylvania Press.

Brontë, C. (1847) *Jane Eyre*. Reprint. Mason, M. (ed.) *Penguin Classics*. New York: Penguin Group, 1996.

Bruhn, J. B. (2013) "Dialogizing adaptation studies: From one-way transport to a dialogic two-way process." In J. Bruhn, A. Gjelsvik, and E. Frisvold Hanssen (eds.) *Adaptation Studies: New Challenges, New Directions*. London: Bloomsbury, 69–88.

Burwick, F. and Powell, M. L. (2015) *British Pirates in Print and Performance*. New York: Palgrave Macmillan.

Cardwell, S. (2002) *Adaptation Revisited: Television and the Classic Novel*. Manchester: Manchester University Press.

Carroll, R. (ed.) (2009) *Adaptation in Contemporary Culture: Textual Infidelities*. London and New York: Continuum.

Cartmell, D. (1999) "The Shakespeare on screen industry." In D. Cartmell and I. Whelehan (eds.) *Adaptations: From Text to Screen, Screen to Text*. London and New York: Routledge, 29–37.

Cartmell, D. and Whelehan, I. (2014) *Teaching Adaptations*. Basingstoke, UK and New York: Palgrave Macmillan.

Coats, K. and Norris Sands, F. (2016) "Growing up Frankenstein: Adaptations for young readers." In Smith, A. (ed.) *The Cambridge Companion to* Frankenstein. Cambridge, UK: Cambridge University Press, 241–55.

Cooper, J. F. (1823) *The Pilot*. Reprint. *The Novels*. 25 vols. New York: The Co-operative Publication Society, 1900.

Cooper, J. F. (1827) *The Red Rover*. Reprint. *The Novels*. 25 vols. New York: The Co-operative Publication Society, 1900.

Corrigan, T. (2007) "Literature on screen, a history: In the gap." In D. Cartmell and I. Whelehan (eds.) *The Cambridge Companion to Literature on Screen*. Cambridge, UK: Cambridge University Press, 29–43.

Cutchins, D. and Perry, D. (eds.) (forthcoming 2018) *The Afterlives of Frankenstein*. Manchester: Manchester University Press.

Cutchins, D., Raw, L. and Welsh, J. M. (eds.) (2010) *The Pedagogy of Adaptation*. Lanham, MD: Scarecrow Press.

Cutchins, D., Raw, L. and Welsh, J. M. (eds.) (2010) *Redefining Adaptation Studies*. Lanham, MD: Scarecrow Press.

Danahay, M. A. and Chisholm, A. (eds.) (2004) *Jekyll and Hyde Dramatized: The 1887 Richard Mansfield Script and the Evolution of the Story on Stage*. Jefferson, NC: McFarland.

Davis, P. (1990) *The Lives and Times of Ebenezer Scrooge*. New Haven, CT: Yale University Press.

Dickens, C. (1838) *Oliver Twist*. Reprint. Tillotson, K. (ed.), The Clarendon Dickens. Oxford: Clarendon Press, 1966.

Dickens, C. (1839) *The Life and Adventures of Nicholas Nickleby*. Reprint. Thorndike, S. (ed.), The Oxford Illustrated Dickens. Oxford: Oxford University Press, 1978.

Elliott, K. (2003) *Rethinking the Novel/Film Debate. Cambridge*, UK: Cambridge University Press.

Fisch, A. A. (2009) *Frankenstein: Icon of Modern Culture*. Hastings, UK: Helm Information.

Forry, S. (1990) *Hideous Progenies: Dramatizations of* Frankenstein *from Mary Shelley to the Present*. Philadelphia, PA: University of Pennsylvania Press.

Geduld, H. M. (1983) *The Definitive "Dr Jekyll and Mr Hyde" Companion*. New York/London: Garland.

Glavin, J. (1999) *After Dickens: Reading, Adaptation, and Performance*. Cambridge, UK: Cambridge University Press.

Glavin, J. (ed.) (2003) *Dickens on Screen*. Cambridge, UK: Cambridge University Press.

Glavin, J. (ed.) (2012) *Dickens Adapted*. New York and London: Routledge.

Griggs, Y. (2016) *The Bloomsbury Introduction to Adaptation Studies: Adapting the Canon in Film, TV, Novels, and Popular Culture*. London: Bloomsbury.

Hitchcock, S. T. (2007) *Frankenstein: A Cultural History*. New York: Norton.

Hoeveler, D. L. (2004) "*Frankenstein*, feminism, and literary theory." In E. Schor (ed.) *The Cambridge Companion to Mary Shelley*. Cambridge, UK: Cambridge University Press, 45–62.

Hoeveler, D. L. (2010) *Gothic Riffs: Secularizing the Uncanny in the European Imaginary, 1780–1820*. Athens, OH: Ohio State University Press.

Hugo, V. (1831) *The Hunchback of Notre-Dame*. Reprint. Liu, C. (trans.) New York: Modern Library, 2002.

Hutcheon, L. (2006) *A Theory of Adaptation*. New York and London: Routledge.

John, J. (2010) *Dickens and Mass Culture*. Oxford, UK: Oxford University Press.

Kerr, J. A. (1826) *The Monster and Magician: or, The Fate of Frankenstein: A Melo-dramatic Romance, in Three Acts*. London: J. & H. Kerr. Reprint. Forry, S. (ed.) (1990) *Hideous Progenies: Dramatizations of* Frankenstein *from Mary Shelley to the Present*. Philadelphia, PA: University of Pennsylvania Press, 205–26.

Koger, A. K. (2012) "Calendar for 1836–1837." In A. L. Nelson, G. B. Cross, and J. Donohue (eds.) *The Adelphi Calendar Project*. N. pag. Web [3 October 2015]. https://www.umass.edu/AdelphiTheatreCalendar/m36d.htm

Koger, A. K. (1992/2012) "Calendar for 1836-1837." In A. L. Nelson, G. B. Cross, J. Donohue (eds.) *The Adelphi Calendar Project*. N. pag. Web. [3 October 2015]. Copyright © 1988, 1992, 2013 and 2016 by Alfred L. Nelson, Gilbert B. Cross, Joseph Donohue. https://www.umass.edu/AdelphiTheatreCalendar/m36d.htm.

Kucich, J. and Sadoff, D. (eds.) (2000) *Victorian Afterlives: Postmodern Culture Rewrites the Nineteenth Century*. Minneapolis, MN: University of Minnesota Press.

Leitch, T. (2007) *Film Adaptation and Its Discontents: From* Gone with the Wind *to* The Passion of the Christ. Baltimore, MD: Johns Hopkins University Press.

Leitch, T. (2011), "Vampire adaptation," *Journal of Adaptation in Film & Performance*, 4(1): 5–16.

Looser, D. (2017) *The Making of Jane Austen*. Baltimore, MD: Johns Hopkins University Press.

Meer, S. (2005) *Uncle Tom Mania: Slavery, Minstrelsy, and Transatlantic Culture in the 1850s*. Athens, GA: University of Georgia Press.

Milner, H. M. (1826) *Frankenstein; or, The Man and the Monster! A Peculiar Romantic, Melo-dramatic Pantomimic Spectacle, in Two Acts*. London: John Duncombe. Reprint. Forry, S. (ed.) (1990) *Hideous Progenies: Dramatizations of* Frankenstein *from Mary Shelley to the Present*. Philadelphia, PA: University of Pennsylvania Press, 187–204.

Murray, C. (2016) "*Frankenstein* in comics and graphic novels," in A. Smith (ed.) *The Cambridge Companion to* Frankenstein. Cambridge, UK: Cambridge University Press, 219–40.

Murray, S. (2012) *The Adaptation Industry: The Cultural Economy of Contemporary Literary Adaptation*. New York and London: Routledge.

O'Flinn, P. (1983) "Production and reproduction: The case of *Frankenstein*," *Literature and History* 9(2): 194–213.

Peake, R. B. (1823a) *Presumption; or, The Fate of Frankenstein*. London: J. Duncombe. Rpt. in Forry (1990): 135–60.

Peake, R. B. (1823b) *Another Piece of Presumption*. Rpt. in Forry (1990): 161–76.

Pedlar, V. (2012) "Opening up the secret theatre of home: Wilkie Collins's 'The woman in white' on the Victorian stage," *The Wilkie Collins Journal* 11: n. pag. Web [22 June 2016]. http://wilkiecollinssociety.org/opening-up-the-secret-theatre-of-home-wilkie-collinss-the-woman-in-white-on-the-victorian-stage/

Pride + Prejudice + Zombies (2016), dir. Burr Steers. Sony. DVD.

Rede, W. L. (1837) *Peregrinations of Pickwick: A Drama, in Three Acts*. London: W. Strange.

Rigney, A. (2012) *The Afterlives of Walter Scott: Memory on the Move*. Oxford: Oxford University Press.

"Robert Louis Stevenson archive: Film versions of Jekyll and Hyde." RLS Website. http://www.robert-louis-stevenson.org/richard-dury-archive/films-rls-jekyll-hyde.html n. pag. Web [19 Apr. 2017].

Robbins, J. (2017) "'It lives!': Frankenstein, presumption, and the staging of Romantic science." *European Romantic Review* 28(2): 185–201.

Rose, B. A. (1996) *Jekyll and Hyde Adapted: Dramatizations of Cultural Anxiety*. Westport, CT: Greenwood Press.

Saggini, F. (2015) *The Gothic Novel and the Stage: Romantic Appropriations*. New York and London: Routledge.

Schober, R. (2013) "Adaptation as connection—transmediality reconsidered." In J. Bruhn, A. Gjelsvik, and E. Frisvold Hanssen (eds.) *Adaptation Studies: New Challenges, New Directions*. London: Bloomsbury, 89–112.

Scott, W. (1814) *Waverly*. Reprint. A. Hook (ed.) Penguin Classics. New York: Penguin Group, 1972.

Shelley, M. W. (1818) *Frankenstein; or, The Modern Prometheus*. Reprint. S. J. Wolfson (ed.). Longman Cultural Editions Series. 2nd edition. Pearson Education, 2007.

Slethaug, G. E. (2014) *Adaptation Theory and Criticism: Postmodern Literature and Cinema in the USA*. New York: Bloomsbury.

Smith, B. H. and Gless, D. J. (eds.) (1991) *The Politics of Liberal Education*. Durham, NC : Duke University Press.

St. Clair, W. (2004) *The Reading Nation and the Romantic Period*. Cambridge, UK: Cambridge University Press.

Stam, R. (2005) *Literature through Film: Realism, Magic, and the Art of Adaptation*. Malden, MA: Blackwell Publishing.

Stirling, E. (1838) *Nicholas Nickleby; or, Doings at Do-the-Boys Hall*. Reprinted as *Nicholas Nickleby: a farce in two acts*, taken from the popular work of that name by 'Boz.' Boston : William V. Spencer, ca. 1858.

Stoneman, P. (2007) *Jane Eyre on Stage, 1848–1898: An Illustrated Edition of Eight Plays with Contextual Notes*. Burlington, NC: Ashgate.

Stowe, H. B. (1852) *Uncle Tom's Cabin or, Life Among the Lowly*. Reprint. A. Douglas (ed.). Penguin Classics. New York: Penguin Group, 1981.

Szwydky, L. L. (2010) "Victor Hugo's *Notre-Dame de Paris* on the nineteenth-century London stage." *European Romantic Review* 21(4): 469–87.

Szwydky, L. L. (2018) "*Frankenstein*'s spectacular nineteenth-century stage history and legacy." In D. Cutchins and D. Perry (eds.) *Adapting Frankenstein: The Monster's Eternal Lives in Popular Culture*. Manchester, UK: Manchester University Press.

Wallings, R. P. (1991) "Richard Bentley." In Anderson, P. and Rose, J. (eds.) *British Literary Publishing Houses, 1820–1880, Dictionary of Literary Biography*, 39–52. Dictionary of Literary Biography Main Series (106), Gale Databases. Web. [18 Apr. 2017]. go.galegroup.com/ps/i.do?p=DLBC&sw=w&u=faye28748&v=2.1&id=HAESHW694952481&it=r&asid=92f5d366251eace782079dd6525903f6.

Zemka, S. (2010) "The death of Nancy 'Sikes', 1838–1912," *Representations* 110: 29–57. Reprinted in John Glavin (ed.) *Dickens Adapted* (2012) Burlington, NC: Ashgate, 397–425.

PART III

Identity

Eckart Voigts

Identity is very closely related to immersions in texts, whether through production and/or reception. The 'Identity' section is concerned with answering the question what happens to the relationship between identity and text once the text itself has been adapted. Judith Butler's notion of performativity has inspired several of the readings in this section. Highlighting the ways in which adaptations always stage already existing material, they question the binary established between source and adaptation and, instead, investigate texts and practices as a site of multiple identities. Adaptations are seen as a vehicle for the display of ever-changing, fluid identity formations, not only with respect to national, ethnic, religious and gender identities – the traditional concerns of cultural studies, but also with respect to adaptive practices themselves. Allowing moments of (self-) reflexivity, this section also addresses the question as to what extent adaptation studies itself shifts, or indeed reiterates its own multiple identities. Several papers intervene in the ongoing fidelity debate in adaptation studies and examine ideas of individuality and originality. It seems that these unresolved conflicts of fidelity and originality still haunt our discourses of identity formation in adaptation studies (see Chapter 3 by Rainer Emig in this volume).

Pamela Demory's contribution is exemplary in focusing not just in queer representation, but also probing the queer nature of adaptation Studies itself. Intervening in the fidelity debate, a queer studies approach necessarily comes down on the side of infidelity. On the one hand, Demory makes the point that texts might be inherently queer, and that adaptation in general might be thought of as a 'queering' of textual identity as a 'straight' text might be 'turned', resulting in a proliferation of diverse and promiscuous transformations. Adaptations can also erase or intensify this textual queerness, as she shows in a long line of adaptations in which the homosexuality of a source text was elided or obscured, using the 2015 film adaptation of *The Danish Girl* as a recent case. On the other hand, Demory also points out many cases in which adaptations brought out an implicit thematic and formalist queerness in a text, as in the case of the adaptation of Michael Cunningham's *The Hours* and many of the recombinant para-adaptations on YouTube. Beyond the text, Demory also discusses queer authorings and 'perverse' readings and welcomes the queer fuzziness in wide notions of adaptation.

Shannon Brownlee brings fidelity discourse to bear on questions of identity formation. Picking up a binary discussed by Mary Ann Doane, she juxtaposes materiality and convention. With Doane, Brownlee opposes essentialist notions of medium specificity, pointing out that a medium is always historically and culturally specific, based largely on conventions that are wrongly taken to be essential to a medium's identity. Brownlee then discusses the specific performances of the racial and gender identity of actors such as Anthony Hopkins, Tilda Swinton or Veruschka von Lehndorff, and points out how identities are unstable and the result of fluid assignations of identity. Her analysis of the film adaptations of Virginia Woolf's *Orlando: A Biography*, or Philip Roth's *A Human Stain* thus refutes Seymour Chatman's supposition that cinematic spectators always read for plot, showing instead the mutability of identity formation on film – additionally diversified in what she terms 'digital mutability'.

Brownlee's essay can thus be usefully linked not just to the essays in the 'Technology' section, but also to the essay by Josh Sabey and Keith Lawrence, who focus on adaptors as spectator–critics. Sabey and Lawrence discuss the work of video essayists such as Catherine Grant, or critics such as George Toles, and conclude that adaptation criticism as itself is a form of adaptation. They productively blur the line between adaptation and criticism and show how critics are experimenters, adaptors, collaborators and co-creators of texts. In this chapter, Sabey and Lawrence create their own hypothetical critique-as-adaptation, recombining existing material in a montage with their own experimental analysis.

In the subsequent essay by Glenn Jellenik, we find a suitable definition of such practice which might be addressed as 'derivative originality'. Jellenik grapples with notions of originality that underpin much of the fidelity discourse in adaptation Studies. He adamantly attacks a 'zombie return' to Romantic identity notions in our field, when adaptation Studies actually should interrogate and problematise such notions. Analysing the Jonze/Kaufman film *Adaptation.* as a linchpin in the field of adaptation studies, Jellenik develops his oxymoronic ideal of 'derivative originality' and, referring to its legal definition, disentangles the epithet from the negative connotation it usually carries. Thus, we can argue that the concept of 'derivative originality' is our field's contribution to questions of originality that have haunted discussions of creative production since the late eighteenth century.

The term 'fluidity' haunts several of the papers in this section, not least in Carol Poole and Ruxandra Trandafoiu's contribution on recent adaptations of *Pride and Prejudice*. Zooming in on P.D. James's *Death Comes to Pemberley* and its BBC screen adaptation, the authors show how a supposedly very socially specific costume drama can be re-functioned and migrated in versatile ways. Thus, Austen is made to speak to a great diversity of social issues – from class-based and feminist to post-colonial renegotiation – spanning a huge variety of historical and cultural spaces.

Katja Krebs opens a set of fascinating cases of identity construction through adaptation. She discusses the success of vaudeville performer William Elsworth Robinson's appropriative masquerade as Chinese Chung Ling Soo. Krebs considers Robinson's show to be an instance of an audience reinforcing their own sense of identity. Vaudeville and music-hall acts of ethnic masquerading analysed by Krebs, such as those of Chung Ling Soo, Haroun-Al-Raschid, Zulieka, the Neapolitan Cabaret or Sigmund Neuberger a.k.a. the Great Lafayette, raise fascinating issues on colonialist adaptation practices in the field of performative mimicry and mimetic comedy. The study of this kind of cultural appropriation thus becomes a testimony to Krebs's plea for adaptation case studies to become "a means to a conceptual end rather than an end in itself."

We can test these general terms and assumptions about the fluidity of identity construction in adaptation by zooming in on the questions of national identity that are the topic of two essays that focus on questions of Australian adaptations. The fluidity of national identity is the central concern of Claire McCarthy's contribution. She argues that Australia's national identity ought to read as an adaptation or as an adaptive palimpsest. Adaptations, according to McCarthy, are central to the formation of Australia's cultural heritage through classic Australian films.

She sketches a move from the nostalgia of post-colonial imagining prior to the multicultural turn that informs adaptations of the 1980s and 1990s, and globalised neoliberalism and globalised environmental concerns that fosters, for instance, the *Mad Max* franchise.

McCarthy thus frames the contribution of Brian McFarlane, who zooms further in on a formative decade of Australian film adaptation. McFarlane argues that a spate of commercially successful and aesthetically adventurous adaptations were crucial in bringing international recognition to Australian cinema. In passing, McFarlane notes that adaptations, as a rule, tended to be adaptations of novels, a phenomenon which invites comparison to the current 'post-literary' phase of film adaptation and its expanded source materials. McFarlane provides an insightful discussion on the role of 'outsiders,' such as British director Nicolas Roeg and Canadian Ted Kotcheff in 1970s Australian adaptation. Touching on the role of the Australian Film Commission, McFarlane's essay throws into sharp relief the function of institutions in addition to landmark productions or auteurs such as Bruce Beresford and Gillian Armstrong in the formation of national cinemas.

13

QUEER ADAPTATION

Pamela Demory

Introduction: what is queer adaptation?

Queer is often used as a handy umbrella term for LGBT identity (as an adjective or noun), but it is also a way of seeing, a way of doing, and a way of being (that is, a verb). In its broadest interpretation, queer denotes "*whatever* is at odds with the normal, the legitimate, the dominant" (Halperin 1995: 62; emphasis in original), but particularly whatever is at odds with normative sexuality and gender identity. And, because sexuality and gender identity are fundamental qualities of being human, queerness has far-reaching implications. Eve Kosofsky Sedgwick (1991: 1) has argued that we cannot understand "virtually any aspect of modern Western culture" unless we also understand how homosexuality and heterosexuality have been constituted and reproduced in our cultural texts. Queer rhetoric scholars Jonathan Alexander and David L. Wallace (2009: 301) argue that "the power of queerness extends beyond exposing and challenging heteronormativity"; queerness "helps us see important connections between our personal stories and the stories that our culture tells about intimacy, identity, and connection" (2009: 303). If, as adaptation scholars, we adopt "the power of queerness," we might identify how adaptation, similarly, both challenges heteronormativity and connects personal and cultural stories about identity.

So what is queer adaptation? Queer adaptation might refer to adaptations that are *about* homosexuality (or non-normative sexual or gender identity more broadly), composed or adapted by queer authors, and consumed by queer audiences and readers – that is, adaptations that explicitly have to do with homosexuality or queer identity in some way. Queer adaptation might also refer to the process of adapting. To queer something may be to reveal the queerness of something apparently 'straight,' or to make something available to a queer point of view. And it may also be to *turn* something, to transform it in some way. To queer, then, may be to adapt.

Queer theory and recent adaptation theory intersect in a number of interesting ways. Judith Butler (1999: 198) argues that gender is constituted through a series of repetitions: "in a sense, all signification takes place within the compulsion to repeat." Individual agency occurs through "variation on that repetition" (198), and it is only "*within* the practices of repetitive signifying that a subversion of identity becomes possible" (199). Using similar language, Linda Hutcheon (2013: 7) defines adaptation as "repetition, but repetition without replication." A given adaptation might be conservative in its approach, but the process of repetition always carries the potential for subversion (see Handyside 2012, for an excellent elaboration of this idea).

Queer theory also aligns with recent adaptation theories in undoing binary thinking. If we think of queer as a verb, as a 'doing' rather than a 'being,' then to queer something is to deconstruct it, to demonstrate the instability of all those apparently obvious oppositions – male/female, gay/straight, homosexual/heterosexual, normal/deviant – that structure our understanding of ourselves and others. To queer is to blur, erase, or trouble the boundaries that would constrain or categorise us. Queer is thus, in Derrida's terms, an "undecidable," that which "can no longer be included within philosophical (binary) opposition, but which, however, inhabit[s] philosophical opposition, resisting and disorganizing it, *without ever* constituting a third term" (Derrida 1981: 43). A similar logic underlies much recent adaptation theory. The oppositions of original/copy, faithful/unfaithful that characterise much popular discourse on adaptation are being dismantled. Instead, scholars discuss adaptation as intertextuality, transtextuality, webs, cross-fertilisation, conversation, and Derridean undecidable (see Hurst 2008). Adaptation studies have also moved well beyond the book-to-film binary, encompassing multiple media, the very promiscuity of which could be said to be queer.

Queer and adaptation are also both intimately concerned with the language of gender and sexuality. Shelley Cobb (2011) has argued persuasively that the value of fidelity – so persistent in popular and academic discourse on adaptation – is rooted in deeply held cultural assumptions about the masculine/feminine gender roles and the 'normality' of the heterosexual romance plot. The language of fidelity constructs a "gendered possession of authority and paternity for the source text within adaptation: the film as faithful wife to the novel as paternal husband" (Cobb 2011: 30). Cobb's feminist critique of the heteronormative assumptions buried in the discourse of adaptation could also be said to be queer.

The characteristics of queerness that are most interesting in relation to adaptation have to do with resistance. As Robert Stam (2005: 11) has pointed out, queer theory has in common with feminist, postcolonial, and race theories an "egalitarian thrust, [a] critique of quietly assumed, unmarked normativities which place whiteness, Europeanness, maleness, and heterosexuality at the center, while marginalizing all that is not normative." He goes on to list several implications of these theories for adaptation studies, including "a revisionist view of the literary canon" that would include more "minority, postcolonial, and queer writers" and "revisionist adaptations" (Stam 2005: 11). Julie Sanders (2006: 98), similarly, groups queer theory with feminist and postcolonial theories, pointing out that adaptations of canonical texts can "write back to an informing original from a new or revised political and cultural position, and … highlight troubling gaps, absences, and silences." She also ties this idea of resistance to what Adrienne Rich (1972: 18–19) describes as "re-vision – the act of looking back, of seeing with fresh eyes, of entering an old text from a new critical direction. … not to pass on a tradition but to break its hold over us." A queer perspective on adaptation can thus be a way of resisting normative ideologies and of revealing the fissures, absences, or silences of canonical texts.

But what characterises queer resistance specifically? For one thing, its focus on sexualities draws attention to particular sorts of absences that other ideological points of view might not catch. There is also a significant strain of queer theory that insists on the value of recognising, even celebrating, queer's negative connotations. Heather Love (2007) argues that if we focus our attention only on the 'progress' that has been made, we are likely to forget the painful – but vital – past and ignore the inequities, violence, and stigmas that remain. To 'queer' something once meant to 'spoil' or 'ruin' it, and to refer to a person as 'queer' was to mark that person as disgusting. In the early 1990s, groups such as Queer Nation and ACT-UP appropriated the epithet, using it as a shock tactic and taking pride in what mainstream society saw as contemptible. 'Queer' came to mean a "militant sense of difference" that asserts the value of the "erotically marginal" (Doty 1993: 3). Now, what Michelle Aaron (2009: 64) refers to as queer's "nasty history … keeps it on its toes, keeps it daring, dancing, and not only astute to the nastiness of the present, but capable of undermining it." For so many years, queer lives were literally 'unrepresented' and 'unrepresentable' in literature and in film. Adaptation has the potential to re-visit and re-present this violent, repressed past.

Queer resistance also manifests itself in positive terms, as openness, fluidity, and erotic pleasure. It signifies, according to Eve Kosofsky Sedgwick (2013: 8), "the open mesh of possibilities, gaps, overlaps, dissonances and resonances, lapses and excesses of meaning." These qualities could be appropriated by adaptation scholars to counter what Julie Sanders (2006: 12) notes as the too-often "linear and reductive" view of adaptation. Queer resists the categories of proper/improper, appropriate/inappropriate, moral/immoral, faithful/unfaithful. As Alexander Doty (1993: 4) notes, "through playfully occupying various queer positions in relation to the fantasy/dream elements involved in cultural production and reception, we (whether straight-, gay- lesbian-, or bi–identifying) are offered spaces to express a range of erotic desire" that is deemed "unacceptable" or "immature" in mainstream culture. Such play is itself a kind of adaptation – we adapt ourselves to identify with a character we see on screen or read in a book, and in the activity of adapting, we play with, or queer, the texts we adapt. Bodily pleasure, promiscuity, infidelity, multiple partners, erotic play are all terms that appear in recent discussions of adaptation. Consider, for example, Robert Stam's description of adaptation as an "amorous exchange of textual fluids" (Stam 2005: 46). Or Gérard Genette's observation that "one who really loves texts must wish from time to time to love (at least) two together" (qtd in Sanders 2006: 7). Sanders (2006: 20) points out that it's "usually at the very point of infidelity that the most creative acts of adaptation and appropriation take place;" Fiona Handyside (2012: 59) suggests that "infidelity can offer new explorations of identity and desire." Promiscuity, group sex, infidelity – all suggest resistance to normative ideas about sexual morality and thus are queer.

Thomas Leitch (2012: 103) concludes his essay on "Adaptation and Intertextuality, or, What Isn't an Adaptation, and What does it Matter?" by suggesting that, "[i]n the spirit of confounding binaries," we defer indefinitely the question of what *isn't* an adaptation – because answering it will always impose disciplinary constraints – and the field "may well flourish more successfully when a thousand flowers bloom." Queer theory is perhaps one species of flower that might add drama and colour to the field of adaptation studies.

Characters and story

One way of thinking about queer adaptation would be to think of queer as something that inheres in a text. We might look, for example, at the relationship between *Maurice*, the 1987 film, and the E. M. Forster novel, or between Sally Potter's 1992 *Orlando* and the Virginia Woolf novel. Or we could examine *Brokeback Mountain* (2005) in relation to the Annie Proulx short story, or analyse the journey of John Cameron Mitchell's *Hedwig and the Angry Inch* from standup comedy act to 2001 film to Broadway musical (2014–5) or Alison Bechdel's *Fun Home* from graphic novel (2006) to Broadway musical (2015–). Or perhaps we might examine the intertextual relationships among the film *Gods and Monsters* (1998), the 1995 novel *Father of Frankenstein*, Mary Shelley's novel, and Director James Whale's Frankenstein films from the 1930s. To consider these as examples of queer adaptation might simply be to acknowledge their 'queer' content; these are stories about queer characters and stories that have been adapted from one medium or form to another. But does that mean that they are 'queer adaptations'? Are they instead simply queer texts that have been adapted? A more productive way of thinking about the queerness of adapted characters and stories would be to look at those texts in which the queerness is either *erased* or *revealed* through adaptation.

Cinema history is full of examples of films that have in some way 'de-queered' their source texts. *Rope* (1948), based on a 1929 play by Patrick Hamilton; *Cat on a Hot Tin Roof* (1958), based on a 1955 Tennessee Williams play; *The Children's Hour* (1961), based on the 1934 Lillian Hellman play; and *Fried Green Tomatoes* (1991), based on the novel *Fried Green Tomatoes at the Whistle Stop Cafe* (1987), are all well-known examples of films that obscured or elided the characters' homosexuality in order to appease censors and appeal to mainstream audiences. Even more recently, the 2015 film *The Danish Girl* has come under scrutiny for its handling of the story of Lili Elbe – the first person to undergo male-to-female gender reassignment surgery. According to *Sight and Sound* reviewer Vadim Rizov (2016: 80), the 2000 novel by David Ebershoff fictionalised important details of the characters' lives, and the adaptation "further tinkers with those spurious details," resulting in a "narrative [that] has been significantly de-queered from the get-go." Analysing this film from the perspective of queer adaptation would mean examining the circumstances of production – how the intersection of still-wary producers, filmmakers, studio heads, and audiences who might find the queerness of these characters unpalatable has produced a film that is shaped according to the ruling conventions of a Hollywood romance narrative.

But the opposite movement is also possible – an adaptation might provide a commentary on the silenced queerness of the source text. As Julie Sanders (2006: 98) points out, adaptations (and appropriations) of canonical texts do not necessarily accept those texts uncritically. When adaptations "write back" to their source texts, they often "have a joint political and literary investment in giving voice to those characters or subject–positions they perceive to have been oppressed or repressed in the original" (Sanders 2006: 98). Examples include Gus Van Sant's 1991 *My Own Private Idaho*, in which Prince Hal, Falstaff and other characters and scenes from Shakespeare's *Henry IV* become gay hustlers in late twentieth-century Portland, Oregon, and Jacob Tierney's 2004 *Twist,* which puts a similar gay spin on Dickens' *Oliver Twist*. Derek Jarman's 1991 *Edward II* does something slightly different in adapting Christopher Marlow's sixteenth-century play; by emphasising the homosexual content of the play and then setting it in the present, he makes a political statement about current events. In the introduction to the published screenplay, Jarman (qtd in Wymer 2005: 147) wrote that the only way to make a "film of a gay love affair" was to "[f]ind a dusty old play and violate it." By dressing Edward III in lipstick and high heels, and by incorporating the "Dance of the Sugar Plum Fairy," a group of gay rights activists, and Annie Lennox singing "Every Time We Say Goodbye," Jarman takes the play's themes of the corruption of power and the fear of sex and creates something utterly modern, particularly during the reign of the then homophobic Tory government (Davidson 2016).

Michael Cunningham's novel *The Hours* provides a great example of a novel-to-novel adaptation that opens up the implicit queerness of Virginia Woolf's *Mrs. Dalloway* and illustrates how an adaptation can voice identities and subject positions that have been silenced. Julie Sanders (2006: 117) points out that Cunningham's novel has a "liberatory treatment of gay rights and politics," and that he "achieves for his characters a freedom of relationships beyond the heterosexually prescribed 'norm' which was well beyond the realm of Woolf's own circumscribed community." In Cunningham's novel, Clarissa is not married to Richard but is instead in a lesbian relationship with Sally, a relationship that "was only hinted at on a subterranean level in *Mrs. Dalloway*" (Sanders 2006: 117). The adaptation is queer in the way it reveals the buried subtext of Woolf's novel.

This kind of reclamation is what most adaptation scholars have in mind when referring to the way queer theory might impact the study of adaptation. But looking at the queer content is just the beginning.

Form

An adaptation might be queer not just in the kind of story it tells, the characters it features, and the themes it addresses; it might also be queer in the way it tells its story. Conventional narrative (particularly in Hollywood film) centres on a heterosexual couple and implies a 'natural' progression from childhood to adulthood, marriage, reproduction, and death. By critiquing or resisting a source text's conventional narrative structure or normative ideologies, by playing with genre, or by relying on non-traditional media, an adaptation might, in the words of Benshoff and Griffin (2004), "explode formal boundaries" and be considered queer.

One area of queer theory that has intriguing implications for adaptation is the idea of "queer time." Jack Halberstam (2005: 1–2) writes: "Part of what has made queerness compelling as a form of self-description in the past decade or so has to do with the way it has the potential to open up new life narratives and alternative relations to time and space." For Halberstam, then, queer time means that those who identify as queer might imagine their lives being organised according to logics other than conventional dictates. Elizabeth Freeman (2010: xxii) has also written extensively on the idea of queer time. She notes that our traditional ways of keeping and marking the passage of time tend to naturalise certain processes that privilege those in power. To deconstruct those assumed processes, she examines "textual moments of asynchrony, anachronism, anastrophe, belatedness, compression, delay, ellipsis, flashback, hysteron–proteron, pause, prolepsis, repetition, reversal, surprise … ." These "queer temporalities" are "points of resistance" that suggest alternative ways of understanding history and narrative (Freeman 2010: xxii).

Neither Freeman nor Halberstam is referring explicitly to the process of adaptation, but following the logic they establish, we could say that a given adaptation might be said to tamper with the temporal dimension of its source text and, in so doing, queer that text. Or possibly the process of adapting is itself queer by virtue of the temporal displacement it produces.

Michael Cunningham's *The Hours* again provides a good example. Julie Sanders (2006: 119) notes the novel's anti-linear narrative structure, pointing out that Cunningham has reclaimed this methodology, long "associated with women's writing," "for another readership and community." For Halberstam (2005: 2), the novel's narrative structure is Cunningham's way of emphasising a "queer rendering of time and space." He points, as an example, to Cunningham's imaginative recreation of Virginia Woolf's writing process:

> Clarissa Dalloway, in her first youth, will love another girl, Virginia thinks; Clarissa will believe that a rich, riotous future is opening before her, but eventually (how, exactly, will the change be accomplished?) she will come to her senses, as young women do, and marry a suitable man.
>
> *Cunningham 1988: 81–2*

In Woolf's (1925: 35) novel, Mrs. Dalloway recalls the "exquisite moment" when Sally Seton "kissed her on the lips," but she has already, as Cunningham writes, "come to her senses" and married "a suitable man" – Richard Dalloway. But in Cunningham's novel, his Clarissa (Clarissa Vaughan) has refused to "come to her senses" and is living in a committed lesbian relationship. In Halberstam's words (2005: 3), "Cunningham's elegant formulation of queer temporality opens up the possibility of a 'rich, riotous future.'" Unlike Woolf's character, his Clarissa is not caught in "the seemingly inexorable march of narrative time toward marriage" (Halberstam 2005: 3).

The 2002 film adaptation of *The Hours* (directed by Stephen Daldry from a screenplay written by David Hare) further manipulates the temporal dimensions of the multiple stories. In *The Hours* DVD extra "The Lives of Mrs. Dalloway," David Hare explains that Cunningham told him that he had taken the three women's stories and "mixed them together in a way that" pleased him. "Now," he said, "you must do whatever you wish with these three stories." Hare acknowledges that he "traveled a long way from Michael's book. But [and he pauses for emphasis] the only way fidelity is achieved in adaptation is through promiscuity." The novel tells the stories of three women in alternating chapters – Mrs. Woolf in the 1930s, Mrs. Brown in the 1950s, and Mrs. Dalloway (Clarissa Vaughan) – each chapter neatly focused on just one character at a time. The film adaptation moves much more fluidly back and forth in time, sometimes using montage sequences to place all three characters in the same 'moment.' In so doing, it emphasises the queer temporal relationships among the three women.

Another way that a film adaptation's narrative form might be queered is through the construction of the gaze. Michelle Aaron notes that *Mansfield Park* is a queer film "[d]espite its achingly straight heritage," not because of the lesbian connotations of Fanny Price's interactions with Mary Crawford (Aaron 2004: 193), but because the film creates a "sexuality of looking [that] is unhooked from its normative roots." The film creates what Aaron (2004: 193) calls "a polyvalency of desire, a general air of erotic objectification available to both the characters and spectators and unhinged from the conventional love-story trajectory." As her key example, she notes the scene when the Crawfords arrive, the way the "starched collars and lowered eyes of gentility" transform into a "swell of deliciously conspicuous lust," effected by a series of "shot-reverse shots released from heterosexual imperatives. The director creates a world in which an idealised heterosexuality, despite being the dominant force of the narrative, ceases to 'own' desire" (Aaron 2004: 193).

Adaptations that feature anti-linear narrative, that resist conventional narrative structures, might be queer; certain genres – horror films, musicals, film noir, and animation – have also been theorised as queer. Benshoff and Griffin (2004: 61) point out that "some cinematic forms (such as the musical and the animated film) invite audiences to glory in the chaotic extravagance that occurs when the rigid social conventions of normality are overturned." What draws "queer spectators" to musicals and to horror films "is their use of excess, drawing in audiences by promising moments that will transgress the 'normal world' (into the abnormal in the horror film, and into the supernormal in the musical)" (61). Most of the films are not explicitly homosexual in theme, yet "the artificiality of the fantasy worlds in these genres, as well as the ineffable quality that seemed to but did not exactly say 'homosexual,' are precisely what the authors find so queer about the genres" (62). *Wicked*, the musical based on Gregory Maguire's Oz books, is a good example. Maguire (2011) says that in his books he purposely left open the question of whether his two female protagonists shared more than friendship, but points out that the musical "stepped even more steeply back from the hint of romantic attraction between the leads … with the effect, some feel, of heightening the possibility of what remains unsaid."

Sean Griffin argues that the same "potential for queerness" lies in animation: "the Carnivalesque nature of animation makes it hard to claim any specific sexual orientation for certain cartoon characters, as well as creating a general sense of fluidity and anarchy that is often expressed through a play with gender and sexuality" (Benshoff and Griffin 2004: 62). SpongeBob Squarepants is a great example – accused of being gay in the right-wing media and embraced by gay fans, his creator Stephen Hillenburg resisted defining his sexuality, calling him "somewhat asexual" – which is still a 'queer' identity. If the cartoon character is queer, then *The SpongeBob Musical*, which made its Broadway debut in the fall of 2017, is certainly a queer adaptation.

Certain media, too, have been identified as 'queer.' A special issue of *Cinema Journal* on "Queer Approaches to Film, Television, and Digital Media" focuses on the "intersection of media and sexuality" (Ahn et al. 2014: 120). In their introduction, the editors explain that "A queer approach to media theory and practice has suggested possibilities for challenging – through critical analysis – overlapping structures of patriarchy, nationhood, citizenship, heteronormativity" (119), all of which has implications for adaptation studies as well. Lynn Joyrich, for example, argues that despite television's history as a mainstream medium, its "anti-teleological temporality makes it … [a] potentially queer medium" (Ahn et al. 2014: 120), an argument that Michelle Aaron (2009) and Holly Furneaux (2010) have also made. Furneaux (2010: 251–2) argues that the television adaptation of *Bleak House* uses the "queer possibilities of serialization to make a case against the heterosexually friendly shape often attributed to the Victorian novel." And Stephen Greer (2015: 50) notes that the BBC *Sherlock*, while not exactly a "queer series," plays with the characters' sexuality, hinting at a romantic relationship between Sherlock and Watson. More important than the characters' identities as gay or straight, however, is the way the show portrays a "post-homophobic world, where the possibility of same-sex desire is not met with disgust."

Queer form might also refer to adaptations in non-traditional (even 'inappropriate') media, including any of the digital forms described under the umbrella 'new media.' Kara Keeling (2014: 152) argues that its "eccentric temporalities and reliance on reading codes" and its interest in "ephemera, publics, viruses, music, and subcultures" marks such media as queer. Cover songs might also be considered instances of adaptation. Gabriel Solis (2010: 297) argues that the practice of "covering" a song in rock music does not simply reproduce a new iteration of an old song but is a "versioning practice" in which the musical texts "gain layers of authorship as they are worked and reworked over time." And cover songs might also be queer; Judith Halberstam (2007: 52) argues that while "there is nothing necessarily queer or alternative about the cover song, the performance of covers can be queered." Fan fiction might also be considered queer adaptation – particularly erotic slash faction, such as Kirk/Spock stories that appropriate the Star Trek universe and imagine romantic and sexual entanglings between the two main characters. Or, to take an even more extreme example, the Barbie Liberation Organization 'adapting' dolls by switching the voice boxes of Barbies and GI Joe dolls (described in Sullivan 2003).

Or consider YouTube videos that take film clips, advertisements, news clips, and other examples of 'straight' media and, through the process of editing and remixing, queer them. Such videos are an example of "disidentification," the process of recycling and recoding that, according to José Esteban Muñoz, "scrambles and reconstructs the encoded message of a cultural text in a fashion that both exposes the encoded message's universalizing and exclusionary machinations and recircuits its workings to account for, include, and empower minority identities and identifications" (qtd in Alexander and Rhodes 2012). Alexander and Rhodes provide the example of a YouTube video in which footage from an old army fitness training film is edited into a queer montage titled "We don't want gays (but we do!!)." While Alexander and Rhodes do not use the term adaptation in their discussion, the activities they describe – recycling, remixing, recoding – are evidence of the same "will to adapt, recycle, and appropriate" (Cartmell and Whelehan 2010: 13) that many adaptation scholars are now studying.

Another example of queer adaptation would be "The Q-Sides," a collaborative art project by film director Vero Majano, photographer Kari Orvik, and DJ Amy Martinez, which "reimagine[s] Latino lowrider culture as inclusive of queer identities" by re-creating a series of classic album covers associated with lowrider culture. Their aim was not to duplicate the album covers in every detail, but to "[open] the door to the possibility of something else. … It's including other stories into this much more traditional narrative" (Kari Orvik, qtd in Kost 2015: E3). In appropriating and queering these traditional artefacts of popular culture, these artists are adapting.

Video games, comics, merchandise, theme park rides – if queer is defined as resistance to the norm, as taking pleasure in what is improper, then all of these non-traditional adaptations might be defined as queer.

Authorship

The idea of queer authorship focuses our attention not on the queerness that might inhere in the text – the source text, the adaptation, or both – but in the author's own intent and/or identity. The idea of the author as the sole arbiter of meaning is outdated – and it is important to avoid what Cartmell and Whelehan (2010: 26) refer to as the "tyranny of intentionalism." But, as Linda Hutcheon (2013: 107) points out, the intentions of authors and adapters (whether novelists, screenwriters, editors, directors, playwrights, graphic artists, or composers) "are often recoverable, and their traces are visible in the text."

So, for example, to discuss the 2011 HBO production of *Mildred Pierce* as an example of queer adaptation might be to pay attention not to the queerness of the text – whether the television production, the 1945 film, or the 1941 James Cain novel – but to the queerness of the auteur, Todd Haynes. Haynes won early recognition for his 1991 *Poison*, a New Queer Cinema classic, and has continued to make films that explore queer sexuality and identity, but not always in a direct or explicit way. In an interview, Haynes explains his view that a "gay film" doesn't have to do with whether or not it's got gay characters or themes, or with some sort of "gay sensibility." To define "gay cinema solely by content … [is] such a failure of imagination" (Wyatt 1993: 26). What Haynes brings to the story of a divorced single mother and her complicated relationship with her daughter is partly an awareness of the status of the 1945 *Mildred Pierce* film as a key text in gay male culture (Halperin 2012: 234). As a gay man, Haynes has access to "subcultural discourses" that allow him "to produce content that is recognizably 'gay/lesbian'" (Dyer 1991: 188). What Haynes also brings to the production is his persistent interest in critiquing "heterosexual, mainstream narrative cinema by making whatever might be familiar or normal about it strange, and in the process hypothesizing alternatives that disrupt its integrity and ideological cohesiveness" (DeAngelis 2004: 42).

Or consider *Gods and Monsters* (1998), a queer film not only because it thoughtfully explores director James Whale's homosexuality in relation to his filmmaking, but also because it is 'authored' by a collaboration of out gay men: director and screenwriter Bill Condon, novelist Christopher Bram, producer Clive Barker, and star Ian McKellen.

Richard Dyer (1991: 186–7), in an exploration of authorship and gay/lesbian identity, argues that while identity is a social construction, "it does make a difference who makes a film, who the authors are. … [O]ne can have no concept of socially specific forms of cultural production without some notion of authorship, for what one is looking at are the circumstances in which counter-discourses are produced, in which those generally spoken of and for speak for themselves," counter-discourses that may be produced through the process of adaptation.

Reception

Rather than inhering in the production of a text – its authorship – queerness might inhere in its reception. In a sense, a queer adaptation is a queer reading or interpretation of a text that – from a straight point of view – might not be queer at all. Because homosexuality was effectively erased from Hollywood film for so long, gay and lesbian audiences became adept at 'reading queerly,' that is, in identifying signs that a character was gay: the Peter Lorre character in *The Maltese Falcon* (1941) fondling his cane and lisping; Joan Crawford stalking around in a black cowboy hat and jeans in the 1954 *Johnny Guitar*; the cowboys playing with their guns in *Red River* (1948).

But 'reading queerly' means more than just identifying possibly gay characters or spotting the erasure of queer content; it means deliberately reading texts outside the matrix of heteronormative convention. Reading against the grain – what Eve Kosofsky Sedgwick (2013: 5) refers to as being "a perverse reader" – is the key to queer spectatorial practices. Alexander Doty (1993: 16), who has explored queer readings of *The Wizard of Oz*, *The Cabinet of Dr. Caligari*, and *Gentlemen Prefer Blondes*, among others, argues that "Queer readings aren't 'alternative' readings, wishful or willful misreadings, or 'reading too much into things' readings. They result from the recognition and articulation of the complex range of queerness that has been in popular culture texts and their audiences all along." In other words, if we cast aside our assumptions about apparently 'straight' texts, we open ourselves – and the texts we consume – to more inclusive, more complex meanings.

But perhaps the most important aspect of queer spectatorial practices for a study of adaptation is the suggestion that reading queerly is a kind of adaptation. In accordance with the tenets of reception theory, meaning does not inhere in the text but is created by the reader or spectator within a particular context. A screening of James Whale's *Bride of Frankenstein* at the Castro Theater in San Francisco will mean something different – and more queer – than a screening of the same film in a high school literature class. Michelle Aaron (2004) argues that there is something queer about spectatorship itself, as it encourages us to cross-identify with characters whose expressions of gender and sexuality may be other than our own. Because of the erotic possibilities that Alexander Doty describes when we "playfully occupy" queer spectatorial positions (1993), any of us might experience "queer moments" when watching a film or play, reading a novel or comic, listening to a woman singing a song written for a man, or inhabiting a video game avatar – and thus adapt that text for ourselves.

Performance

One of the basic tenets of queer theory is that identity is performance. "'Queer,'" says Eve Kosofsky Sedgwick (2013: 9), "seems to hinge much more radically and explicitly [than the terms 'gay' or 'lesbian'] on a person's undertaking particular, performative acts of experimental self-perception and filiation." This conception of the way we "act" our identities has "had radical implications," as James Loxley explains, "for how we might think about the relation between theatrical performance and the apparently real or serious world offstage, implications that performance theorists have themselves sought to spell out in recent years" (2007: 3). And it has, I would add, implications for the relationship between an adaptation and its source text as well, particularly in theatrical adaptations, but potentially for any adaptation.

An adaptation – particularly a stage adaptation – is a kind of drag. Performers use costume, makeup, wigs, and mannerisms to perform identities that may or may not align with their own gender expressions. Drag is also used (by queer theorist Judith Butler) as an analogy for gender performance. As Butler (1999: xxiii–xv) explains, gender performance is not the same as doing drag – we perform our gender identities unconsciously. But we adopt certain clothing, hairstyles, manners, ways of speaking that we associate with a particular gender – and these can be changed. Gender is not a fixed, unchangeable 'reality.'

To consider adaptation as queer performance, we can examine how queer actors or performers affect meaning in a given adaptation. Douglas McGrath's *Nicholas Nickleby* is queer, according to Holly Furneaux, largely because McGrath cast queer actors (Nathan Lane, Alan Cumming) and actors associated with queer roles (Charlie Hunnam and Jamie Bell). Anna Blackwell (2014: 351) notes that "[a]n actor-based approach to adaptation studies … allows us to refocus the critical gaze by considering how the actor shapes the adaptation, not only in their performance but through their intertextual physicality." One example of this intertextual physicality would be the radical 1991 staging of *Streetcar Named Desire*, titled *Belle Reprieve*, written and performed by gay and lesbian actors, which "realigned the traditional gendered roles of man and woman as they were represented in *Streetcar* and in the 1951 film version of Williams's play. In the process, *Belle Reprieve* interrogated the political and sexual implications of portraying woman as the object of male desire and man as the indomitable agent of that desire" (Kolin 2000: 121). Another example is offered by Megan Brodie, writing about her production of *Orlando*, based on Sara Ruhl's 1998 adaptation of Virginia Woolf's 1928 novel. Brodie realised that she could manipulate the gender dynamics of the play beyond what Woolf or Ruhl had indicated in their texts by casting any character of any sex to play any role. She could even change the casting from performance to performance, thereby "fruitfully destabiliz[ing] and further queer[ing] an already queer text" (Brodie 2014: 167).

Same-sex performances of *Romeo and Juliet* (which of course are in some ways more 'faithful' to the Shakespearean stage, where female roles were played by men, than most contemporary stagings) or any adaptation that plays with gender roles and/or sexuality might have similar radical implications for our understanding of gender and sexuality.

Thomas Leitch (2012: 99) has argued that adaptation itself might be defined as performance. This would mean treating "the performance text … as a recipe for a new creation rather than a court that has issued a restraining order anticipating any possible infractions by future realizations." And while such a broad definition might become somewhat messy – in a sense every performance of anything becomes an 'adaptation' – perhaps it would be a productively queer mess.

Process

What I've tried to suggest in this chapter is the queerness of adaptation itself; to queer and to adapt are (or can be) processes of reading, of resisting, of destabilising notions of authorship, authority, ideology. Queer theory can thus do more than help us understand the buried stories and silenced voices of people whose sexuality or gender expression does not conform to conventional narrative patterns. It can help us understand the nature of adaptation itself.

Works cited

Aaron, M. (2004) "The new queer spectator," in M. Aaron (ed.) *New Queer Cinema: A Critical Reader*. Edinburgh, UK: Edinburgh University Press, 187–200.

Aaron, M. (2009) "Towards queer television theory: Bigger pictures sans the sweet queer-after," in G. Davis and G. Needham (eds) *Queer TV: Theories, Histories, Politics*. New York: Routledge, 63–76.

Ahn, P., Himberg, J. and Young, D.R. (2014) "In focus: Queer approaches to film, television, and digital media," *Cinema Journal* 53(2): 117–121.

Alexander, J. and Wallace, D.L. (2009) "The queer turn in composition studies: Reviewing and assessing an emerging scholarship," *College Composition and Communication* 61: 300–20.

Alexander, J. and Rhodes, J. (2012) "Queer rhetoric and the pleasures of the archive," *Enculturation: A Journal of Rhetoric, Writing, and Culture* 13: n. pag. Online (accessed 29 April 2016).

Benshoff, H. and Griffin, S. (2004) "General introduction," in H. Benshoff and S. Griffin (eds) *Queer Cinema: The Film Reader*. New York: Routledge, 1–15.

Blackwell, A. (2014) "Adapting *Coriolanus*: Tom Hiddleston's body and action cinema," *Adaptation* 7(3): 344–52.

Brodie, M. (2014) "Casting as queer dramaturgy: A case study of Sarah Ruhl's adaptation of Virginia Woolf's *Orlando*," *Theatre Topics* 24(3): 167–74.

Butler, J. (1999) *Gender Trouble: Feminism and the Subversion of Identity* (second edition). London: Routledge.

Cartmell, D. and Whelehan, I. (2010) *Screen Adaptation: Impure Cinema*. London: Palgrave Macmillan.

Cobb, S. (2011) "Adaptation, fidelity, and gendered discourses," *Adaptation* 4(1): 28–37.

Cunningham, M. (1988) *The Hours*. New York: Farrar, Straus, and Giroux.

Davidson, A. (2016) "Derek Jarman: Five essential films." *BFI Film Forever*, British Film Institute, 25 January: n. pag. Online. (accessed 6 February 2016).

DeAngelis, M. (2004) "The characteristics of new queer filmmaking—case study: Todd Haynes," in M. Aaron (ed.) *New Queer Cinema: A Critical Reader*. Edinburgh, UK: Edinburgh University Press, 41–52.

Derrida, J. (1981) *Positions*. Trans. Alan Bass. Chicago, IL: University of Chicago Press.

Doty, A. (1993) *Making Things Perfectly Queer: Interpreting Mass Culture*. Minneapolis, MN: University of Minnesota Press.

Dyer, R. (1991) "Believing in fairies: The author and the homosexual," in D. Fuss (ed.) *Inside/Out: Lesbian Theories, Gay Theories*. New York: Routledge.

Freeman, E. (2010) *Time Binds: Queer Temporalities, Queer Histories*, Durham, NC: Duke University Press.

Furneaux, H. (2010) *Queer Dickens: Erotics, Families, Masculinities*. Oxford: Oxford University Press.

Greer, S. (2015) "Queer (mis)recognition in the BBC's *Sherlock*," *Adaptation* 8(1): 50–67.

Halberstam, J. (2005) *In a Queer Time and Place*. New York: New York University Press.

Halberstam, J. (2007) "Keeping time with lesbians on ecstasy," *Women and Music* 11: 51–8.

Halperin, D. (1995) *Saint Foucault: Towards a Gay Hagiography*. New York: Oxford University Press.

Halperin, D. (2012) *How to Be Gay*. Cambridge, UK: Harvard University Press.

Handyside, F. (2012) "Queer filiations: Adaptation in the films of François Ozon," *Sexualities* 15(1): 53–67.

Hurst, R. (2008) "Adaptation as an undecidable: Fidelity and binarity from Bluestone to Derrida." in D. L. Kranz and N. C. Mellerski (eds) *In/Fidelity: Essays on Film Adaptation*. Cambridge, UK: Cambridge Scholars Publications.

Hutcheon, L. (2013) *A Theory of Adaptation*. 2nd edition. New York: Routledge.

Keeling, K. (2014) "Queer OS," *Cinema Journal* 53(2): 152–7.

Kolin, P.C. (2000) Williams: *A Streetcar Named Desire*. Cambridge, UK: Cambridge University Press.

Kost, R. (2015) "Old soundtrack gets a more inclusive spin," *San Francisco Chronicle* 15 June: E1, 3.

Leitch, T. (2012) "Adaptation and intertextuality, or, what isn't an adaptation, and what does it matter?" in D. Cartmell (ed.) *A Companion to Literature, Film and Adaptation*. Malden, MA: Blackwell, 87–104.

"The lives of Mrs. Dalloway" (2003) DVD extra, *The Hours*, dir. Stephen Daldry, Paramount, DVD.

Love, H. (2007) *Feeling Backward: Loss and the Politics of Queer History*, Cambridge, UK: Harvard University Press.

Loxley, J. (2007) *Performativity*. The New Critical Idiom. London: Routledge.

Maguire, G. (2011) "Friends of Dorothy: How gay was my Oz?" *Huffpost Queer Voices*, *The Huffington Post*, 31 October 2011: n. pag. Online (accessed 6 February 2016).

Rich, A. (1972) "When we dead awaken: Writing as re-vision," *College English* 34(1): 18–30.

Rizov, V. (2016) "The Danish Girl," review, *Sight and Sound* (January) 26(1): 80.

Sanders, J. (2006) *Adaptation and Appropriation*, The New Critical Idiom. New York: Routledge.

Sedgwick, E.K. (1991) *Epistemology of the Closet*. Berkeley, CA: University of California Press.

Sedgwick, E.K. (2013) "Queer and now," in D. Hall and A. Jagose (eds) *The Routledge Queer Studies Reader*. New York: Routledge, 3–17.

Solis, G. (2010) "I did it my way: Rock and the logic of covers," *Popular Music and Society* 33(3): 297–318.

Stam, R. (2005) "Introduction: The theory and practice of adaptation," in R. Stam and A. Raengo (eds) *Literature and Film: A Guide to the Theory and Practice of Film Adaptation*, Malden, MA: Blackwell, 1–52.

Sullivan, N. (2003) *A Critical Introduction to Queer Theory*. New York: New York University Press.

Woolf, V. (1925) *Mrs. Dalloway*. San Diego, CA: Harvest.

Wyatt, J. (1993) "Cinematic/sexual: An interview with Todd Haynes," *Film Quarterly* 46(3). Reprinted in J. Leyda (ed) *Todd Haynes: Interviews* (2014), Jackson, MS: University of Mississippi.

Wymer, R. (2005) *Derek Jarman*, British Filmmakers, Manchester, UK: Manchester University Press.

14

FIDELITY, MEDIUM SPECIFICITY, (IN)DETERMINACY

Identities that matter

Shannon Brownlee

Two principal threads in adaptation theory are the discourses of fidelity and medium specificity. These intersect when we ask how particular media can support or foreclose fidelity in adaptation. Adaptation theorists have charted their frustration with the moralistic ideal of fidelity, but generalisations about what specific media 'can' and 'cannot' do can be similarly confining, especially when media conventions are conflated with the wider possibilities that materials of expression can offer. Both theoretical approaches intersect and gain a political dimension when we ask how either faithful or unfaithful adaptations in different media can or do construct identities related to questions of social justice, such as race, gender, and age. By attending to questions of fidelity when adaptations test the limits of medium specificity, we can gain a better sense of the different ways identities are constructed, naturalised, and denaturalised, and propose ways of representing and imagining identity that may challenge oppressive social regimes.

(In)fidelity and medium (non-)specificity

The prejudice that a text ought to be faithful to its progenitor has been unpopular with adaptation theorists for decades. Nonetheless, in her excellent overview of the place of fidelity discourse in post-millennial criticism, Casie Hermansson makes a compelling case for its ongoing utility as one tool among many. First, the rejection of fidelity may lead to an overvaluation of *in*fidelity which, as Hermansson points out, is merely the other side of the fidelity coin (2015: 152). Looking only to unfaithful adaptations for adventurous or interesting ideas can be limiting, especially in discussions of identity. Certainly, activist artists have intelligently critiqued oppressive constructions of identity through deliberate infidelity. For example, Patricia Rozema's 1999 screen version of Jane Austen's 1814 *Mansfield Park* interpolates dialogue, sounds, images, and events to draw explicit parallels between the commodification of white women and African people on the principle family's Caribbean plantation. However, we restrict ourselves if we only look for insights and transgressions in overtly unfaithful adaptations with explicit political agenda. The stage puppetry of Nick Stafford's *War Horse* has embodied the character of Joey the horse from Michael Morpurgo's 1982 novel in a way that not only solves the technical challenge of representing an equine character in the theatre but also brilliantly exploits the role of sentimentality in representations of non-human subjects. Likewise, Lego versions of *Star Wars* characters in toys or video games can tell us much about how racial and gender identities get caught up in layers of corporate branding. The political activism of Rozema's *Mansfield Park*, although admirable, does not necessarily yield more insights about constructions of social identities than do the magical stagecraft of *War Horse* or the humour of Lego *Star Wars*; both fidelity and infidelity can teach us a great deal about identity.

Furthermore, while moralistic discussions of fidelity can indeed be "tiresome" (Andrew 2000: 31), the approach remains important to a great number of people and, as such, should not be dismissed entirely. Although fidelity discourse has traditionally been aligned with an elitist preference for literature over 'mass' media, Colin MacCabe has argued more recently that academic discourse must not "seal itself hermetically off" from this "highly evaluative" form of popular discussion of adaptation (2011: 9). MacCabe's injunction to attach "equal importance" to both adaptation and adapted text rather than "promote one over the other" (2011: 8) may not always be consistent with popular discourse, but it is a welcome amendment to fidelity criticism in retaining the potentially useful emphasis on likeness while rejecting the moralistic preference for the 'original.' However, it also begs the question of whether fidelity discourse is moralising by definition or whether this is merely an unfortunate feature that can be cut away. Shelley Cobb's (2010) analysis of the patriarchal rhetoric of both fidelity criticism and its detractors demonstrates that ideological biases are not so easily expunged. Moreover, to ignore people's passions and evaluations is to ignore some of the most informative aspects of their responses, as we will see in case studies below.

Fidelity discourse also raises the important issue of what 'sameness' and 'difference' mean, given the constant need to account for medium specificity. No single definition of a 'medium' can fully capture the term's complexity, but Mary Ann Doane writes lucidly that we generally think of a medium as "a material or technical means of aesthetic expression (painting, sculpture, photography, film, etc.), which harbours both constraints and possibilities, the second arguably emerging as a consequence of the first" (2007: 130). When we analyse adaptations, we address a wide range of materials of expression, from recorded images and bodies moving through space to recorded sounds, live sounds, paints, canvas, and many others. Medium specificity, however, is not simply the presence or absence of certain material or technical means of expression. At its most dogmatic, medium specificity is "yoked to a notion of Greenbergian formalism wherein every authentic work of art is caught in a self-reflexive spiral, referring only to itself and its own conditions of existence (in painting, for example, flatness)" (Doane 2007: 131). In adaptation theory, medium specificity has often played a happier role, contesting the fidelity discourse by valuing the "different but equal" status of various media. Ironically, Hermansson writes, this can bring us right back to fidelity when "appreciation of each medium's particular strategies for telling stories … remains fidelity by other (medium-specific) means" (2015: 152). In other words, analyses of the specificities of different media can sometimes be seen as (in)fidelity criticism drained of its moralising. Although this is not always true of adaptation theory addressing medium specificity, questions of fidelity and medium are crucially linked in their attention to formal similarities and differences in the communication of narratives.

Doane raises a question that productively complicates previous key texts on adaptation: the relation between materiality and convention. Citing Rosalind Krauss's definition of a medium not as the materials of expression alone but as a set of conventions that emerge from these materials, Doane emphasises the differences between convention and materiality, as well as their interrelation (2007: 131). In adaptation studies, however, these have frequently been conflated. George Bluestone begins *Novels into Film* with the question of how the two media "both join and part company" (1957: 1) perceptually. "Seeing" through language, he writes, is different from seeing through the eyes, although both media use imagery. He attends to the optical process viewing a film (1957: 14), but quickly asserts that this material is always embedded in a "set of fluid, but *relatively homogeneous*, conventions" (1957: 5, emphasis added). André Bazin (1967) attends to the ontologically realist specificity of the photographic image in a 1945 essay, but his essays on Theatre and Cinema from 1951 contrast the conventional anthropocentrism and logocentrism of the classical theatre with photographic realism and de-emphasis on the human in cinema. This approach must have foreclosed his consideration of some very interesting work circulating in Paris at the time he wrote, such as mime artist Marcel Marceau's wordless mimo-dramas (which included adaptations of literature) or Alain Resnais's film of Marceau's mimo-drama in his early short, *La Bague* (1946). Both Bluestone and Bazin understand materials of expression ontologically – as a 'given' – while larger units are understood as a series of conventions to which a similar degree of givenness is nevertheless ascribed. However, while individual elements signify in the context of larger units of meaning, that ought not to make materiality identical to convention.

Studies based in semiotics are similarly divided between the individual signifier and the signifying system. Dudley Andrew's overview of semiotic approaches to adaptation notes that fidelity is usually framed in relation to the "letter" or the "spirit" of the adapted text (2000: 31). Fidelity to the letter implies that elements such as lines of dialogue can be mechanically transferred from one medium to another, prioritising content and deprioritising materials of expression. Conversely, fidelity to the spirit addresses the specificity of larger signifying systems. Andrew finds that, for many, meaning in one semiotic system can more or less 'match' that in another insofar as the signifiers can be arranged in relationships that approximate the overall effect of the adapted text. Concepts of fidelity to the 'letter' and to the 'spirit' both suggest that signifiers in different media can be associated with similar signifieds, but fidelity to the 'spirit' generally addresses larger narrative blocks and signifying systems than the former, which emphasises the material of expression (or irrelevance thereof) of individual signifiers.

Such semiotic studies, especially those that also draw on narratology, regularly find themselves unable to address texts whose materials belong to the medium but whose broader units of narration do not fall within what is narrowly defined as its conventions. For example, in "What Novels Can Do That Films Can't (And Vice Versa)," Seymour Chatman (1980: 126) mentions some films that depart from fairly conventional cinematic language. However, he argues that film audiences watch primarily for the plot, and that this mode of spectatorship is appropriate to the medium because on-screen events "move too fast" (1980: 126) for the audience to absorb all the visual details. Furthermore, less plot-oriented cinematic moments are mere departures that cause problems (like a ticking taxi metre) for the audience (1980: 130). This claim is dubious: for example, fandom of individual theatre or film performers prioritises the intertextual or extradiegetic element of star persona rather than plot. Chatman's prioritisation of plot because everything else moves too fast is more understandable coming from the era before pause and rewind functions, and Laura Mulvey (2006) and others have analysed how home viewing technologies have radically changed narrativity and pleasure. However, it also demonstrates that, as Doane argues, our understanding of any medium needs to be historically and culturally specific, not essentialist. If a medium cannot be reduced to its materials of expression, neither can it be reduced to the conventions by which its materials are combined. Mime, Aboriginal storytelling, and flash mobs can be 'theatre' as much as Shakespeare; Paul de Man's concept of "literariness" (1997), the opacity of verbal language, can be found not only in conventional novels but also in instruction manuals (especially those in translation!) and tweets. A major problem of medium specificity is that the larger narrative conventions can come to stand for the medium as a whole in a way that forecloses the analysis of many texts.

Performativity and (in)determinacy

In a rejection of modernist forms of medium specificity as well as cultural elitism, several theorists in the twenty-first century have turned to cultural and ideological dimensions of adaptation. Robert Stam and Linda Hutcheon, in particular, have offered categories with broader reach than individual media. Hutcheon (2006) uses the categories of telling, showing, and interacting, not to deny the role of materiality, but to re-insert the often sadly neglected role of the audience member in understanding adaptations as such. The case studies below demonstrate how useful audience responses can be in teasing out multiplicity of signification. Stam's "single-track" versus "multitrack" approach (2000: 56) attends to the materials of expression of different media without drawing absolute lines between and among them; media can share tracks, such as words, as well as the number of tracks by which they communicate. Both Hutcheon and Stam balance hybridity with specificity. However, these approaches can sometimes dwell on these large-scale concepts in a way that re-inscribes the apparent ideological 'neutrality' or givenness of individual materials: the 'nature' of the visual image, word, and sound.

In recent decades, though, debates originating outside of adaptation studies about construction versus essentialism of identity have taught us not to see materiality as 'natural' in any way. As Judith Butler (1990, 1993) writes, performance of gender, race, and sexuality constitute identity rather than reflecting essence. This is not a denial of materiality. The bodies that Butler discusses exist in space and time, speak, gesture, and feel pain, but these bodily experiences are circumscribed by discourse. One way in which adaptation studies can contribute to these debates and expose the naturalisation of materiality is by comparing materials of expression to show that identities are not inevitably or consistently embodied in particular words, images, sounds, or objects. Rather, in a given cultural–historical context, material signifiers are culturally intelligible and "contoured" (Butler 1993: 17) in a finite number of ways through regulatory, discursive regimes such as colonialism and normative heterosexuality. "Contouring" is not determination; both discourse and the materials of expression have an impact on, but do not determine how we interpret signifiers. Taking materials of expression seriously requires situating them within cultural, discursive contexts but not reducing them to homogeneous units that work harmoniously within a larger system of meaning. In other words, we need to reject a master narrative of the 'nature' of image and word as well as the 'nature' of femininity, masculinity, whiteness, and so on.

The brief case studies in the following section focus on a particular contrast between verbal and visual media that is often taken for granted: the amount of detail possible for each and their potential for abstraction. Chatman notes that literary language can limit the number of details by using fewer words, whereas the number of details in a film image is "indeterminate" (1980: 125). Film semiotician Christian Metz (1974) similarly discusses the impossibility of finding the smallest unit of cinematic meaning, since a single frame includes multiple objects and codes (lighting, composition, etc.). On the other hand, verbal language may be more abstract than a visual image. Stam writes: "Flaubert never even tells us the exact colour of Emma Bovary's eyes ... A film, by contrast, must choose a specific performer" (2000: 55). Similarly, a literary character may remain uncharacterised in terms of categories of identity such as race, age, sexual orientation or, very occasionally, gender, but a character performed in the flesh is beholden to the physical specificity of a performer, and further contouring by lighting, prosthesis, etc., only changes the morphology rather than increases its abstraction (although an audience's knowledge of the performer outside the role may confer duplicity on such contouring). Even a line drawing would be hard pressed to be as abstract as a word can be, as the received wisdom goes.

The tension between potential and practice, however, is important. The prevalence of gendered pronouns in those languages that use them make it likely that literary characters' identities are gender specific, and the first-person narrators in Jeanette Winterson's *Written on the Body* (1993) and Sarah Caudwell's mystery novels (1982, 1986, 2002a, 2002b) are rare examples of gender indeterminacy in English. And, although racial identity is not tied to pronouns in the same way (proper nouns are more likely to connote race and ethnicity), writing and reading practices may make 'unspecified' racial identities nothing more than 'unmarked' identities that conform to the dominant norm within a given community of interpretation. The case studies below cannot do full justice to the rich texts, nor do they intend to; instead, they start from the perspective of individual signifiers rather than larger units of meaning in order to interrogate the received wisdom concerning the determinacy of the visible and the indeterminacy of the verbal. Specific aspects of materiality may not have the same degree of semiotic importance in every instance. Here, the shape of a body may be of paramount importance in an audience's interpretation of gender; there, that body's gestures may carry more weight. In short, these case studies argue that visual specificity and verbal indeterminacy are 'constructed' within particular texts instead of being 'determining' factors.

Pronouns and embodiment

Philip Roth's novel, *The Human Stain*, takes pronouns seriously. At a crucial moment, its protagonist feels "Free to enact the boundless, self-defining drama of the pronouns we, they, and I" (2001: 109). Coleman Silk, a light-skinned African American, passes as a white Jew for most of his adult life. His decision is rooted in a desire for self-determination in the face of what he perceives to be the strictures of the "we" of African American group identity. He does not realise "that he [is] a Negro" (2001: 108) until he attends Howard University; then, at once, "the raw I was part of a we with all of the we's overbearing solidity" (2001: 108). It seems that Coleman exploits linguistic indeterminacy in order to be a white "I" instead of an African American "we" in the same way that Flaubert never tells us the exact colour of Emma Bovary's eyes. However, Coleman's conscious disidentification with the African American "we" has created an unconscious conflict that erupts in racialised epithets such as "lily-white" (2001: 81) and – costing him his academic job – "spooks." Thus, the novel both exploits the fact that pronouns "lack the standardised and continuous significance of other linguistic terms" (Silverman 1983: 44) to avoid essentialising race, and also demonstrates how cultural–historical discourse unconsciously circumscribes racial identity.

Some of the reception of Robert Benton's 2003 film adaptation of *The Human Stain* essentialises racial identity in a way that seems to affirm the concreteness and specificity of the visual image. Two actors play the role: the white, blue-eyed Anthony Hopkins is the older Coleman, while the mixed-race, hazel-eyed Wentworth Miller is the younger.[1] David Stratton sees Hopkins's performance to be strong, but writes that many will see it to be "utterly preposterous" (2003: 4), a highly "wayward example of casting" (2003: 4), for Hopkins to play an African-American. Miller "convinces as a light-skinned African-American in a way Hopkins never does" (2003: 5). Here, Coleman is constructed as 'essentially' African-American; Miller's racial identity is naturalised on-screen and off and, by contrast, Hopkins's is naturalised off-screen by being denaturalising on-screen. Erica Abeel also criticises Hopkins's performance as "the priapic Silk" for being too "donnish" (2003: 102). Abeel's diction has racial subtext, especially in light of her approval of Miller, a "buff, libidinous" and "green-eyed sizzler" (2003: 102): the characterisation of Black masculinity as "buff," "libidinous," "priapic," and "sizzling" is rife with racial stereotypes, while "donnishness" connotes whiteness in Anglophone culture. Both responses construct Coleman as 'really' African-American, despite his performance of Jewishness, and Hopkins's and Miller's racial identities in similarly essentialist terms. The apparent specificity of the image, highlighted by visual contrasts between the two performers, is caught up in this essentialism.

Other responses recognise a potential for indeterminacy in the visual performance text that echoes the pronominal fluidity of the novel. Roger Ebert (2003) asks: "Does Hopkins look as if he 'could' have been black? How can you answer that question about a man who successfully passes for white?" The casting of a white-identified actor could thus be seen as an affirmation of the power of Coleman's performance of whiteness over an essential racial category. While Stratton emphasises colouring over performance, Talise D. Moorer does the opposite, applauding Hopkins's "natural affinity" for the role. Writing for the Black-owned and -operated *New York Amsterdam News* and addressing a racially marginalised reader, she finishes: "[b]eyond leaving the cinema entertained by this film, one leaves better educated as to how painful it is to be cast down merely because of the colour of *your* skin" (2003: 18, emphasis added). Hopkins's powerful performance makes him part of the African American 'we' on-screen even if he is not identified as such off-screen. By prioritising performance, Moorer suggests that gesture, facial expression, and other aspects of the 'movement' of the body, rather than skin and eye colour, are the salient visual signifiers. Together, Abeel, Stratton, Ebert, and Moorer's responses show that audiences generally take account of both social discourse and the materiality of bodies (both their shape and their movements). However, the differences among them, including their "highly evaluative" (MacCabe 2011: 9) observations, demonstrate that audiovisual signification can be ambiguous – perhaps not in the same way as the verbal pronoun is, but to a similar degree.

While *The Human Stain* focuses on the plural pronoun versus the singular, Virginia Woolf's 1928 *Orlando: A Biography*, interrogates gender pronouns and exploits the ease with which, on the page, 'he' can be transformed into 'she.' The opening sentence is a masterpiece of prevarication disguised as assertion: "He – for there could be no doubt of his sex, though the fashion of the time did something to disguise it – was in the act of slicing at the head of a Moor which swung from the rafters" (1973: 3). Orlando's gender is specific, as there is "no doubt" of his sex, but not stable. The young favourite of Queen Elizabeth I ages with magical slowness, falling into a trance in the eighteenth century only to wake and stand "upright in complete nakedness before us, and while the trumpets [peal] Truth! Truth! Truth! we have no choice left but confess – he was a woman" (1973: 88). Only a few paragraphs on does Orlando's narrator begin to use "she" instead of "he" "for convention's sake" alone (1973: 89). Of course, taking the narration at face value fails to account for Woolf's powerful irony, through which one could plausibly read Orlando to be initially masquerading as a man (Hovey 1997: 399). However, the abstraction of verbal language also leaves room for an interpretation less shackled to biology in a "reconciliatory fantasy of transcending sexual difference and reaching the neutralisation of sex" (González 2004: 83). Gender is socially and legally powerful in the novel, as the female Orlando loses her ambassadorial position and estate, but it is accidental in relation to personality: after the transition, "in every other respect, Orlando remained precisely as he had been" (Woolf 1973: 89). The power of the pronoun is socially and economically coercive, but irrelevant to psychology and, perhaps, biology. In short, literary indeterminacy enables a change in gender and leaves the nature of that change open to interpretation.

In her 1992 film adaptation, director Sally Potter strategically affirms the relative immutability of the body in comparison to the word. While two actors play Coleman Silk to represent different ages and performances of race, in *Orlando*, the female-identified, albeit famously androgynous, Tilda Swinton plays both genders of the titular character. In the transformation sequence, in a moment of fidelity to the letter of the novel, she appears naked for the first and only time. The film thus uses Swinton's female body to assert the 'truth' of the sex change; simultaneously, Swinton's identity off-screen constantly undercuts the character's apparent diegetic maleness, as Woolf, on one view, undercuts the 'doubt of his sex' by protesting too much about its stability. However, while Woolf's protest is effectively about the 'accidental' nature of any gendering,

Swinton's bodily presence grants femaleness a different, more 'real' status than maleness. Its feminism rests on strategic essentialism, enabled by the materiality of the body, and represented visually in contrast to the abstraction of the word. Both novel and film use medium specificity to feminist ends, overtly linking their materials of expression to critiques of patriarchy and normative heterosexuality, and their essentialism (or lack thereof) is strategic.

While Potter's *Orlando* asserts visual concreteness and specificity, another film uses images to achieve an effect closer to Woolf's verbal ambiguity. Ulrike Ottinger's *Dorian Gray im Spiegel der Boulevardpresse* is an avant-garde 1984 mashup of Oscar Wilde's *Picture of Dorian Gray* and Fritz Lang's three *Dr. Mabuse* films (1922, 1933, and 1960). While Swinton's nakedness in *Orlando* makes the performer's biological sex part of the film's visual text, the body and gender of the titular character in *Spiegel* is highly indeterminate. The story is a battle of wills between Dorian (supermodel Veruschka von Lehndorff) and Frau Dr. Mabuse (Delphine Seyrig). Media magnate Mabuse is an amalgam of Wilde's Lord Henry Wotton, the hypocritical upper-class man who leads Dorian astray, and Fritz Lang's all-seeing, all-knowing villain. Seeking material for her sensationalist press, Mabuse leads the virtuous young Dorian into well-documented debauchery and ultimately orchestrates his death to capitalise on his funeral; however, Dorian simultaneously massacres Mabuse and her minions. The film's double ending is not entirely successful, but the intention is clearly to offer two narratively incommensurable outcomes.

The film is far more successfully duplicitous in its performance of gender. Alice Kuzniar writes that the main question it raises is "how to read the opaque signs of Dorian's gender" (2000: 146). Frau Dr. Mabuse says that Dorian is a "beautiful and somewhat dull and inexperienced young *man*" (emphasis added), and narratively he is a fairly traditionally masculine man who pursues a woman, vanquishes the villain, etc. Visual signification is a different, more contradictory matter. Von Lehndorff's athletic, six foot tall figure, short, slicked hair, and men's clothing all signify masculinity, although the androgynous effect references the legacy of Wilde's 'effeminate' dandy very closely. The gap between character and performer is emphasised by the untrained von Lehndorff's lack of technical clarity (Brecht 1996: 148–53), and the makeup darkening Dorian's upper lip and eyebrows is unconvincing. Furthermore, the performer is best known to film audiences as the model in a five-minute, sexually charged photo shoot sequence in *Blow Up* (Antonioni, 1966); although von Lehndorff is not nude in this scene, she is scantily clad and objectified enough that we know what she 'really looks like' under her costume. Ottinger's film references this explicitly since this Dorian is also photographed and exploited for profit. The director's reputation as a lesbian filmmaker also frames Dorian – the object of desire – as female. Finally, the film's postmodern style and double ending discourage straightforward, cohesive interpretation. Ultimately, the signification of gender is gleefully contradictory and multiple.

This multiplicity is confirmed by diversity in the film's reception. Roswitha Mueller (1985) does not distinguish greatly between the modes in which Dorian and Frau Dr. Mabuse are performed, although they are assigned different gender pronouns within the diegesis; for her, both are women taking over men's roles in the adapted texts. In contrast, Kuzniar writes of the "blindness" and "captivation" she experienced on first viewing, not realising that Dorian was performed by a female-identified actor: "I was profoundly disturbed every time his/her image appeared on the screen. I was confused by the appearance of the five-o'clock shadow and anxiously looked for the tell-tale swelling of the breasts, which I didn't find" (2000: 152). Although she writes that "visible contradictions in Dorian's gender are erased" (2000: 154), why was she "confused"? Why did she look for breasts? Not only did the cinematic signifier construct an ambiguous gender for Kuzniar, but even this ambiguity was uncertain. For myself, this 'uncertain' ambiguity is precisely the pleasure of this text. Even though 'women play men' in this film, 'woman' and 'man' do not become reified categories, nor do the performers' bodies become reified as a 'natural' female ground on which performance signifiers are inscribed. The denaturalisation of gender and biological material is enabled simultaneously by the uses of the audiovisual materials of expression and by intertextuality, both adaptation and other forms of reference such as Ottinger's, von Lehndorff's, and Wilde's bodies of work. *Dorian Gray im Spiegel* is perhaps the clearest instance in which visual signification matches the indeterminacy and abstraction of the gendered third person pronoun.

Age and visual ambiguity

In contemporary Western culture, most of us experience identities of gender and race as relatively stable – at least in contrast to our age identity. People are 'expected' to change from young to old as they are not expected to change from white to Asian or female to male, though we are increasingly recognising such possibilities for mutability. Woolf's *Orlando* illustrates this difference: Orlando's aging process is as unusual as the gender transformation, but the latter happens only once in a substantial narrative passage devoted to it while the character's magically slow aging is mentioned periodically and casually. Aesthetic conventions also represent age identity differently from race or gender. While most literary characters are identified by gendered pronouns and many are assigned race through marked or unmarked categories, the age of many is only roughly drawn. Similarly, the performance of age identity is conventionally much less rigidly policed than we have seen identities of gender and race to be: twenty-somethings regularly play teenagers; thirty-somethings play characters whose career status would be more plausible for a fifty-year-old. Thus, age identity has a very different relation to the materials of expression than gender and race do.

One body of texts that highlights the intersection of fidelity, medium (non-)specificity and age identity is that surrounding *Alice's Adventures in Wonderland* by Lewis Carroll (pen name for Charles Dodgson). In the famous 1865 novella of a girl's dream world adventures, word play and nonsense poetry abound. Absurdity illuminates the arbitrariness of social convention and the violence of pressures towards conformity. Alice, Carroll tells us, is seven years old, an important time for learning these social rules. She balances delicately between a childlike fascination with the strangeness of Wonderland's rules (and, by extension, of socialisation in general), and an incipient, adult impulse to assert the rules that her own waking world tells her to follow. Thus, her age identity is crucial to the novella's logic games and absurdity.

As Kamilla Elliott (2003) points out, Carroll's novellas have never been 'pure' from the point of view of medium. Carroll illustrated his first, handwritten manuscript, and John Tenniel's drawings of a blonde girl in a pinafore are globally recognisable. Medium specificity continues to play a role amidst hybridity. Verbally, Elliott shows, Mock Turtle Soup works through absence: the reader must infer that a Mock Turtle is neither a turtle nor the calf whose flesh masquerades as turtle (the word "calf" being absent from the name of the dish). However, Tenniel's illustrations of the Mock Turtle as part calf and part turtle "undo his verbal emaciation, creating a plump, cacophonous semiotic surplus" of "hooves, shell, tail, ears, and flippers – a both/and rather than a neither/nor figure" (Elliott 2003: 214–15). Here, contrast between verbal language and image follows received wisdom: the verbal is identified with ambiguity and absence, while the image is highly specific in its plenitude.

When it comes to signifying age, however, the word is precise where visual culture is chaotic and multifarious. A number of visual and audiovisual adaptations have raised Alice's stated age, either within the diegesis or only extradiegetically, at the level of the actor. Furthermore, sexualisation of the protagonist seems tied to the growing perception in the twentieth century of Dodgson's paedophilia (cf. Leach 1999: 11–12, 35–43, 58–59; Brooker 2004: 20–22, 51–65). Visual culture plays a role here, as the photos Dodgson took of Alice Liddell and other young girls are frequently given as evidence of his lust. Whatever the 'truth' of Dodgson's desire, this has had a powerful influence on both close and loose adaptations. Sometimes the sexualisation occurs within the adult world. The numerous examples include video games (2000, 2011) in which Alice is 18 or 19 years old and porn films, such as *Alice in Wonderland: The X-Rated Musical Comedy* (Bud Townsend, 1976) and *Alice* (Erica McLean, 2010), in which actresses' nudity leaves no doubt of her post-pubescence. Such adaptations are unfaithful to the letter of Carroll's novella, but faithful to a perception of its sexual charge; in these cases, the visually present, adult female bodies are haunted by the child body Dodgson is believed to have loved. Thus, these adaptations contextually signify two age identities, one present and 'appropriate' as an object of adult male desire, the other absent and inappropriate.

Other (audio)visual adaptations have been more ambiguous in their processes of signification. The first film adaptation of *Alice in Wonderland* (Hepworth and Stow, 1903) featured a visibly post-pubescent May Clark, but Hepworth's insistence "that the images stay faithful to the drawings of Sir John Tenniel" (Brown 2014) encourages the audience to interpret Alice as a child. This follows the stage performance tradition, in which teenaged actresses played a character much younger. Isa Bowman was a "rather mature and curvy" (Leach 1999: 248) fourteen-year-old, and presumably Dodgson's mistress when he secured the stage role for her, although she signified a child much younger for Victorian audiences. Finally, graphic novels have exploited the medium's potential for visual abstraction in order to blur the boundaries between adult and child. Alan Moore and Melinda Gebbie's *Lost Girls* (2009) and *Miyuki-chan in Wonderland* (2001) by manga collective CLAMP both feature Alice in sexually explicit, sadomasochistic relationships. The verbal signifiers echo Carroll's specificity: in *Lost Girls,* we learn that Alice is fourteen when first coerced into sex, and *Miyuki-chan* tells us she is school-aged. Visually, however, the message is more mixed. The curves of the Alice figures in both graphic novels suggest post-pubescence at odds with the way the conventionally sylph-like limbs connote extreme youthfulness. Like Tenniel's Mock Turtle's semiotic surplus, these illustrated Alices exploit the difficulties of pinning down age identity based solely on visual cues; the illustrations present a Frankenstein-like collage of physical features associated with different age groups. The body of *Alice* texts is so enormous that most of us can think of additional examples that may complicate the case further; however, even these few demonstrate that the performance of age does not only confirm the determinacy and specificity of visuals against the potential indeterminacy of verbal signifiers.

Conclusion

The preceding case studies suggest that Bluestone was half right when he stated that visual and verbal – and, we could add, auditory, haptic, olfactory, and gustatory – materials of expression "both join and part company" (1957: 1) in a "set of fluid, but relatively homogeneous, conventions" (1957: 5). The conventions are fluid but not homogeneous, as becomes clear if we start from the perspective of individual signifiers rather than larger units of narration. The lack of homogeneity is particularly conspicuous if, following Hutcheon's lead, we ascribe real value to the community of interpretation in which those signifiers provoke such diverse responses. This lack of homogeneity does not mean that the concept of medium specificity is useless or outmoded. As Hermansson finds the fidelity discourse to be useful if approached as one tool among many, the concept of medium specificity can be very useful as long as axiomatic assumptions about what particular media 'can' and 'cannot' do are not allowed to obscure our observations. The question is: what *have* particular media done?

In this regard, digital media might offer a welcome addition to rather than a rebuttal of medium specificity. The digital turn is often equated with the end of medium specificity, as images, words, and sounds are all technically rendered by zeros and ones, but Doane (2007) and Charlotte Brunsdon (2012) both write that medium specificity still has traction as long as it is culturally contextualised. Brunsdon delineates institutional and textual similarities and differences between television and film to argue that "medium specificity may be more nationally specific than much contemporary theorisation suggests" (2012: 457). Doane similarly argues that media must be historically situated, writing that, although the digital offers a "*fantasy* of immateriality" (2007: 148, emphasis added), the "challenge of digital media, in its uses and theorisation, is that of resisting … the digital's subsumption within the dream of dematerialisation and the timelessness of information, returning history to representation and reviving the idea of a medium. Making it matter once more" (2007: 148). In short, for both Doane and Brunsdon, the modernist emphasis on ontology can give way to a postmodern emphasis on discourse without requiring us to abandon the concept of medium altogether. This is apt, since the digital also raises the question of scale; although we know that zeros and ones are 'behind' the images, words, and sounds, they are literally and sometimes metaphorically undetectable. Cultural discourse is the microscope that allows us to see this materiality, and conversely, the digital may inject a much-needed cultural dimension into medium specificity.

The question of medium specificity and materiality is particularly pressing when we consider instances of identity related to questions of social justice and accompanying questions of racism, gender, sexual normativity, and the age-appropriateness of sexual acts. When we look to an adaptation for a new cultural and social imaginary, a new way of conceiving of and performing the self, we are not only 'reading for the plot.' We are also looking for ways of engaging materiality. Again, the digital is provocative. Digital media offer materials of expression, from online avatars to digital aging and rejuvenating 'makeup,' that may seem entirely new. However, such technological determinism may not do justice to the variety of performed identities, either now or in the past, a variety that individual signifiers in adaptations can help us to see. If the pause and rewind functions demonstrate that Chatman's plot-oriented cinematic spectator is historically contingent (and probably never was so single-minded), perhaps digital mutability can show us that the constraints that seem to define media are, as Doane writes, "after all, not very constraining" (2007: 131).

Note

1 I do not mean that Hopkins 'is' white and Miller 'is' mixed race (or, below, that Tilda Swinton and Veruschka von Lehndorff 'are' female) in an essentialist manner; I only mean that these identities are consistently, discursively assigned to them in film reviews, interviews, and other aspects of public discourse.

Works cited

Abeel, E. (2003), "The Human Stain (film)," *Film Journal International* 106(10): 102.
Alice (2010), dir. Erica McLean. CalVista. DVD.
Alice in Wonderland (1903), dirs. Cecil M. Hepworth and Percy Stow. Hepworth & Co. Film.
Alice in Wonderland: The X-Rated Musical Comedy (1976), dir. Bud Townsend. Essex Pictures. DVD.
Alice: Madness Returns (2011), Spicy Horse/Electronic Arts. Video Game.
American McGee's Alice (2000), Rogue Entertainment/Electronic Arts. Video Game.
Andrew, D. (2000), "Adaptation," In: James Naremore (ed.), *Film Adaptation*, New Brunswick, NJ: Rutgers University Press, 28–37.
Austen, J. (2005), *Mansfield Park*, Cambridge, UK: Cambridge University Press.

La Bague (1946), dir. Alain Resnais. Unifrance.org: n. pag. Web [15 May 2016].

Bazin, A. (1967), *What Is Cinema?*, Gray, H. (trans.), Berkeley, CA: University of California Press.

Blow-Up (1966), dir. Michelangelo Antonioni. MGM. DVD.

Bluestone, G. (1957), *Novels Into Film*, Berkeley, CA: University of California Press.

Brecht, B. (1996), *Brecht on Theatre: The Development of an Aesthetic*, London: Methuen.

Brooker, W. (2004), *Alice's Adventures: Lewis Carroll in Popular Culture*, New York: Continuum.

Brown, S. (2014), "Alice in Wonderland (1903)," *Screenonline*: n. pag. Web [29 Sept. 2015].

Brunsdon, C. (2012), "'It's a film': Medium specificity as textual gesture in *Red Road* and *The Unloved*," *Journal of British Cinema and Television* 9(3): 457–79.

Butler, J. (1990), *Gender Trouble: Feminism and the Subversion of Identity*, New York: Routledge.

Butler, J. (1993), *Bodies That Matter: On the Discursive Limits of "Sex,"* New York: Routledge.

Carroll, L. (2009), *Alice's Adventures in Wonderland and Through the Looking-Glass and What Alice Found There*, New York: Oxford University Press.

Caudwell, S. (1982), *Thus Was Adonis Murdered*, London: Penguin.

Caudwell, S. (1986), *The Shortest Way to Hades*, London: Penguin.

Caudwell, S. (2002a), *The Sibyl in Her Grave*, London: Robinson.

Caudwell, S. (2002b), *The Sirens Sang of Murder*, London: Robinson.

Chatman, S. (1980), "What novels can do that films can't (and vice versa)," *Critical Inquiry* 7(1): 121–40.

CLAMP (2001), *Miyuki-Chan in Wonderland*, Tokyo: Kadokawa Shoten.

Cobb, S. (2010), "Adaptation, fidelity, and gendered discourses," *Adaptation* 4(1): 28–37.

Doane, M. A. (2007), "The indexical and the concept of medium specificity," *Differences: A Journal of Feminist Cultural Studies* 18(1): 128–52.

Dorian Gray im Spiegel der Boulevardpresse (1984), dir. Ulrike Ottinger. Ulrike Ottinger Filmproduction. VHS.

Dr. Mabuse: The Gambler (1922), dir. Fritz Lang. DVD.

Ebert. R. (2003), "The Human Stain," *Rogerebert.com* 31 Oct: n. pag. Web [31 July 2015]

Elliott, K. (2003), *Rethinking the Novel/Film Debate*, Cambridge, UK: Cambridge University Press.

González, E. S.-P. (2004), "'What phantasmagoria the mind is': Reading Virginia Woolf's parody of gender," *Atlantis* 26(2): 75–86.

Hermansson, C. (2015), "Flogging fidelity: In defense of the (un)dead horse," *Adaptation* 8(2): 147–60.

Hovey, J. (1997), "'Kissing a negress in the dark': Englishness as a masquerade in Woolf's *Orlando*," *PMLA* 112(3): 393–404.

The Human Stain (2003), dir. Robert Benton. Miramax. DVD.

Hutcheon, L. (2006), *A Theory of Adaptation*, London: Routledge.

Kuzniar, A. (2000), "Allegory, androgyny, anamorphosis: Ulrike Ottinger's *Dorian Gray*," In: *The Queer German Cinema*. Stanford, CA: Stanford University Press, 139–56.

Leach, K. (1999), *In the Shadow of the Dreamchild: A New Understanding of Lewis Carroll*, London: Peter Owen.

MacCabe, C. (2011), "Bazinian adaptation: *The Butcher Boy* as example," In: MacCabe, C., Murray, K., & Warner, R. (eds.) *True to the Spirit: Film Adaptation and the Question of Fidelity*, Oxford: Oxford University Press, 3–25.

de Man, P. (1997), "The resistance to theory," In: *The Resistance to Theory*, Theory and History of Literature, Vol. 33, Minneapolis, MN: University of Minnesota Press, 3–20.

Mansfield Park (1999), dir. Patricia Rozema. Miramax. DVD.

Metz, C. (1974), *Film Language: A Semiotics of the Cinema*, Taylor, M. (trans.), New York: Oxford University Press.

Moore, A. and Gebbie, M. (2009), *Lost Girls*, Mariette, GA: Top Shelf Productions.

Moorer, T. (2003), "'The Human Stain': A masterful audit of self-deception," *New York Amsterdam News* 94(45): 18.

Morpurgo, M. (2011), *War Horse*, London: Collins Education.

Mueller, R. (1985), "The mirror and the vamp," *New German Critique* 34: 176–93.

Mulvey, L. (2006), *Death 24x a Second: Stillness and the Moving Image*, London: Reaktion Books.

Orlando (1992), dir. Sally Potter. Sony. DVD.

Roth, P. (2001), *The Human Stain*, New York: Vintage International.

Silverman, K. (1983), *The Subject of Semiotics*, New York: Oxford University Press.

Stam, R. (2000), "Beyond fidelity: The dialogics of adaptation," In: Naremore, J. (ed.) *Film Adaptation*, New Brunswick, NJ: Rutgers University Press, 54–76.

Stratton, D. (2003), "The Human Stain," *Variety Movie Reviews* 280(43): 4–5.

The Testament of Dr. Mabuse (1933), dir. Fritz Lang. Nero Film. DVD.

The Thousand Eyes of Dr. Mabuse (1960), dir. Fritz Lang. DVD.
War Horse (2007), adapted by Nick Stafford. National Theatre. Stage Production.
Wilde, O. (2006), *The Picture of Dorian Gray*, Oxford: Oxford University Press.
Winterson, J. (1993), *Written on the Body*, New York: Knopf.
Woolf, V. (1973), *Orlando: A Biography*, Boston, MA: Mariner.

15

THE CRITIC-AS-ADAPTER

Josh Sabey and Keith Lawrence

> There remains much to learn of his relation to the world, and that it is not to be learned by any addition or subtraction or other comparison of known quantities, but is arrived at by untaught sallies of the spirit, by a continual self-recovery, and by entire humility … [T]here are far more excellent qualities in the student than preciseness and infallibility.
>
> *Ralph Waldo Emerson,* Nature

We begin with Emerson's words – partially because his flamboyant, intuitive thinking remains at odds with contemporary orthodoxy. The century-long consideration of the nature and role of adaptation criticism, together with the responsibilities or capacities of the adaptation critic, has traditionally centred on questions (and assumptions) about how the adaptation should be approached, carried out, evaluated, or appreciated. Early on, such questions quite naturally evoked critical responses grounded in fidelity, given that adaptation is, in contrast to other art forms, defined by its imitative and translational nature. Especially during the late 1980s through the early 2000s, as J.D. Connor suggests (2007), fidelity criticism was denigrated as overtly moralistic (Robert Stam), boring (Robert B. Ray and Dudley Andrew), culturally problematic (Christopher Orr), and naïve (Linda Hutcheon). Yet 'fidelity' persists to this day as an element of scholarly as well as popular adaptation criticism, even enjoying a mild resurgence during the past fifteen years as literary and cultural studies absorbed the influences of such movements as neo-formalism and constructivism and responded to new debates about evaluative aesthetics, anti-realism aesthetics, and intentionality (cf. Bender 2005: 80–98; Thomson-Jones 2009: 131-41). Somewhat rhetorically, Connor asks how fidelity-centred adaptation studies has "resisted such an onslaught – not simply of Hutcheon, Stam, Andrew, Orr, Naremore, Ray, and McFarlane, but also of Irigaray, Kristeva, Foucault, Derrida, Bakhtin, and Barthes?" He answers his own question in at least three ways: first, "whether something is faithful seems an easier question to settle than whether [something] is better"; second, in order to objectively justify her evaluation of a given adaptation, the lay person quite naturally looks to questions of fidelity; and finally, in a paraphrase of Kamilla Elliott, fidelity discourse reminds 'cinema of its literariness.'

As cultural theory has been integrated with traditional adaptation theory over the past forty years and more, the adaptation critic still engages in comparative analysis of subject adaptations and source texts – but such analysis has widened to include social, political, economic, philosophical, and other intersections or interactions among subject and source texts and the larger culture. The logicality of comparative discourse, whether centred in aesthetics or culture, does not, in the minds of many adaptation scholars, grant it a meaningful *raison d'être*; and the scramble to discover workable post-comparative or post-cultural approaches to adaptation proceeds unabated. More than a decade ago, Thomas Leitch (2003: 149) asserted in "Twelve Fallacies in Contemporary Adaptation Theory" that because adaptation studies lacks a satisfactorily nuanced "presiding poetic," the discipline is not moving forward with appropriate theoretical rigour. Three years before Leitch, Robert Ray declared:

> Without the benefit of a presiding poetic, film and literature scholars could only persist in asking about individual movies the same unproductive layman's question (How does the film compare with the book?) getting the same unproductive answer (The book is better).
>
> *Ray 2000: 44*

Attempting to create new conversations, Linda Hutcheon (2006: 170–1) turns adaptation criticism in on itself to incorporate "academic criticism and reviews of a work" within what she calls a "continuum" of adaptations – "versions, revisions," "revisitations," "re-productions," "translations," "transcriptions," "condensations," "spin-offs," and "expansions." To see the critic as an artist or the artist as a critic is an old trick.[1] But to see the critic as adapter – as Hutcheon suggests in passing – is not, and our purpose here is to show that, in deliberately assuming the title of the adapter, the adaptation critic potentially accesses a richly expansive understanding of adaptation studies and a vision of the liberating "artistic scholarship" we call criticism-as-adaptation.

What or who, then, is the critic-as-adapter (CAA)? She is, most simply, a performance artist who defines or establishes the nature of her own artistic work as she critically adapts the art of others. As a performance artist, she writes – or performs – for an audience; she is, therefore, amenable to principles undergirding or governing performed art. Pioneering performance theorist Richard Schechner (1988: xi) asserts that there are only two kinds of performance. The first – represented by a soccer match, for example, or a jam session – might be called 'spontaneous' performance, where both details and outcomes are unknown and unpredictable to performers as well as observers, unfolding live. The second – typified by a film, stage play, or Beethoven symphony – is a set enactment of a representational text, a text embodying its creator's impression or understanding of (a) a select glimpse of the real world and (b) how the performance itself will unfold and what each performer will do. At least twice removed from reality, this might be called 'adaptational' performance. Naming his performance categories, respectively, "*lila* – sports, play" and "*maya* – illusion," Schechner (1988: xi) insists that "so is all life *lila* and *maya*" and that all performance is, therefore, "an illusion of an illusion," or an adaptation of an adaptation. In light of Schechner's argument, any critic is by definition an adapter, transforming her perception of a given subject text (which, in turn, is a reflection or adaptation) into a performed narrative that provides an audience access to the subject text and to her impressions of it.

Perhaps the clearest personifications of the CAA can be seen in non-traditional critic–scholars like Catherine Grant, whose cutting-edge 'film essays' may be almost entirely comprised of fragments from the cinematic works she critiques – scenes mashed together, frames spliced or otherwise manipulated, projection speed changed. Grant (2011, 2013) knowingly blurs lines between scholarship, prosumerism, and artistry, defining her work as a "creative critical" means of delivering "haptic and sensuous criticism," wordless criticism, and thesis-less criticism.[2] Quite predictably, we argue that Grant's essays are adaptations – not merely because, in each, Grant literally adapts a film text as its own criticism, but because she sutures her own artistic and critical vision to it.

Viewing adaptation criticism as itself a form of adaptation – and thus of art – may profitably encourage the critic-as-adapter to consider his potential roles as he critically adapts a given text – written, filmed, or performed. In recognising that he is, among other things, a collaborator, creator, and experimenter alongside the artist–adapter whose work he critiques, the CAA may employ such tools as spontaneity, doubt, and humility to create new, intriguing critical forms.

The CAA as collaborator

Pioneering French film critic Christian Metz famously encouraged previous generations away from "descriptive theory" and toward analysis of "filmic fact" (1974: 12)—toward a semiotics of film critique marked by critical objectivity, precision, and "systemisation" of film structure and content (1974: 14–21, 287). Metz tended to romanticise the scholarly critic as a kind of academic equivalent of the title character of *Shane* (1953), the reticent stranger who employs objective, hard-nosed common-sense to create order from chaos. Pitting himself against artwork if not artist, the objective critic adopted a posture of distrust, distance, and competition vis-à-vis his subject and his onlookers. At his best, such a critic successfully stared down his film–text subject and, by virtue of his disinterested, outsider status and steely-eyed divination, secured and then revealed to an awestruck but seemingly irrelevant audience an untainted perspective of the subject's methods and message.

Some thirty-five years after Metz's pronouncement, George Toles (2010: 49) would mock the notion that "tenderness itself must be opposed and rooted out so that the power balance can shift back to the beholder." From the perspective of the critic-as-adapter, it is virtually impossible (and highly undesirable) to separate the critic and criticism from the adapter and adaptation. To even consider doing so is to ignore the fact that the traditional critic has long shared elements of the artistic adapter's role. Both the adapter and the traditional critic effect original translations of source texts. And in creating their translations, both selectively use source materials; both employ the tools of their respective disciplines to manipulate, excise, enhance, or reconfigure source and original elements as they develop coherent 'narratives' conveying desired themes, feelings, perspectives, or ideals. And depending on personal allegiance to a variety of fidelities – to source material, to theoretical or aesthetic ideals, to the broadly defined languages and conventions of their respective media, to audience or consumer expectations, to whomever finances their respective endeavours – both the artist–adapter and traditional critic may be similarly restrained from expressing themselves as fully as they otherwise might.

But connections between the traditional adaptation critic and the artist–adapter go beyond shared roles. The adaptation critic has always been primarily an adapter; foundational to the criticism she writes are the intersecting internal adaptations she creates beforehand, together with the ideals or standards they suggest. That is, as she reads a novel or story or as she experiences another kind of source text, she creates an internal adaptation of it in her mind. And when she experiences a film or other adaptation of a familiar source text, she creates another internal adaptation, one that may be more consciously constructed than was her internal adaptation of the source text, and one that is likely even more idealised. This adaptation will borrow from the film all that she liked in it, but elements perceived as flawed or unacceptable will be excised or replaced. Understandably, her idealised adaptation of the film will be influenced by her internal adaptation of the relevant source text – even when she may prefer certain elements of the film to corresponding source-adaptation elements. Eventually, her written critique of the film will constitute yet another adaptation, a narrative performance that – regardless of her larger focus or purposes – implicitly reveals the interplay of the film adaptation, the critic's internal adaptations, and (possibly) the source text.

Perceiving the adaptation critic as himself an adapter not only encourages but necessitates connectivity between critic and subject text. Such connectivity may logically be expanded to include the source text as well as the makers of the source and subject texts. Indeed, as the CAA shapes his critique of the subject text, he appropriately sees himself as co–collaborator with the makers of that text. Like them, he stretches toward an ideal: he wants to realise a version of the source text that is enlivened by the vitality and meaning attached to valued personal experiences with the text. Like them, he seeks to measure the success of the subject text as a realisation of the ideal. These are vital connections, but the collaborative role of the CAA potentially extends further still.

Assuming that, at least on one level, art is a medium of communication and that an artist may communicate with an audience through her art, and assuming a highly qualified validity of the well-known declaration of Knapp and Michaels (1982: 724) that "the meaning of a text is simply identical to the author's intended meaning," the CAA may rely on the subject text and pertinent resources to hazard suppositions about the vision of the makers of the text – their responses to and collaborative vision of the source text, their manipulation of source–text elements, their formulation of the subject text as both an adapted and an independent work. Having made such suppositions, the CAA may effectually collaborate with the subject-text makers to learn how the text works – independently as well as in aesthetic, cultural, historical, and other contexts – and how and why it accrues or communicates meaning.

Devoid of Roy Andersson's statement at the 2014 Venice Film Festival that his film *A Pigeon Sat on a Branch Reflecting on Existence* (2014) "had been inspired by Italian director Vittorio De Sica's 1948 film *Bicycle Thieves*" (*BBC News*), the CAA might not recognise the highly original and achingly funny *A Pigeon* as an adaptation, much less an adaptation of De Sica's dark film. Suspicious yet interested, the CAA learns what Andersson and his col-laborators have said about their film; she is especially intrigued by an interview Andersson granted *Svenska Dagbladet*, a Stockholm newspaper, in January 2013 – more than a year before he finished filming *A Pigeon*. In the interview, Andersson says his film's title was inspired by Pieter Brugel's fifteenth-century painting, *The Hunters in the Snow*, which shows birds in a tree watching the men below them. "I see the birds looking at the village men," Andersson says, "and I wonder what the birds are thinking. My title is a kind of euphemism, then, for 'What are we *really* doing?'" (Spektra 2013).[3]

As she more carefully considers *A Pigeon* alongside *Bicycle Thieves*, the CAA notes the episodic structure of each, a structure underscoring the sense of futility and disconnection in the respec-tive films. She perceives the humour of *A Pigeon* as counterpoint to the grim inevitability of *Bicycle Thieves* – but each as the ironic signifier of emptiness and waste, of the human tendency to pursue wrong objectives. She determines to write a critique narrating her experiences with the two films, and she wants to involve Andersson himself. She locates an email address; Andersson generously allows her limited access to a portion of his screenplay for *A Pigeon*, sends photos, tells her what he most admires about De Sica's film. All these she includes in her narrative, now a collaboration; Andersson himself writes a brief response to their text, which they include as an epilogue. Showing her own movement from suspicion to awareness and Andersson's from auteur to critic–artist, their narrative is a sophisticated celebration of two critics' capacities to learn and to be surprised – and to engender similar learning and surprise in their readers.

There are countless ways the CAA might collaborate. For example, he might initiate a col-laborative critique through a blog posting, a posting wherein he introduces his subject text, an adaptation of a well-known source text. He invites his online friends or followers to experience the subject text and then, by a set deadline, to post their responses to it – with the caveat that they may not refer to the source text in their responses. In a follow–up blog after the response deadline, the CAA asks his respondents to assess their earlier experience. Were they frustrated or energised by the assignment? What did they focus on – and why? Then, eschewing the kind of statistical analysis implied by the project and concentrating instead on the "revised subject text" implied by his blogs and the responses they generated, the CAA finds creative, engaging ways to highlight primary modes or concerns of popular adaptation criticism.

Or perhaps the CAA wants to re-create a stylised ten-minute segment from a recent film adaptation shown at a local film festival, and then to learn whether her re-creation forces recon-sideration of parallel moments in the subject film as well as the source text. Because she has connections, she invites the film's screenwriter and the original text author to assist her in plan-ning and creating the segment; they also collaborate on an accompanying narrative about the significance of the project.

Collaborative criticism-as-adaptation may originate among scholars in a variety of disciplines – music, film, economics, culture, literature, politics, or science – or among a variety of trained professionals – screenwriters, editors, bloggers, photographers, cinematographers, directors, pro-ducers, marketers, musicians, actors, writers, artists, or popular critics. This fact alone forever separates the work of the CAA from traditional critique. Indeed, there assuredly will be advances in adaptation theory and practice as the critical field itself is expanded and deepened.

The CAA as creator

A work of adaptation criticism is seldom praised for its artistry or style, for its plurality of mean-ings or the way it resists a single interpretation. Given that scholars and journals alike champion the clear and rigorous claim that sets up all that follows – neatly ordered, cleanly constructed, and cleverly stated – how is the CAA to address films like Terrence Malik's *Tree of Life* (2011; adapted autobiography) or Ingmar Bergman's *The Seventh Seal* (1957)? How does she create a single argument about either film, much less assign the film itself a focused 'meaning?' More crucially, why would she want to do so, knowing that neither film could continue to live, intact and recognisable, enduring such bondage?

Here is the same problem stated in a slightly different way. In his rather remarkable close reading of 'fragments' from William Wellman's *Other Men's Women* (1931), George Toles (2010. 165) includes this description of a scene from very late in the film. Bill, the guilt-ridden pro-tagonist, has just received an unexpectedly warm invitation from Lily, the widowed wife of his deceased best friend, Jack:

> We then watch from behind [Bill] as he clambers to the top of the carriage and pro-ceeds to run the entire length of the train, a tiny figure almost vanishing in the distance. Two-thirds of the way through his dash, he pauses for an instant, deep in the frame, raises his arms, and leaps in jubilation. He does not turn back to look at Lily, but we feel he is aware, somehow, that she might be looking at him. … [T]his concluding image of the train traveling outward and elsewhere seems to link Bill's precarious, uninhibited leap into life with the dreamy self-containment of Jack's dark passage to death. The beckoning future is viewed from behind, available in a swift, telling glimpse, but still visibly exceeding our grasp; we can't slow the future down, make it linger and attain better focus. It is all about transitory proximity in the midst of rushing separa-tion: moving toward something that is hauntingly in back of you, awaiting your arrival.

This is focused, evocative writing, and Toles skilfully implies parallels between Bill's emotional and physical 'journeys.' An allusive adaptation, the passage remakes Wellman's original scene as a devotional text, a paean to human generosity and to hope itself. And in the sentence that concludes his essay, Toles (2010: 166) effectually remakes himself as a neophyte CAA who writes to "unsettle the process of knowing." The problem is that, in the balance of his essay, Toles remains a largely traditional evaluator – too focused, too narrowly driven by his own aim to situate "'lost' fragments" in "communion with each other" and to "conjure up alternative homes for them" while showing "perfect movie fusion" (Toles 2010: 161, 166). The CAA recognises that, even in isolation, the scene that Toles describes is suggestive of far more than Toles allows. On the one hand, Bill's giddy run along the top of a moving train suggests his having surmounted the very things that killed his friend Jack; on the other, the scene underscores Bill's fragility and inconsequentiality. And regardless of what future Bill is plunging toward, he is – at least for the present – going there without Lily, who is rapidly receding into a past as blurred as his own future.

Maude Fulton wrote the original screenplay for *Other Men's Women*, but she may have been influenced by the real-life love triangle involving Harry Thaw, Stanford White, and Evelyn Nesbit, Thaw's wife. When Fulton was 25, and on the night of her Broadway debut in *Mam'zelle Champagne* (1906), White and the partially deranged Thaw were in the audience, seated at separate tables near the stage. Without warning, Thaw jumped to his feet during the show's finale, scrambled toward White's table, and shot him point-blank in the face, declaring: "You've ruined my life!" (Nevius and Nevius 2009: 197). Undoubtedly scarred by this scene that surreally played itself out in front of her, and by the lurid 'Trial of the Century'[4] that followed, Fulton – as she worked on her screenplay some twenty-five years later –may have consciously changed the character of the triangle's husband to make him generous and self-sacrificing, thereby reshaping scandal-sheet catastrophe as redemptive tragedy. And as she refined the details of her script, she may have been helped by the triangle in the William Faulkner short story, "Honor" (1930), which – like her own – also had a generous husband and a conscience-stricken friend at its centre.

If the cinephile is to attend to "rescuing fragments" – as Toles advises in the title of his essay – and if he is to do so as a critic-as-adapter, he must find ways to narrate multiplicities. Figuratively, he will assert (as did the possessed man to Christ), "My name is Legion, for we are many" (Mark 5:9). In creating a response to Toles' fragment from *Other Men's Women*, the CAA might begin somewhat traditionally, deftly summarising the fragment from various perspectives and playing the collective possibilities against the film's larger plot. From that point forward, however, the CAA creates his own form and voice as he broadens the reach of the fragment/film as adaptation. Here is a rough, unstructured outline of such a presentation:

> *Thaw-White-Nesbit reconfigured as redemption. Faulkner and dishonour. Engineers versus pilots, forgiveness versus revenge, humilityversus fear. The prodigal son and his brother; Abel and Cain. Repeated sorrows; recurrent joys. Life shot through with irony; resilient faith. The taste of jealousy, the shape of honor, the sound of hope.*

Matthew Wickman (2016) observes that the academic projects garnering the most acclaim tend to be those that "conjure an air of ultimacy: we never say of them that they are merely 'interesting.' This makes them more important – and, occasionally, insufferable." Wickman (2016) thus warns that "unless an air of ingenuity leavens [our projects], unless they bear an aesthetic as well as an ethical appeal, they can become sanctimonious, bloviating." The traditional critical argument is an anatomist's knife, opening the body of the text and exposing its workings, its mysteries. The critic-as-adapter, whose argument is a canvas on which she paints the narrative of her sapience, is concerned with the living soul – with the body's beauty and secrets, certainly, but also with its actions and interactions, its receptivity and influence, its vitality, reality, and potential.

Well-written criticism-as-adaptation brings into close proximity multiple and divergent elements, scenes, perspectives, texts, or impressions, enabling them to speak meaningfully together. While rigour will always remain a crucial standard by which academic criticism is measured, according it pre-eminence underserves intuition, experimentation, and exploration. By de-privileging rigour, especially that rigour attached to single-strain or correct-versus-incorrect interpretations or argumentation, the CAA naturally adopts a non-competitive, playful, and fulfilling relationship with the texts she critiques. She is far more exploratory and tentative in her thinking and writing than she is definitive or declaratory. Her criticism-as-adaptation is rich and incorporative; it may also be refreshingly disruptive, inchoate, and inconclusive.

Because adaptation itself is defined by the reworking of form, the 'creator CAA' will allow his critiques to assume new configurations or appearances, to be expressed through different genres – films, screenplays, blogs, paintings, fan-page or social-media postings, photo-essays, songs, narratives, poems, videos, or plays. He will embrace modification and change as inherent to adaptation itself, remembering that the root of the term, *ad aptus*, means 'to fit;' thus, he will use comparative analysis not as the basis for cut-and-dried value judgments but as the genesis of understanding and appreciation. While recognising that "the old verities and truths" will never become irrelevant, that "the problems of the human heart in conflict with itself alone can make good [art]" (Faulkner 1950), he will eschew the easy answer; indeed, he will value above any answer "the straightforward question worthy of the attention of a wise one" (Morrison 1993). His most significant creative impulses may be stirred by the messiest, most obscure, or least extractable elements of his texts.

If the CAA justifiably believes, for example, that the language of Jeremy Podeswa's *Fugitive Pieces* (2007) – adapted from Canadian poet Anne Michaels' novel of the same title (1996) – is its most exquisitely necessary element, she will confront the question of how one uses personal language to talk about the language of others. She may decide one does not. Instead, others' language must do its own speaking, and she can only interweave her words,[5] telling how it was

to sit in arthouse dankness, quiet, pre-titles scene unfolding over sharp memories of
novel's opening lines, toes pulled tightly in as
This isn't a joke, Jakob, listen. Don't open the door to anyone. Promise! And be quiet
there, Jakob's noiseless self-excavation doubling his wallpaper birth-from-death
NO ONE IS BORN JUST ONCE…IF YOU'RE LUCKY, YOU'LL EMERGE AGAIN IN SOMEONE'S ARMS
reduced to mere political risk
Shut up! Shut up! You're coming with us. You'll get us killed. You understand?
and dwarfed, in turn, by unsaid guilt, by unrelenting haunting of family no more
BUT WORSE THAN THOSE SOUNDS WAS THAT I COULDN'T REMEMBER HEARING BELLA
AT ALL.
FILLED WITH HER SILENCE
yet ever-present, flitting hovering brooding possessing
To live with ghosts requires solitude.
I know only fragments of what Athos contained
to prove that mortals do not dispel ghosts, but, bound, silent, only endure
MY DEEPEST STORY MUST BE TOLD BY A BLIND MAN, A PRISONER OF SOUND
to endure;
and Spielberg, Benigni, and eventually dozens more never can make voice enough
for millions silenced
While I was learning Greek and English, learning geology and poetry, Jews were filling the
corners and cracks of Europe …
…

In his ground-breaking article, "A Certain Tendency of the French Cinema" (1954), François Truffaut (2008: 11) noted the inclination of certain of his fellow filmmakers working in adaptation to divide a literary source text into "filmable scenes and unfilmable scenes" – and in the latter case, "to invent *equivalent* scenes, that is to say, scenes as the novel's author would have written them for the cinema." Truffaut (2008: 11) continues, "What annoys me about this famous process of equivalence is that I'm not at all certain that a novel contains unfilmable scenes, and even less certain that these scenes, decreed unfilmable, would be so for everyone." A paraphrase of this declaration is, we believe, a worthy cry for the critic-as-adapter to take up, especially when she embraces the creator's role: 'I'm not at all certain that a film contains unwriteable elements, and even less certain that such elements, decreed unwriteable, must be so for me.'

The CAA as experimenter

Genuine art is perceived as experimental; traditional criticism – bound to expectations of form and content – is not. But if, as we have argued, the critic-as-adapter is by definition an artist, it follows logically that her writing also can and should be experimental. In contrast to the 'creator,' who, however inspired, works deliberately and from careful plans toward known or expected outcomes, the 'experimenter' courts spontaneity, flirts with doubt, and teases failure. Avant-garde film critic Catherine Grant (2013) – referenced in our introduction – describes her work as "mad poetry," a term she borrows from Adrian Martin, and says her methodology depends on "a kind of intense theorizing, something that's coming from somewhere that feels less consciously controlled than the academic rigor that we've [held to] in the past" (ibid.). She goes on to say that she makes artistic decisions on hunches spontaneously and "without knowing why," (ibid.) guided by intuition and, in no small part, the desire to be different.

There are experimentalist parallels in the work of Manu Yáñez. In the introduction to his audiovisual essay, "Thought, Action and Imagination," Yáñez (2014: ¶ 3) celebrates the fact that the film essay is still in "embryonic development," and still "very open to the formulation of questions and hypothesis, as well as fertile relation to its possibilities." Defining his own work as "marked by an air of uncertainty, and by a great number of intuitions," Yáñez (2014: ¶ 3, 10) asserts that "doubts have always been present, circling the minds of those critics who resist the siren call of ultra-dogmatic criticism." His claim is simple enough: doubt is not the by-product of experimentation, but one of its crucial tools. Certainly, then, the critic-as-adapter should doubt tradition and, "without fearing the rupture of some established model" (Yáñez 2014: ¶ 17), liberally embrace the unknown. But she should also doubt herself, never becoming safely lodged in her own way of seeing and doing. Each time she begins to write, the CAA may remind herself of the original meaning of 'essay,' recognising that her best work potentially begins as an attempt, as experimentation.

In *The Learning Paradigm College* (2003), John Tagg (2003: 54) asserts that learning, "from its very beginnings, entails a process of courting failure and learning to play with it." He continues:

> None of us would be walking or talking – and certainly not reading, writing, and calculating – had we not embarked at an early age on the systematic project of doing things that were definitively impossible for us and repeatedly failing at them for an extended period of time. … Toddlers … embrace learning goals with gusto. And this principle of trying out the currently unachievable as an intrinsically interesting endeavor is a principle that drives successful enterprise at every stage of life.

Discovering first-hand what does not work may be fully as important as pioneering what does, especially as one expands the boundaries and promise of one's field. And intelligent failure holds far more significance than shallow success. Gus Van Sant's *Psycho* (1998), a shot-for-shot remake of Hitchcock's original masterpiece (1960), was widely denounced as an artistic failure. One of the many popular critics to pan the film was Roger Ebert (1998), yet he nevertheless called it an "invaluable experiment in the theory of cinema." Without it, he declared, we might never have known that "a shot-by-shot remake is pointless" – or that "genius apparently resides between or beneath the shots, or in chemistry that cannot be timed or counted." Ebert's implicit argument, of course, is that devoid of artists willing to experiment, to take crucial risks, there would be no "invaluable" platform from which to film or paint or write or dance – and hence there would be no genuine art.

When working with an obviously experimental text like Spike Jonze's *Adaptation.* (2002), the critic-as-adapter might naturally rely on an intuitive approach to scholarship, an approach indirectly bound to spontaneity, doubt, and risk. He might question how (for example) the dark romanticism of this quintessential postmodernist film is bound to its tone, pacing, and style – or how Jonze adapts his own metatextual screenplay to create life-giving metacinematic dissonance in the film's characters and structure. The more demanding challenge, perhaps, is to apply an intuitive approach to an established film – a film like *Citizen Kane* (1941), for instance, that continues to enjoy the loftiest of reputations. Orson Welles's first masterwork has a popular reputation as a "mystery drama" (Wikipedia), as "Drama, Mystery" (IMDb), as the "epic tale of a publishing tycoon's rise and fall" (Rotten Tomatoes, Critics' Consensus). Leonard Maltin (2010) asserts that *Citizen Kane* "broke all the rules and invented some new ones." Forty years earlier, Pauline Kael (1971) infamously called it a "*shallow* masterpiece" that "manages to create something aesthetically exciting and durable out of the playfulness of American muckraking satire." Through the years, however, the film has accrued a reputation expressed through very different adjectives (joining 'rule-breaking' and 'inventive') – gritty, profound, hard-edged, true, shocking, devastating, elemental.

The CAA might well wonder how to engage an established film on an experimental level or, after nearly eight decades of ongoing discussion and analysis, what there possibly might be left of the film to experiment with. But she has long been intrigued by what she sees as the film's blatant sentimentality, and she has learned from her research that not many fellow critics address this element of the film. If they do, they call it by another name (like 'mystery' or 'twist' or 'epic' or 'drama') that automatically makes it a non-issue. The CAA is pleased to find an incidental reference to *Citizen Kane* in a book by Jerome McGann (1996: 119) on lesser-known British Romantic poets, a reference describing *Kane* as "a nostalgic film—in terms of Schiller's famous naïve/sentimental formulation, a sentimental film." McGann even suggests that the film's "sentimentality is directed at 'the sentimental' itself" (ibid.). Yet by the end of his paragraph, McGann backpedals, arguing that Welles' treatment of the "psycho-social structure" of "unspoiled childhood represented as a kind of dream memory" is "at once sympathetic and critically reflective" – to the end that "the film desires a 'naïve,' an entirely sympathetic absorption in the Rosebud world; but its self-conscious awareness of the fantastic character of that world blocks any simple, naïve engagement" (ibid.).

Wishing that McGann had stopped halfway through his argument, the CAA is persuaded that the film's sentimentality is not erased through 'critical reflection,' 'self-consciousness,' 'inventiveness,' 'grittiness,' 'hard edges,' or anything else. It is, she believes, far more than the emotional crux of the film. The casual perusal of sponsored and unsponsored film blogs has led her to believe that, for many lay critics, the final minutes of the film determine its greatness. Before that point, the film is very good indeed – but, for such critics, the revelation of "Rosebud" – whether as 'twist,' surprise, shock, or heart-wrencher – makes the film great. On one level, the CAA considers the image of Rosebud, over the course of the film, to be milked drier than Jack-in-the-Beanstalk's cow. She wonders how the sentimentality of the film escapes notice or, perhaps more accurately, why it works. On another level, she wonders not simply how it works, but why it may be necessary to the film. For it to work, she realises as she moves into the project, especially for the mature or jaded critic, it cannot call attention to itself, and it must embody more than simple innocence – the pure, sweet joy of childhood untouched by the drives that clutter adult life: power, sex, money, influence, achievement, escape. She hypothesises that its value to the film may have something to do with moral compasses and sympathising with a character otherwise undeserving of sympathy, but she determines to put rationales on hold until she has set up and carried out her experiment.

The role of 'CAA as experimenter' carries inherent dangers. Perhaps the most sinister of these is the inclination or temptation to shock or surprise readers – a phenomenon related to adolescent experiences in high school chemistry lectures where the best experiments always ended in explosions. As Wickman (2016) has warned, the "attraction" of less desirable academic projects "primarily consists in their power to surprise." Rather than attempting to surprise, the wise CAA will instead be surprised by the texts he encounters, by the texts he responds to. Catherine Grant (2013) exemplifies this attribute perfectly when she confesses in one of her films that "the feeling I engender in the video is the feeling I had watching it."[6] While she uses the word "video" to describe her own mashup and "it" to refer to the subject film, she creates dynamic confusion by apparently collapsing the meaning of subject, object, and adaptation through what might be called a syntactical palimpsest. Simultaneously, she measures her original experience by its capacity to surprise her, and she continues to be surprised through the moment of sharing her experience with readers.

Because experimentation preserves opinions and beliefs as nascent and nebulous, as "easily reversible" (Durgnat 2006), to experiment is, in very real ways, to yield control. This can be an uncomfortable fact for the traditional critic, who 'wins' by making rock-solid cases for his focused arguments, his brilliant readings unlocking 'meaning.' The CAA, in contrast, is energised by openendedness. The more meanings, the merrier, especially when they are in richly resonant conversations with one another and surrounding elements. The yielding of control may be especially important to the CAA exploring a subject text that resists paraphrase and deduction, a text like Christopher Nolan's *Memento* (2000). A quick survey of blogs about the film shows that at least nine of ten bloggers are fascinated by the film's 'meaning,' by what really happens, by the 'true story.' Only about one in ten appreciates the film for its refusal to mean. The CAA initially considers acting as 'creator' – and learning why 'truth' is the object of so many fans of the film.

But he is also intrigued by ways the film does and does not fit the American horror genre. He considers doing a kind of comparison with a favourite slasher film, but then begins to consider other questions: How is horror generated in the film? How is it experienced by characters in the film? By the viewer? To what ends? He decides to experiment with tentative answers to such questions, approaching *Memento* as an after-postmodernist[7] exploration of horror. He begins by considering the kinds of horror faced by Leonard, the film's protagonist. Most obviously, the horror of his wife's rape and murder compels Leonard to avenge himself on those responsible. On another level, Leonard's horror is the possibility that he killed his own wife and invented an alternate history to cloak his guilt. On yet another, his anterograde amnesia is his horror – his not remembering or knowing, his existence as a non-agent. These all seem fairly obvious claims, however, and so he will not write about them. But he uses them as paths to other considerations.

As he begins considering the kinds of horror experienced by the viewer, beginning with visceral responses to the violent, chaotic events of the film, he realises this is the material that will yield results. Ironic cultural implications of his own visceral responses lead the CAA to perceive a darker horror: because the viewer never receives a reliable backstory, it makes little difference that the events of the main story are, by the end of the viewing experience, more or less clear. The larger rationale of the narrative, including its logic and ethics, is both obscure and relative. This leads the CAA to recognise an even more compelling horror, one deriving from the facts that, regardless of backstory or rationale, the viewer naturally submits to Leonard in sympathy, ethics, and purpose – and that, at film's end, the viewer (unlike Leonard) cannot plead insanity. The CAA devotes most of his response to putting into play the resonances and implications of this last claim.

The collaborator and creator proceed with some degree of knowledge, foresight, and fore-recognition, with a tentative plan. Within the critic-as-adapter herself, these roles are balanced by the experimenter – who proceeds unknowingly and humbly. When properly understood, knowledge and doubt cannot only coexist, but can afford the CAA buoyancy and flexibility in her research and thinking, enabling her to plumb the depths of texts as well as to map their sometimes intractable surfaces. Together with spontaneity and the risk of failure, critical doubt draws her attention to what is uncertain, untranslatable, and overlooked, and yet significant.

By focusing on criticism-as-adaptation, the critic is liberated to adapt new subjects, forms, elements, and ideologies. In doing so with both critical and artistic integrity, the CAA will be engaged, as both observer and participant, in significant conversation with her subjects and with adaptation itself. The critic-as-adapter thus writes at the edge of understanding – always projecting herself and her expectations into the darkness, but always surprised by and enamoured of the irreducible texts before her.

Notes

1 This discussion is traceable back to Immanuel Kant's aestheticism as popularised by Oscar Wilde in "The Critic as Artist" and, more recently, of George Bluestone's (1968: 6–20) groundbreaking assessment of the critic's relation to the novelist or the film adapter, Roland Barthes' (1981: 78–9, 96–7) striking erasure of boundaries separating criticism from art, and Robert Stam's (2005: 8) description of film adaptation as "a form of criticism."
2 Grant's self-describing phrase "creative critical" comes from "How Long is a Piece of String?" (2013); "haptic and sensuous criticism" is from "Touching the Film Object?" (2011). The other descriptors in this sentence are paraphrases of artistic ideals covered in both texts.
3 The webpage in question was accessed using Google Translate (Swedish to English); the English here is corrected slightly from the Google 'translation.'
4 So designated by contemporary news reporters across the United States.
5 In this opening segment of a hypothetical critique-as-adaptation, the CAA's words are in standard typeface; these words are interwoven with speeches from Podeswa's screenplay (italics) and lines from Anne Michaels' original novel (all-caps).
6 This line comes from a segment of Grant's mashup "How Long Is a Piece of String?" (2013)—the segment labeled "Video 8: Extract from *Unsentimental Education*"—and is part of a written headnote to the incorporated video extract.
7 Given that deconstructionist analysis yields no historical understanding whatsoever.

Works cited

Adaptation. (2002), dir. Spike Jonze. Columbia. DVD.
Barthes, R. (1981) *Camera Lucida: Reflections on Photography*, Trans. Richard Howard, New York: Hill and Wang.
BBC News (2014) "Roy Andersson film scoops Venice Golden Lion award," in *BBC News Entertainment & Arts*. Online. Available at: http://www.bbc.com/news/entertainment-arts-29098103 (accessed 17 March 2016).

Bender, J. (2005) "Aesthetic realism 2," in J. Levinson (ed.) *Oxford Handbook of Aesthetics*, New York: Oxford University Press, 80–98.

Bicycle Thieves (2007), dir. Vittorio De Sica. Ente Nazionale Industrie Cinematografiche, 1948, Criterion. DVD.

Bluestone, G. (1968) *Novels into Film*, Berkeley, CA: University of California Press.

Connor, J.D. (2007) "The persistence of fidelity: Adaptation theory today," *M/C Journal*, 10(5): n.p.

Durgnat, R. (2006) "Culture always is a fog," *Rouge*, 8. Online. Available at: http://www.rouge.com.au/8/interview.html (accessed April 7, 2016).

Ebert, R. (1998) "Psycho," in *RogerEbert.com*. Online. Available at: http://www.rogerebert.com/reviews/great-movie-psycho-1960 (accessed 10 October 2015).

Emerson, R.W. (1849) *Nature*, New York: J. Munroe.

Faulkner, W. (1950) "William Faulkner – Banquet Speech," Stockholm. Online. Available at: http://www.nobelprize.org/nobel_prizes/literature/laureates/1949/faulkner-speech.html (accessed 5 November 2015).

Fugitive Pieces (2007), dir. Jeremy Podeswa. Maximum Film. DVD.

Grant, C. (2011) "Touching the film object? Notes on the 'haptic' in videographical film studies." *FilmAnalytical*. Online. Available at: http://filmanalytical.blogspot.de/2011/08/touching-film-object-notes-on-haptic-in.html (accessed 4 November 2015).

Grant, C. (2013) "How long is a piece of string? On the practice, scope, and value of videographic film studies and criticism," Audiovisual Essay Conference, Frankfurt Filmmuseum/Goethe University, November 23–4, 2013. Conference presentation, Goethe University. Frankfurt Filmmuseum, Frankfurt am Main, Germany. Online. Available at: http://reframe.sussex.ac.uk/audiovisualessay/frankfurt-papers/catherine-grant/ (accessed 21 March 2017).

Hutcheon, L. (2006) *A Theory of Adaptation*, New York: Routledge.

Kael, P. (1971) "Raising Kane – I," *New Yorker*, 20 February: n. pag.. Online. Available at: http://www.newyorker.com/magazine/1971/02/20/raising-kane-i (accessed 20 July 2015).

Knapp, S. and W.B. Michaels (1982) "Against theory," *Critical Inquiry* 8(4): 723–42.

Leitch, T. (2003) "Twelve fallacies in contemporary adaptation theory," *Criticism* 45(2): 149–71.

McGann, J.J. (1996) *The Poetics of Sensibility: A Revolution in Literary Style*, Oxford: Oxford University Press.

Maltin, L. (2010) "Leonard Maltin movie review, *Citizen Kane*," TCM. Online. Available at: http://www.tcm.com/ tcmdb/title/89/Citizen-Kane/ (accessed 10 May 2015).

Memento (2000), dir. Christopher Nolan. Summit Entertainment. DVD.

Metz, C. (1974) *Language and Cinema*, The Hague, the Netherlands: Mouton.

Michaels, A. (1996) *Fugitive Pieces*, Toronto, ON: McClelland & Stewart.

Morrison, T. (1993) "Toni Morrison – Nobel lecture," Stockholm. Online. Available at: http://www.nobelprize.org/nobel_prizes/literature/laureates/1993/morrison-lecture.html (accessed 13 August 2015).

Nevius, M. and J. Nevius (2009) *Inside the Apple: A Streetwise History of New York City*, New York: Free Press.

A Pigeon Sat on a Branch Reflecting on Existence (2014), dir. Roy Andersson. Roy Andersson Filmproduktion AB. DVD.

Psycho (1960), dir. Alfred Hitchcock. Universal. DVD.

Psycho (1998), dir. Gus Van Sant. Universal. DVD.

Ray, R. (2000) "The field of 'literature and film,'" in J. Naremore (ed.), *Film Adaptation*, New Brunswick, NJ: Rutgers University Press, 38–53.

Schechner, R. (1988; rev. Routledge Classics ed. 2003) *Performance Theory*, New York: Routledge.

The Seventh Seal (2009), dir. Ingmar Bergman. AB Svensk Filmindustri, 1957; Criterion. DVD.

Shane (1953), dir. George Stevens. Paramount. DVD.

Spektra, T.T. (2013) "Roy Andersson siktar *på Guldpalmen.*" *Svenska Dagbladet*, "Culture," 27 January: n. pag. Online. Available at: https://www.svd.se/roy-andersson-siktar-pa-guldpalmen/om/kultur (accessed 4 September 2016).

Stam, R. (2005) "Introduction: The theory and practice of adaptation," in Robert Stam and Alessandra Raengo (eds) *Literature and Film: A Guide to the Theory and Practice of Film Adaptation*, Malden, MA: Blackwell, 1–52.

Tagg, J. (2003) *The Learning Paradigm College*, Boston, MA: Anker.

Thomson-Jones, K. (2009) "Formalism," in P. Livingstone and C. Plantinga (eds) *Routledge Companion to Philosophy and Film*, New York: Routledge, 131–41.

Toles, G. (2010) "Rescuing fragments: A new task for cinephilia," *Cinema Journal* 49(2): 159–66.

Tree of Life (2011), dir. Terrence Malick. Fox Searchlight. DVD.

Truffaut, F. (2008) "A certain tendency of the French cinema," in B.K. Grant (ed.) *Authors and Authorship: A Film Reader*, Malden, MA: Blackwell, 9–18; originally published as "Une certaine tendance du cinéma français," *Cahiers du Cinéma* 31 (January 1954), 15–29.

Wickman, M. (2016) "Anomalous? Ultimate? Why do we care about scholarly projects?" in *BYU Humanities Center*. Online. Available at: http://humanitiescenter.byu.edu/ultimate-or-anomalous-why-we-care-about-scholarly-projects/ (accessed 18 October 2016).

Yáñez, M. (2014) "Thought, action and imagination. [Frankfurt Papers]," in *The Audiovisual Essay*. Online. Available HTTP: http://reframe.sussex.ac.uk/audiovisualessay/frankfurt-papers/manu-yanez/ (accessed 4 January 2016).

16

ADAPTATION'S ORIGINALITY PROBLEM

"Grappling with the thorny questions of what constitutes originality"

Glenn Jellenik

Do I have an original thought in my head? My bald head.

Charlie Kaufman

The chimera of originality

Adaptation has an originality problem

By that, I do not mean that the act of adaptation has an originality problem, but rather that the systematic act of reading adaptations, or adaptation studies, has an originality problem. The zombie return of fidelity criticism to our field signals a lapse back into a traditional literary studies, recycling and internalising of Romanticism's critical definitions of originality, and correlating definitions of genius and art. Such definitions clash with many of the productive postmodern critical urges and potentials of adaptation and often lead critics blindly down fallacy-strewn alleys. Thomas Leitch (2003: 163, emphasis in original) peers down one such alley in his seminal catalogue of critical fallacies that riddle and stunt the field: "[Fallacy] 9. *Source texts are more original than adaptation* … It is much easier to dismiss adaptations as inevitably blurred mechanical reproductions of original works of art than to grapple with the thorny questions of just what constitutes originality." Leitch (2003: 163) argues that adaptations represent a fundamental critical problem, because they "raise questions about the nature of authorship." Rather than engage and confront the rich and messy critical implications and conceptual ruptures opened by adaptation's complication of traditional notions of authorship and originality, some scholars have begun to fall back to critical models, such as fidelity, which recycle and re-affirm Romanticism's ancient definitions. Books such as Colin MacCabe's *True to the Spirit* (2011), which includes contributions by Dudley Andrew, Tom Gunning, Frederic Jameson, Laura Mulvey and James Naremore, and David Krantz, and Nancy Mellerski's *In/Fidelity: Essays on Film Adaptation* (2008) attempt to re-centre recently marginalised fidelity models. Cassie Hermansson (2015: 147) points to such texts as evidence of "a post-millennial resurgence in fidelity criticism" in order to "make a (final?) recuperative claim for fidelity as one essential tool in the intertextual toolbox of adaptation studies." To be fair, 'new' fidelity attempts to distance itself from the traditional fidelity urge. Andrew (2011: 38) insists on the possibility of what he labels "genuine fidelity," a move away from the fidelity concept of adaptation as the translation of a literary source through filmic equivalents, or what Andrew (2011: 38) calls "vain and simple-minded matching." He posits a shifted "deeper fidelity," achieved when an adaptation is "[s]ubservient to [its] source, but not slavishly mechanical in rendering it" (Andrew 2011: 38). Clearly, the 'new' fidelity is just as dependent as the 'old' fidelity on subjective, evaluative, and critically problematic terms such as 'deeper,' 'genuine,' 'essence,' and 'spirit.' Furthermore, such vague theoretical steps back to (new) fidelity models cannot escape inherently constructing outdated, backward-gazing compare/contrast feedback loops wherein adaptations are viewed as secondary ("subservient") and, as such, capable of limited/prescribed originality.

Linda Hutcheon (2012: 3–4) recognises both this dynamic and the fundamental problem it poses for adaptation studies: "It is the (post-) Romantic valuing of original creation and of the originating creative genius that is clearly one source of the denigration of adapters and adaptations." The implied hierarchy constructed by Romanticism's definitions of originality, and our culture's subsequent internalisation of those definitions, dictates adaptation as secondary, lesser. Such critical assumptions lie at the heart of fidelity discourse. In order to circumvent fidelity while maintaining a focus on the act's twice-told essence, Hutcheon (2012: 7) configures adaptations as "repetition, but repetition without replication." She positions adaptations as deconstructive, rendering the traditional literary studies' original/derivative and primary/secondary binaries unproductive: "[A]n adaptation is a derivation that is not derivative – a work that is second without being secondary. It is its own palimpsestic thing" (Hutcheon 2012: 9). This opens the door to viewing adaptation as more than a generic shift, more than a movie made from a book; Hutcheon's palimpsest-formula echoes Brian McFarlane's suggestion that adaptation can be read to function as its own multi-layered thing, "as an example of convergence among the arts" (McFarlane 1996: 10). The adaptation functions, potentially, as a dialectic; not a novel made into a movie, but rather the discrete product of a convergence of novel and movie.

This essay picks up these critical threads to argue that adaptation studies represents an ideal critical ground from which to question, confront, and begin to move beyond Romanticism's outmoded definitions of originality. A close look at the act and performance of adaptation offers the critical opportunity, as Leitch puts it, to grapple with the thorny questions of just what constitutes originality. That necessitates a move away from well-landscaped definitions of originality that rely on binary rhetoric: source/copy, original/derivative, pure/contaminated. The fidelity-niche of adaptation studies is the last academic holdover of the New-Critical urge to read literary texts (in the case of adaptation studies, 'sources') as transcendental signifieds, to position texts as fundamentally authentic, original, self-identical. Postmodern theoretical concepts have thoroughly un-wound the structural justifications and underpinnings of Romanticism's fixed, clean, and clear concepts of originality. According to Derridean deconstruction, "All of our notions of authenticity, originality, and the like that were sustained by the fiction of a truth completely separate from the structure of signification have to be rethought" (Rivkin and Ryan 2004: 261). Within that radically shifted semiological economy, "the authentic will itself prove to be an artifact, the original derivative, the self-identical a double, the natural itself contrived and conventional" (Rivkin and Ryan 2004: 261). Deconstruction hollows out and collapses the original/derivative binary, clearing a space to develop the concept of derivative originality, or more specifically, of the act of adaptation – Hutcheon's repetition without replication – as fundamentally creating conditions that can lead to derivative originality.

The traditional critical positioning of the spirit underwritten and fueled by originality was shaped and developed in the Romantic period. William Hazlitt (1825: 149) opens his *Spirit of the Age* portrait of Lord Byron by calling him and Sir Walter Scott, whose portrait immediately precedes Byron's, "the greatest geniuses of the age." Yet Hazlitt (1825: 149) divides the pair by pointing out that they "afford a complete contrast to each other." The distinction that fleshes out this contrast revolves around issues of originality: "[Byron] is, in a striking degree, the creature of his own will. He holds no communion with his kind … He cares little of what he says, so that he can say it differently from others" (Hazlitt 1825: 150, 152). Thus, Byron embodies the Romantic ideals of organic and essential originality: (self) invention, innovation, and individuality. Scott veers sharply from Byron's Romantic ideal in that he is *"servile to nature and to opinion"* (Hazlitt 1825: 153; emphasis in original). Thus, Hazlitt (1825: 153) concludes: "The genius of Sir Walter is imitative … that of Lord Byron is self-dependent." That differentiation defines Scott's drastic limitations, as according to Hazlitt's system: "He is just half what the human intellect is capable of being … He shudders at the shadow of innovation … He has either not the faculty or not the will to impregnate his subject by an effort of pure invention" (Hazlitt 1825: 123, 127). Hazlitt posits Scott as an incomplete representative of the Romantic zeitgeist – *the spirit* only at half capacity. That minimisation derives from his choice and treatment of sources: "He has taken his materials from the original authentic sources, in large concrete masses, and not tampered with or too much frittered them away. He is only the amanuensis of truth and history" (Hazlitt 1825: 133–4). Scott is an adapter. Specifically, his innovation of the historical novel is adaptation. Rather than celebrate that genre's potential to construct derivative originality, Hazlitt posits adaptation as an act of midwifery. This explicitly denies Scott originality; his inability to reproductively create ("impregnate") codes him as the secretary ("amanuensis") who merely takes history's dictation. The highest recommendation of Scott's historical novel, according to the zeitgeist of the original genius, is that it remains faithful to its source – its generic transfer doesn't bastardise the history it copies. And the language of Hazlitt's marginalisation of Scott for his crimes against originality is strikingly similar to subsequent literary critical marginalisations of adaptation.

Hazlitt's definitions of originality and genius became internalised in the culture and subsequent criticism to such an extent that when William Wordsworth published his five-act tragedy *The Borderers* in his collected works, he prefaced the play with an extraordinary short confession: "Readers already acquainted with my Poems will recognize, in the following composition, some eight or ten lines which I have not scrupled to retain in the places where they originally stood. It is proper, however, to add, that they would not have been used elsewhere, if I had foreseen the time when I might be induced to publish this Tragedy" (Wordsworth 1910: 54). Yes, Wordsworth apologises for recycling eight lines of his own poetry almost 50 years after its composition, as if he had been caught plagiarising himself. Further, his use of the word "scrupled" casts his artistic act in decidedly ethical terms. Originality is virtuous, derivation dishonest.

That ethos carries over into traditional adaptation criticism, which, as Stam (2005: 3) points out, employs rhetoric that conflates the textual and ethical: "The [field's] conventional language … has often been profoundly moralistic, rich in terms … like 'infidelity,' 'betrayal,' 'deformation,' 'violation,' 'bastardisation,' 'vulgarisation,' and 'desecration' … each word carrying its own specific charge of opprobrium." And as Stam (2005: 3) observes, in his treatment of Spike Jonze and Charlie Kaufman's 2002 dissertation on the act of adaptation, *Adaptation.*,[1] "Even the metaphor of murder is invoked. 'We have to kill him,' the Susan Orlean character says of her adapter, 'before he murders my book.'" Clearly, the hyperbolic ethical positioning of the act of adaptation, as well as the false original/derivative dichotomy, is not critically productive. In place of hyperbole and false dichotomy, I suggest the deconstructive dialectic enabled by the concept of derivative originality.

The act of adaptation functions as a direct questioning (and problematising) of Romanticism's originality ethos. To theorise and flesh out the concept of derivative originality, I will explore *Adaptation.* through the lens of the film's own direct engagement and grappling with the thorny questions of just what constitutes originality. Viewed through that lens, the film enacts a dialectical debate on issues of originality and derivation, a perpetual struggle between cultural, artistic, and personal pressures to conform to either end of the original/derivative spectrum. Out of that dialectic springs both the concept of derivative originality and the film itself. That is, in its rehearsal of a dialectic debate between originality (thesis) and derivation (anti-thesis), *Adaptation.* produces a third possibility: derivative originality – something not new, yet new; derived, yet not derivative; original, yet not originary, Hutcheon's repetition without replication. Derivative originality functions as a fundamental questioning of Romanticism's definitions of originality, and an exploration of its implications allows adaptation critics to reconstitute concepts of originality and the capacities of the act of adaptation.

Adaptation. is a profound process ...

Adaptation. introduces screenwriter Charlie Kaufman (Nicholas Cage) as a character whose initial overall understanding of art and originality derive straight from Wordsworth and Hazlitt. The film opens on a black screen with Charlie's neurotic stream-of-consciousness voiceover: "Do I have an original thought in my head?" The evocation of artistic angst as the conceptual introduction to a comic plot suggests that such Romantic-period definitions of originality are still active in our culture, and that they are still very much on the surface of our discourse on adaptation. Indeed, the generative creative conflict in *Adaptation.* centers on the act of adaptation's fundamental (perceived) originality problem, as well as a productive tension between theory and practice. We soon learn that Charlie's original question: "Do I have an original thought in my head?" is the neurotic residue of his writer's block, and that the block stems from the fact that Charlie has agreed to adapt a book for the screen. Simply, Charlie's crisis is constructed by the notion of the act of adaptation as fundamentally derivative.

Far from sanctioning Charlie's creative angst (and thus accepting Romanticism's definition/valuation of originality), *Adaptation.* introduces a deconstructive foil, Charlie's artistic binary opposite: his twin brother (and aspiring screenwriter) Donald (also Nicholas Cage). Where Charlie anxiously avoids influence and nebbishly obsesses over originality, Donald blithely represents a derivative/formulaic approach to writing. The first half of *Adaptation.* rehearses this theoretical clash between Charlie and Donald. However, rather than come down on either side of the constructed spectrum, the film enacts a collision between Charlie's literary originality-thesis: "[A writer's] goal is to do something new ... Writing is a journey into the unknown" and Donald's filmic derivation-antithesis (borrowed from screenwriting guru Robert McKee): "[W]e have to realize that we all write in a genre, and we must find our originality within that genre."

This debate centres on the constitution of the act of adaptation. Theoretically, Charlie explains, adaptation offers the opportunity for artistic fulfilment through originality: "The script I'm starting, it's about flowers ... Nobody's ever done a movie about flowers before. So there are no guidelines." The project's lack of guidelines, formulae, or recipes squares with Charlie's Romanticism-driven artistic ethos, in which "writing is a journey into the unknown" where the writer should always have the "goal to try to do something new." Practically, however, Charlie's journey into the unknown hits a snag; he spirals into an artistic and existential crisis when the challenges of the specific act of adaptation become manifest.

In the end, the film's dialectic produces what I call *derivative originality*. Charlie's writer's block in *Adaptation.* parallels a reader's block within areas of the field of adaptation studies. Some of the same artistic prejudices that paralyse Charlie – his acceptance of and adherence to a critically impossible definition of originality and genius – block the field. Specifically, recent attempts to critically re-centre fidelity marginalise the agency of the act of adaptation, containing an adaptation's potential for originality within the borders of its source and prompting critics to engage in backward-gazing reading strategies. Yet *Adaptation.*'s dialectic construction of derivative originality recommends the field as the natural place to interrogate issues of originality. Adaptation studies is not bound to recycle and reify retrograde concepts of originality; in fact, it clears a natural theoretical space to rehearse and push on new definitions of it.

Obviously, a film titled *Adaptation.* will draw the attention of adaptation critics. Robert Stam uses the film as his point of departure for the introduction to *Literature and Film* in 2005, a reworking and expansion of his seminal essay "Beyond Fidelity: The dialogics of adaptation" (Stam 2000). Stam (2005: 1–3) suggests that the film's treatment of adaptation (its reflexivity, complex (and hilarious) engagement with issues of fidelity, radical intertextuality, and use of Darwinian adaptation as a metaphor for textual adaptation) underscores a productive and largely untapped set of tropes, with which the field of adaptation studies might position adaptations going forward. Antonella D'Aquino's 2007 essay "The Self, the Ideal, and the Real" considers *Adaptation.* centrally concerned with artistic access to reality, specifically with the ways in which it rehearses the tensions between art and life. For D'Aquino (2007: 562), who reads Donald as an imagined (fictional) character who exists as a part of Charlie's split self, Charlie's block stems from his "mistaking art for life ... [his inability] to distinguish between what's real and what's imagined." Kamilla Elliot's "Postmodern Screened Writers" engages the movie's play with reality in order to counteract the field's fidelity urges, pointing out that *Adaptation.* "foregrounds concepts of authorship and identity" while simultaneously denying its meta-character "an autonomous, individual, core, stable identity" (Elliot 2013: 23). Such a fragmented and unstable representation accords with a Barthesian death of the author. Elliot (2013: 23) uses that fracturing to argue that the film "shifts the focus from adaptation as product to adaptation as process; it sidelines questions of how true [it is] to the book it adapts, displacing them with questions of how true the film [is] to the life of its writer; it foregrounds questions of which medium is closer to life or reality, books or films." Vartan Messier's "Desire and the 'Deconstructionist': *Adaptation* as Writerly Praxis" also attempts to use *Adaptation.* as an avenue by which to circumvent the field's discourse on fidelity: "By approaching adaptation as an intertextual intermedial process, the shortsightedness of an essentialist, centric approach bound to fidelity can be swept away" (Messier 2014: 67). Messier (2014: 68) argues that *Adaptation.* functionally illustrates this liberation by deconstructing and destabilising the structural underpinnings of fidelity: "The film overturns the hierarchies between text and adaptation, source and copy, author/auteur and reader/viewer."

In addition to the rich set of Derridean deconstructions traced by Messier, I add the film's deconstruction of the original/derivative binary. The driving constructive tension of *Adaptation.* is (ironically) Charlie's writer's block, occasioned by his attempt to adapt Susan Orlean's *The Orchid Thief* (1998) into a screenplay.[2] Messier (2014: 71) configures Charlie's conflict as a fidelity issue: "His perceived incapacity to faithfully reproduce the book [is] due to his flawed initial approach. In other words, the narrative *errancy* of fidelity." For Messier, the film deconstructs what Stam (2000: 14) refers to as the "chimera of fidelity." However, I argue that Charlie's block is constructed by the internal clash between his (perceived) debt to originality and the derivative nature of adaptation. That is, Charlie is blocked by the chimera of originality. Additionally, the film's diegetic negotiation of that block forms a dialectic that clears a critical space in which to question traditional definitions of originality.

In addition to adapting Orlean's book, the film fictionalises screenwriter Kaufman's actual experience. That is, the screenplay is also (reportedly) based on the story of its own construction. Hired to adapt *The Orchid Thief*, Kaufman discovered that the book lacked a coherent narrative structure. In order to write himself out of an ensuing block, Kaufman wrote himself into the story, morphing Orlean's book from a work about flowers into a screenplay about an adaptation of a book about flowers. In focussing on the derivative originality of Kaufman's workaround, I argue that his performance simultaneously participates in, fundamentally questions, and ultimately enables adaptation studies to shift the critical notions of adaptation and its relationship with originality.

Early in his project to adapt Orlean's *Orchid Thief* into a screenplay, Charlie foregrounds the opportunity to "journey into the unknown" and grow as a Wordsworthian artist. At his pitch meeting for the project, he considers his source's abstract nature an enabling asset: "I think it's a great book … It's great, sprawling *New Yorker* stuff, and I'd want to remain true to that. You know? I'd want to let the movie exist, rather than be artificially plot driven." However, the adapter soon collides with the fact that Orlean's book contains very few linear plot elements. Rather than deliver on the promise of artistic freedom and originality, the constraints of adaptation threaten Charlie's concept of himself as an artist. His writer's block sends him to his agent, Marty (Ron Livingston), in an effort to escape the project. To prove his source's un-adaptableness, Charlie reads Marty the (actual) *New York Times* book review: "'There's not nearly enough … to fill a book, so Orlean … digresses in long passages … no narrative really unites these passages.' I can't structure this. It's that sprawling *New Yorker* shit." In the move between book and script, between adaptive theory and practice, "great sprawling *New Yorker* stuff" becomes "that sprawling *New Yorker* shit." Charlie is stymied by the fact that "The book has no story. There's no story!"

The generic implications of Orlean's frustration of narrative are highlighted later by screenwriting guru Robert McKee's reaction to Charlie's project. After hearing Charlie's synopsis of his source, McKee answers, "That's not a movie." That reaction reflects and rhymes with film theorist Christian Metz's concept of film narrativity as a simple abstraction: "Film tells us continuous stories; it 'says' things that could be conveyed also in the language of prose; yet it says them differently … The basic formula, which has never changed, is the one that consists of making a large continuous unit that tells a story and calling it a 'movie.' 'Going to the movies' is going to see this type of story" (Metz 1974: 44–5). Charlie's block, then, functions as a struggle with the generic pressure to impose an artificially plot-driven narrative structure on the book, to sacrifice his project's potential for originality by adhering to generic formula. Marty proposes just such a work-around: "Make [a story] up. I mean, nobody in this town can make up a crazy story like you. You're the king of that." Here the materialist agent evokes Charlie's originality, his trademark inventiveness. Rather than view that capacity as artistically liberating, however, Charlie folds back on originality again to point out that the specific constraints of the act of adaptation theoretically forestall such originality. "No, I didn't want to do that this time. It's somebody else's material. I have a responsibility to Susan." Thus, *Adaptation.* floats the possibility that the act of adaptation centers on a fundamental originality crisis.

However, even as the film offers such theory/practice and original/derivative divides, it questions and complicates, rather than reifies, definitions of the act of adaptation and of originality. In that questioning and complication, the adaptation also provides an example and a space in which adaptation critics can systematically question and grapple with issues of originality, that simultaneously structure and limit a cultural understanding of the act of adaptation. Specifically, it questions whether or not the act of repetition forestalls originality. If not, what does repetition without replication look like? Does originality generate from the basic system of filmic equivalents, as suggested by Jean-Luc Godard's concept that "something filmed is automatically different from something written, and therefore original" (Godard 1972: 200)?

The configuration of adaptation as the finding of filmic equivalents for a source's literary devices is a traditional adaptation studies theory that casts adaptation as an act of translation. The problems of such a system are all too evident, as Hutcheon (2012: 16) points out: "[I]n most concepts of translation, the source text is granted an axiomatic primacy and authority, and the rhetoric of comparison has most often been that of faithfulness and equivalence." Rather than negotiate and license originality, such a theory negates its possibility; if an adaptation is tasked merely with translating a literary source through functional equivalents, it lacks the capacity to construct meanings beyond those previously constructed by that source. Joseph Strick's 1967 New-Wave version of Joyce's *Ulysses* works as an example of a functional-equivalent adaptation. Despite employing a broad range of avant-garde filmic devices, "Strick presented his role as being to serve the original author without interposing himself into the work" (Geraghty 2008: 66–7). As a consequence, "*Ulysses* is generally now seen as a complete failure" (Geraghty 2008: 66). Yet from a structuralist perspective, as traced by Brian McFarlane in *Novel to Film*, film adaptations certainly offer cinematic equivalents to the literature they adapt (McFarlane 1996: 194–202).

The question, perhaps, is whether the system of filmic equivalents represents a lack of originality or an opportunity for a repositioned concept of originality. *Adaptation.* represents an example of the work and creative/original potential of non-translational functional equivalents. One of the film's central thematic elements is the linking and suturing of Orlean's prose treatment of orchids and the act of adaptation (both filmic and biological). At one point, the film discusses the radical variety found in orchids as a species. To do so, the filmmakers turn to a moment in Orlean's prose: "There are more than thirty thousand known orchid species … Orchids are considered the most highly evolved flowering plants on earth … One species looks just like a German shepherd dog with his tongue sticking out, one looks like an onion … one looks like a monkey" (Orlean 1998: 42–3). Orlean explains that the biological diversity and complexity of orchids stems from their reproductive reliance on cross-pollination rather than self-pollination. The film re-iterates this content, but within an altered context. As Charlie, researching orchids and attempting to break his writer's block, wanders alone through a sun-bathed orchid show in Santa Barbara, the audience hears a voiceover from Susan (Meryl Streep): "There are more than thirty thousand known orchid species. One looks like a turtle. One looks like a monkey. One looks like an onion. One looks like a German shepherd." The quote from Orlean's text is just part of the functional equivalent, as shots of orchids accompany the words, adding the missing visual to prose that labours to capture the colour, richness, and strangeness of the orchids. Thus, the film completes and adds a missing visual component to Orlean's project.

Further, the setting of the flower show calls back to two previous flower-show moments in the film: one in which John Laroche (Chris Cooper) first impresses Susan by seductively and poignantly disserting on the miraculous "love making" between the insect and its "soul mate" flower, and a moment in which Charlie and Alice (Judy Greer), a waitress at a local diner, leave the flower show for a romantic tryst in the sun-dappled woods. That moment is broken by a smash cut that places Charlie back in his dark bedroom. And it turns out that he is not at the flower show at all, but rather lying in bed, masturbating to an imagined tryst at an imagined flower show. The scene amounts to a fantasy of cross-pollination with Alice. The re-use of the setting provides an opportunity to connect Susan and Charlie's writing process to the concept of pollination. Susan, the productive writer, engages with cross-fertilisation with her subject (Laroche), while Charlie, the blocked writer, is forever closed off from others, left to unproductively self-fertilise/masturbate.

Beyond that, the adaptation overtly extends the metaphor beyond biological diversity, using orchids as a direct metaphor for relationships, desire, and love. The film also shifts out of adaptation proper and into the stories of the inserted character of Charlie and a (fictional) romantic relationship between Susan and Laroche. As Charlie wanders alone through the flower show, Susan's voiceover on orchids dissolves in the middle of an observation: "One looks like …" and is sound-bridged into a voiceover from Charlie: "… a schoolteacher … one looks like a gymnast

… one looks like that girl from high school with the creamy skin. One looks like a New York intellectual with whom you do the Sunday Times crossword puzzle in bed … One has eyes that dance. One has eyes …" Now each "flower" becomes a correlating woman at whom Charlie gazes. And the final sentence fragment ("One has eyes") is completed with a final cut from the flower show to Susan's dark apartment and her eyes as she types alone. Charlie's voiceover ends "One has eyes … that contain the sadness of the world." That conflation of orchids and love, of biological and textual adaptation functions as a central theme of the adaptation.

Some of the tension that arises from the use of translation as a trope for adaptation revolves around the chimera of originality. According to translation theorist Albert Braz (2007: 17), "If the translator becomes creative to the point of ignoring the original work … no translation can take place." At the centre of Braz's observation is the idea that there is a tipping point of "creative" interaction at which a text ceases to function as a translation and becomes … original. Within that formula, Kaufman's translation of *The Orchid Thief* might be seen as too creative to function as an adaptation. Yet I argue that, rather than threaten to shed his source, Kaufman's screenplay produces a derivative originality that works to complete his source.

What we talk about when we talk about originality

And here we come to a prime example of the dialectic that produces derivative originality, as well as its creative capacity. At its midpoint, *Adaptation.* questions and destabilises one of the fundamental assumptions of adaptation studies: the completeness of the source. Traditional adaptation studies assumes the completeness of the source text, that perspective licenses and fuels fidelity discourse, wherein an adaptation is measured by its ability to approximate that completeness. Yet *Adaptation.* illustrates that adaptation (both biological and textual) functions not as a reflection and repetition of completeness, but as a step in an on-going process of completion. Adaptation is not the mere story or echo of reproduction, it is an enabling device that creates the conditions for reproduction. Without adaptation, reproduction is not possible, and the story remains necessarily incomplete. From a narrative perspective, the claim of Orlean's book's incompleteness acts as a pivot-point from Charlie and Donald's dialectic debate on originality and art, which dominates the film's first two (static) acts, to the McKee/Metz-driven third act, which delivers the type of story the audience recognises as a 'movie.' The first accusation of *The Orchid Thief*'s unfinished status comes in the form of McKee's aforementioned summary judgment of Orlean's book: "That's not a movie." McKee's claim positions Charlie's source material as incomplete, as an organism constitutionally unable to adapt to the new environment of film–story.

As if taking its cue from McKee, the film drops Orlean's book-proper here. Charlie abandons his theory of art and enlists his brother's help. From that moment, Donald hijacks the story/screenplay, adding all the elements that Charlie's pitch-meeting manifesto sought to avoid: "I don't want to cram in sex or guns or car chases or characters learning profound life lessons or growing or coming to like each other … the book isn't like that and life isn't like that. It just isn't." Donald (via McKee) reminds the audience that movies are not books or life, and they are like that, as the last act of *Adaptation.* proves. To diegetically flesh out McKee's claim that Orlean's book lacks something essential, Donald's initial analysis of *The Orchid Thief* also zooms in on lack: "[The secret of flowers is] not in the book, Charles." Not only does the book lack material for a movie, but Donald suggests that the verbal story it purports to tell is incomplete; it contains a secret only revealed in the subtext. The film then sets out to complete the book, to flesh out its subtext. Thus, *Adaptation.* posits adaptation not as a re-telling of a story but as an engagement with it, a furthering and completion of its narrative urges. The film argues that change is not a choice in adaptation, that originality is actually a pre-requisite and a necessity.

In Donald's first plot act, he replaces Charlie, pretends to be him and interviews Susan in order to unearth the subtextual secret that will allow Charlie to complete an adaptation of an incomplete book. Donald theorises that not only is *The Orchid Thief* incomplete, but that Susan is also missing, or rather not revealing, something essential. During the interview, he mentions that, in reading *The Orchid Thief*, he detected a subtext of romantic attraction between Susan and Laroche, and asks if she has kept in contact with him. Susan answers calmly: "Well, our relationship was strictly reporter–subject. I mean, certainly an intimacy does evolve in this kind of relationship … By definition, I was so interested in everything he had to say. But the relationship ends when the book ends." No such meeting between Charlie Kaufman and Susan Orlean ever occurred. The film's third act completely separates it from Orlean's book and transforms it into the exact thing that Charlie's inner Wordsworthian artist struggled to avoid: "an orchid heist movie … changing the orchids into poppies and turning it into a movie about drug running … cram[ming] in sex … guns … car chases … characters learning profound life lessons." From that perspective, Donald's interview of Susan represents another of Kaufman's forays into creative fiction, a necessary adaptive departure from the source text in order to bring a narrative-challenged source into line with Metz's concept of a "movie" as "a large continuous unit that tells a story." Yet that interview between the fictional Donald and the fictional Susan actually appears in Orlean's non-fiction book – though not at all in the form offered in the film.

A Reader's Guide was added to the 2000 edition of *The Orchid Thief*, containing excerpts of an interview between Orlean and writer Tim McHenry. During the interview, McHenry asks Orlean if she has kept in touch with Laroche since the book was published; McHenry explains his question by suggesting a subtext of attraction in the text. Orlean (2000: 292) answers: "Our relationship was strictly reporter and subject. It is certainly true that you develop a kind of intimacy with someone you are writing about … By definition, everything he had to say was interesting to me, because that is what I was there to do: to find out about him … [But] the relationship ends when the book ends … There was never any flicker of romantic interest on my side, and I suspect on both sides." The similarity between the McHenry/Orlean interview and the Donald/Susan interview suggests that, rather than invent the scene, Kaufman's adaptation re-purposes and re-contextualises it. The product of that recycling of the Reader's Guide interview works as a prime example of the creative capacity of derivative originality.

Within the original interview, Orlean (2000: 292) shuts the door on the question of her involvement with Laroche, and she does so by positing her book as complete: "the relationship ends when the book ends." In *Adaptation.*, Susan delivers the same line; however, Donald's reading of *The Orchid Thief* as incomplete licenses his refusal to accept her answer. In a film that regularly associates authors and their texts, Donald naturally transfers the book's incompleteness to its author. The audience has already seen Susan as fundamentally incomplete, as lacking something vital. Though a successful and productive (and original-ish) writer, she exists in the film as a character with a void at the centre of her life: though an excellent detached observer, she longs to be a wanter, someone like Laroche. Yet she is incapable of fulfilling her desire to desire by herself. Significantly, her first moment of satisfaction, of ease, comes in a moment of collaboration, when, during a phone conversation, she and Laroche each sustain a note in order to produce the sound of a dialtone. This introduces and develops a theme of fulfilment through cross-pollination/collaboration that the film consistently privileges, pushes, and pursues.

Completion through collaboration extends beyond the diegesis of the film adaptation, as well. The re-contextualisation of the McHenry/Orlean interview functions as a springboard for the insertion of a fictional Susan/Laroche love affair. In a sense, Kaufman Hollywoods

Orlean's "sprawling New Yorker" book by heeding Marty's advice and "make[ing] up a crazy story." But such derivative originality also proves David Hare's claim that "The only way to achieve fidelity in adaptation is through promiscuity" (Hare 2002). When discussing the radical plot changes wrought upon her book, Orlean (2008) observes: "[The movie] was truer to the spirit of the book than a conventional [adaptation] could have been … Charlie unearthed themes that were not explicit … I saw, sort of, qualities of the book emerge through the film that I hadn't really even had a grip of myself." Here, a real-life author acknowledges something along the lines of Stam's observation that "[a]uthors are sometimes not even aware of their own deepest intentions" (Stam 2000: 57–8). This acknowledgement of the intentional fallacy licenses an adaptation to function less as a "copy" than as a reading or interpretation of its source. Further, Orlean's comments contradict Frederic Jameson's odd fidelity-driven concept that an adaptation that seeks to match its source's originality must "breathe an utterly different spirit altogether" (Jameson 2011: 218). According to Orlean, the adaptation's added plot lines do not "breathe a different spirit"; rather, they enrich the source and flesh out unexplored and untapped potential.

It is vital to note that only part of this process involves invention-proper. Kaufman invents a love affair, a fictional romance between reporter and subject. Yet the seeds of that invention emerge from a subtextual reading of the book. Further, the film's first accusation of that relationship is recycled. Kaufman's dialogue is 'borrowed' rather than 'original' – it is, literally, derived from his source. Yet it is not. Here we see derivative originality, as well as a complex actualisation of Hutcheon's concept of adaptation as repetition without replication. Kaufman's arrangement is simultaneously derivative and original – derivative originality. The deconstructive tension between this term's seemingly opposite components is built into *Adaptation.*, as well as our cultural concept of adaptation.

Within the economy of literary criticism (driven as it is by Romanticism's definitions of originality), the term "derivative" carries negative connotations: "When a lesser product comes of literary theft, we call that work derivative; when a masterpiece results, we excuse [the theft]" (Urgo 1997: 7). With that in mind, perhaps the term "derivative" should be avoided by adaptation critics. Yet from a legal perspective, an adaptation is considered (and protected as) a "derivative work," which, according to the United States Copyright Act (Title 17, Ch. 1), "is a work based upon one or more preexisting works, such as a … fictionalization, motion picture version … or any other form in which a work may be recast, transformed, or adapted. A work consisting of editorial revisions, annotations, elaborations, or other modifications which, as a whole, represent an original work of authorship, is a 'derivative work.'" No evaluation exists in copyright law – "derivative" implies no lesser status, merely the preexistence of a source. And interestingly, the signifier "original work of authorship" structures the legal definition of a "derivative work." Thus, the seeming contradiction in the term derivative originality begins to appear less manifest.

So while the term causes some inherent friction, where better than adaptation studies to, as Julie Grossman (2015: 1) puts it, "invite a kind of critical thinking that moves viewers and readers beyond their comfort with inherited boundaries and preexisting patterns." In her book, Grossman associates adaptation with Mary Shelley's tropes of hideous progeny and monstrous birth. She does so not to marginalise the act and product of adaptation; rather, she appropriates those terms to remind readers that concepts such as 'hideous' and 'monstrous' are mental and cultural constructions rather than objective identities, and the same holds true for the cultural understanding of adaptation. It is in that spirit that I employ the term "derivative originality." For in derivative originality, we find the nexus of Braz's concept of "creativity" in translation and Orlean's concept that Kaufman's creativity "unearthed" hidden themes that cleaved to the spirit of the book. The potential of derivative originality to produce meaning in adaptation allows us to re-examine, question, and redefine our fundamental notions of what we talk about when we talk about both adaptation and originality.

Notes

1   The regularity with which critics incorrectly write the name of this film is disturbing. Kaufman/Jonze, et al. punctuate the title with a period. It is a specific rhetorical choice that should not be ignored – by anyone – but particularly not by the very people tasked with looking closely at and into and processing these texts.
2   The amount of times Orlean's non-fiction book is referred to as a "novel" in the existing criticism on the film is disturbing.

Works cited

Adaptation. (2002), dir. S. Jonze. Columbia. DVD.

Andrew, D. (2011) "The economies of adaptation," in C. MacCabe, K. Murray, and R. Warner (eds) *True to the Spirit: Film Adaptation and the Question of Fidelity*, New York: Oxford University Press, 27–40.

Braz, A. (2007) "The creative translator: Textual additions and deletions in *A Martyr's Folly*," in N. Cheadle and L. Pelletier (eds) *Canadian Cultural Exchange/Echanges Culturels au Canada: Translation and Transculturation/Traduction et Transculturation*, Waterloo, ON: Wilfred Laurier University Press, 15–28.

D'Aquino, A. (2007) "The self, the ideal, and the real – the artistic choice of three creative minds: Fellini, Allen, and Kaufman," *Italica*, 84(2/3): 556–77.

Elliot, K. (2013) "Postmodern screened writers," in S. Wells-Lassagne and A. Hudelet (eds) *Screening Text: Critical Perspectives On film Adaptation*, Jefferson, NC: McFarland, 22–42.

Geraghty, C. (2008) *Now a Major Motion Picture: Film Adaptations of Literature and Drama*, Lanham, MD: Rowman & Littlefield.

Godard, J.-L. (1972) *Godard on Godard*, ed. J. Narboni and T. Milne, New York: Viking.

Grossman, J. (2015) *Literature, Film, and Their Hideous Progeny: Adaptation and ElasTEXTity*, New York: Palgrave Macmillan.

Hare, D. (2002) in "DVD Extras," The Hours, dir. S. Daldry. Miramax. DVD.

Hazlitt, W. (1825) *The Spirit of the Age, or Contemporary Portraits*, London: Colburn.

Hermansson, C. (2015) "Flogging fidelity: In defense of the (un)dead horse," *Adaptation*, 8(2):147–60.

Hutcheon, L. (2012) *A Theory of Adaptation*, 2nd edn, New York: Routledge.

Jameson, F. (2011) "Afterword: Adaptation as a philosophical problem," in C. MacCabe, K. Murray, and R. Warner (eds) *True to the Spirit: Film Adaptation and the Question of Fidelity*, New York: Oxford University Press, 215–33.

Krantz, D.L. and Mellerski, N.C. (eds) (2008) *In/Fidelity: Essays on Film Adaptation*, Newcastle, UK: Cambridge Scholars.

Leitch, T. (2003) "Twelve fallacies in contemporary adaptation theory," *Criticism*, 45(2): 149–71.

MacCabe, C., Murray, K., and Warner, R. (eds) (2011) *True to the Spirit: Film Adaptation and the Question of Fidelity*, New York: Oxford University Press.

McFarlane, B. (1996) *Novel to Film: An Introduction to the Theory of Adaptation*, New York: Oxford University Press.

Messier, V. (2014) "Desire and the 'deconstructionist': *Adaptation* as writerly praxis," *Journal of Adaptation in Film and Performance,* 7(1): 6582.

Metz, C. (1974) *Film Language: A Semiotics of the Cinema*, Chicago, IL: University of Chicago Press.

Orlean, S. (1998) *The Orchid Thief*, New York: Ballantine.

Orlean, S. (2000) "*The Orchid Thief*: A reader's guide," in S. Orlean, *The Orchid Thief*, New York: Ballantine, 285–96.

Orlean, S. (2008) "*The Orchid Thief* author Susan Orlean discusses *Adaptation*.," on *YouTube*. Online. Available HTTP: https://www.youtube.com/watch?v=bUwrIeEB9-Y> (accessed 11/14/2016).

Rivkin, J and Ryan, M. (2004) *Literary Theory: An Anthology*, 2nd edn, Malden, MA: Blackwell.

Stam, R. (2000) "Beyond fidelity: The dialogics of adaptation," in J. Naremore (ed.) *Film Adaptation*, New Brunswick, NJ: Rutgers University Press, 54–76.

Stam, R. (2005) "Introduction: The theory and practice of adaptation," in R. Stam and A. Raengo (eds) *Literature and Film: A Guide to the Theory and Practice of Film Adaptation*, Malden, MA: Blackwell, 1–52.

Stam, R. and Raengo, A. (eds) (2005) *Literature and Film: A Guide to the Theory and Practice of Film Adaptation*, Malden, MA: Blackwell.

United States Copyright Act. Online. Available HTTP: https://www.law.cornell.edu/uscode/text/17/101 (accessed 11/14/2016).

Urgo, J.R. (1997) "Introduction: Reiving and writing," *Faulkner Journal*, 13(12): 3–14.

Wordsworth, William (1910) *The Poetical Works of Wordsworth*, London: Oxford University Press.

17

MIGRATION, SYMBOLIC GEOGRAPHY, AND CONTRAPUNTAL IDENTITIES

When death comes to Pemberley

Carol Poole and Ruxandra Trandafoiu

Published in 2011 as a crime sequel to *Pride and Prejudice, Death Comes to Pemberley* was adapted for *BBC One* and first aired on Christmas Day, 2013. It subsequently aired on *PBS* in the United States in November 2014 and has had several re-runs on the British-based free to view *Drama* channel. P.D. James, one of the most successful and revered British crime writers of the twentieth century, produced a worthy homage to her favourite writer (James 1999: 227), published in time to celebrate the two-hundred-year anniversary of *Pride and Prejudice* in 2013.

As a multi-genre sequel, P.D. James's novel mixes Austen's comedy of manners and love story with a crime caper and whodunit, police procedural and journalistic court reporting, infused with Gothic elements and fan fiction tributes. The result is a multi-authored, fluid text that should be read in the context of authorship and genre polygamy, whereby both novels, together with their adaptations for television and cinema, co-exist and travel in multiple fields of meaning, defined by in-betweenness. Although the ensuing analysis will focus on *Death Comes to Pemberley* and its screen adaptation for the *BBC*, we will inevitably and simultaneously refer implicitly or explicitly to Austen's work and its numerous television and film adaptations, thus offering a contextualised socio-historical approach.

Any text that has endured various adaptations, sequels, and mash-ups, can be considered a migrant on a journey of exile, resettlement, adaptation, and cultural translation, resulting in "contrapuntal" (Said 1994: 36) co-presences, multiple embodiments, and shifting identities. It is therefore appropriate to use diasporic theory, as described in the following section, to better understand the text's journey of migration from one space to another, from page to screen and back again. Identities are shaped by departures and arrivals, emigration and immigration, and in-between experiences, and we borrow from the arsenal of diasporic theory concepts such as alienation–adaptation, journey–capital, multiplicity–hybridity, change, place polygamy, symbolic geography, homeness, trauma, and memory. These concepts expand on already existing theories in adaptation studies, such as Stam (2005), Hutcheon (2006), and Sanders (2006), who examine hybridity and transtextuality through authors from Bakhtin to Kristeva, and Deleuze to Genette. However, while these approaches still put the text at the centre of a mainly literary analysis, diasporic concepts offer new tools for a socio-historical analysis and a sociology of adaptation, which is a more fitting method of investigating identity, change, and multiplicity.

A sociology of adaptations: migration, change, and belonging

The novels and screen adaptations considered in this chapter offer multiple research opportunities, in relation to romance and crime in particular. However, our aim is to shift the emphasis from a more traditional literary perspective to a socially oriented one. While most adaptation studies focus on textual analysis, our approach is socio-cultural and historical and opposed to formalism; we aim to place the texts in their social contexts, thus illuminating social change and identity shifts.

Our sociologically oriented approach not only differs from traditional textual approaches in adaptation studies, but also from the recent endeavour to "showcase a broadly sociological approach to adaptation, foregrounding … the industrial structures, interdependent networks of agents, commercial contexts, and legal and policy regimes within which adaptations come to be" (Murray 2012: 6). Although Murray uses the term 'sociological,' her otherwise worthwhile industry-based approach only marginally touches upon socio-historical aspects, being mainly concerned with economic and political capital. Her approach is a "deliberately *con*textual" one (Murray 2012: 7; emphasis in original), but of a different kind to ours. However, we share with Murray (2012: 4) a desire to move from a "comparative aesthetic evaluation" towards "scrutinising adapted texts for their critical reworking of power structures often only covertly registered in source texts." We use Bourdieu's classic concepts 'capital' and 'field,' already appropriated by diasporic theory (see Meinhof and Triandafyllidou 2006), to illuminate some of these hidden social structures. In particular, in the second part of the chapter we look at the way James's mash-up and its screen adaptation have engaged with social change, women's status, class rigidity, and symbolic social spaces, and have been influenced by the author's own political identity.

Austen's novel has been described as her "most socially idealistic book" (Amis 1990: 102), primarily because Darcy is ultimately ruled by impulse, and feelings bridge class hierarchies. Idealism becomes the tool that Austen deploys to provide us with a subtle commentary on the rigidity of the English class system. Although by no means a radical, Austen is, with hindsight, an innovator. P.D. James, on the other hand, one of the more conservative writers of detective fiction, establishes a natural coupling between social transgression and punishment. For her, crime itself is the result of a system that becomes unbalanced; crime happens when people trespass into social and geographical spaces that are alien to them, where they do not belong. However, it would be wrong not to recognise James's ability to enmesh valuable social debates within this otherwise reticent approach to social change.

The application of socio-cultural and historical analysis to a sociology of adaptation allows us to focus on identity, otherness, and change in relation to the characters' psychologies and their 'journeys,' and provides us with much needed alternative methodologies to traditional textual analysis (Murray 2012: 16). This is a more rewarding undertaking, both in terms of reading/viewing pleasure and critical potential, since from a literary perspective, P.D. James's crime story dries up under the pressure of mimicking Austen's style and maintaining the characters' credibility, thus taking James away from what she usually excels at: the careful plotting of the crime and its psychological justifications. A sociology of adaptation is not an entirely new approach, with Said and Rushdie, respectively, offering similar readings to literary texts (most notably Austen's *Mansfield Park*) and films (such as *Slumdog Millionaire*, 2008). Nor is it alien to filmmakers like Gurindher Chadha, whose *Bride and Prejudice* (2004) is written from a post-colonial perspective and contributes to what Stam (2005: 11) calls "an egalitarian thrust" in adaptation studies connected to the closer exploration of identity and oppression.

Edward Said (1994) applies his personal migratory experience to an analysis of nineteenth-century novels in imperial France and Britain to provide a "contrapuntal" reading of literary texts. As defined by Roger Silverstone (2007: 85) in his key text, *Media and Morality*, the contrapuntal "speaks to the inevitable, continuous and significant juxtaposition of elements and threads in a life, a text, a history … the presence in a single discourse of more than one voice." Using *Mansfield Park*, among other examples, Said (1994) delivers a subtle decoding of social and historical references, particularly class, slavery, and the empire, which illuminate but at the same time complicate the storyline. The contrapuntal thus reveals the hidden tension that can under-line one's socio-historic identity while that identity is in flux, on a journey of migration and change. The co-presence of at least two cultures or stories – Austen's well-known family story and the implied larger history that Said brings into focus to discuss historical class hierarchies – reveals multiple spaces of existence between "here" and "there," existence and absence, what is visible and what is invisible. This "double consciousness," itself a key theme in diasporic theory (see Du Bois qtd in Gilroy 1993: 112), creates contrapuntal effects of adaptation and rejec-tion, acceptance and alienation, consonance and dissonance, evident in the characters' dilemmas and tribulations.

James's mash-up can be read in a similar vein. Both the novel and its screen adaptation become part of a contrapuntal continuum or field of multiplicity. They sit in tension to one another, but also to previous creative and critical takes on both Austen and P.D. James, and, through micro-stories, offer insights into larger national and literary histories. Similarly, in his review of *Slumdog Millionaire* (2008), Salman Rushdie (2009: 2) appropriates the metaphor of migration to define screen adaptations as "translation, migration and metamorphosis, all the means by which one thing becomes another." While we observe the characters' identities on a journey of change, we can also reflect on the hybrid and shifting nature of a text's identity, as it is operating among and across spaces of hybridity.

Chadha's appropriation of Austen in *Bride and Prejudice*, on the other hand, provided the opportunity to replace the class discourse, dominant in Austen, with an ethnic and racial one, derived from globalisation concerns over the negotiation of cultural difference. In P.D. James's whodunit, appropriation became likewise a conduit for exploring social and political concerns from a different historic and ideological positioning to that of Austen. This has resulted in an interesting symbolic dialogue between the two authors about, among other things, the status of women and the significance of tradition, which we will explore more closely in the next sec-tion through the theoretical lenses of concepts such as alienation, journey, capital, and hybridity.

The characters' shifting social identities are explored under the first signs of historical upheaval, which often results in bringing alterations in social conditions and women's status, resulting in changed identities and new social practices. It is possible that P.D. James moved the action to 1803–4 (six years into the marriage of Elizabeth and Darcy) because it is a time more clearly defined by change. The opportunity to explore, albeit briefly, Britain's incipient indus-trialisation and its effects on class, the relationship between landowners and servants, women's emancipation, and legal reforms in the judicial system and policing, is possible for James because there is actually more to Austen than romance and comedy of manners. The backdrop in *Pride and Prejudice* is a complex social context, which P.D. James is able to tease out into the socio-political arena by problematising themes that Austen was not interested in exploring fully herself (James 2012: 329).

From the point of view of social change, the historical moment is defined by fluidity and in-betweenness: the golden age for landowners is coming to an end; the judicial system is under scrutiny; the class system is under pressure. Fluidity (in a Deleuzian/Guattarian sense) and liquidity (Bauman 2005) define identity – that of the characters, and that of the work, which is "the sum of its versions" in this "fluid-text approach" (Bryant 2013: 47). Issues of identity thus affect the macro level, such as authorship or genre, always relevant in the case of a sequel or a mash-up, but also the micro level – characterisation, plotting, and setting. Approached from the angle of identity, hybridity is one of the dominant notions emerging, tying together the concepts of journey, change, and multiple spaces (physical and symbolic/social).

It is the symbolic social journeys, through which 'capital' is acquired and the historical and social spaces that are being traversed, intersected, or reformed, that are of particular interest. Characters are given certain challenges: Darcy (played by Matthew Rhys) has to reconcile his marriage to Elizabeth (Anna Maxwell Martin) – which brings Wickham (Matthew Goode) back into his life – and his duty to his ancestors, family, and estate; Elizabeth must resolve her class complex and take on the challenge of solving the crime (on screen, not so much in the book); Georgiana (Eleanor Tomlinson) must choose between love and duty; the young law-yer Alveston (James Norton) and Wickham have to navigate a legal system which is under critical scrutiny. Consequently, their identities change as they are moulded by traumatic events and radical choices.

As Rushdie (2009: 4) warns us, those who "cling too fiercely to the old text, the thing to be adapted, the old ways, the past, are doomed to produce something that does not work, an unhappiness, an alienation, a quarrel, a failure, a loss." By providing a parallel between cultural and artistic adaptation, he concludes that "the process of social, cultural and individual adapta-tion, just like artistic adaptation, needs to be free, not rigid, if it is to succeed" (Rushdie 2009: 4). James's characters are like migrants on a journey that could result in survival or failure, gain or loss; they engage in an evolutionary process resulting in the survival of the fittest (see also Stam 2005: 3). Words such as 'journey', 'change,' or 'capital,' from the arsenal of cultural and migra-tion theory can become useful in this context. In applying diasporic and migration theories to the field of adaptation studies, Rushdie raises inevitable questions around power, control, and ultimately ownership, about adaptation as a critical process with its internal–external, inclu-sion–exclusion dynamics.

Key concepts in diasporic theory can illuminate the text from a fresh perspective, reveal-ing what might have stayed hidden: the threat of an unhappy adaptation, as Rushdie warns; contradictions that appear out of collectiveness and multiplicity, with various texts (off and on screen) in dialogue with each other (see also Cartmell 2010: 126); and the resulting accumula-tion of symbolic capital. Schober (2013: 105) calls this process "intermediality," "understood as a dynamic and highly complex network in which media are in constant reciprocal interaction, not only with each other, but also with their recipients, various social and cultural forces and … within the intricate network of remediations." How we read a text constitutes a political act (Silverstone 2007: 88), and if we move "away from what adaptation *is* and towards what adapta-tion *does*" (Dicecco 2015: 163; emphasis in original), aesthetic debates can be replaced by politi-cal ones (see Dicecco 2015: 173), allowing us to move forward from textual analysis towards a more sociologically based one.

Changed identities and fluid symbolic geographies

Austen has been accused of ignoring historical events and providing us with a "timeless pre-sent" (Tanner 1972: 7), but this is not entirely true. Austen's main characters ask, even though obliquely, fundamental questions about the nature of happiness, and the answers that Austen provides for them have often revolved around attaining both love and money. Although not explicitly addressed, love and security are socio-economic issues and the paraphernalia of war, politics, and philosophy. While Austen's position can be gauged to some extent from the novels, P.D. James's views are more obviously disclosed in autobiographical reflections.

James, from the viewpoint of another literary genre and another period, is comfortable plac-ing her characters in a more politicised setting, at a time when the seeds of social upheaval and change, with the advent of industrialisation, were being sown. Her novel is a murder mystery, but also a police procedural and a critical commentary on judicial reforms, the class system, and the rights of women, issues that have informed her personal (conservative) politics.

P.D. James was familiar with William Godwin's first ever mystery novel in the English language. Published in 1794, it established a strong link between crime and social upheaval. Although James (1999: 17) found it "unreadable," she acknowledges that Godwin uses the detective novel formula "to say something about society." More importantly, Godwin was Mary Wollstonecraft's husband. Not by accident, the famous women's rights advocate is the only historical figure (apart from Napoleon) mentioned in James's novel. By including Wollstonecraft at the centre of a significant dialogue, James made a point about women's rights and also, more generally, about social injustice and its possible effects. Through suitor Alveston's voice she wrote: "we do not need to be a disciple of Mrs. Wollstonecraft to feel that women should not be denied a voice in matters that concern them. It is some centuries since we accepted that a woman has a soul. Is it not time that we accepted that she also has a mind?" (James 2012: 142). James transforms petty squabbles between suitors into a conduit for exploring the first traces of feminism. Thus, her story becomes history.

We know that the journey of change is still embryonic because it is still a man (Alveston) who voices concerns about women's rights. He speaks, while the women stay silent. However, with two novels written by women and Juliette Towhidi's script, we end up with a story dominated by strong female characters – Elizabeth and Georgiana – who accomplish what they set out to do. As a mash-up, *Death Comes to Pemberley* presented the *BBC* with a dilemma. The screen production had to resolve the tension between viewers' expectation of romance in Austen and costume drama more generally, and James's feminism and social interests. In the end, with Lizzie, the woman who ultimately manages to have it all and saves the day, we have a very modern heroine indeed, that both contemporary audiences and P.D. James, herself an independent working mother, would have empathised with. To accomplish this, the *BBC*'s production had to select from a number of social and political concerns evidenced in James's book, primarily one: women's emancipation, told via a story of love rivalry.

Adaptation selectivity has precedence in other Austen-inspired work. To give just one example, *ITV*'s 1996 production of *Emma* provided some historical contextualisation and implicit references to the need for change in terms of servants' status, poverty and crime, and the formality of rank, but they were swiftly brushed aside for the final celebratory scenes. Andrew Davies's harvest feast "smacks too much of 'merrie England' to ring true" just after the bad harvest of 1812, enclosures, and unrest among low-paid agricultural workers, and not long before the Peterloo massacre and the passing of the Corn Laws to protect landlords' interests (Brown 2015: 229), but it provided the happy and romantic ending viewers expected. In *Death Comes to Pemberley* we have a more satisfying resolution. Romance dominates, but does not overwhelm, and leaves some room at least to note other impending social reforms.

James's views are not radical; on the contrary, she retains reticence vis-à-vis social change. Detective fiction often sees crime as being motivated by social and political change. Nordic noir is often the preferred example of such a link, starting in the 1960s with the work of Sjöwall and Wahlöö, who wrote from a clear left-wing position. However, with her concerns about change, especially change that happens too quickly and threatens tradition, James (a Conservative Peer in the House of Lords) most likely wrote from the standpoint of British conservatism. As Horsley (2005: 58) observes: "For James, one of the most conservative of British crime writers, individual evil would seem to be a consequence of the growing secularisation and fragmentation of modern society … and her high-church Anglicanism is joined politically to Thatcherite Conservatism."

The unhappiness, fear, and doubt that permeate relationships in the book can also be read as an ideological metaphor for the state of a society threatened by change. However, James's more conservative approach also means that she successfully picks up on imbalances in the sexist and class-bound social system of Austen's time. The crime itself is the outcome of a split, unbalanced society; when worlds collide, tragedy happens. The screen version makes a return to romance, but, in doing so, loses some of the more interesting social issues raised in the novel.

To a certain extent, P.D. James deals with issues of class, money, marriage, and honour with even less romanticism than Austen did two hundred years ago. She seems critical of the radicalism of the French Revolution and hints at Darcy's weakness for marrying down. When Darcy and his cousin, Colonel Fitzwilliam (played on screen by Tom Ward), keep an overnight vigil in the aftermath of the murder, James (2012: 125) writes: "He [Colonel Fitzwilliam] would never have married an Elizabeth Bennet, and Darcy occasionally felt that he had lost some respect in his cousin's eyes because he had placed his desire for a woman above the responsibilities of family and class." The word 'desire' rather than 'love' hints at a much baser emotion and is derogatory. The allusion here is that love carries the danger of alienation; it puts tectonic plates in motion and it destabilises the known order. There is a similar hint in the love story between Louisa and Wickham. In the book, their love child pays the price of the ill-thought affair, although on screen consequences are mollified for, presumably, more liberal audiences.

While Austen created heroines who "through their desirability" managed to "dangerously tempt men away from the fixed social order" (Blum 2003: 165), P.D. James desexualises Elizabeth in her sequel, transforming her into someone bound by tradition, class, and etiquette (Craig 2011), possibly to elevate her to Darcy's level, which is a rather disparaging treatment. James's Elizabeth Darcy is so unlike Lizzie Bennet, that screenwriter Juliette Towhidi (2014) had to return to Austen to rediscover her beloved heroine: "I went back to the original. And there was Lizzy Bennet, in all her timeless wit, irreverence and sense of mischief." It is telling that James hints again at the outcome of migration between traditionally fixed social spaces: alienation and lack of integration.

We find Elizabeth mostly indoors or in a carriage, often chaperoned though she is now a married woman, a reversal of the previously independent and outdoorsy Elizabeth who displayed the ability to escape the social conventions of the time. The use of space is therefore evocative of the character's social and cultural position. In Austen, Lizzie walked three miles to Netherfield to visit her ill sister, an expression of sexual energy, rebelliousness, and love of nature (according to Sue Birtwistle, the producer of the 1995 *Pride and Prejudice*, qtd in Parrill 1999). In *Death Comes to Pemberley*, she seems burdened by duty, framed by windows or doors, while looking upon a wet and windy landscape, although the screen version empowers her to some extent, as she is given the role of crime solver.

Physical spaces allude to the social and mental spaces inhabited by Austen's characters (Tanner 1972: 31), so it is no surprise that James dwells a great deal on the geography of the Pemberley estate and especially the location of the murder. Finding the place where the murder took place is key to solving the mystery. In the meantime, being in or out, part of or banished from, also tells a social story. As James (1999: 16) explains, she uses "place" to create an atmosphere, aid characterisation, give credibility, and as a symbolic device. In her author's note at the end of the novel, she emphasises this when she observes that place and setting are a key consideration, particularly in a crime novel, because of the "contrast between peace, order and beauty and the contaminating eruption of violent death" (James 2012: 329).

The Pemberley estate and its various component parts present us with a symbolic geography. Place is "polygamous," to adopt another term from the arsenal of diasporic theory (Georgiou 2006): it is physical, but also social; it speaks of inclusion and exclusion, "here" versus "there"; it is divided by time, then and now, with flashbacks being used to travel between spaces of inclusion and spaces of alienation, clearly changed by the passing of time. Locations such as Pemberley, the forest (with its gravestone, coachman's house, clearing, carved tree, and ruins), as well as the pub and the courts in town, punctuate the development of the plot and become important clues, but they are also markers of social identities. They conjure momentary "third spaces" defined by "in-betweenness" (see Bhabha 1994; 1996), when normal roles and hierarchies are suspended to allow identities to be renegotiated. The court, for example, is the place where Wickham plays either the villain or the hero and where Darcy suffers the humiliation of being reduced to a mere witness, thus prompting him to revisit the consequences of his marriage. In Episode Three (2013), we see the Penny Press being sold in Derby, which adds local and historical colour and interest, but it is also a sign of how times are changing. We are told indirectly that literacy and the popular press are becoming a catalyst for social change.

Change is a recurring theme that provides an important narrative thread. Through it, James alludes to a range of social problems. Her stance is at times conservative (appreciative of the aristocracy) and at times in favour of reform (police and courts). Through lawyer Alveston's voice, for example, she writes: "there is one reform which we would like to see: the right of the prosecuting counsel to make a final speech before the verdict should be extended to the defence" (James 2012: 165).

The comparative device that is the French Revolution allows James to view positively the slow, managed, and controlled change taking place in British society at the time. This is a period when social and political institutions are beginning to come under attack in the early phase of the industrial revolution, and there is the constant threat of a French invasion. The dangers may have felt real at the time, but the impact on the upper classes would have still been small. In the book, Elizabeth ruminates: "*outside there is another world which wealth and education and privilege can keep from us, a world in which men are as violent and destructive as is the animal world. Perhaps even the most fortunate of us will not be able to ignore it and keep it at bay forever*" (James 2012: 55; emphasis in original). It is significant that James puts this excerpt in italics. Through such reflections, characterisation, and dialogue, James comments on the way identities would be affected, as if she were debating with herself the virtues and dangers of social change. The crime itself produces the kind of trauma that results in questions about change, adaptation, and identity, with Darcy reflecting: "The natural order which from boyhood had sustained him had been overturned and for a moment he felt as powerless as if he were no longer master in his house" (James 2012: 66).

It is not by chance that James breaks a key cardinal convention in crime writing – the killer shall never be a servant (Van Dine 1928, rule 11) – because it allows her to connect poverty and crime. Moreover, through Will Bidwell's sister Louisa (played on screen by Nichola Burley), she has the opportunity to emphasise the moral pitfalls that might await working class women and the plight of unmarried mothers. By telling the story of women like Louisa, James explores the larger social story of illegitimacy and its consequences for mothers and children.

In the screen version we see Louisa nursing her baby. Breastfeeding is not an image often seen in costume drama, but it provides Elizabeth with a clue regarding the truth. Illegitimacy was handled by Austen only when some degree of respectability had been recovered, usually through the offer of protection by a pillar of society. We see this both in *Emma* and *Sense and Sensibility*. In P.D. James the harsh consequences of illegitimacy are clear to see when Louisa's baby is sent away (another sign of conservative attitudes), although they are mollified in the screen version, where Louisa is allowed to keep the baby in the end. In Episode Three (2013), Mrs. Reynolds (Joanna Scanlan), the housekeeper, asks why children must always pay the price for the sins of their fathers, a comment that helps address a real social issue in a more direct way, and hints at a possible change of attitude. Significantly, it is women, who, as first-hand witnesses of the consequences of birth outside wedlock, are more ready for change.

The screen adaptation further highlights such debates, notably through dramatising the relationship between the spouses, Elizabeth and Darcy, who find themselves at odds over Georgiana's choice of husband. We also have an interesting dialectic throughout the television series with Colonel Fitzwilliam and Sir Selwyn Hardcastle (Trevor Eve) representing the old, and Elizabeth and Alveston (eventually Darcy and Georgiana) the new, via their attitudes towards gender equality and marriage. Tradition would dictate that Georgiana married her cousin, Colonel Fitzwilliam, higher in rank and wealth than lawyer Alveston and a choice Darcy would initially prefer. However, Elizabeth takes Darcy on a journey, at the end of which he is happy to see Georgiana marrying for love. Darcy and Georgiana may have had liberal inclinations (Darcy's marriage is a case in point), but these are tempered by the burden of responsibility. Not unlike migrants on a journey of adaptation, before their identities become clear, they must grow and develop as human beings, through torment and trauma, acquiring new capital while remaining the pillars of nobility that P.D. James seems to value.

Georgiana would already have a clear example of equality within marriage in Darcy and Elizabeth. During the murder inquiry Darcy says, as he draws his wife near, "Forgive me, Sir Selwyn, my wife and I have no secrets." In Episode Two (2013), the exchange between Darcy and Elizabeth regarding the Colonel's proposal to Georgiana also shows an equal marriage footing, when Elizabeth retorts, "We discussed this, Darcy, and we agreed." The screen version plays cleverly with the tension between old and new, tradition and modernity. Alveston is the character given the task of voicing liberal views based on the writings of the time, like Wollstonecraft, and possibly Locke and Hume. The Colonel, his rival for Georgiana, chooses to interpret his views as radical when they are just progressive. Dr. Clitheroe (character absent from the screen version), also describes Alveston as "something of a radical" when "[t]he peace and security of England depends on gentlemen living in their houses as good landlords and masters … If the aristocrats of France had lived thus, there would never have been a revolution" (James 2012: 216–7). Alveston courts and later proposes to Georgiana directly, unlike the Colonel. Geography, space, and mapping are used metaphorically when readers and viewers are asked to imagine the life Georgiana would have in London, with Alveston, in opposition to the presumably miserable existence in the Colonel's remote Northern castle, a picture of which Georgiana and Alveston find in a book. There is again a hint here of multiple and contrapuntal existences, which depict once more a world on the cusp of change, fluid and in flux.

Death Comes to Pemberley continues the tension between duty and desire, already present in *Pride and Prejudice*, but also magnifies it. Old traditions clash with new ideas and, on screen in particular, those characters representing tradition are represented in a poorer light than those representing change. However, change is never radical. Georgiana still marries aristocracy (Alveston inherits a baronetcy) and Darcy's main roles as owner of Pemberley and carrier of the family line are cemented through another heir. So although the Colonel says to Darcy: "You sound like a radical … I had not realized that you had such an interest in the law or were so dedicated to its reform" (James 2012: 166), change is limited and the status quo is, for the best part, maintained. It is the trespasser of social conventions, the 'other' (Wickham), who has to exit this rigid social space in order for the equilibrium to be resumed. He has always been an outsider, and this ultimately leads to an adaptation failure and symbolic punishment.

Class transgressions are viewed particularly negatively by the old guard, with the Colonel commenting on Wickham's upbringing: "it was dangerous for him to enjoy a privilege which, once boyhood was over, he could not share … Changes in his status and expectations were perhaps made too drastically and too suddenly" (James 2012: 32). This is echoed by Hardcastle

(James 2012: 96) and given as a motive for the probability that Wickham is the murderer. Through Wickham and Lydia escaping the straightjacket of the British class system by departing to America, there is, nevertheless, a hint at a more egalitarian future. Alienation can give way to integration within the right milieu, a society that is ready to accept migrants or "hybrids" like them, who belong nowhere in particular.

Additional seeds of change are evident in the way Elizabeth and Darcy's relationship with their servants is becoming warmer, more fluid, and based on mutual respect and partnership. Yet, rather than being a critique of the rigid class system, this is an endorsement of a well-managed and mutually beneficial relationship between masters and servants. In her friendly rapport with the housekeeper, Mrs. Reynolds, Elizabeth shows how master–servant partnerships can define a community of mutual interest. It is mainly on screen, when Elizabeth visits Will Bidwell (Lewis Rainer) in the cottage, that we clearly see relaxed class barriers, magnified in comparison to the novel. The lady of the manor and the coachman's ill son hold each other's hands in an act of real emotional exchange, while Elizabeth remarks: "Will, for all my fine clothes I can't do anything about what is happening to you right now, but I offer what I can, which is my friendship" (2013). Such moments describe a contrapuntal relationship between the individual and fixed social identities.

The screen version tends to be less conservative and, in the wake of Joe Wright's 'modern' 2005 Austen adaptation, there are modernising and egalitarian attempts. Except for Lady Catherine (played by Penelope Keith), the costumes lack frills or puffed up sleeves; the simple linens and muslins enable Elizabeth to move more freely, to grab and hug her child or snatch her coat to take lunch with Darcy at the assizes. Casually, she seems to be exposing herself to the rough and tumble of the local town, more like a modern–day aristocrat. For her this is more readily possible, because she has 'migrated' socially already, she has acquired the right capital which results in a hybrid identity, though, of course, she still has to overcome an inferiority complex.

Yet any journey is defined not just by looking forward but also by looking back. References to tradition are still made in the screen version, to echo the more numerous references in the book. Darcy's first appearance on screen is to remonstrate that his peace is disturbed by the hustle and bustle of the arrangements for the ball. He alludes to the tradition that the man of the house is always present on the day of the ball, to which Elizabeth responds, "perhaps some traditions need updating … all good things come to an end." This early suggestion of change, even inside this hierarchical and relatively cloistered environment, is significant. However, change does not happen straight away. Georgiana's remark that she would not marry without Darcy's approval, exemplifies the inherent contrapuntal tension between the old and new order, between individual desires and social constraints. Eventually though, she frees herself from the burden of personal self-sacrifice to marry the man she loves.

Paul Gilroy (1993: 19) explains in one of his most influential texts that in terms of identity, the term 'routes' is more significant than the homonym 'roots.' The journey of change is undertaken within an in-between space or a "scape" (Said 1994: 1–15) that leads to self-actualisation, defined by movement, occasional placelessness, and repositioning. It is interesting to see so many characters out of place in this Austen sequel. Darcy and Georgiana need to find the right "home" for their changing identities. Wickham starts his journey unhomed but, through migration, might become homed. His experiences on and off the battlefield have earned him the symbolic capital needed in the New World. Perpetually dislocated in terms of social standing, Wickham would only thrive where class is irrelevant. The home becomes a metaphor for morality, leading to the establishment of a moral geography. Stigma, marginalisation, 'us' and 'them,' are also about belonging (or lack of), with misfits like Wickham often shamed or punished, which also shows a dose of conservatism on James's part.

The Colonel, an aristocrat who has been exposed to the realities of the battlefield, has no problem doing questionable business with Mrs. Younge (played by Mariah Gale), Georgiana's former rogue minder, but Darcy finds this repugnant. Darcy may run a large estate, but his morality and liberal views come from idealism rather than having been pressurised by the external realities. The murder and its inescapable impact lead him to confront the seedier side of life. Alveston is a realist because of the work he does at the Old Bailey, yet his ethics and views on reforming the judicial system and women's rights are shaped by his strong moral compass and in opposition to harsh realities. These characters operate between complex spaces: Pemberley, the forest, the pub, the abbey ruins, the courts and prisons. These spaces are used metaphorically in various ways: as protective fortresses, as agents of corruption, as a cause for change.

Space was also important for Austen. In his analysis of *Mansfield Park*, Said (1994: 102) is right to emphasise the importance of space in the relationship between home, Europe, and the empire, but also the importance of home as the domestic private space, colonised, we might add, by women's lives and love stories. Space is also an important indicator of status (Pemberley versus Longbourn or Cheapside) and Austen provided us with a symbolic map of fashionable places for the aristocracy of the time (London, Bath, Brighton). The landscape aids characterisation, explains character relationships, and emphasises class distinctions (Parrill 1999: 33). The colonisation of space by women is also significant for gender relations. "The Library at Pemberley was as freely open to her as it was to Darcy" (James 2012: 15), reflects Elizabeth at the beginning of James's novel, and further on remarks: "It had been a revelation to Elizabeth that there were men who valued intelligence in a woman" (James 2012: 16). When Darcy is irked that the Colonel insists on the two men taking position in the library in the aftermath of the murder (James 2012: 122), even though Darcy would have preferred the comfort of his wife's bed, we have a clear separation between old fashioned chivalry and new attitudes for couples. Elizabeth interprets it (rightly) as an attempt to create a wedge between the spouses, allowing for manipulation and control. Metaphorically, it speaks of the individual's alienation in the face of an oppressive social system.

The tension between the inside and the outside served several purposes in Austen and also for James. The Pemberley estate is about status and family line in Austen, but needs to serve primarily as the location of a crime in James and the scene of (transitory) marital discord in the screen version. Spaces and locations transition and are being constantly reconfigured. The weather is used, typically of a crime story, to set up the atmosphere, but also to delineate various spaces and to contain and separate characters. Being inside or outside becomes a way of alluding to secrets, separation, and difference. When Elizabeth observes Georgiana walking with Colonel Fitzwilliam in the garden, for example, we are aware of the incongruity between the two.

"Pemberley, deep in the Derbyshire countryside, resembles the other isolated communities in James's oeuvre, placing its inhabitants under unusual strain and showing them in a new light because of it," observes Craig (2011). He also notes the similarities between James's Darcy and Adam Dalgliesh, the detective–poet from her crime novels. Indeed, James is mostly at home in enclosed spaces: an island in *The Lighthouse* (2009a), a manor house in *The Private Patient* (2009b), or a clinic in *A Mind to Murder* (2010). In the screen version of *Death Comes to Pemberley*, the use of colour, the subtle interplay between light and shadow and the use of 'Gothic' spaces, like the abbey ruins for the ethically problematic transaction of a baby, serve to further emphasise the characters' identity and moral traits. Gothicism is used to hint once again at alienation, the incongruity between individual freedoms and social constraints, two parallel stories which become contrapuntal in the depiction of a certain historical moment.

The title of the sequel, *Death Comes to Pemberley*, is significant. Pemberley is the place where secrets and knowledge coexist, where past, present, and future come together, the turning point for both the action and the lives of the main characters. Pemberley is the place where, in Austen, there is a significant arc in the story (Elizabeth and Darcy come to an unspoken understanding), where characters change direction (Darcy has changed his manners and Elizabeth has changed her mind), and where text and subtext, the known and the hidden, run in parallel (Elizabeth and Darcy's shared knowledge of Wickham's misrepresentations, though hidden from view, remains a constant source of tension and alienation).

P.D. James appropriates the notion that at Pemberley appearances can hide certain secrets and transforms it into the source of a greater mystery, a crime. The ambiguous outer landscape, beautiful but threatening, matches the inner and multifarious landscape that defines the characters' identities. The house itself, with its corridors, staircases, doors, and locks, as well as the surrounding landscape (park, gardens, forest, hills, and gorges), becomes another character in the plot and visually assists many plot developments. The landscape itself becomes a clue. This is obvious from the opening scene, which sees two servant girls wandering into the woods. Frightened by a ghostly encounter and accompanied by the sound of screeching birds, the girls scream as they run from the woods towards the house. The juxtaposition is thus created: Pemberley (safety, family) versus woods (danger, sin), the inside versus the outside, appearance and reality, truth and lies. It is another instance of contrapuntal tension and a sign of hybridity and fluidity (anything can change, at any time).

The metaphorical journey of change towards new identities is mainly told through location and space, but also, to a certain extent, through the subject of trauma.

Before the characters move forward, they have to look back to resolve previous traumatic events. The discovery of the body is only one such event, copiously exploited on screen through mise-en-scène, the use of colour and music, to recreate the impact on Darcy in the novel: "a mental earthquake in which he no longer stood on firm ground and in which all the comfortable conventions and assumptions which since boyhood had ruled his life lay in rubble round him" (James 2012: 124).

Memory in the novel and flashbacks on screen are effectively used to merge the two stories (*Pride and Prejudice* and *Death Comes to Pemberley*) and give us backstory. Flashbacks are used, for example, to explore the relationship between Darcy and Wickham from boyhood and visualise in particular one event that they shared as equals: the hanging of a young peasant for theft. This is related to a backward–forward movement. The flashbacks depict the bond between Darcy and Wickham in the past, but also offer us clues about Wickham's possible future, as they are accompanied by musical motifs and images associated with the wood (the scene of the crime), his arrest, and potential hanging.

When Wickham recounts his version of events to Sir Hardcastle, the magistrate in the case, flashbacks focus on the verbal exchange at the inn between Wickham and the victim, Captain Denny (Tom Canton). They are both a clue and a red herring: a clue because this serialised flashback eventually reveals the truth, but also a red herring because Denny's accusation that Wickham knows nothing of how women feel could easily refer to his wife Lydia (Jenna Coleman), rather than Louisa Bidwell, his lover.

Elizabeth's prejudice, her initial rejection of Darcy, and her infatuation with Wickham are also reiterated to us by flashbacks, giving us an insight into the character's past and present identities. In Episode One (2013), we sense Elizabeth's regret at perpetuating Wickham's prejudice regarding Darcy. In Episode Two (2013), we have Elizabeth's flashback to Darcy's first proposal and a previous ball where she overheard comments on her fine eyes, the sparse wealth of her father's estate, and her sister's marriage to Darcy's steward's son. Both flashbacks spark fear that Darcy might be regretting his choice of bride. Darcy himself has a flashback to the payment he had to make to Mrs. Younge in order to find Lydia and Wickham in London.

These flashbacks act as a reminder of the characters' history and the *Pride and Prejudice* story, but also as clues. In *Death Comes to Pemberley*, it is Wickham and Mrs. Younge who try to pay Louisa off. In Episode Three (2013), the past catches up with Wickham. His flashbacks to the hanging of Patrick Riley and the events in the woods construct him as a more complex character, while also serving to fill in the gaps. The use of memory or screen flashbacks are not just filmic techniques – interestingly, little used in period drama – but also a social commentary alluding to poverty and class differences as a possible cause for crime. The flashbacks also reveal a long process of alienation, which explains Wickham's ultimate predicament. His upbringing may have given him some advantages (he does not end up like Riley), but has put him at odds with his own milieu: he becomes an outcast, never to be an insider, thrown in between social spaces. Flashbacks also speak metaphorically about the hybrid nature of identity, and the outcome of trauma, which is change.

The journey of change, the navigation of liquid or symbolic spaces and geographies, as well as negotiating trauma, are an important part of migration, but also of textual adaptation. Through these devices, identities are reconfigured: the identity of the text and its simultaneous embodiments, as well as the identities of characters and plots. The text is thus contextualised, allowing for more nuanced and multiple meanings. Most importantly, the text acquires a moral identity through social mapping and the politicisation of time and space. *Death Comes to Pemberley*, already a mash-up, a hybrid, offers an appropriate example in our attempt to propose a sociology of adaptation. In its various incarnations, off and on screen, we witness the contrapuntal negotiation between the characters' stories and the larger social and historical canvass, a process influenced by the author's own political stance. Most importantly, this example shows that social issues such as women's emancipation and class hierarchies remain in constant renegotiation, across various historical and cultural spaces. This speaks for the potential of costume drama to be constantly renewed for contemporary audiences. Identity will never cease to be a point of contention.

Works cited

Amis, M. (1990) "Miss Jane's Prime," *The Atlantic*, 262: 100–2.
Austen, J. (1972) *Pride and Prejudice*, 1813, London: Penguin.
Austen, J. (1996) *Mansfield Park*, 1814, London: Penguin.
Austen, J. (2012) *Sense and Sensibility*, 1811, London: Penguin.
Bauman, Z. (2005) *Liquid Life*, Cambridge, UK: Polity.
Bhabha, H.K. (1994) *The Location of Culture*, London: Routledge.
Bhabha, H.K. (1996) "Culture's in-between," in S. Hall and P. du Gay (eds) *Questions of Cultural Identity*, London: Sage, 53–60.
Blum, V.L. (2003) "The return to repression: Filming the nineteenth century," in S.R. Pucci and J. Thompson (eds) *Jane Austen and Co. Remaking the Past in Contemporary Culture*, Albany, NY: State University of New York Press, 157–78.
Bride and Prejudice (2004), dir. G. Chadha. Miramax. DVD.
Brown, J.P.C. (2015) "Screening Austen: The case of *Emma*," *Adaptation*, 8(2): 207–36.
Bryant, J. (2013) "Textual identity and adaptive revision: Editing adaptation as a fluid text," in J. Bruhn, A. Gjielsvik, and E.F. Hanssen (eds) *Adaptation Studies: New Challenges, New Directions*, London: Bloomsbury, 47–67.
Cartmell, D. (2010) *Screen Adaptations – Jane Austen's* Pride and Prejudice: *The Relationship between Text and Film*, London: Methuen.
Craig, A. (2011) "Death Comes to Pemberley," *New Statesman*, 14 November. Online. Available at: http://www.newstatesman.com/books/2011/11/pemberley-darcy-death-wickham (accessed 19 April 2017).
Death Comes to Pemberley (2013), dir. D. Percival. BBC. DVD.
Dicecco, N. (2015) "State of the conversation: The obscene underside of fidelity," *Adaptation*, 8(2): 161–75.
Emma (1996), dir. D. Lawrence. ITV. DVD.

Georgiou, M. (2006) *Diaspora, Identity and the Media: Diasporic Transnationalism and Mediated Spatialities*, Cresskill, NJ: Hampton Press.

Gilroy, P. (1993) *The Black Atlantic: Modernity and Double Consciousness*, London: Verso.

Godwin, W. (2005) *Caleb Williams*, 1794, London: Penguin.

Horsley, L. (2005) *Twentieth-Century Crime Fiction*, Oxford: Oxford University Press.

Hutcheon, L. (2006) *A Theory of Adaptation*, London: Routledge.

James, P.D. (1999) *Time To Be in Earnest: A Fragment of Autobiography*, London: Faber & Faber.

James, P.D (2009a) *The Lighthouse*, London: Penguin.

James, P.D (2009b) *The Private Patient*, London: Penguin.

James, P.D (2010) *A Mind to Murder*, 1963, London: Faber & Faber.

James, P.D (2012) *Death Comes to Pemberley*, London: Faber & Faber.

Meinhof, U.H. and Triandafyllidou, A. (2006) "Beyond the diaspora: Transnational practices as transcultural capital," in U.H. Meinhof and A. Triandafyllidou (eds) *Transcultural Europe: Cultural Policy in a Changing Europe*, Basingstoke, UK: Palgrave Macmillan, 200–22.

Murray, S. (2012) *The Cultural Economy of Contemporary Literary Adaptation*, London: Routledge.

Parrill, S. (1999) "What meets the eye: Landscape in the films *Pride and Prejudice* and *Sense and Sensibility*," *Persuasions*, 21: 32–43.

Pride and Prejudice (1995), dir. S. Langton. BBC. DVD.

Pride and Prejudice (2005), dir. J. Wright. Working Title. DVD.

Rushdie, S. (2009) "A fine pickle," *The Guardian Review*, 28 February: 2–4.

Said, E. (1994) *Culture and Imperialism*, London: Vintage Books.

Sanders, J. (2006) *Adaptation and Appropriation*, London: Routledge.

Schober, R. (2013) "Adaptation as connection – transmediality reconsidered," in J. Bruhn, A. Gjielsvik, and E.F. Hanssen (eds) *Adaptation Studies: New Challenges, New Directions*, London: Bloomsbury, 89–112.

Silverstone, R. (2007) *Media and Morality: On the Rise of the Mediapolis*, Cambridge: Polity.

Slumdog Millionaire (2008), dir. D. Boyle. Celador. DVD.

Stam, R. (2005) "Introduction: The theory and practice of adaptation," in R. Stam and A. Raengo (eds) *Literature and Film: A Guide to the Theory and Practice of Film Adaptation*, Oxford: Blackwell, 1–52.

Tanner, T. (1972) "Introduction," in J. Austen, *Pride and Prejudice*, 1813, London: Penguin, 7–46.

Towhidi, J. (2014) "Interview with Juliette Towhidi," in *BBC Media Centre*. Online. Available at: http://www.bbc.co.uk/mediacentre/mediapacks/pemberley/towhidi (accessed 1 December 2015).

Van Dine, S.S. (1928) "Twenty rules for writing detective stories." Online. Available at: http://gaslight.mtroyal.ca/vandine.htm (accessed 1 December 2015).

18

ADAPTING IDENTITIES

Performing the self

Katja Krebs

As has been observed on numerous occasions, adaptation studies has broadened its scope of enquiry for better or worse from the relatively narrow field of literature to film adaptations. This Companion is a case in point where the chapters in sections such as history, identity, reception and technology go far beyond the limited novel-to-film model. Adaptation studies now encompasses studies that engage with rewriting, appropriation, intertextuality and intermediality as examples of processes and products of adaptation. For some scholars, such proliferation is a step (or three) too far and they offer a convincing argument why adaptation studies needs to undergo a process of disambiguation and pay more attention to the development of a more particular definition of adaptation (see, for example, Cardwell in this collection) in order to establish more clearly a much-needed analytical aesthetics of adaptation. Furthermore, adaptation studies has yet to excise its, arguably, over-reliance on the case study at the expense of more detailed consideration of the conceptual framework within which we read adaptations. And while the case study is a well-established and rigorous methodology, it will only move the field of adaptation studies on if it becomes a means to a conceptual end rather than an end in itself.

Being fully aware of the dangers inherent in the broadening of a concept of adaptation to contain elements of intertextuality, a position which could lead to an adaptation studies without focus, I would also like to embrace the scope that an adaptation studies which refuses to settle on a narrow definition of its field brings with it. As a result, what follows is a consideration of a series of case studies, arguably more intertextual than adaptational by nature, in order to develop further the concept of adaptation rather than concentrate on the specifics of the case studies themselves only. Such an analysis has the potential to offer not only a detailed consideration of processes of adaptation in themselves but, importantly, processes and positions of identity construction through adaptation, performance and reception. As such, this chapter's aim is to contribute to the development of adaptation studies through a discussion and analysis of notions and displays of identity through adaptation.

While looking for suitable examples of adaptation of identity and adaptation as identity in the Theatre Collection at the University of Bristol as long ago as 2014, I came across boxes of so-called miscellaneous content, ordered in appearance yet un-catalogued beyond assigning letters of the alphabet to its material: a–c, d–f and so forth. And within those boxes were hidden stories of performers whose acts were based on various forms of identity construction through adaptation. A number of these performed identities seemed based upon constructions of cultural identities through the adaptation and performance of national identity. And the very fact that their content became known by happenstance as miscellaneous items rather than through cataloguing is noteworthy in terms of the position the archive plays vis-à-vis non-canonised popular performance and adaptation practices. While the role the archive plays is important to our understanding, appreciation and categorisation of adaptation, original, popular and canonical texts, to list but a few, the confines of this chapter mean that such concerns can only be alluded to. Let us return to the content of these boxes and the display of adaptation of/as identity within them.

While I am sure that instances of adaptation of/as identity as were preserved in those boxes are not necessarily exclusive to the first half of the twentieth century nor to the Anglo-American stage, they happen to coincide with "the heyday of interculturalism in the west – at least as far as it involved contact or integration with, or appropriation of, the cultural forms of the 'other' (Indigenous, 'oriental', or other non-western peoples)" (Knowles 2010: 11). These boxes in front of me contained some intriguing examples of adaptations of identity. Grudgingly I paid little attention to the French Poodle impersonator[1] in an effort not to muddy the waters between impersonation and adaptation; instead, I focused on the adaptation, through performance, of cultural and national identities. The most intriguing of all the cases I came across is Chung Ling Soo a.k.a. William Elsworth Robinson (1861–1918).

Famous as a Chinese magician and best known for his trick of catching marked bullets fired at him, he died on stage in 1918, at the age of 57, after said trick went tragically wrong. What is most fascinating is that all throughout his British career, the public at large insisted on believing this white American to be Chinese. Robinson was born in New York in 1861 and, after an early stage career in the US, he relocated to London in 1905. Adaptation in terms of Robinson's stage persona is twofold: not only does he present himself to be Chinese and thus adapt Western assumptions of the East, he also adapts the act (and name) of Ching Ling Foo, a Chinese performer who toured the US from 1898 before performing in London himself.

Robinson's presentation of self, his performance of a cultural identity, on and off stage was so tightly constructed that he even dressed up his female sidekick, Soo-Suee-Seen a.k.a. Olive Path, and child, and, at the height of his fame, appeared in public with an interpreter at his side at all times, pretending only to speak Mandarin. Adaptation and identity are so closely knit in this case that this active adaptation of the imagined other became known to, or rather was acknowledged by, the public at large as adaptation not until after his death. Of course, Robinson's appearance as Chung Ling Soo as well as an audience's readiness to accept his adapted body as that of the cultural other perpetuates the colonial project. Looking at photographs of Robinson/Soo it becomes apparent that this adaptation of the culturally other relies on very specific instances of costume. While make-up plays some part in Robinson's/Soo's process of adaptation, it is the readiness of an audience to ignore facial features and instead embrace costume as a guarantor of authenticity that is noteworthy here. As his stage persona, Robinson appeared with a Manchu queue, itself intertwined with the Qing Dynasty (1644–1912) which imposed this hairstyle on the Han Chinese. This braided hairstyle is specifically male and appears in Western depictions of China as early as the seventeenth century. We can find them, for example, in paintings of Chinese performers by Johan Nieuhof (1618–1672). Nieuhof worked for the Dutch East India Company, and his account of the Dutch Embassy to China during the seventeenth century "remained the authority on China into the mid-nineteenth century" for English readers (Pagani 1998: 30). Western imaginations of the Chinese other are closely intertwined with this partly shaved and partly braided hair, and Robinson used such a hairstyle as a method of adaptation on and off stage. Together with wearing long tunics or jackets, imposed on court and government officials also as part of the Qing Dynasty, an audience literate in the reading of certain visual clues was willing to suspend disbelief to the extent that Robinson's North-American ancestry came as an apparent revelation after his death.

This kind of adaptation then offered seemingly authentic access to the other through the adaptation of an imagined self: "Only in particular circumstances is nationality self-consciously and actively imagined, usually in circumstances where the borderlines between 'us' and 'them' can be not only marked but negotiated and brought into being" (White 2001: 109). Through the act of adaptation Robinson/Soo actively imagined, and thus engendered a public imagining or rather confirmation of such imagining, of what is meant to be 'Chinese'. He enacted the borderlines between 'them' and 'us', whether by performing the act of translation with an interpreter by his side, or by constructing and enacting notions of authenticity through adaptation. And his adaptation of the notion of the Chinese magician was deemed more authentic than the Chinese magician this was based upon himself, namely Ching Ling Foo who was born in Beijing in 1854 and died in Shanghai in 1922 after touring the US and Europe. Adaptation then becomes an integral part of the "cannibalisation of forms without respect for the cultures that produced them" and it participates actively "in the west's colonisation of the world's cultures and peoples" (Knowles 2010: 12). Crucially, the source of this particular adaptation is not only a specific native Chinese magician, but more generally the Western ideal of the Chinese other. The pleasure of adaptation, which according to Hutcheon (2006) is firmly related to an audience's knowledge of the source, in this case consists of the pleasure of having one's own cultural assumptions confirmed. What comes together here is Hutcheon's notion of the pleasure of adaptation with Laura Mulvey's notion of the pleasure of the gaze where "[t]he determining male gaze projects its phantasies on to the female figure which is styled accordingly" (Mulvey 1999: 837). In this case, the gaze determines the Chinese figure in terms of colonial phantasies.

Robinson's/Ching Ling Soo's fame is not to be underestimated, and he continues to be part of a public imagination and collective cultural memory as a quick search on YouTube demonstrates. We can find a murder mystery narrated by Boris Karloff which lays the blame for Robinson's/Soo's death firmly at his estranged wife's door, numerous short documentaries and magic circle discussions, and, more recently, an opera based on his life: 2006 saw *The Original Chinese Conjuror* composed by Raymond Yiu with a libretto by Lee Warren and performed at Southwold Pier, Suffolk, UK, and the Almeida Theatre in London. In other words, Robinson/Soo still captures the imagination of a Western audience. And the material available seems to have proliferated to a large extent since I first opened that box marked 'miscellaneous' in the Theatre Collection, University of Bristol.

Robinson played an undeniably active part in this adaptation that is the enactment of Chung Ling Soo, yet the audience was and continues to be complicit in the process of such adaptation. As if to confirm the centrality of the audience's complicity, since his death, his enduring audience seems more interested in Robinson's adaptation of the imagined other, enacted by his performance on and off stage and made visible by his tragic death, than his skills as a magician. And such adaptation of the imagined other, as is the case with Robinson/Soo, becomes central in an audience's construction and confirmation of their own identity. This specific act of identity construction through adaptation becomes an act of confirmation of the audience's understanding of their notion of self.

It is the comprehensive nature of this particular adaptation of identity that captures the imagination. This instance of adaptation, while it is certainly recognisable as adaptation, strives to be invisible[2] and the audience is complicit in such pretence. It is not the relationship between the source and its adaptation that necessarily provides the audience with the pleasure inherent in adaptation (see Hutcheon 2006). Arguably, it is so-called identity maintenance vis-à-vis an audience's notion of themselves that takes on a more central role here and contributes immensely to the pleasure of adaptation. Sanders talks about the pleasure of adaptation in terms of its participation in canon formation and "activating and in some cases reactivating the profile and popularity of certain texts" (Sanders 2006: 29). Robinson's/Soo's performance reactivates, and continues to reactivate even after his death, the popularity of imagining the Chinese other, whereby the Chinese other is constructed in relation to an understanding of self. Sanders argues further that "familiarity with the source [is not] necessary, but the experience [of an adaptation] is certainly altered by that stance of familiarity" (Sanders 2006: 29). The source of this particular adaptation – the collectively imagined other – does not necessarily precede the experience of Robinson's/Soo's adaptation but rather exists simultaneously with it as well as complicates it. The identity of the audience is confirmed, constructed and enacted through this performance and adaptation of the imagined other. In this case, source and adaptation are a multi-layered structure of multiple texts which is not necessarily governed by a hierarchical or temporal relationship. Yet, it allows insight into an Anglo-American audience's notion of identity of self as well as into the imagined identity of the other.

The story of William Ellsworth Robinson/Chung Ling Soo helps conceptualise the way we can think of identity in terms of adaptation or, indeed, adaptation in terms of identity, and what follows is an attempt to unpick the interwoven nature of adaptation, imitation, mimicry and the performance of (imagined) identity further.

Mimicry and imitation of well-known or imagined identities were, at the time of Robinson's performance of Chung Ling Soo, not an unknown form of popular performance: especially in comedy, mimicry was a conventional style, enacted particularly by female performers. As Susan Glenn observes:

> Though male performers were also implicated in the epidemic of [mimicry], imitations, especially imitations of well-known performers, were largely the province of female comics. … These women were part of what might be called a mimetic moment in American comedy in the years between 1890 and end of the 1920s. On the popular stage of vaudeville and musical revue every conceivable kind of comic imitation was in full flower: blackface, minstrelsy, gender impersonation, burlesque, parody and ethnic caricature.
>
> *Glenn 1998: 48*

While Glenn makes this point with particular reference to the American stage, parody, gender impersonation and burlesque were also very much *en vogue* on most European and especially British vaudeville and music hall stages. Arguably, this kind of performance can be classified as what Julie Sanders terms "mere imitation" (2006: 12). Performers like Robinson, however, went far beyond notions of mimicry and imitation; Robinson's rewriting of identity can be understood as adaptation, as his performance can be read as more than "replication as such, but rather complication, expansion rather than contraction (Andreas 1999: 107)" (Sanders 2006: 12) as we have seen above. In short, it complicates and expands an audience's understanding of self.

Mimetic comedy of the kind referred to by Glenn – the comedy of imitation and parody – thrived in the tension between nineteenth-century bourgeois fascination with imitation and the early twentieth-century "modernist" intellectual glorification of "authenticity" (Glenn 1998: 54). Such glorification of authenticity goes hand in hand with global modernity and is, for example, also expressed by Robinson's/Soo's audience's insistence on believing his performance of the other to be authentic.

> The entertainment industry, of which theatre and tourism have historically been integral and interrelated components … is, then, a form of "poetic world-making" that

works across the terrain of fantasy and materiality, fabricating "intersensory illusionings of passing fancies", capital momentum, and concrete communities.

Werry 2005: 357 citing Boon 2000: 524–556

Robinson's enactment of the imagined other is an example of "poetic world-making" but it also functions in relation to the establishment and confirmation of concrete communities: them and us. The tension between imitation and authenticity Glenn talks about very much exceeds the realm of mimetic comedy and can be observed in the modes of presentations of the self in performance outside of the confines of comedy. One such example is Haroun-al-Raschid, whose real identity I have not been able to confirm, yet photographs of his performance persona, which I came across in that box of miscellaneous content in the Theatre Collection at Bristol University, indicate that he performs with reference to *One Thousand and One Nights* and presents a performative imagining of the literary adaptation of Haroun-al-Rashid, the fifth Abbasid Caliph, who is considered "synonymous with a golden age of medieval Islamic civilization" (El Hibri 1999: 17). This performance of Haroun-al-Raschid includes a turban, ear-rings and makeup, yet it is the spectacles the performer wears which locate him temporally as well as geographically and make visible the adapted nature of the performance. An audience is not, as was the case with Robinson/Soo, able to pretend they are watching an authentic performance and, as is the case with Robinson/Soo, it is not so much the identity of Haroun-al-Raschid that is of interest; instead, what is of importance is to what extent such a performance of adaptation contributes and shapes the identity of an audience. Edward Said has argued in his seminal *Orientalism*, "European culture gained in strength and identity by setting itself off against the Orient" (1979: 3), and in these cases an audience's sense of identity is strengthened through the performance of the imagined other, a performance of what the audience considers to be identifiably oriental. The source of these adaptations then, are the coexistent, simultaneous imaginings of the other. And who better to embody the oriental other, against whom the identity of a European or western self can be constructed, than the figure of Haroun-Al-Rashid.

Further examples of the simultaneity of source and adaptation, performance of the other and constructions and maintenance of identity, include the levitating miracle that is Zulicka, whose real name I have not been able to determine, and the Anglo-Neapolitan Cabaret, who under the guidance of Guiseppe Ceci offer 'authentic' Italian routines to be performed in restaurants, ladies' festivals, concert parties and the like. Flyers, documents and, in some cases, photographs can be found in the miscellaneous boxes in the Bristol Theatre Collection. While it has proven difficult if not impossible to find any proof as to the 'real' identities of the performers who present themselves as embodying such imagined identities, the audience seems prepared to accept such adaptations of identity as part of the appeal of these acts. Zulieka perpetuates the imagined identity of the oriental woman: claiming to be from Khartoum, Sudan's capital, and witness to battles involving British and Egyptian forces in the latter half of the nineteenth century, the posters accompanying her performance contain a sketch of Zulieka levitating above a small crowd. The imagery taps into visual depictions of the contemporaneous trends in spiritualism, table-turning in particular.[3] Zulieka, the description on the poster claims, is also known as Thauma, the eighth wonder of the world, and she lives, moves, breathes and swings her bodyless form in mid-air. Wearing an off-the-shoulder bodice and handing out playing cards to male members of the audience, the exotic meets popular performance in a display of an imagined identity of the other.

Similarly, the Neapolitan Cabaret, which is not dependent on notions of the non-western other but instead perpetuates the notion of the romantic Italian ideal, adapts the audience's imagination of what is meant to be Italian, or rather Neapolitan. The Cabaret consists of four female performers, all swinging tambourines and dressed in identical costumes complete with head-coverings which are most akin to Wilhelm von Gloeden's painting from 1870 entitled *Costumes of the Romana and Neapolitan Countryside: Woman in the Traditional Costume.*[4] The poster further announces an act which displays up to date costumes, is snappy and full of talent, has originality and punch and is delightfully graceful. For an audience which cannot afford the great Italian tour, an authentic adaptation of that experience is available for hire.

Of course, all these examples are based on imagined sources, and thus are essential in the construction of the identity of self and identity maintenance, as well as the construction and maintenance of the imagined other. In other words, the sources for these adaptations exist in the communal and collective imagination of the audience at the moment of encounter, and they do so in relation to an audience's understanding of their own identity. William Ellsworth Robinson adapts a collectively imagined identity of the other when he performs Chung Lin Soo and his body becomes the medium of adaptation. The source for his adaptation is not a specific magician nor performance nor text but instead an imagined identity of the other which exists only in relation to the self. Similarly, Zulieka as well as the Neapolitan Cabaret are not based on specific experiences but rather adapt and consequently embody the imagined other. The body, through which adaptation of identity is enacted here, is a reflection of a particular notion of identity: a specific cultural, national, gender identity in which it is embedded rather than necessarily that which it adapts. Adaptation of an imagined identity, in these particular cases, represents identity within the adapting culture rather than offer a glimpse of the identity of the imagined other. Colette Conroy discusses the relationship between the body of the spectator and the body on stage, and it becomes clear that the performed body is always necessarily understood in terms of the body of the self:

> The analysis of the performing body also tells us something about the spectating body. Whenever I watch or analyse a piece of theatre I occupy a physical perspective, and I rely on my own body as the vantage point of my analysis. So my analysis is always subject to the restrictions or possibilities that my own body imposes or opens up.
>
> *Conroy 2010: 6*

Bodies that present adaptations or ideals of identity, as is the case with all the examples so far, do so in relation to the body of the spectator, or rather the spectator understands the performing body necessarily in relation to her or his own body. And our own body "mediates the relationship between self-identity and social-identity: consequently, the social meanings attached to bodily display and expression are an extremely important factor in an individual's sense of self" (Coupland and Gwyn 2003: 2).

Less specific in his adaptation of a national identity than Robinson, Haroun-Al-Raschid, Zulieka and the Neapolitan Cabaret, and more concerned with the adaptation of a cultural identity is Sigmund Neuberger/the Great Lafayette. Arguably the Liberace of his day, several stories of Neuberger's/Lafayette's authentic self were in circulation which contributed very much to his construction of an, arguably, "unmoored self" (Schweitzer & Guadagnolo 2012: 154). Depending on which newspaper article you consult, the Great Lafayette a.k.a. Sigmund Neuberger was born in the US in 1871 to Polish and Jewish parents and studied art in Italy before working as an entertainer in mining camps in the Western US. Apparently, his first appearance in London was at the Alhambra Theatre in 1892 before returning to the States to develop his acts and take London by storm under his pseudonym the Great Lafayette when performing at the London Hippodrome in 1900.

Alternative accounts tell of Sigmund Neuberger being born in Munich, the son of a wealthy German silk merchant. As the story goes, one day he brought home a dog, a stray no less. When his father ordered the dog out, Sigmund "hurt and unhappy, walked out, taking the dog with him. He went to America cutting himself off entirely from his family" (Anon. 15 February 1955). Apparently, so this romantic version of events goes, his family only found out what happened to him when reading a newspaper article about the Great Lafayette and his dog.

There is more to his likeness to Liberace than his fondness of dogs: one article in 1955, long after his death in 1911 – which demonstrates, just as is the case with Chung Ling Soo, that he was very much part of a collective cultural memory – quoted him as saying that he enjoyed "the endearing elegance of female friendship" but it was animals that he cared for the most (Jackson 1955). Sigmund Neuberger/Great Lafayette lived in an extravagant house in Tavistock Square in central London, adorned with murals of his dogs, as a full-page picture spread in *The Sketch*[5] (Anon 17 May 1911) attests; the house was equipped with a bathroom for his dogs and an internal staircase garden. Only one of the seven photographs depicts Neuberger's/Lafayette's stage show: his personal life seemed to capture the imagination of his audience to a much larger degree. In addition to the dogs' bathroom and staircase, we get to see Lafayette in his bed accompanied by two of his four-legged friends. Also, a photograph of his dog Beauty whose "parting has caused a wound that never can be healed" (Anon 17 May 1911: 171) and the Great Lafayette's very own banknote is included. A close friend of Harry Houdini's, Neuberger/Lafayette is undoubtedly an act of comprehensive adaptation where even the glimpses into an apparently authentic self are themselves acts of adaptation. It is as if intertexuality itself is embodied here as multiple layers of adaptation come together in the construct that is the Great Lafayette. He died on 9 May 1911 in a fire during one of his performances at the Empire Theatre in Edinburgh (see Mc Kerracher 1985: 166–171). His body was found near the remains of the lion which appeared in his show and "it was thought Lafayette had died trying to save the beast" (Jackson 1955: n.p.).

It is his embrace of the 'unmoored self' similar to Chung Ling Soo, that makes him an interesting case study in terms of adaptation of the other and identity of the self. While Chung Ling Soo enacted and embodied pseudo-translation – with a translator at his side he convinced audiences that he could speak only (an imagined) Mandarin – Lafayette revels in the mystery of an anonymous source. What both have in common is that their performance acts and personas are adaptations of an imagined other. Adaptation and source exist simultaneously and are inextricably linked as well as contributing to notions of identity within their audience. What these performances are based upon "are two experiences of performing identity: departing from one's own identity and staging the fantasy about the other; and resuming one's own identity through faking otherness" (Soares 1998: 294).

While Chung Ling Soo/William Elsworth Robinson is an extreme example of the former – that is, departing from one's own identity and staging the fantasy about the other – the Great Lafayette a.k.a. Sigmund Neuberger is the embodiment of the latter: resuming one's own identity through faking otherness. Soares, in his article "Staging the self by performing the other: Global fantasies and the migration of the projective imagination" (1998) has identified these two issues of the presentation of the self as a characteristic of the late twentieth-century phenomena of globalisation; arguably these adaptation practices of the self which we can observe at the end of the nineteenth and beginning of the twentieth century are much earlier examples of a performance of a global imagination and identity construction. The Chung Ling Soos and Great Lafayettes of the performance circuit are examples of the embodiment and thus performance of an imagined other, where the nineteenth-century fascination with imitation meets the twentieth-century one with authenticity; an authenticity that is firmly bound up with constructs of identity. As Vannini and Williams argue, "Authenticity is not so much a state of being as it is the objectification of a process of representation" (2009: 3) and firmly linked to identity construction and identity maintenance. In the examples cited above, the process of representation which Vannini and Williams talk about, is based upon adaptation, whether that is the adaptation of historical and literary figures, that is, Haroun-al-Raschid, or the imagined other as is the case with Chung Ling Soo.

As argued above, the imagined other is, of course, not imagined or constructed by the performer alone, but part of a collective imagination, without which the acts described above would not have excited and enthralled audiences to the extent they did. And an audience also has a collective memory (see Erll 2005) of these constructed and constructing identities. Dante, international man of mystery, makes such collective memories central to his adaptation of identity. While the examples so far tend to adapt collective imaginings of the other, Dante offers a doubling of such adaptation: rather than adapting a collective imagining of the other, his source are the performances based upon such collective imagining and identity construction in that he offers an ironic, at once celebratory and critical engagement with notions of identity bound up in such memories of imagined otherness.

Harry August Jansen a.k.a. Dante, born in Copenhagen in 1883, emigrated to the US with his family in 1889 and died in 1954. Dante prided himself in his performances being funny and suitable for all the family. He also presented himself very much as an international performer, who impressed with his skills the imagined originators of specific tricks such as fakirs and the like. And it is the international encounters and destinations which are the focus of posters for his performances: in these early years of globalisation and tourism, the figure of Dante is superimposed on what could at first sight look like a collage of holiday snaps on a hand-drawn map. We can see Dante in Japan, Holland, Germany, Sweden, France, England, Canada, Argentina, Russia, Italy, China, Norway, Denmark, Spain, the US, Australia, Africa, Brazil, Venezuela, Uruguay and Austria. A programme for his performances at the Alhambra Theatre in London between 1936–1937 adds cities to the list of countries, and we can count at least 30 ranging from Hamburg to Cape Town and Buenos Aires to Norfolk.

Dante's performances not only present himself as multiple imagined identities, but he also imitates, in a comic vein, performers such as Chung Ling Soo, the Great Lafayette and notions of authenticity. During his act, costume changes allow him to take on the appearance of a Chinese magician, similar to Chung Ling Soo, as well as that of a non-specific other, such as the Great Lafayette. The comedy inherent in his performance depends on a knowingness between him and his audience: Dante relies on his audience to recognise his sources of adaptation, namely the likes of the Great Lafayette and Robinson/Soo. With such recognition comes a humour which is self-deprecating: the audience's memories of these performances of the other, and their subsequent confirmation of self, form an essential part of Dante's comedy. Dante is only funny if one recognises the act as an adaptation of these other performances and the audience's relationship to them.

Dante's programme notes include a description of his trick 'Fountania' which reads:

> Here Dante reproduces with real oriental splendour a correct facsimile of the great Chinese wonder worker, personating realistically the great Ching Lee Foo in his speciality: Fountania. Over 500 gallons of water are necessary to present this bewildering novelty. Originated by Confuse-Us, 5000 B.D. (Before Depression) portrayed by Dante as the Chinese necromancer, Foo Ling Yu, and his able asso-see-ates (Chinese for assoshates), Chee-ting-yo, Won-lung-gon, Long-tie-pin, Won-fat-Cow, Sing-so-low, Won-long-tack, and Tu-Fat-Yet.

V.P. – Very important. These titles are the result of more than 15 minutes of endless research into the Chinese ancestry in all the chop suey restaurants of the worlds, and are therefore not authentic. Pardon Us!!!"

These programme notes are performative in their own right; the humour depends on a speaking aloud of the names listed in order to hear 'Fooling You', 'Cheating You', 'One Long Gone' and so forth. They also reference the relationship between a confirmation and maintenance of self and the colonial notion of the other. Thus, Dante not only demonstrates the extent to which performers such as Chung Ling Soo, Lafayette and the like entered a collective consciousness and certainly formed part of a collective memory, but his humour is possible only because of a collective fascination with an imagined other, maintaining the self and notions of authenticity as performed by Chung Ling Soo and others.

Conclusion

Adaptation in all these case studies becomes a dramaturgical means of the performance of the self and the imagined other whether on stage only or beyond. Furthermore, an important element of these adaptations is identity maintenance, that is the confirmation and maintenance of the identity of the self for an audience vis-à-vis a representation of the other. Dante's humourous referencing of these acts of adaptation must be understood as part of a collective memory of the performances of self upon which Dante's act are based. The sources of Dante's adaptation also all form part of what Bryant identifies as a form of writing as cultural event; while Bryant may understand 'writing' to be very much the production of 'text' as in the authored 'written word', it might be helpful to think of such writing as not only a cultural event, but also a practice enacted on the body and the imagination which goes beyond a 'written text'. In his "fluid text approach" to adaptation "a *work* is the sum of its versions; *creativity* extends beyond the solitary writer, and *writing* is a cultural event transcending media." (Bryant 2013: 47) Once we go beyond writing as producing written text and instead encompass enacted text, Bryant's position is very helpful in as much as it supports an attempt to conceptualise adaptation as collective dramaturgical 'writing' which allows insight into and becomes symptomatic of constructions and maintenance of identity.

All these examples above then are not merely curiosities, which one happens to come across in un-catalogued, miscellaneous boxes in archives, but instead they allow us to see modern examples of performance where "the construction of the self becomes a residual derivative of the process of imagining the other, or performing imaginary otherness" (Soares 1998: 288). Importantly, Soo and the others also engender a broadening of our understanding of adaptation which goes beyond the written text and instead has to be understood as a dramaturgical practice of performance as much as processes of identity construction.

Adaptation as an act of identity construction and maintenance provides insights into specific moments of such maintenance and construction which, in the cases above, are part of colonial and globalising processes. And while the colonial process may be understood as a form of power relationship between nation states, it is also at its core a one-sided, democratic, creative and communal process. Such democracy, creativity and community is located within the seat of power, which it displays, and cannot be shared with the imagined other. It depends upon an audience to claim ownership over its existence. It questions the very idea of singular authorship and chronological relation of texts. This is never more obvious than in the theatre, where ownership and authorship is shared and embodied by both the makers and the spectators of the performance.

The not unproblematic pleasure inherent in such ownership and authorship may be the reason for adaptation's popularity and economic success. It certainly formed an important part of the allure of the performances discussed above, which are all adaptive events; adaptive events which exist only as a multiplicity of texts. What adaptation in the theatre makes very clear is that adaptation is what Regina Schober calls "a process of forming connections" (Schober 2013: 91), and anything but a stable and fixed entity. Performance and source, text and rewriting, watching and re-watching make up a dynamic and reciprocal network of identity construction and maintenance which, in these cases, is based upon holding on to the binary that is 'them' and 'us'.

Notes

1 London-based theatrical agent Charles Lauri (1860–1903) was famous for his animal impersonations, and he performed regularly as part of the pantomime season at Drury Lane, London.
2 Such invisibility of adaptation may make the product and process more akin to translation where the translator's invisibility (see Venuti 2008) is still upheld as the ideal.
3 Table-turning was a popular element of the séance where participants placed their hands on a table waiting for it to rotate through the power of spirits.
4 Von Gloeden's painting is part of the Alinari Collection, Florence.
5 According to the British Newspaper Archive, "The Sketch … described itself as a 'A Journal of Art and Actuality'. It was published weekly and was for 'the cultivated people who in their leisure moments look for light reading and amusing pictures, imbued with a high artistic value'" (https://www.britishnewspaperarchive.co.uk/titles/the-sketch).

Works cited

Anon. (17 May 1911), "'The Great Lafayette and Beauty' – and his last scene", in *The Sketch*, 171.
Anon. (17 February 1955), 'One man and his dog – the life and death of the Great Lafayette', in *The Evening News*, n.p.
Bryant, John (2013), "Textual identity and adaptive revision: Editing adaptation as fluid text", J. Bruhn, A. Gjelsvik and E. F. Hanssen (eds.), *Adaptation studies: New Challenges, New Directions*, London & New York: Bloomsbury.
Conroy, Colette (2010), *Theatre and the Body*, Basingstoke, UK: Palgrave Macmillan.
Coupland, Justion & Richard Gwyn (2003), "Introduction", in J. Coupland and R. Gwyn (eds.), *Discourse, the Body, and Identity*, Basingstoke, UK: Palgrave Macmillan, 1–18.
El-Hibri, Tayeb (1999), *Reinterpreting Islamic Historiography: Hārūn al-Rashīd and the Narrative of the 'Abbāsid Caliphate*, Cambridge, UK: Cambridge University Press.
Erll, Astrid (2005), *Kollektives Gedächtnis und Erinnerungskulturen: Eine Einführung*, Stuttgart, Germany: J. B. Metzler.
Glenn, Susan (1998), "'Give an imitation of me': Vaudeville mimics and the play of the self", in *American Quarterly* 50: 1, pp. 47–76.
Hutcheon, Linda (2006), *A Theory of Adaptation*, London & New York: Routledge.
Jackson, Dorothy (1955), "The world's strangest stories no. 168", in *The Evening News*, n.p.
Knowles, Ric (2010), *Theatre and Interculturalism*, Basingstoke, UK: Palgrave Macmillan.
McKerracher, A. C. (1985), "The last illusion", in *The Scots Magazine*, 166–171.
Mulvey, Laura (1999), "Visual pleasure and the narrative cinema", in Braudy and Cohen (eds.), *Film Theory and Criticism: Introductory Readings*, New York: Oxford University Press, 833–844.
Pagani, Catherine (1998), "Chinese material culture and British perceptions of China in the mid-nineteenth century", T. Barringer and T. Flynn (eds.), *Colonialism and the Object: Empire, Material Culture and the Object*, London & New York: Routledge, 28–40.
Sanders, Julie (2006), *Adaptation and Appropriation*, London & New York: Routledge.
Schober, Regina (2013), "Adaptation as connection: Transmediality reconsidered", in J. Bruhn, A. Gjelsvik and E. F. Hanssen (eds.), *Adaptation studies: New Challenges, New Directions*, London & New York: Bloomsbury.
Schweitzer, Marlies and Daniel Guadagnolo (2012), "Feeling Scottish: Affect, mimicry, and vaudeville's 'inimitable' Harry Lauder", in *Journal of Dramatic Theory and Criticism* 26:2, 145–160.

Soares, Luiz E. (1998), "Staging the self by performing the other: Global fantasies and the migration of the projective imagination", in *Cultural Values* 2: 2–3, 288–304.

Vannini, Phillip & J. Patrick Williams (2009), "Authenticity in culture, self and society", in P. Vannini and J. P. Williams (eds.) *Authenticity in Culture, Self and Society*, Farnham, UK: Ashgate, 1–20.

Venuti, Lawrence (2008), *The Translator's Invisibility: A History of Translation*, London and New York: Routledge.

Werry, Margaret (2005), "'The greatest show on earth': Political spectacle, spectacular politics, and the American Pacific", in *Theatre Journal* 57: 3, 355–382.

White, Richard (2001), "Cooees across the strand: Australian travellers in London and the performance of national identity", in *Australian Historical Studies* 32: 116, 109–127.

19

ADAPTATIONS DOWN UNDER

Reading national identity through the lens of adaptation studies

Claire McCarthy

Australian adaptations *Picnic at Hanging Rock* (Weir 1975), *Strictly Ballroom* (Luhrmann 1992), and the Mad Max films (Miller 1979; 1981; 1985; 2015) are key markers in popular representations of Australian heritage, nationality, and identity within Australia and internationally. They are all in the top 100 highest earning Australian feature films (*Screen Australia* 2016b), and they have each received extensive critical attention and acclaim (O'Regan 1996; Powers 2015). Drawing on Monika Pietrzak-Franger's (2012) discussion of adaptations as palimpsests, this chapter examines these films as adaptations. In particular, the way in which they adapt history, national stereotypes, and genre to represent new versions of Australia's 'national story.' This chapter argues that reading *Picnic at Hanging Rock* (1975), *Strictly Ballroom* (1992), and *Mad Max: Fury Road* (2015) as adaptations reveals the ongoing construction of nation, heritage, and ethnicity in Australia. It demonstrates that through the process of adaptation these texts exist as layered versions of themselves and their surrounding contexts. They are palimpsests conjuring an image of Australian national identity at the same time as they deconstruct the notion of Australian-ness they are deemed to represent.

In her authoritative work, *A Theory of Adaptation*, Linda Hutcheon (2006: 8–9) defines adaptation as:

> An acknowledged transposition of a recognizable other work or works;
> A creative *and* an interpretive act of appropriation/salvaging;
> An extended intertextual engagement with the work.

Each of the adaptations listed above represents aspects of this definition. Adapted from Joan Lindsay's 1967 novel (reprinted in 2009), *Picnic at Hanging Rock* (1975) re-imagined Australia's colonial era as a vision of women in white dresses threatened by a sexualised landscape. It produced images of sanitised, white Australian heritage in contrast to the violent dispossession of Australia's indigenous people during British colonisation. I will discuss this in more detail with reference to 'retrocolonialism' developed from Imelda Whelehan's (2000) concept of "retrosexism."

Strictly Ballroom (1992) is an adaptation of Baz Luhrmann's 1984 play of the same name, and was most recently made into a musical that is touring the United Kingdom (*Strictly Ballroom the Musical*). The film is also an interpretive and salvaging act, incorporating Australia's multicultural policies that were adopted by the Federal Government in the 1970s and 1980s. Multiculturalism followed the final abolition of the White Australia Policy in 1975. The White Australia Policy was established in 1901 to restrict immigration to Australia on the basis of race (Jupp 2007). *Strictly Ballroom* (1992) is one of a group of 1990s Australian films influenced by officially condoned or bureaucratic multiculturalism, and that acknowledge and represent the cultural diversity of the Australian population. It re-negotiates Australia's national story to include reference to the nation's migration history and ethnicity.

Mad Max: Fury Road (2015), the fourth film in the Mad Max franchise (1979–2015), represents the Australian adaptation of Hollywood genres, the character of Max and environmental issues. It produced a cosmopolitan version of Australian identity for national and international audiences and achieved massive financial success. As of May 2016, *Fury Road* had made over $153 million in Australia, contributing to 2015 being the most financially successful year the Australian screen industry has ever seen (*Box Office Mojo* 2016; Screen Australia 2016a). In this latest incarnation, Max is played by a British actor, Tom Hardy, and the film was shot in Namibia, Africa, instead of the Australian desert. However, *Fury Road* (2015) is still known as an 'Australian' film because of its status as an adaptation. The intertextual engagement it has with the series of Mad Max films produces a collective understanding that the Mad Max franchise is an Australian story.

In her treatise on adaptation, Hutcheon (2006: 9) goes on to describe the "palimpsestic" nature of adaptations, which is something that I want to develop along with Pietrzak-Franger's (2012: 85) extension of this concept using Derrida's theory of "hauntology." By extending the definition of adaptation to consider each text as a layered entity the process of construction is revealed. Pietrzak-Franger (2012: 85) argues that studying adaptations in this way foregrounds "the constant dialogue between the past, present and future." The past is not present in these conversations but rather manifests as a ghost that frames how we interpret the current time and plan for the future. Analysing adaptive texts as palimpsests, therefore, provides a window into representations of the past through the glass of contemporary culture. The view distorts in accordance with our current cultural preoccupations and frames of reference.

As a metaphor for adaptation, a palimpsest refers to the layering, repetition, and reuse of material that is inherent in the adaptation process. A palimpsest is a parchment that has been used again and again. The pages reveal the most recent text as well as hints and traces of writing from before. As a concept in adaptation studies, this metaphor allows us to consider the intertextual signs that cluster and manifest in the adapted entity. It also assumes an atmosphere or location that surrounds the adaptation and the process of its construction. The concept of a palimpsest is, then, pertinent to a discussion of national identity because it dismisses all thought of a faithful rendition (Pietrzak-Franger 2012: 85). This framework could be a chance to bury what Eckart Voigts-Virchow (2009: 137) calls the "undead hand" of fidelity that casts its lingering shadow over the discipline. If there can be no "faithful rendition," any attempt at, or claim to, faithfulness is futile. The palimpsest or adaptation's capacity to haunt is not predicated on bringing the past to life but instead relies on acknowledging the process of adaptation in the production of meaning.

Derrida (1994: 202) writes: "To haunt does not mean to be present," but argues "it is necessary to introduce haunting into the very construction of a concept." The ghost or spectre of the past, conjured for new audiences, is how meaning is constructed and is at the heart of the adaptation process. As nostalgia reinvents images of the past it does so retrospectively, removing the sting as it paves the way for future representations built on a history of intertextual meaning. In the following analysis, I will focus on a close reading of *Picnic at Hanging Rock* (1975), *Strictly Ballroom* (1992), and *Mad Max: Fury Road* (2015) because they are the best known textual examples. They occupy a central place in the constellation of adaptations. However, the chapter also draws attention to the adaptation process and how it relates to the construction of national identity by examining the multiple versions that exist for each adaptation and how this relates to representations of nation, heritage and ethnicity.

Picnic at Hanging Rock (Lindsay 2009; Weir 1975) is the story of three schoolgirls and a teacher who go missing at the now infamous Hanging Rock in rural Victoria. Only one of the girls is ever found. In a prefatory statement in the book it says:

> Whether *Picnic at Hanging Rock* is fact or fiction, my readers must decide for them-
> selves. As the fateful picnic took place in the year nineteen hundred, and all the
> characters who appear in this book are long since dead, it hardly seems important.
>
> *(Lindsay 2009: 7)*

This statement creates ongoing suspense in the text as to whether or not the tale is fact or fiction. On the one hand, it describes the participants as characters, which provides a reference to fictional norms. On the other, it declares that they are all dead, which makes the narrative impossible to authenticate. It also frames the story as a legend by representing it as paradoxically historical yet unverifiable. Through repetition and popular knowledge of the text, *Picnic at Hanging Rock* (1975) contributes to a collective sense of Australia's cultural heritage and mythology.

The film version was critically and popularly successful and helped to develop Weir's inter-national standing as a director (*NFSA* 2016). The romantic feel of the images was assisted by a bridal veil draped over the lens of Russell Boyd's camera (Byrnes 2015), which created an arthouse aesthetic similar to a series of 1970s and 1980s films funded by the Australian Film Commission (AFC). These included *Sunday Too Far Away* (Hannam 1975), *My Brilliant Career* (Armstrong 1979), and *We of the Never Never* (Auzins 1982). All set in rural Australia, these films were dubbed the 'AFC-genre' (Dermody and Jacka 1988: 32). Because of the government's investment, the AFC-genre films were seen as cultural markers of national values, heritage, identity, and purpose (Adams 1984: 71; Turner 1989: 103; Couzens 2016: 46). Brian McFarlane (1993: 90) suggests that *Picnic at Hanging Rock* (1975) ushered in a series of literary adaptations concerned with representations of Australian heritage and marked the start of a new Australian cinema or New Wave cinema that was recognised internationally.

The film begins with white text on a black background:

> On Saturday 14th February 1900 a party of school girls from Appleyard College pic-
> nicked at Hanging Rock near Mt Macedon in the State of Victoria.
> During the afternoon several members of the party disappeared without a trace.
>
> *(Weir 1975)*

The next shot is of the towering pillars of Hanging Rock looming out of a dark, grey mist. The title of the film appears in a flourish of white cursive letters over the landscape. Then, over a shot of a breeze blowing dried out thistle heads, which is a gesture to 'time,' one of the film's central themes, a young woman says: "What we see and what we seem are but a dream. A dream within a dream" (Weir 1975). The camera fixes on an establishing shot of stately Appleyard College before cutting to a close-up of Miranda waking up in her bed (Weir 1975). Miranda is one of the girls who goes missing. She is also the most beautiful and angelic of all the students. She is dressed in a white nightgown reminiscent of the British heritage adaptations (Weir 1975). The opening scene is complemented by an ethereal pan flute score by Gheorghe Zamfir (Weir 1975). The combination of music and images foregrounds a sense of nostalgia for things that are irretrievable: dreams, the missing girls, and spotless images of Australia's colonial past.

Nostalgia is further conveyed through the characters' frequent reference to time and to symbols of renaissance culture. The girls are celebrating Valentine's Day in the Roman tradition. As Miranda walks away from the picnic she tells the group, "We shall only be gone a little while." Mademoiselle, watching her go, says, "I know that Miranda is a Botticelli angel" (Weir 1975). When the picnickers wake from their naps they realise their watches have stopped and that the girls have not returned (Weir 1975). This disruption and obsession with time and high culture speaks to Pietrzak-Franger's use of the concept of hauntology. While the past is in conversation with the present and the future, the past itself can never be realised; rather, it can only manifest as a spectral return (Pietrzak-Franger 2012: 79). As the search for the girls plays out, Miranda's only return is a spectral one through the repetition of the wistful scene where she and the others walk away from the group towards Hanging Rock.

To discuss this sense of nostalgia further, I will now introduce the concept of 'retrocolonialism.' I have adapted this term from Imelda Whelehan's (2000: 11) "retrosexism," which she coined in her work on feminism. Whelehan (2000: 11), who has also written extensively on adaptations, defines retrosexism as "nostalgia for a lost, uncomplicated past." Retrosexism implies a reimagining or conjuring that "[o]ffers a dialogue between the past and the present" given rise to by a change in the status quo of the gender order or from fear that those changes will occur (Whelehan 2010: 11). Retrocolonialism refers to a simplified construction of the past, a retrospective envisioning that offers an image of white Australian heritage that does not reference Aboriginal people or dispossession, let alone ongoing controversy over land rights or migration.

Retrocolonialism is a process of nostalgic representation. It differs from postcolonialism because it does not refer to an entire discourse or the ongoing or aftereffects of colonialism on contemporary societies. Postcolonialism is what Pramod Nayar (2010: 1) defines as "the academic, intellectual, ideological and ideational scaffolding of the condition of decolonization." On the day of the picnic, the girls are reminded by their teacher Ms McCraw that the "rock is extremely dangerous" (Weir 1975). She bans them from "tomboy foolishness even on the lower slopes" and warns them about "venomous snakes and poisonous ants" (1975). Yet the images as they approach their picnic destination are bucolic. Rural Australia passes by the windows of the carriage in faded tones. Dry yellow fields appear paradoxically lush, even though Australia is regularly in drought. It is a celebration of the Australian landscape connected to white heritage. Yet, knowing as we do that the landscape too is a palimpsest, we can recognise a latent danger in this mythology.

As the carriage approaches, Mr Hussey tells the girls that Hanging Rock is more than a thousand years old. Ms McCraw corrects him, saying it is millions of years old: "Quite a recent eruption really." One of the girls peers up at the rocky outcrop and says, "Waiting a million years, just for us" (Weir 1975). Ms McCraw is somewhat of a geology expert but it does nothing to save her from her mysterious fate on the rock. Through a postcolonial lens and awareness of alternative histories of Australia these statements register the absence of the nation's first people from the text. More than forty years since the film was released, these gaps can be read as a naivety or as inciting ownership attributed to the dominant white Australian community. The characters romanticise place through their awe over the longevity of the landscape, and by imagining they are the only ones in it. In this version of the national story, the landscape has been waiting for them. This is itself an adaptation of *terra nullius* – the legal doctrine that claimed the land was empty – on which the country was founded.

The looming mass of Hanging Rock, its age, the anxiety it inspires about the land, and the mysterious disappearance of the girls all suggest a specific, retrospective white imagining of the colonial era. This imagining is not aware that indeed the land was not or is not solely the resource or product of white settlers. Ross Gibson (1992: 69) argues that *Picnic at Hanging*

Rock and other AFC films that focus on landscape became statements of Australian identity by presenting the land "as generically Australian." Gibson (1992: 69) argues that they became "the projective screen for a persistent national neurosis deriving from the fear and fascination of the preternatural continent." The suggestion that the rock has been waiting for the girls the whole time speaks to the inherent dangers in the land that are both real and imagined. It also invests the landscape with a sense of white Australian ownership.

Australia is prone to bushfires and drought. The snakes are poisonous and the bites from the ants sting, but there is also a metaphoric risk that retrocolonialism helps identify. This risk comes to bear through the Land Rights movement and greater awareness of the circumstance of British invasion and resettlement of the land under false pretences of *terra nullius*. In this context, and with the bicentenary celebration of British arrival coming up in 1988, inventing or perpetuating a white Australian heritage through films such as *Picnic at Hanging Rock* (1975) was of utmost ideological importance.

When *Picnic at Hanging Rock* (1975) produced these retrocolonial images of white Australian heritage, the Aboriginal Land Rights Movement was in full swing, the White Australia Policy was finally abolished, and multiculturalism was being introduced (Collins and Davis 2004; Jupp 2007: 12). However, *Picnic at Hanging Rock* (1975) was fixated with representing images of white colonial heritage imbued with a sense of nostalgia for the intangible past. These representations were adapted from literature and history to produce a sense of Australian national identity linked to white heritage situated in the Australian landscape. Even Lindsay's novel published in the late 1960s is an adaptation of the imagery of the Australian colonial era in the style of neo-Victorianism. Louisa Hadley (2010: 2) defines neo-Victorianism as literature set in the Victorian era but written in the 1960s or later. Through the process of adaptation, *Picnic at Hanging Rock* (1975) offers a spectral return of an imagined past that provides lineage to Australia's heritage through colonial and white history.

Weir's film (1975) offers a romantic yet hostile relationship to the landscape symptomatic of real fears about the ongoing adaptation of Australia's national identity to recognise Aboriginal rights and to include cultural diversity brought to bear through migration. The site where *Picnic at Hanging Rock* is set is in the land of the Wurundjeri Aboriginal nation, but Aboriginal people are not mentioned in any of the adaptations that I am aware of, other than instances where the Rock is described as a tourist destination. Instead, the mystery of the missing girls is romanticised through retrospective nostalgia for a white, colonial era. The adaptation reflects a desire to shape a sense of cultural heritage connected to the Australian landscape but without reference to the history or continued presence of Australia's Aboriginal people.

Considered with this retrocolonial lens, *Picnic at Hanging Rock* has many similarities with British heritage adaptations of the 1980s. "Heritage" was produced as a concept in Britain in the 1980s, and is closely associated with identity formation (Childs 2012: 89). It is not history handed down from the past to the present, but appropriated representations of the past that are "superimposed by a modern generation" (Childs 2012: 89). In Britain, the term "heritage film" came to address issues of national, ethnic, cultural, class, and gender ideologies that are crucial to the analysis of these films (Voigts-Virchow 2007: 123). Analysis of these films drew attention to the way the heritage of certain groups of people took possession of the past for that group. Voigts-Virchow (2007: 123) argues:

> Heritage industries (film, novels, tourism, theme parks, etc.) re-establish the past as a property or possession, which by 'natural,' or better, 'naturalized' right to birth, 'belongs' to the present or to be more precise, to certain interests or concerns active in the present.

As a heritage film, *Picnic at Hanging* constructs a lineage or reference point for contemporary Australia to a cultural history that privileges the white settler experience and literary history, even in the context of greater recognition of Australia's first people and the introduction of multiculturalism.[1]

The heritage represented in *Picnic at Hanging Rock* is related inextricably to the Australian landscape and the mystery and allure of the bush. While the characters might be long since dead, they have been incarnated through adaptations, retellings, and remakes. As the spectre of *Picnic at Hanging Rock* remains in the cultural imaginary, so do assumptions about an explicitly white cultural heritage refracted through images of women in white dresses and rural vistas that lie beneath the ominous shadow of the rock as the narrative adapts into an ever-expanding palimpsest. In 1980, Yvonne Rousseau published an unsolicited sequel into Lindsay's novel *The Murders at Hanging Rock,* which speculated on what happened to the girls. Weir's producer Patricia Lovell called her autobiography *No Picnic* in reference to the film (Byrnes 2015). In the 1990s, Australian garage rock band, Magic Dirt, appropriated the imagery of Miranda walking into the bush for a film clip for their song "She Riff."

In 2016, Tom Wright adapted *Picnic at Hanging Rock* for the stage in conjunction with Malthouse Theatre and Black Swan State Theatre Company. On the Malthouse Theatre website, the play is promoted as a "chilling adaptation" and claims the text "has haunted the Australian psyche for over a century" *Malthouse Theatre* 2016). A further adaptation was developed by New York-based singer–songwriter Daniel Zaitchik and playwright Jordan Harrison for the Lincoln Center (*Picnicathangingrock.com* 2016). Hanging Rock is also currently advertised as a popular tourist destination in the Macedon Ranges. In this adaptation the rock itself is the protagonist and storyteller:

> Hanging Rock has fitted in so much over 6 million years … from a sacred place for local indigenous people and backdrop to Joan Lindsay's book Picnic at Hanging Rock and Peter Weir's film of the same title, to a colourful host of music concerts, markets and popular horse races, the Rock tells many tales about the history of the Macedon Ranges.
>
> *Visit Macedon Ranges 2016*

Additionally, in a film review of *The Virgin Suicides* (Coppola 1999), Roy Rodenstein argues that "in many ways" it is an updated version of *Picnic at Hanging Rock* (2000), although he suggests that where Weir "took full advantage of the disappearances' unexplainable nature, Coppola's film is too fickle and heavy handed to draw sustained strength from the girls' suicides" (Rodenstein 2016).

The National Film and Sound Archive posit "a key factor in *Picnic at Hanging Rock*'s enduring appeal is that the mystery of the disappearing girls is never solved" (*NFSA* 2016). However, in Lindsay's initial manuscript the mystery was, if not solved, then explained. Spoiler alert, but prior to publication, the editor removed a final chapter that explained the girls as leaping from the rock into a time warp. Despite the publication of this additional material in 1987 in *The Secret of Hanging Rock,* the popular version of the narrative remains the one in which the girls are never found, their disappearances never explained, and the land remains ominous. The mystery that surrounds the popular version continues to be adapted in order to fuel the legend's longevity in the cultural imagination. *Picnic at Hanging Rock* has become an Australian classic through a long tradition of adapting the text and the sheer existence of so many versions. Read in this way, the adaptations of *Picnic at Hanging Rock* duplicitously produce and deconstruct any notion of an essential Australian identity, and as such become important sites of analysis in their own right.

In contrast to *Picnic at Hanging Rock*'s representation of white Australian heritage, *Strictly Ballroom* redefines "Australianness" in relation to official multiculturalism. I use the word 'official' to refer to the specific brand of state-sanctioned cultural diversity policy introduced in

Australia at a similar time to when the AFC was funding literary adaptations such as *Picnic at Hanging Rock*. Multiculturalism began to be discussed in the 1970s by then Minister for Immigration, Al Grassby. In 1977 the Australian Ethnic Affairs Committee defined multiculturalism for the first time as "'cultural pluralism' based on the principles of social cohesion, equality of opportunity and cultural identity" (Australian Government Department of Social Services 2015). This marked a shift in policy frameworks for cultural diversity in Australia, although the National Agenda for a Multicultural Australia was not formally introduced until 1989, but it was not until the 1990s that multicultural themes started to appear in popular Australian film.

Examples include *Strictly Ballroom* (Luhrmann 1992), The *Heartbreak Kid* (Jenkins 1993), *Head On* (Kokkinos 1998), and *Looking for Alibrandi* (Woods 2000). The *Heartbreak Kid* (1993), based on the play by Richard Barret (1987), is about a Greek-Australian teacher (Claudia Karvan) who starts a soccer team in an Australian Rules-dominated school and has an affair with one of her Greek-Australian students. It was later made into a spin-off series, *Heartbreak High*, an Australian version of Degrassi High. *Head On* (1998) is adapted from Christos Tsoilkas's novel *Loaded* (1995). It is about a young, Greek-Australian man exploring his homosexuality and rejecting his migrant parents' aspirations for him. *Looking for Alibrandi* (2000) was adapted from a young adult novel by Melina Marchetta (1992), in which high-school student, Josie, comes to terms with her Sicilian-Australian heritage amid the turmoil of her final exams. *Strictly Ballroom* is about Scott Hastings, a champion ballroom dancer, learning "new steps" with second-generation migrant, Fran, who is initially cast as an outsider because of her Spanish ethnicity.

In keeping with the didactic nature of Australian multiculturalism, all of these films except *Head On* (1998) are set in learning institutions: *Strictly Ballroom* (1992) in a dance studio and the other two in high schools. This is significant because it reflects the newfound cultural tolerance espoused by Australian multiculturalism as distinct from former White Australia policies. The educational settings suggest not only a coming of age but acceptance and tolerance of multiculturalism as something the Australian nation needed to or needs to learn. These films represent a new Australian multiculturalism (Hage 1998: 201; Simpson, Murawska and Lambert 2009: 23) predicated on a "coming of age through ethnicity" (Simpson, Murawska and Lambert 2009: 34). In line with government policy, these films convey a lesson in cultural difference and tolerance more than a full embrace of cultural diversity.

Strictly Ballroom (1992) was the most financially successful of these adaptations clearing $21 million at the domestic box office (*Screen Australia* 2015). The film is set in Sydney's infamous Kings Cross, marked by the iconic Coke sign that sits on the roof of the building where Scott and Fran dance around the Hill's Hoist[2] washing line to "Time After Time" by Cyndi Lauper. It launched director Baz Luhrmann's career and was the first of the so-called "red curtain trilogy" that includes *Romeo and Juliet* (Luhrmann 1996) and *Moulin Rouge* (Luhrmann 2001) (Cook 2010). Luhrmann later went on to make the grandiose, epic *Australia* (2008) starring Nicole Kidman and Hugh Jackman, which is relevant to this topic but outside the scope of this current chapter.[3] As mentioned in the introduction, Luhrmann more recently made *Strictly Ballroom* into a musical, which premiered in 2016 at the West Yorkshire Playhouse in Leeds. This is significant because it demonstrates the longevity of the adaptation and its ongoing representation of Australian multiculturalism and its transcultural appeal.

Strictly Ballroom (1992) is the story of Scott Hastings learning to dance "new steps." Scott is a young Australian ballroom dancer who, with the help of dance partner Fran, learns the *pase doble*, a traditional Spanish dance, and performs it with her at the Pan-Pacific ballroom dancing championships. Fran is the daughter of poor Spanish migrants. She lives with her non-English-speaking father and grandmother in a small shop on the edge of the railway line that is literally on the wrong side of the tracks. Scott and Fran learn the dance on a specially built wooden dance floor at the back of the shop, which provides a sanctuary from the hostile and culturally insensitive world of Australian ballroom dancing. Jon Stratton (1998: 149) argues that the fore-grounding of ballroom dancing in the film naturalises "the experience of culture as spectacle." The viewing of this spectacle occurs comically when considering the context of mainstream Australia that does not participate in the sequined extravagance of ballroom dancing. More importantly, it casts Fran's migrant family as exotic outsiders in contrast to mainstream Australia. Scott's actions are against the wishes of his dance teachers, who are scandalised by his desire to dance steps that are not "strictly ballroom." Despite their objections, Scott learns the dance from Fran's father, who wears a sequined jacket. Fran's grandmother watches on as she hems a red flamenco dress that belonged to Fran's mother and that Fran later wears at the competition. When Scott struggles to pick up the rhythm, Fran's grandmother, whose lines are subtitled, beats her palm against Scott's chest to show him the beat of the dance is in time with his heart.

The juxtaposition of Spanish and Australian culture also reveals a gaucheness and small-mindedness of Australian culture in relation to migrant authenticity. Australian identity in *Strictly Ballroom* is defined in contrast to this essential Other, and later, when first Scott's father and then the whole room applaud Scott and Fran for their dance, it is redefined in relation to the Other as part of a broader multicultural story. Ghassan Hage (1998: 202) claims the film is "based on an opposition between an archaic, restricting Anglo-Australian culture and a living, promising White cosmo-multiculturalism." The power struggle in the film occurs between the white Australians, while the ethnic characters enable Scott to "embrace ethnic difference and to use it to 'beat' his archaic and unsophisticated Australo-British opponent" (Hage 1998: 202). Fran and her extended family serve to make Scott and the other white Australian characters more cultured and empathetic. The Flamenco rises from the annals of European culture, whereas Australia is positioned as uncouth and uncultured, obsessed with the arbitrary and rigid rules of competitive ballroom dancing. The rich tradition of Spanish culture is pitted against the kitsch naivety of Australia. Spanish culture is represented as authentic and essential. The fluoro colours of the Pan-Pacific Championship held in a Returned Servicemen's League (RSL), which is typically a club full of poker machines, represent Australia. The images are of the self-service buffet, women wearing too much makeup, and sleazy, boozy men.

Strictly Ballroom is an adaptation of Luhrmann's play (1984), but it is also an adaptation of official multiculturalism and a representation of new Australian identity seeking to incorporate cultural plurality into a new hierarchy of Australian values. At the start of the film, Fran's bad skin and poor dance skills, coupled with her angry and protective, non-English-speaking father, are set in direct opposition to Australian culture. But like Cinderella at the ball, it becomes clear that Fran and her family hold the key to true passion and life fulfilment for the Australian characters. Fran translates an expression for Scott from Spanish, "*A vivir con miedo es una vida media vivido/* A life lived in fear is a life half-lived." As the narrative unfolds Scott takes on this advice refusing to dance with the safe, Anglo-Australian partner his family has arranged for him. This message is emphasised when it is revealed that Scott's father did not ruin his career by dancing new steps as Scott is first told, but limited his life by being too afraid to. The adaptation of multiculturalism into Australian cinema in this way represents the incorporation and appropriation of ethnicity into broader Australian culture.

Considering *Strictly Ballroom* as a palimpsest draws attention to the difference or alternative national story presented through an adaptation and inclusion of multiculturalism in Australian film. However, Elder (2007: 141) argues that "the multicultural story is often deployed by Anglo-Australians to find out more about themselves rather than to encourage intercultural exchange." Seen in this light, Fran's role is to exploit difference through representations of the exotic Other without actually challenging dominant narratives (Hage 1998: 201). However, it can also be read with regard to the process of adapting Australia's national identity. While Fran turns into a beautiful Spanish dancer joining Scott to dance the *pase doble* at the Pan-Pacific championships, it is Scott who is liberated from confines of Australian society. This could be read as Fran merely reaching her authentic, ethnic potential. However, when *Strictly Ballroom* is juxtaposed with *Picnic at Hanging Rock*, both of which are considered to be classic Australian films, the concept of palimpsestuous adaptations reveals the adaptation of shifting political and cultural views that require the representation of new versions of the national story.

Strictly Ballroom speaks to the adaptation of multiculturalism into Australian identity and the legacy of Australia's ongoing migration programs. Keith Jacobs (2011: 105) argues "that the foregrounding of migrant experiences and the rites of passage" in films like *Strictly Ballroom* "is symptomatic of wider changes in Australia that stem from processes set in train by globalization." In contrast to Hage, Jacobs (2011: 105) argues that cinema provides an

> important medium for migrant representation because of the opportunities it affords to subvert traditional Anglo-Celtic narratives that house, support and rehearse discriminatory or biased forms of national identity.

Whether or not these biased or dominant narratives of national identity are challenged successfully, the airing of alternative stories reveals the process by which Australianness is constructed. It demonstrates how it is made with reference to the past, but for a contemporary audience. As established ideas about preordained cultural hierarchies are eroded, adaptation studies is well placed to research the political beliefs and ideologies that form part of the adaptation process as well as the social and historical context that surround them (Hassler-Forest and Nicklas 2015: 2–3). Examining *Strictly Ballroom* as a palimpsest in the broader context of Australian society demonstrates the role that adaptation has played in popularising positive, multicultural narratives as part of the national story.

Telling a different story altogether, the Mad Max franchise is an Australian adaptation of popular Hollywood Western, action, and Sci-Fi genres. Now boasting four feature films, as well as comics, graphic novels, and video games, Mad Max is at the height of Australian popular culture, well known domestically and abroad. The Mad Max franchise began in 1979 at a time when the AFC-genre was at its peak, but *Mad Max* did not receive government funding. It was privately funded to the tune of $400,000 and projects a very different view of Australia to the ethereal tones of *Picnic at Hanging Rock*. Despite its lack of critical attention, *Mad Max* (1979) stayed in the *Guinness Book of World Records* for almost twenty years as the most profitable film in relation to its budget. Overdubbed with American voices for the United States market, which is itself an example of adaptation, *Mad Max* is the story of police officer Max Rockatansky of Main Force Patrol fighting to protect his wife and child from a violent biker gang in the Australian desert set, "A few years from now." A global oil crisis has devastated the world, resulting in the dystopian Australia Max finds himself in.

This setting reflected the social and political context of the time of the film's production. The 1970s saw significant changes in the Australian economy, reacting to international volatility following the 1973 oil price shock and the doubling of inflation to 12.9 per cent between 1973–1974, closely followed by stagflation the year after as a result (Treasury). *Mad Max* (1979) was distinctly Australian, represented by Australian accents and the landscape, but it also spoke to the global issues of fuel shortages and environmental disaster. In *Mad Max 2: Road Warrior* (Miller 1981) the setting shifts from oil crisis to post-apocalypse. Society is in decay and Max serves as a sheriff to protect fellow survivors seeking refuge in the wastelands against marauders. By the time *Mad Max 3: Beyond the Thunderdome* (Miller 1985) was released, a third world war has occurred in the backstory and the world is in the aftermath of a nuclear winter. The transition in this third adaptation of Mad Max followed the floating of the Australian dollar in 1983, allowing it to vary with supply and demand against other currencies as part of a global economy.

Analysing the Mad Max adaptations reveals their appropriation of environmental and economic issues that were current at the time of each film's release. The plot device of nuclear war contributed to the film's relevancy in a global context. It was produced towards the end of the Cold War, when nuclear disarmament was a hot topic. *Mad Max 3: Beyond the Thunderdome* also features American pop star Tina Turner. Turner's character connects the film through her famous face to a wider international audience. Aunty Entity, Turner's character, is the ruler of Bartertown, which is fuelled by methane gas captured from pigs that are kept in factory stalls underneath the settlement. Max prevails, but the final scene depicts Sydney in ruins, recognisable nationally and internationally because of the Sydney Harbour Bridge.

The Mad Max franchise represents an alternate facet of Australian identity, that of an international citizen, increasingly aware of the global nature of war and environmental disaster. This represents growing environmental awareness in Australia and gestures to the continent's isolation – geographically and politically from global centres – making it both a refuge from disaster and a world unto its own. Over the course of the Mad Max films, Miller has adapted action, thriller, and Western genres, all within a futuristic Australian outback setting. The fourth film, *Mad Max: Fury Road*, was released to critical acclaim at Cannes in 2015 and saw astronomical box office success pushing Australian film across the profit line of its previous record, $63.4 million, set in 2001 (*Screen Australia* 2015). *Fury Road* made $45 million domestically in its first weekend (*Screen Australia* 2015).

While the debate over whether *Fury Road* (2015) is a feminist film is beyond the scope of this chapter, what is relevant is the adaptation of an Australian classic into a Hollywood blockbuster, and the extent to which *Fury Road* (2015) remains an Australian film. While homage is still paid to iconic Australian character, Max, albeit in this incarnation played by a British actor, it seems that Max and the film in general works for national and international audience in distinct ways. *Fury Road* was nominated for ten Academy Awards, including Best Director, and has been hailed as one of Australia's most successful films (*If.com.au* 2016). However, as noted in the introduction, the extent to which *Fury Road* (2015) is an Australian film is complicated by the fact that it was shot in Namibia because the land around Broken Hill in rural Australia was too green to represent the desolation that is so much a part of the script. Tom Hardy played Max, alongside co-star Charlize Theron (Furiosa), instead of Australian actors. While *Fury Road* (2015) was a grossly successful "Australian film," any essential Australian identity is undermined by the cosmopolitanism status of its adaptation, production, and release. While the adaptation of dystopia from oil crisis to nuclear winter serve the purpose of the Mad Max genre, they also provide insight into how adaptation in accordance with specific social and political factors serves to popularise certain films and representations.

The social and political context of the making and consumption of adaptations has a "material, public and economic" effect beyond or alongside their status as cultural products (Hutcheon 2006: 28). Hutcheon (2006: 28) argues:

> even in today's globalised world, major shifts in a story's context … for example, in a national setting or time period can change radically how the transposed story is interpreted, ideologically and literally.

It is not the national setting that changes in Mad Max, but the environmental and economic situation to make each adaptation even more relatable for its contemporary audience.

The franchise demonstrates an appetite to adapt Hollywood genres in an Australian context, and in the national image.

Considering each of the Mad Max films as adaptations in their social and political context demonstrates the way in which case studies of adaptations understood as palimpsests can be used to examine "larger ideological shifts" (Hassler-Forest and Nicklas 1). While this is true of film theory generally, "ideological criticism" moves beyond close reading to question the broader political and economic context the film constructs and challenges and "the critical role ideologies play in culture" (Pramaggiore and Wallis 2006: 329). However, analysing these films as palimpsests demonstrates the way each one conjures a sense of national identity like a spectral return in constant conversation with the past, present, and future. The adaptation of Mad Max is an example of the effect of globalisation on Australian culture and cultural products, making it increasingly diverse, cosmopolitan, and transnational. Mad Max is distinctly Australian and simultaneously redefines Australia as a general member of the West. National cinemas by definition exist nationally and internationally in relation to each other and in terms of dissemination and consumption (French and Poole 2013: 86). *Fury Road* (2015) is an Australian product intended for a global audience. Only through the process of adaptation, as each sequel is built on the narrative that preceded it, and Mad Max is recognised as a palimpsest, is its identity as an Australian text realised. Through adaptation, *Fury Road* is an example of the effect of globalisation on Australian cultural products, making it more diverse, cosmopolitan, and transnational, or, at the least, more commercially viable in a Hollywood dominated film industry.

By appropriating Hollywood genres and funding models, *Fury Road* represents an internationalising of Australian culture. Building on the work of Saskia Sassen, Tom O'Regan, and Anna Potter (2013: 7), I argue that as globalisation proceeds, its operating parts become more difficult to see because of their reach into all aspects of the film industry, from creation, production, distribution, ownership, and audience. As the Mad Max films were adapted, they were subject to the dynamics and constraints of globalisation, as well as Australia's national film industry. While *Fury Road* (2015) as a genre film would never be held to the same standard of the AFC-genre to represent the nation, it is nevertheless a marker of Australian identity. What makes *Fury Road* an Australian film is not its location or the accent of its lead character, but the knowledge of its status as an adaptation, and as such its characterisation as a palimpsest. Its Australian legacy lies just beneath the surface.

In recent years, adaptation theory has broadened the study of adaptations within their social and political context to provide insight into how adaptations shape and reflect assumptions about nation, heritage, and ethnicity. Australian cinema provides a case study that shows how some adaptations hold a central place in the development of Australia's cultural heritage and the recognition of classic Australian films. Viewed in this way, Australia's national identity has been informed by a nostalgic reimagining of a bloodless colonial past, multicultural aspirations of the 1980s and 1990s, and globalising imperatives of a free market and environmental issues. This has occurred in the political service of nationhood in adaptations such as *Picnic at Hanging Rock* and other AFC genre films. It also manifests in response to official multiculturalism in adaptations such as *Strictly Ballroom*. In comparison, the Mad Max adaptations mark a shift away from representing Australia through its landscape or people. Instead, *Fury Road* adapts Hollywood genres that reflect cosmopolitan identities and international stories. In the case of Mad Max, this manifests in people looking for a home wherever they might find one free of environmental devastation.

This is evidence of adaptation embedding itself in a transnational conversation where, as humans do, cultural products also adapt to ensure the longevity of an Australian film industry within a broader Hollywood environment. In becoming more cosmopolitan, transnational, and global, a question for further research is are Australian adaptations still helping to define the nation and its heritage, or are they having an homogenising effect that makes Australian film more profitable transnationally? Comparing recent box office figures indicates that, despite the ongoing adaptation of national tropes in some films, the "Australianisation" of Hollywood (or is it the "Hollywoodisation" of Australia?) is what the people want.

To conclude, Pietrzak-Franger (2012: 85) argues that thinking about adaptations with reference to hauntology allows us to see adaptation "as palimpsestuous" and presupposes an impossibility of a faithful rendition of an adapted text" because this can only every appear through "spectrality." In the same way, essential representations of Australian identity are rendered implausible through their sheer variation. This is evident in the brief close reading conducted in this chapter. The tropes, stereotypes, myths, and heritage loom like spectres, but they also undermine any true notion of Australianness. Reading Australia's national identity as an adaptation or as an adaptive palimpsest demonstrates this. Pietrzak-Franger (2012: 85) argues:

> Such a theorization also makes it possible to highlight the constant dialogue between the past, present and future as central to adaptation … it prioritizes an ethically creative and critical engagement with adapted texts. It throws into relief not the similarities between the text adapted and its adaptation but privileges difference and ineluctable otherness as fundamental to the process of thinking of adaptations as adaptations.

So much of Australian Studies is taken up with critiquing clichéd images of Australia in film and literature. Perhaps adaptation studies can provide a new lens, through which the ongoing construction of myths, legends, and heroes that make up the land down under are revealed.

Notes

1 In 1967 a national Referendum was held to grant Aboriginal Australians citizenship rights. This contrasts with women achieving the vote around Australia by no later than the 1920s, and the introduction of citizenship legislation in 1948 that could be accessed by migrants despite the White Australia policy.
2 A Hill's Hoist is a rotary clothesline. It was deemed emblematic enough of Australian domestic culture to be included as a prop in the Opening Ceremony of the Sydney 2000 Olympics along with lawn mowers and white picket fences.
3 *Australia* combines the genres of historical romance, Western, and war movie to tell an outback adventure story that reinscribes the history of the stolen generations as part of Australia's national identity. It adapts popular Australian imagery of the Drover and is based on Xavier Herbert's *Capricornia* (1939) and *Poor Fellow My Country* (1975), as well as *Gone with the Wind* (dir. Victor Fleming 1939) and *The Wizard of Oz* (dir. Victor Fleming 1939).

Works cited

Adams, P. (1984) "Two views," *Cinema Papers* 44–45: 70–72.
Australian Government Department of Social Services (2015) *Settlement and Multicultural Affairs*. Available at: https://www.dss.gov.au/our-responsibilities/settlement-and-multicultural-affairs/publications/the-people-of-australia-australias-multicultural-policy (accessed 3 November 2015).
Box Office Mojo (2016) "Mad Max: Fury Road (2015) – Box Office Mojo," *in Boxofficemojo.com*. Online. (accessed 11 July 2016).
Byrnes, P. (2015) "Why *Picnic at Hanging Rock* changed Australian cinema and still mystifies." *The Sydney Morning Herald*. Online. (accessed 11 July 2016).
Childs, P. (2012) "Cultural heritage/heritage culture: Adapting the contemporary British historical novel," in Pascal Nicklas and Oliver Lindner (eds) *Adaptation and Cultural Appropriation: Literature, Film, and the Arts*, Berlin and Boston: De Gruyter, 89–100.

Collins, F. and T. Davis (2004) *Australian Cinema after Mabo*, Cambridge, UK: Cambridge University Press.

Cook, P. (2010) "Transnational utopias: Baz Luhrmann and Australian cinema," *Transnational Cinemas* 1(1): 23–36.

Couzens, A. J. (2016) "Recalling romance and revision in the film adaptations of *Robbery Under Arms* and *The Chant of Jimmie Blacksmith*," *Adaptation* 9.1 46–57.

Derrida, J. (1994) *Spectres of Marx: The State of the Debt, the Work of Mourning and New International*. Trans. Peggy Kamuf. New York and London: Routledge.

Elder, C. (2007) *Being Australian: Narratives of National Identity*, Crows Nest, UK: Allen and Unwin.

French, L. and M. Poole (2013) "Internationalising Australian film and television: The AFI and an Australian Academy of Cinema and Television Arts (AACTA)," *Metro* 176: 86–91.

Hadley, L. (2010) *Neo-Victorian Fiction and Historical Narrative: The Victorians and Us*. St. Martins: Palgrave Macmillan.

Hage, G. (1998) *White Nation: Fantasies of White Supremacy in a Multicultural Society*, Annandale, NSW: Pluto Press.

Hassler-Forest, D. and P. Nicklas (2015) *The Politics of Adaptation: Media Convergence and Ideology*, London: Palgrave Macmillan.

Head On (1998), dir. Ana Kokkinos. Dist. Strand Releasing.

Heartbreak Kid (1993), dir. Michael Jenkins. Village Roadshow.

Hutcheon, L. (2006) *A Theory of Adaptation*, New York and London: Routledge.

If.com.au (2016) "Mad Max: Fury Road Claims Australian Record for Most Academy Award Nominations," in *If.com.au*. Online. Available at: http://if.com.au/2016/01/15/article/Mad-Max-Fury-Road-claims-Australian-record-for-most-Academy-Award-nominations/NKRGMYEBLJ.html (accessed 12 July 2016).

Jacobs, K. (2011) *Experience and Representation: Contemporary Perspectives on Australian Migration*, Surrey, UK: Ashgate Press.

Jupp, J. (2007) *From White Australia to Woomera: The Story of Australian Immigration*, Cambridge: Cambridge University Press.

Lindsay, J. (2009) *Picnic at Hanging Rock*, Melbourne: Penguin Group.

Looking for Alibrandi (2000), dir. Kate Woods. Robyn Kershaw Productions.

McFarlane, B. (1993) "The Australian Literary Adaptation: An Overview," *Literature/Film Quarterly* 21(2): 90–100.

Mad Max (1979), dir. George Miller. Kennedy Miller Productions, Crossroads, Mad Max Films.

Mad Max: Fury Road (2015), dir. George Miller. Warner Bros., Village Roadshow Pictures, Kennedy Miller Mitchell, Rat-Pac Dune Entertainment.

Mad Max 2: Road Warrior (1981), dir. George Miller. Kennedy Miller Productions. Dist. Warner Bros.

Mad Max 3: Beyond the Thunderdome (1985), dir. George Miller and George Ogilvie. Kennedy Miller Productions. Dist. Warner Bros.

Malthouse Theatre (2016) "Picnic at Hanging Rock · Malthouse Theatre," in *Malthousetheatre.com.au*. Online. (accessed 12 July 2016).

Marchetta, M. (1992) *Looking for Alibrandi*, Melbourne: Penguin Books Australia.

My Brilliant Career (1979), dir. Gillian Armstrong. Peace Arch Production Company. Dist. Analysis Film Releasing Corporation.

Nayar, P. K. (2010) *Postcolonialism: A Guide for the Perplexed*. London, New Dehli, New York, Sydney: Bloomsbury Publishing.

NFSA (2016) "Picnic at Hanging Rock: 40 Years of Mystery // National Film and Sound Archive, Australia," in *Nfsa.gov.au*. Online. Available at: http://www.nfsa.gov.au/visit-us/exhibitions-presentations/picnic-at-hanging-rock-exhibition/ (accessed 11 July 2016).

O'Regan, T. (1996) *Australian National Cinema*, London and New York: Routledge.

O'Regan, T. and A. Potter (2013) "Globalisation from Within? The De-Nationalising of Australian Film and Television Production," *Media International Australia Incorporating Culture and Policy*, 149: 5–14.

Picnic at Hanging Rock (1975), dir. Peter Weir. Picnic Productions.

Picnicathangingrock.com (2016) "Picnic at Hanging Rock – Daniel Zaitchik," in *Picnicathangingrock.com*. Online. (accessed 12 July 2016).

Pietrzak-Franger, M. (2012) "Conversing with Ghosts: Or, the Ethics of Adaptation," In Pascal Nicklas and Oscar Lindner (eds) *Adaptation and Cultural Appropriation: Literature, Film and the Arts*, Berlin, Boston: De Gruyter, 70–87.

Powers, J. (2015) "Postcard from Cannes: *Mad Max* Will Astound You," in *Vogue*. Online (accessed 11 July 2016).

Pramaggiore, M. and T. Wallis (2006) *Film: A Critical Introduction*, Boston: Pearson Allyn and Bacon.

Rodenstein, R. (2016) "FILM REVIEW HHH: The Virgin Suicides – The Occasional Golden Moment – The Tech," in *Tech.mit.edu*. Online (accessed 12 July 2016).

Screen Australia (2015) "Media Releases: 2015, a Record-Breaking Year for Australian Film," *in Screen Australia*. Online (accessed 3 November 2015).

Screen Australia (2016a) "MPDAA Confirm 2015 Best Year Ever for Australian Film," in *Screen Australia*. Online (accessed 11 July 2016).

Screen Australia (2016b) "Top Films at The Box Office – Australian Films – Cinema – Fact Finders – Screen Australia," in *Screen Australia*. Online (accessed 11 July 2016).

Simpson, C., R. Murawska, and A. Lambert (2009) *Diasporas of Australian Cinema*, Bristol, UK: Intellect Books.

Stratton, J. (1998) *Race Daze: Australia in Identity Crisis*, Smithfield, NSW: Pluto Press Australia.

Strictly Ballroom (1984), dir. Baz Luhrmann. Available at: https://www.ausstage.edu.au/pages/event/1488 (accessed 3 November 2015).

Strictly Ballroom (1992), dir. Baz Luhrmann. M&A Productions.

Strictly Ballroom the Musical (2015) "Home," in *Strictly Ballroom the Musical*. Online. Available at: http://www.strictlyballroomthemusical.com (accessed 3 November 2015).

Sunday Too Far Away (1975), dir. Ken Hannam. South Australian Film Corporation. Roadshow (Australia) and Columbia-Warner (UK).

The True Story of the Kelly Gang (1906), dir. Charles Tait.

Turner, G. (1989), "Art Directing History: The Period Film," in Albert Moran and Tom O'Regan (eds) *The Australian Screen*, Melbourne: Penguin.

Visit Macedon Ranges (2016) "Hanging Rock – Visit Macedon Ranges," in *Visit Macedon Ranges*. Online. (accessed 12 July 2016).

Voigts-Virchow, E. (2007) "Heritage and Literature on Screen: *Heimat* and Heritage," in Deborah Cartmell and Imelda Whelehan (eds) *The Cambridge Companion to Literature on Screen*, Cambridge: Cambridge University Press, 123–137.

Voigts-Virchow, E. (2009) "Metadaptation: Adaptation and Intermediality – Cock and Bull," *Journal of Adaptation in Film and Performance*. 2(2): 137–152.

We of the Never Never (1982), dir. Igor Auzins. Adams Packer. Hoyts Distribution. Umbrella Entertainment.

Whelehan, I. (2000) *Overloaded: Popular Culture and the Future of Feminism*, London: The Women's Press.

20

ADAPTATION AND THE AUSTRALIAN FILM REVIVAL

Brian McFarlane

When people started to speak excitedly about the 1970s' Australian film 'revival'[1] (even the term 'renaissance' was sometimes invoked), they tended to have in mind the films made from well-loved novels. These could be 'classics' such as *My Brilliant Career* (1979) or they could have been merely 'popular', such as *Picnic at Hanging Rock* (1975), but, whatever their provenance, the films thus derived seemed to bestow a touch of class on the long-awaited upsurge in local film-making. There had been box-office successes such as *The Adventures of Barry McKenzie* (1972), *Stork* (1971) and *Alvin Purple* (1973), with their adroit mixtures of raucous comedy and risqué sex, but it was really the films with literary affiliations that earnt the critical seal of approval – and that of 'serious' filmgoers.

It is likely that all commercial cinemas draw heavily on literary sources. So what is it – particularly in the case of Australian filmmakers in the 1970s – that makes this such a prevalent phenomenon? It almost seemed as if, in the burgeoning Australian cinema of that time, prestige in one narrative medium might rub off on another – and the value of a pre-sold title was not to be overlooked. Whatever the thinking behind them, the rush of adaptations undoubtedly helped to make Australian cinema not merely a force to contend with at home but brought it to international attention for the first time in decades – or, perhaps, ever. In 1979 a writer claimed that, since the inception of the Academy Awards in 1927–8, "more than three-fourths of the awards for 'best picture' have gone to adaptations … [and that] the all-time box-office successes favour novels even more" (Beja 1979: 78). Perhaps the combination of quality and commercial elements hinted at here was at work in the preponderance of adaptations in the major titles that contributed to the emergence of the Australian film revival of the 1970s.

Based on …?

The adaptations so prominent in this revival were overwhelmingly derived from novels. In the decade from 1971 to 1981, there were several that had their origins in plays, a few derived from television series and even one (*The Adventures of Barry McKenzie*) from a comic strip. And a year later, one of the most financially successful, albeit intellectually bereft, films of the period, *The Man from Snowy River* (1982), took A.B. 'Banjo' Patterson's famous bush ballad as its starting point. The television spin-offs were as negligible as this species of adaptation has usually been: the mildly agreeable *Country Town* (1971), from the long-running 'soap', *Bellbird* (1967–77), and *Number 96* (1974) and *The Box* (1975), based on two semi-salacious series that had taken advantage of the relaxation in censorship in the decade. All three of these have sunk without a trace in the history of the national cinema's development.

The plays included David Williamson's *Stork*, *The Removalists* (1975), *Don's Party* (1976) and *The Club* (1980), the latter two directed by Bruce Beresford, one of the key figures in the revival, whose compelling *'Breaker' Morant* (1980) was also based on a play (by Kenneth Ross). Williamson was of course a major player in Australian drama and film at the time and for several decades later when other of his plays were adapted to the screen, but only the two directed by Beresford received much critical acclaim, though *Stork* had secured its place among the 'ocker' comedies[2] so enjoyed by audiences and reviled by critics. The only other play-derived films of the period were John Duigan's *Dimboola* (1979), from Jack Hibberd's comedy about a small-town wedding, and Tim Burstall's tale of conflict in a remote drilling-site, *The Last of the Knucklemen* (1979), from John Power's three-act play.

The two play-derived films that have had a more enduring life are *Don's Party* and *'Breaker' Morant*, partly because of the vitality and relevance of the theatrical text, and perhaps even more so because their director Bruce Beresford emerged as arguably the most protean craftsman of the revival. He had survived the popularity of the 'Barry McKenzie' films (there was a sequel in 1974, *Barry McKenzie Holds His Own*) to go on to more serious fare, both in Australia and abroad, and some of his most memorable films are adapted from plays or novels.

It needs to be made clear that there were also some films from original screenplays that made their mark then and that have proved to have staying power. Three of the most notable of these were Ken Hannam's *Sunday Too Far Away* (1975), Fred Schepisi's *The Devil's Playground* and George Miller's *Mad Max* (both 1979). The first, an eloquent study of outback shearers at work, appeared in the same year as *Picnic at Hanging Rock*, and it is not an exaggeration to say that deriving from them came a new awareness of an Australian quality cinema. (Both were listed by Melbourne's leading daily newspaper reviewer as among the year's top ten films (Bennett 1975), a state of affairs without precedent in living memory.) The second is a still poignantly acute chronicle of a frustrated childhood in a seminary, and *Mad Max*, an imaginative thriller-cum-road movie, went on to spawn three sequels, most recently *Mad Max: Fury Road* (2015), certainly the most popular franchise yet developed here. But in the decade under consideration now, these three − along with some smaller-scale realist pieces such as *The F.J. Holden* (1977) and *Mouth to Mouth* (1978) − are the exception rather than the rule. The adaptations of novels dominated the production output of this watershed decade, and probably the public perception of its achievements.

From page to screen

Two for the road

Released within months of each other in 1971 were two novel-derived films, both of which, in retrospect, were of a quality that might have launched the 'revival'. The fact that they did not is worth thinking about. They were *Walkabout*, based on James Vance Marshall's 1959 novel, and *Wake in Fright*, adapted from Kenneth Cook's 1961 novel. They were both directed by 'outsiders': respectively, British cinematographer-director Nicolas Roeg and Canadian Ted Kotcheff, who had worked mainly in Britain, and perhaps their 'outsider' connections contributed to their status at the time. Also, they were both largely set in remote and uninviting environments, and they both uncomfortably articulated critiques of certain Australian values. Perhaps these descriptors did not constitute a recipe for commercial success, and neither, despite their excellences, enjoyed this réclame, though their reputations have survived their contemporary rejection by audiences. The authors of the indispensable *Australian Film 1900–1977* write that *Walkabout*, in spite of some positive reviews, "did not fare well commercially" (Pike and Cooper 1980: 258–9), and of *Wake in Fright*: "although critics were unanimous in their support, publicity was poor and the public stayed away" (Pike and Cooper 1980: 259).

To look at *Wake in Fright* again after several decades is to wonder at its being almost a lost film over that period. In its restored form, the work of the Australian National Film and Sound Archive, it found much more receptive audiences in 2009, confirming what some of us had always believed: that is, this 1971 film could easily have been the banner announcing the revival. It emerges as a tough, uncompromising adaptation of Cook's short novel, and, in its way, it now looks to have more in common with the later products of the 1970s, adaptations and otherwise. By this, I mean that there was a recurring sense of 'coming of age', of *Bildungsroman*, about a good deal of the local production, perhaps most vividly felt in the versions of novels such as *The Getting of Wisdom* (1977), but also in films working from original screenplays, such as *The Devil's Playground*. It is tempting – and I'm not prone to resist the temptation – to see this element in relation to what was happening in Australian life at the time, when, under the leadership of Labor Prime Minister Gough Whitlam, a new, more sophisticated sense of nationalism was making itself felt. For instance, imperial honours were done away with and Australia devised its own system of rewarding achievement, but there was too a wider feeling of emergence from a long adolescence marked by undue deference to a remote Britain. The setting up by the Whitlam Labor government (1972–5) of the Australian Film Commission in 1975, replacing earlier funding bodies, no doubt gave a boost to production – and perhaps to the more prestigious national cinema of which literary adaptations were in the vanguard.

As to *Wake in Fright*, the film hews closely to the novel's narrative contours, but my point is that, even when this is so in the adaptations of the period, the *choice* of text, at the outset, suggests awareness of major change in the national life. The protagonist of *Wake in Fright* is John Grant, a young teacher who has been sent to a tiny outback[3] school in a deserted landscape in which only the school, the pub and the Tiboonda railway siding break the desolate emptiness of the scene. Significantly, Grant is played by an English actor, Gary Bond, and this, along with the freshness of his youth, contributes to his subsequent sense of being an outsider in the crude world in which misadventure will deposit him. At the end of the school term, he plans to fly to Sydney where there is the promise of surf and a girl called Robyn. First, he takes a train to the country town, Bundunyabba ('The Yabba' to its residents), where he will next morning board the train to Sydney. His plans go seriously awry, however, when he becomes involved in a 'two-up' gambling game and the company of a gaggle of boozy locals. He loses his money, drinks himself into a stupor, engages in a kangaroo-shoot of mindless cruelty with some of his drinking companions, and later, even drunker, is the victim of sexual assault by Doc Tyden, an alcoholic medico who listens to opera in his disgusting shack. Doc is played by an English actor, Donald Pleasence, noted for his skill in purveying corruption of various kinds, and when Grant is saved from suicide and returns to Tiboonda, it is as though he has shaken off this Old-World vice and reached a new sense of maturity.

It is not, of course, just Old-World vice he's had to deal with: outback Australia has confronted him with plenty of its own. The film's depiction of a repulsive male culture in which boisterous drinking-into-insensibility is the order of the day, or gathering to watch the inane two-up pastime that incites the men of The Yabba to risk losing everything and from which women are totally excluded (as they are in the crowded pub scene) was not likely to appeal to those audiences who, a few years later, would flock to more decorous adaptations. When one of the boozers, Dick (a young Jack Thompson), asks rhetorically, when Grant is standing to one side talking to the daughter of the house that is their current roistering site, "What's the matter with him? Rather talk to a woman than *drink*?", we are aware of the film's sharp critique of what was often depicted as good, manly fun in Australia. Grant's trajectory from Tiboonda school to Tyden's shack is something like a descent into hell from which he escapes by the skin of his teeth to begin the next stage of his growth. Cook's novel went through seven reprints in the wake of the film, and it is satisfying to note that the film has now undergone resuscitation in public and critical esteem.

The other film adaptation released at almost the same time, Nicolas Roeg's *Walkabout*, constituted a similar challenge for audiences. Roeg, had been one of Britain's most respected cameramen for twenty years before co-directing *Performance* (1970), a skilful but controversial potpourri of sex, violence and rock music. He may have seemed an odd choice to direct *Walkabout*, since the James Vance Marshall novel is essentially a book for children, but in the event he made something haunting and memorable from Marshall's tale of children dumped in the Australian outback, coming to terms with some of life's more alarming encounters and growing from the experience. Roeg's film, working from a screenplay by the controversial dramatist Edward Bond, has taken little more than the central informing idea – two white children lost and helped by an Aboriginal boy on 'walkabout', itself a coming-of-age ritual – and made something memorable but tonally very different from the original. One would not have expected Bond to settle for anything comforting; and the novel, described by one commentator as "amazing garbage" (Hutton 1985: 334), needed to be re-imagined considerably if it was to make any appeal to adult audiences.

In a 1994 interview, in answer to a question as to what had drawn him to the novel, Roeg had this to say:

> I liked the setting; I liked the idea of being lost – of being lost in life … I never felt it was important to be so faithful to the actual written description of what is happening. I was caught by the idea of two children lost in a desert. Children are lost people anyway, trying to search for themselves; they are trying to grow up and become people.
> *McFarlane 1997: 489*

Roeg's own words reinforce the idea that the novel offers little more than a starting point for an ambitious and demanding film that enforces consideration of the physical cruelties and racial schism that are part of Australian history. Though some progress has been made in addressing these issues since 1971, there are enough traces remaining to render Roeg's film a still-challenging experience.

If the novel appears simplistic to adult readers in its black/white contrasts and confrontations, as well as over-written for young readers, it does share with the film a certain arbitrariness in placing its youthful protagonists in the harsh expanses of the Australian desert. Almost too arbitrarily for easy credibility, John and Mary in the novel are made victims of an air crash that leaves them alone and without provisions in the middle of nowhere. Roeg's answers this by starting the film in Sydney with "a series of sounds and images which suggest that modern man has not transcended the rituals of the tribe but merely perverted them" (Dawson 1971). Images of bustling streets, a class of schoolgirls doing breathing exercises and a soulless-looking apartment block shortly give way to that of a father driving his two children into the desert for a picnic. While the children are setting up the picnic, he sets fire to the car and then shoots himself. Certainly, there is still something arbitrary about this as narrative starter, but there is also an unnerving suggestion that he has been brought to this point by some dislocation in his city life. And the film will end bleakly back in 'civilisation' with the girl now married and living in the sort of apartment we saw earlier, with memory flashes of how she, at fourteen, along with her six-year-old brother, had survived being 'lost' with the help of an Aboriginal boy of her age.

In his novel, Marshall names the white siblings, but in the film the three young protagonists are listed in the cast simply as Girl, White Boy and Black Boy, as if to draw attention to their archetypal significance. Roeg's is a poetic conceptualising of the situation, not a realist one, and this may help to explain the film's failure to find substantial audiences at the time of release. This was not perhaps the 'Australia' that audiences wanted to see, one in which the inadequacy of 'white knowledge' to get the children to safety on their own and the self-destruction of the Black Boy (David Gulpilil, in his first film), whose courtship dance terrifies the Girl (Jenny Agutter), do not offer narrative comfort.

This is a very rich film, both thematically and visually (thanks to Roeg's superb camera work, which does justice to both the threat and harsh grandeur of the landscape), and deserves much more detailed treatment than offered here. The two aspects I stress here have to do with its place in the 1970s revival. Like its contemporary adaptation, *Wake in Fright*, it offers a harsh insight into ways in which white Australia has failed to come to terms with both the promise and the dangers of what the country has had to offer. As one commentary puts it: "In this film, the interior is still brim full of spiritual infinitude which answers the needs of Eurocentric mankind escaping from a shrinking Europe" (Dermody and Jacka 1988: 81). These are 'Answers' which many in 1971, and perhaps even now, are not anxious to accept. The other recurring element of the Australian cinema of the time, especially manifest in its literary adaptations, is that of coming-of-age. Whereas in many of the films of the period this may be seen in a positive light, this is not really the case here: the Girl, tentatively aware of her own sexuality, is forced into a mothering role with her little brother and deeply disturbed by the Black Boy's growing interest in her. She *does* mature as a result of her experiences, but this does not make for happiness, as the poignant last moments imply. Return to civilisation has not been able to offer the sense of freedom, even of exhilaration, that she'd found in moments in the desert.

Two directors and two girls making their way

Two directors who made their names with their 1970s Australian films and subsequently worked internationally were Bruce Beresford and Gillian Armstrong, and the films they directed are now strongly redolent of what the new cinema was offering then. Beresford made his first short films while still a university student, and there were more than a dozen of these, some made here, some in the UK, before he began his feature career with the 'Barry McKenzie' films. But his serious reputation as a filmmaker really got underway with his adaptations of David Williamson's play, *Don's Party* (1976), and Henry Handel Richardson's famous novel, *The Getting of Wisdom*.[4] The latter is my focus here, along with Armstrong's *My Brilliant Career*, derived from Miles Franklin's well-loved classic. Armstrong, a graduate of the Film and Television course at Swinburne Technical College in 1971, was the first woman to direct a feature film in Australia for forty years. I treat these two adaptations together because the protagonist of each is a girl on the brink of adulthood, dealing with a range of conventional restraints that may have belonged to the periods in which each was set (early twentieth century, when the Federation – achieved in 1901 – was still very young) but which struck resonant chords in the pro-feminist 1970s.

The Getting of Wisdom, like much of Richardson's fiction, is strongly autobiographical. In her memoir, *Myself When Young* (1948), she makes clear that the experiences that constitute the episodic plot of *The Getting of Wisdom* are based on memories of her own schooldays. She is quite explicit in the memoir about how she wants us to see Laura at the end of her schooldays and at the end of the novel, more or less as a portrait of the artist as a young girl. She is also quite clear about the real-life basis for Laura's infatuation with the older girl, Evelyn: "The attraction this girl had for me was so strong that few others have surpassed it" (71, endnote 12). These are the two strands of the novel as they are dealt with in Beresford's film that I want to consider here. These two aspects of Laura's growth – the one imaginative/intellectual, the other sexual – both contribute to the novel's somewhat bleak final pages, when she runs down an avenue: "Then came a sudden bend in the long, straight path. She shot round it and was lost to sight" (Richardson 1961: 237). At this point Laura has come to some self-knowledge, and there is a suggestion of possible fulfilment at some future date, but it is not enough to offset the pervasive bleakness of the rest of the novel. It is as a potential artist, one standing solitary, partly because of her calling as an artist, a creator of narratives, partly because of her temperament, that we are required to see here. And like her own creator, she will be an artist whose work, however imaginatively detailed, must be rooted in autobiographical 'fact'.

Beresford's film adaptation was one of a batch ushered in by *Picnic at Hanging Rock*, which set a seal of approval on new Australian cinema. As suggested above, these films had in common a preoccupation with growing up, a setting in Australia's past, a generally episodic treatment derived from similarly episodic novels, and a curiously decorous tone (at least superficially) at odds with the national myths and with the preceding 'ocker' comedies. Several decades later, *The Getting of Wisdom* looks like one of the more tough-minded of those period films, even if its toughness works differently from the novel's. The novel itself, of course, was also a period piece when it appeared in 1910, based on the author's schooldays in the 1880s. What one of the criticisms heard casually about these flagship films of the new Australian cinema, once the initial enthusiasm had died down, was that they were all comfortably set in the past, as though a newly emergent cinema should be grappling with contemporary issues in contemporary settings. I would argue that there is no reason why a film set in the past, or a film adapted from a literary classic of almost seven decades earlier, should not reflect as much of the time of its *making* as it does of its *setting*.

When asked what had attracted him to filming this novel, Beresford said:

> I think, you know, it was the story of the girl being an outsider, and I'd always felt that this was my position, which was probably not true. … I think it was just a role I liked to cast myself in. Also, I thought it was a very good novel, much more concise than any of her other novels to put it mildly. And it had a very well-observed, very cleverly delineated range of characters. It was based on her own experiences at school, but I'd also based the film-script on her memoir, *Myself When Young*, which covered the same period.
>
> *McFarlane 2007: 113*

The film chooses, as it has every right to do in pursuit of its own coherence rather than in slavish 'fidelity' to its antecedent, to diverge from Richardson's conception by turning Laura into a budding concert pianist at the end. At the school's final Speech Day, she plays to an appreciative audience and the headmaster announces that she has won a scholarship entitling her to two years' study in Leipzig, which was what happened to Richardson herself. As Beresford went on to say, "it was a matter of its being much more filmable. It meant we had scenes of Laura sitting at the piano and outraging people by playing Thalberg, whereas it would have been much more difficult to make it interesting to show her writing" (McFarlane 2007: 113). This seems a convincing and sensible reason for 'adapting' one of the novel's central concerns, without in any necessary way undermining the idea of Laura's move towards a new stage in her life. The film's final run down the avenue has a zest not felt in the novel's ending.

Writing about the film eight years later, the English critic Robin Wood (1983: 200) noted that, though the film had been "accorded 'sensitive-study-of-young-girl-growing-up' treatment, not a single review I read made any reference to lesbianism, or to the fact that the 'growing up' expresses itself centrally in the acceptance of a lesbian identity". In the then-popular coming-of-age film sub-genre, this particular aspect of the process in Laura's case was perhaps critically subsumed into the larger context of coming-of-age. It now helps to account for the film's on-going power to compel attention.

The other literary 'classic' of a young woman determined on making her own way was Miles Franklin's *My Brilliant Career* (1901). This (again semi-autobiographical) novel pre-dates the 1970s feminist movement by seven decades, but anticipates it in a protagonist, Sybilla Melvyn, who asserts her own identity and independence, and Gillian (or 'Gill' as she was listed on the credits) Armstrong's 1979 film underlines this resolve even more forcibly than the novel does. In the novel's last chapter, its narrator–heroine writes: "What the future holds I know not, and am tonight so weary that I do not care" (*My Brilliant Career* 230). Armstrong's film closes with Sybilla (Judy Davis) renouncing the handsome, tolerant man, Harry Beacham (Sam Neill), who loves her and whom she may indeed love, in order to take her future into her own hands, epito-mised by her consigning the manuscript of the book she has been writing to Blackwoods [*sic*] Publishers, Edinburgh.

Just as it seemed that Australian audiences might have had enough of period pieces or deco-rative adaptations of varying degrees of distinction, *My Brilliant Career* came as a shot in the arm. Thematically it belonged with those other literary-based films which explored coming-of-age in various settings, and with the feeling the decade had generated of a shift in the national culture. By chance, the only other film of 1979 that made a major impact, critically or commer-cially, *Mad Max*, injected another contrary, very potent, stimulus into the revival's progress. And each launched a major star who would go on to international success: Mel Gibson in *Max* and Judy Davis in *Career*, two more-than-usually-forceful screen personalities (both NIDA-trained stage actors) who had each made only one little-noted previous feature appearance.

If the taste for literary adaptation had declined after bringing some prestige to the Australian film industry, *My Brilliant Career* could be seen as representing the sub-genre at its best. With its woman-dominated credentials – director Armstrong, producer Margaret Fink, associate pro-ducer Jane Scott and screenwriter Eleanor Witcombe at work on its female author's original – it could be seen as embodying some of the 1970s feminist thinking, especially in its repudiation of a conventional romantic conclusion.

Like *The Getting of Wisdom*, Armstrong's film also opens in rural Australia with its heroine involved in the process of fiction-making, but, whereas *Wisdom*'s Laura's is an oral fiction which she is telling to her younger sister, Sybilla is first seen writing at a table. Armstrong traces her heroine's journey towards her coming-of-age via the moves she makes from the genteel poverty of her parents' home through a series of locations (all in outback New South Wales) which the mise-en-scène depicts with more concern for meaning than for mere pictorialism. By the end of the film, Sybilla has moved further along what would be the trajectory of her life than Laura had in *The Getting of Wisdom*, but they have in common a determination to be true to their own insights into their own natures and the possible places they might find for these in the worlds they would go on to inhabit.[5]

Other rites of passage

In the four films considered in some detail so far, the sense of place is important in determining the crucial steps towards a new, more mature understanding of self: Grant in his remote outback school and the crude rural town in which accident locates him; the Girl in *Walkabout* whose experiences in the Australian desert will stay with her well into adulthood; Laura, who will have to leave her tiny rural township; and Sybilla who will, for the film's length, be confined to various rural settings but knows at its end that her future lies elsewhere. And this is the case with several other adaptations of the period that chart the rites of passage of their protagonists: it is almost as if, for some, coming to terms with a harsh environment will be a source of their inner maturity, while for others getting away from it will be the necessary hurdle.

Peter Weir's adaptation of Joan Lindsay's *Picnic at Hanging Rock* may well be the film most firmly and fondly associated with the revival. This largely unremarkable novel was reprinted more than a dozen times between 1975 and 1982, no doubt in response to the popular success of the film. It is the story of the disappearance of a party of private-school girls during the eponymous outing, and Lindsay, somewhat disingenuously, suggested that whether "[it] is fact or fiction, my readers must decide for themselves. As the fateful picnic took place in the year nineteen hundred, and since the characters who appear in this book are all long since dead, it hardly seems important" (Lindsay 1997: 6). Set in the Australian bush, the film establishes very early, through Russell Boyd's wonderfully evocative cinematography, the two monoliths between which the action will take place. These are the Rock itself, with the vast, threatening faces it presents to the world, and the granite façade of Appleyard College. The latter may appear to be trying to impose its decorums on a landscape indifferent to it, but the Rock, seemingly impervious to the humans who venture upon it, will be more than a match for the school's influence.

The film's subsequent strength will lie not in its development of a tightly organised narrative but in the evocation of an atmosphere, and its coherence will lie in the way it explores the conflict implied imagistically in its two key icons – the Rock and the College – and for its handling of the rites-of-passage motif, characteristic of all these adaptations. Having written about *Picnic* in detail on several occasions,[6] I want now only to refer briefly to this particular aspect. Peter Weir was interviewed (1981: 325) a few years after the film's release. When questioned about the film's exploration of a smothered sexuality at work in the depiction of these teenage girls, he claimed: "I was never really interested in that side of the film … For me, the grand theme was Nature, and even the girls' sexuality was as much a part of that as the lizard crawling across the top of the rock." Perhaps so, but one of the film's most vivid images is that of Irma, the one girl who survives the vanishing on the Rock, as she returns to the College after having been nursed back to health. She left the College to go on the picnic dressed in filmy white as all the girls were; when she comes back, to say goodbye to the remaining school pupils, who are engaged in some sort of eurythmic exercise in the gym, she is a very different figure. She is now dressed in a long crimson cloak with hat to match, quite different from her school uniform. But there is also an air to the way she carries herself and speaks that suggests she is no longer the girl who went missing some months before. Whether or not this interested the director, there is a distinct aura of sexual coming-of-age here. Whatever has happened to her on the Rock (and the film, like the novel, maintains its mystery about this), she has left girlhood behind her. The other girls, sensing some new awareness in Irma, crowd around her hysterically, as if trying to share this experience.

Two other adaptations of the decade, with male protagonists who undergo their own rites-of-passage experiences, are Kevin Dobson's *The Mango Tree* (1977) and Fred Schepisi's *The Chant of Jimmie Blacksmith* (1978). Each is based on a prestigious novel of the decade, the former on Ronald McKie's 1974 novel, the latter on Thomas Keneally's of 1972.[7] *The Mango Tree* signals its approach to its source novel in its opening images and accompanying voice-over. A steam train is speeding off, and as a young man makes his way to his seat he imagines/remembers a woman's voice saying to him: "You're going away from this town. You must. You know that. You'll decide what you'll do – write, fly a plane, who knows? Remember I love you. I'll never forget you." An older woman's voice takes over the soundtrack to say, "Remember, Jamie, we're not all sinful. The church claims we're all innocent, but some are more innocent than others." We are set up to ask: what kinds of experience have brought the boy to this point? The two voices will prove to be those of Miss Pringle (Diane Craig), Jamie's French teacher who initiates him into the ways of sexual love, and of his Grandma Carr (Geraldine Fitzgerald), who, his parents having died, has provided him with a home and wise counsel.

There are several problems with this film, produced and written by actor Michael Pate and starring his son, Christopher Pate, as Jamie. It may seem to announce initially that we are in for another rites-of-passage piece, and so, up to a point, this is what we get. Not only does Jamie undergo his sexual orientation, but he also intervenes to stop a teacher from bullying one of his classmates and, importantly, twice challenges the grandmother he loves and to whom he owes so much: first, he tells her "I won't be going" to church when she calls him to accompany her, and, second, despite *her* fear of flying, he accepts against her wishes a flight from visiting war ace, Bert Hinkler. Jamie is taking off in several senses.

But the film doesn't find enough confidence in his coming-of-age as a means of holding our interest, not even when it is seen in the context of the new maturing of the nation at large in the light of Australia's contribution to the British forces in the World War, which comes to an end in the course of the film's narrative. Michael Pate's screenplay (working from McKie's sprawling novel) drags in too many side issues and too many characters, along with their relationships and activities, without exploring any of these in detail and without their having any real bearing on what we take to be the central concern of Jamie's move towards maturity. A stronger central performance, as distinct from the mere amiability Christopher Pate offers (and he is too old to be a convincing schoolboy) might have helped. As it is, the Irish-American actress Geraldine Fitzgerald's wonderfully humane inhabiting of the Grandma Carr role is the strongest element in the film, even though this doesn't seem to be its rallying point.

The last of these adaptations that I want to refer to is *The Chant of Jimmie Blacksmith*. Although each of these films in some way highlights the rites-of-passage theme, *The Chant of Jimmie Blacksmith* differs radically from the others in the sense that its focus is on the hideous matter of racial hatred and, in turn, the hatred that this instils in the eponymous Jimmie (Tommy Lewis). He aspires to marriage, getting a job, generally making his way in the white community, having been encouraged by the clergyman, Mr Neville (Jack Thompson), who has taken Jimmie under his wing. He marries a white girl (Angela Punch), believing her to be pregnant with his child, only to find the child cannot possibly be his. This discovery leads him to go on a murderous rampage, shouting, "I declared war; that's what I've done", taking revenge on "everyone that's done me wrong". There is talk of 'marauding blacks', but there is also talk of the whites who "took away a way of life", bringing to the Aborigines "a whole host of improvements", including alcohol abuse and syphilis, among others. The film ends tragically for Jimmie, and, set as it is at the turn of the century, there is the ironic notion of a nation coming of age, amid "all this Federation nonsense".

Jimmie Blacksmith was not likely to please audiences who had welcomed the decorative period pieces of the revival. It never deals in nostalgia for aspects of Australia's past; in fact, it offers an abrasive critique of a key aspect of this nostalgia. American critic Pauline Kael (1980: 204) wrote: "This great Australian film, *The Chant of Jimmie Blacksmith*, which was made in 1978 and has finally opened here [US, 1979], has almost nothing in common with the other Australian films of recent years."[8] The rites-of-passage idea, at work in all these adaptations, is here savagely mocked, from the early moment when some of his tribesmen are forcing grog down his throat, so as "to make you a man, make you a father of men". The film, written, produced and directed by Schepisi, may have had disappointing box-office returns at the time, but over thirty years later it carries a charge that perhaps only the abrasive *Wake in Fright*, made at the earlier end of the decade, can still deliver.

Retrospect and hindsight

Looking back on this very exciting period of Australian filmmaking, a few recurring elements now emerge with some clarity. The sheer preponderance of adaptation – and I have here concentrated only on a handful of *novels*-into-film – now seems surprising in the light of the prevailing 'ocker' titles that had set the revival in motion. Adaptation has been an ongoing feature but without dominating perception of the national cinema's subsequent output. For instance, the 1980s saw respectable if not greatly adventurous dealings with Christina Stead's *For Love Alone* (1986, directed by Stephen Wallace), and Tim Burstall grappled with D.H. Lawrence's *Kangaroo* (1987), but the decade is more likely to be remembered for the box-office successes of the *Mad Max* films of 1981 and 1985 and the hugely popular '*Crocodile' Dundee* (1986) than for its cinematic dealings with literature. And during the last years of the century, films such as *Strictly Ballroom* (1992), *Muriel's Wedding* (1994) and *Shine* (1996) are more quickly recollected than, say, *Country Life* (1994), Michael Blakemore's under-valued relocation of Chekhov's *Uncle Vanya* to outback Australia, or Armstrong's *Oscar and Lucinda* (1997), a reworking of Peter Carey's eccentric novel. In the new century, apart from a daring *Macbeth* (2006) set in Melbourne's gangland, the adaptations tended to be of contemporary novels, such as Ana Kokkinos's *Head On* (1998) from Chris Tsiolkas's novel, *Loaded*, which dramatises without compromise the explosive conflicts of its youthful, gay, Greek protagonist.

This essay has focussed on those adaptations which helped to shape the all-important revival of Australian cinema in the often-tumultuous 1970s. These titles gave a sort of coherence to the national output; they were all essentially in the realist mode, as were such other key films of the period as *Sunday Too Far Away* (1975) and *The Devil's Playground* (1979). And like so many revival films, adaptation or otherwise, they tapped into a sort of national coming-of-age. Most of their directors – Weir, Armstrong, Beresford and Schepisi, for instance – went on to work internationally, sometimes to award-winning effect, but their reputations were originally and memorably made with their work in literary adaptation in the Australian cinema they helped to create, and to which they have all intermittently returned. Re-watching their 1970s films, one registers that they cut their teeth on some potent material and in the process brought something equally potent of their own to bear on it.

Notes

1 Also known as the "Australian New Wave", the Revival is typically dated from 1970 to 1985. It was spurred in part by government funding of the Australian Film Institute in 1970, and the establishment of the Australian Film Development Corporation in 1970, and the Australian Film Commission in 1975. Several of the films discussed below were funded, at least in part, by these bodies.

2 'Ocker' is a term referring to an Australian who speaks and acts in an uncultured manner, using a broad Australian accent (or 'Strine,' according to *Wikipedia*). It carries a masculinist, hedonistic resonance, at odds with more obviously cultivated tastes.

3 *Outback* was the title given to the film when shown internationally.

4 'Henry Handel Richardson' was the name under which Ethel Lindesay Richardson wrote.

5 There were of course more films derived from novels than I can deal with here, both in the decade of the revival (think of Henri Safran's *Storm Boy* (1976), from Colin Thiele's popular children's book of 1964) and subsequently. Some of the better-known ones include Peter Weir's *The Year of Living Dangerously* (1982, from Christopher Koch's 1978 novel), Carl Schultz's *Careful He Might Hear You* (1983, from Sumner Locke Elliott's 1963 novel), Richard Flanagan's *The Sound of One Hand Clapping* (1998, from his own 1997 novel) and Fred Schepisi's *The Eye of the Storm* (2011, from Patrick White's 1973 novel). My emphasis in this study is on the place of the literary adaptation in the crucial years of the Australian film revival of the 1970s.

6 For example, in *Words and Images: Australian Novels into Australian Films* (Melbourne: Heinemann, 1983) and more recently in "The Long Shadow of *Hanging Rock*," *Screen Education*, No. 75, Spring 2014.
7 McKie won the Miles Franklin Award; Keneally's novel was nominated for the Booker Prize.
8 "New Yorker" (1980) reprinted in A. Moran, and T. O'Regan (eds) *An Australian Film Reader*, Paddington, NSW: Currency Press, 204.

Works cited

Beja, M. (1979) *Film and Literature*, New York: Longmans.
Bennett, C. (1975) "Triumph clouded by future fears," Melbourne: *The Age*, 14.
Dawson, J. (1971) "*Walkabout*," *Monthly Film Bulletin*, 38 (454), 227–8.
Dermody, S., and Jacka, E. (1988) *The Screening of Australia*, vol. 2, Sydney: Currency Press.
Hutton, A. (1985) "Black Australia and film: Only if it makes money," in A. Moran, and T. O'Regan (eds) *An Australian Film Reader*, Paddington, NSW: Currency Press.
Kael, P. (1980) "New Yorker," reprinted in A. Moran, and T. O'Regan (eds) *An Australian Film Reader*, Paddington, NSW: Currency Press, 204–10.
Lindsay, J. (1997) *Picnic at Hanging Rock* (1967), Camberwell, Victoria: Penguin.
McFarlane, B. (1997) "Nicolas Roeg," in B. McFarlane *An Autobiography of British Cinema*, London: Methuen/bfi publishing.
McFarlane, B. (2007) "From Ockers to Oscars: An interview with Bruce Beresford," *Metro Magazine: Media & Education Magazine*, 154: 110–5.
My Brilliant Career (1979), dir. G. Armstrong. Greater Union Organisation, Margaret Fink Productions, New South Wales Film Corp. DVD.
Pike, A., and Cooper, R. (1980) *Australian Film 1900–1977*, 1998, rev. edn, Melbourne, VIC: Oxford University Press.
Richardson, H.H. (1948) *Myself When Young*, London: Heinemann.
Richardson, H.H. (1961) *The Getting of Wisdom*, London: Heinemann.
The Chant of Jimmie Blacksmith (1978), dir. F. Schepisi. Perf. Freddy Reynolds, Angela Punch McGregor. The Film House, Victorian Film Corporation. DVD.
The Mango Tree (1977), dir. K. Dobson. Perf. Geraldine Fitzgerald, Robert Helpmann. Pisces Productions. DVD.
Wake in Fright (1971), dir. T. Kotcheff. Perf. Gary Bond, Donald Pleasence. NLT Productions, Group W. DVD.
Weir, P. (1981) "Peter Weir: Towards the Centre," interview by B. McFarlane and T. Ryan, *Cinema Papers*, 34: 322–9.
Wood, Robin (1983) "Quo Vadis Bruce Beresford?" in A. Moran, and T. O'Regan (eds) *An Australian Film Reader*, Paddington, NSW: Currency Press.

PART IV

Reception

Dennis Cutchins

Reception, simply put, is the recognition of the necessary role an audience, a reader, an experiencer, or an interpretive community plays in interpreting any text. In textual studies, reception is often a game-changer, since it constantly forces scholars to shift their focus away from texts and the structures and patterns revealed by close reading, and toward the dynamic and unpredictable relationships that exist between texts and those who experience them. These relationships are always present, though they are sometimes ignored, but they become even more unpredictable when a text is transplanted to a different time or place. Since adaptation studies often deals with texts that have been transplanted across cultures, through time, or from one medium to another, reception should play a significant role in the field. Even astute theorists of adaptation, however, have sometimes ignored reception, treating both source texts and adaptations as monolithic entities whose relationships are static and can be mapped. This is often productive, but always dangerous, since anyone who experiences a text will always do so from a slightly different angle.

This principle is important for many reasons, not the least of which is that all adapters begin, at least, as people experiencing a text. Imagine a screenwriter assigned to write a screenplay for a new adaptation of Herman Melville's *Moby Dick*. She may begin by reading the novel and creating an abstract, a detailed plot outline or treatment. If the production team approves of it, that treatment becomes the source for the new script. The screenwriter may have understood and abstracted *Moby Dick* as a story about a man seeking vengeance against a whale, as an allegory about the dangers citizens face when putting their fate in the hands of corrupt leaders, as a story about a lowly sailor who somehow survives the destruction of a whaling ship, or in any number of other ways. In the words of Obi-Wan, each of these abstractions is true, "from a certain point of view," but each would very naturally lead to distinctly different interpretations of the novel. More to the point, the different abstractions/interpretations would tend to result in very different adaptations.

The next six essays explore these dynamics of reception as they play out in different ways and across diverse media. Amanda Ruud's "Embodying change: Adaptation, the senses, and media revolution" explores adaptations as sensory experiences that are inscribed on literal bodies. By recognizing this physicality, she observes, "adaptation makes a mediated experience of a cultural text both sensuous and sensible." Using examples from early silent films, Ruud makes complex theory seem like commonsense. In a somewhat less theoretical vein, Bradley Stephens's "Great voices speak alike: Orson Welles's radio adaptation of Victor Hugo's *Les Misérables*" analyzes an early Welles radio adaptation in light of the famously intimate relationship radio listeners have with the medium. Stephens argues that Welles found in Hugo a kindred spirit, and in *Les Misérables* an echo of depression-era problems. Suzanne Speidel also writes about radio adaptations in "Lux presents Hollywood: Films on the radio during the 'golden age' of broadcasting," but in this case they were radio adaptations of Hollywood films, often created using the same actors. Speidel reminds readers of the economic and practical reasons these adaptations were popular, and the role they played in keeping Hollywood on the minds of listeners. In "Reconfiguring the Nordic Noir brand: Nordic Noir TV crime drama as remake," Yvonne Griggs points out that some adaptations adapt the style of a genre, rather than any particular text. She writes about the worldwide phenomena of Nordic Noir television, even in productions that have nothing to do with Scandinavia. Anna Blackwell's "Tweeting from the grave: Shakespeare, adaptation, and social media" is a delightful treatment of the social media tempest in a teapot surrounding the rediscovery of the bones of Richard III in 2012. Blackwell reminds us that interpretive communities thrive on Facebook and Twitter and have the potential to shift public opinion, even about texts that are more than 400 years old. In "Adaptation, fidelity and reception" Katy Meeks and Dennis Cutchins argue for the primary role reception theory should play in adaptation studies. By sharing their own, very personal, experiences with texts they work to shine a new light on familiar topics like fidelity.

21

EMBODYING CHANGE

Adaptation, the senses, and media revolution

Amanda Ruud

While scholarship on reception has made important gestures to epistemology and cognition, it has often been relatively blind to the senses it invokes. Is adaptation a kind of transformation we recognize by sight? Or is it something we feel? When we encounter adaptations that surprise us, repel us, or invite us in, do we experience that as a kind of touch? It is possible that it is all of these, and more? And yet, our scholarship resists addressing these sensible and affective encounters with adaptation beyond their ability to be transformed into epistemological metaphors. More often than not, an adaptation—say, a Shakespeare play produced as a silent film, or a novel turned into a ballet—will signal itself in part by a change in address to the senses. They ask their audiences to experience visual, aural, and even haptic transformations to a narrative with which we have familiarity and often even emotional attachments. Just as transformations to familiar texts can potently move an audience (Who hasn't heard the lay critic's frustrated cry: 'It wasn't like that in the book!'?), so too do transformations in the media by which we experience them. New technological advances, such as the invention of 'moving pictures', have forcibly affected spectators in ways that cold terms such as 'recognition' and 'reception' don't seem properly to name. Indeed, Alfred Gell needed the term 'enchantment' to describe this moment of perception (1992: 211). Gell argued that wonder, fascination, and pleasurable incomprehension were affects almost bound to accompany an audience's experience of a new technology or medium. Though Gell's argument is concerned with significant evolutions in technology, it seems to me that the 'enchantment' he describes may also be broadly applicable to the experience of adaptation. Do we not also experience sensations of surprise, of visual wonder, of haptic desire when we encounter alterations that are less extreme? A new staging of a familiar opera might still have the capacity to make its audience gasp with surprise or grimace at a new characterization—all moments of experience that register in a feeling body. It is to this somatic experience of adaptive change that I wish to draw our attention.

Increasingly, adaptation studies scholarship has called on us to attend to the manifold ways in which audiences experience adaptation. To date, "experiencing adaptation" has largely been a subject of epistemological study—asking how we distinguish between knowing and unknowing audiences, or how viewers or readers work to fill in the gaps that adaptations inevitably possess (Hutcheon 2013: 120–128). Here, I hope to posit a new approach that has its roots in the phenomenology of perception, one that takes the whole of embodied experience—and the existential stakes of that experience—as the conditions under which adaptation

245

can be received. What if we started experiencing adaptations as 'experience expressing experience'? That is, as sensory events that themselves accomplish two things: first, they reflect on the experience of the adapted thing (novel, film, etc., filtered through the memory and subjectivity of the adaptor) while offering that experience to another audience, and second, they reflexively comment on the means by which that experience can be conveyed (the medium). To do so would, I believe, bring into play the wider array of sensory transformations that take place in adaptive behavior and make each one of those transformations signify—reflecting on and articulating what it means to receive a text, a film, a play, a novel, etc. in a living body. These bodies, I suggest, do not always receive adaptations and media transformations as epistemological puzzles; rather, they come under the spell of experience offered to them visually, aurally, sensibly. The purpose of this chapter is to test the possibilities of this phenomenological approach to the study of adaptation.

Experience expressing experience

Let us begin, then, with two definitions and an axiom. The definitions I draw from two theorists who engage adaptation—and particularly film adaptation—as a kind of lived experience. The axiom is my own attempt to point to an ever-present, if underappreciated, aspect of adaptation studies: sensation.

Our first definition comes from Linda Hutcheon. In *A Theory of Adaptation* (2013) Hutcheon provides one of the most comprehensive recent approaches to articulating adaptation's means and its ends. Describing adaptation as a process of reception, she offers this definition: "adaptation is a form of intertextuality: we experience adaptations (*as adaptations*) as palimpsests through our memory of other works that resonate through repetition with variation" (Hutcheon 2012: 8). Though Hutcheon's dominant metaphor is always textual—asserting adaptation's role as a signifying act—I want to note the sensory mélange that her definition invokes. A palimpsest, that layering of text upon text that indicates both engagement and revision, is a visual metaphor. Recognizing a palimpsest requires a *viewer* to notice the layering effect of multiple overwritten sentences, not just as a linguistic challenge but also as a function of *sight*. And yet, Hutcheon's next sentence offers the evocative term 'resonate,' remembering that we receive adaptations not just as viewers, but as hearers too. This shuttling between seeing and hearing should not, I think, strike us as a poorly mixed metaphor. Rather, it should remind us that receiving adaptations is an act of complex coordination that centers not only in the minds and memories of a receptive audience, but on their bodies too. If we are going to consider adaptation as a kind of reception, as Hutcheon rightly suggests we should, then we should acknowledge that adaptation *embodies* that reception in two important ways: it gives a form (a body) to a particular moment of reception, and it delivers that experience to another set of bodies to be received yet again. To fully articulate what it means to receive an adaptation, it seems necessary that we focus our attention on the second set of bodies: the audience, spectator, reader, player, etc. If adaptation studies are going to come to a robust understanding of adaptation as a process of reception, then we will need to take these sensory metaphors seriously, remembering that experiencing adaptation is a bodily act.

This leads us to our second definition. Quoting Maurice Merleau-Ponty, Vivian Sobchack begins *The Address of the Eye* thus: "What else is film if not an 'expression of experience by experience'? And what else is the primary task of film theory if not to restore to us, through reflection upon that experience and its expression, the original power of the motion picture to signify" (1992: 3–4). Sobchack goes on to posit that film's superlative claims to communicative power lie in its ability to make itself "sensuously and sensibly manifest." That is, a film is itself an act of seeing or hearing that can be seen and heard, and an act of reflecting on experience that can be felt by a reflecting and understanding audience. For Sobchack, film's definitive function is a phenomenological one. Though Sobchack makes these claims for all of cinema, I want to posit that this "expression of experience by experience" is a particularly keen definition of adaptive behavior. Adaptations—and not only cinematic adaptations of texts, but all forms of inter- and intra-medial adaptation—possess this phenomenological potential. They make a lived experience of a cultural unit (a text, a joke, a comic book, a film, a novel, etc.) manifest to experience again. In other words, adaptations *produce* experiences at the same time as they *reflect on* experiences. And importantly, if we are going to map Sobchack's phenomenological interests onto our exploration of adaptation, then the ends of phenomenological inquiry might also apply. Those ends are intersubjective. Insofar as adaptations seek to generate community around reflection on lived experience, they serve as mediating objects in the gathering of subjects. In short, expressing experience by experience allows multiple subjects to meet at the point of sensation.

This then, is the deceptively simple axiom I have to offer: adaptation makes a mediated experience of a cultural text both sensuous and sensible, and can always be experienced as such. Adaptation is experience expressing experience.

While this axiom does not claim to be a complete definition of adaption, it does offer some helpful support to the ongoing methodological soul-searching in adaptation studies. For instance, in a more recent attempt to posit a systematic approach to adaptation studies, Patrick Catrysse offers three desirable characteristics of an adaptation studies methodology. A favorable approach to adaptation should be 1) "descriptive rather than prescriptive," 2) "target [oriented] rather than fidelity based," and 3) "trans-individual, systemic, and corpus-based" (2014: 15). These characteristics are meant to mirror the general conclusions of the recent history of adaptation scholarship, but at least two of them also manage to reflect the virtues of approaching adaptation as experience expressing experience. Echoing the formula Catrysse recommends, this approach encourages adaptation scholars to describe the how and the why of an adaptation, but makes the likes and dislikes of the analyst a moot point. If an adaptation expresses the *experience* of a text rather than the text itself, then the response 'you should have better experienced it another way' is an irrational commentary. One may acknowledge that they experienced the 'original' cultural text differently, but that has little to do with their encounter with another experience: that which is occasioned by the adaptation. Likewise, this approach to the study of adaptation helps to liberate scholarship from the legendary bogeyman of fidelity criticism. What is at stake in experience expressing experience is not a mimetic *representation* of a text or cultural object, which could be more or less correct. Rather, the experience of that cultural object becomes a source of further experiential reflection, a reorientation that at least decenters—if not elides—the issue of fidelity to another text.

But besides responding to these (perhaps over-rehearsed) critiques of early adaptation criticism, approaching adaptation as an embodied and sensory reflection on experience can offer multiple new benefits. First, it ameliorates the defensive impulse of adaptation scholarship to counter iconophobia with a histrionic show of semiotic prowess. That is, it admits that adaptations 'signify' in complex and interesting ways without submitting fully to a linguistic model of criticism. Second, it builds into its methodology a reflection on changing media: technological transformation brings with it a new experiential apparatus which adaptations can both employ and interrogate. And third, it proposes an end for adaptive behavior that is essentially humanist and potentially political: that end is an intersubjective encounter across disparate moments of sensory experience. In what follows, I emphasize these benefits while turning to my experience of the earliest Shakespeare films to mediate my argument.

247

Narrative and iconophobia

Perhaps without intent, adaptation scholarship has historically accepted an opposition between the (sensory) pleasures of receiving adaptation and the scholarly pleasures of making adaptation signify. This chasm between the sensual and the sensible has, I think, been a primary cause of our hesitation to consider adaptation from the perspective of the body. Thus, when Robert Stam sought to explain the sources of a perceived academic denigration of adaption, he named "iconophobia and logophilia" as its twin sources (2000: 58). Citing "iconophobia," Stam's argument assumes that there is an implicit alliance between the pleasures of visuality and the pleasures of adaptation. To adapt a text—in the case of Stam's work, a novel made into a film—is to make the narrative *visible* through the cinematic apparatus. But the appeal to the eyes, Stam seems to concede, is a kind of plebeian pleasure—one that ostensibly comes at the cost of narrative complexity. In response, Stam, among others, took up the challenge of proving adaptation's semiotic capabilities. Though expressed by means of a primarily visual/aural display, adaptation is, he argued, structurally complex, multi-vocal, and politically significant. Thus, while adaptation has consistently seemed to have visual pleasure as one of its aims, scholarship has tended to disavow that pleasure and, in effect, cede the territory of pleasure from the senses over to the more cerebral territory of semiotics and narrative complexity. Ironically, even this dichotomy finds itself challenged when we consider the sensory address that underlies even a motive as ascetic as "logophilia." For signification itself depends on an ability to perceive difference, either by sight (noting the visible difference between a 'b' and a 'd') or sound (hearing the distinction between a sound articulated by the lips and the tip of the tongue). Considered this way, the very grounds of adaptation scholarship's claim to academic significance have always already been rooted in moments of sensory experience, whether we allow ourselves to revel in those senses or not.

More importantly, as Christa Albrecht-Crane has noted, allowing adaptation scholarship to remain a debate over narrative competence and adaptations' power to signify risks essentializing the media that adaptations employ, effectively treating them all as narrative machines. Instead, she suggests, "film [and, I would add, video games, comic books, and other adaptive apparatuses] might form an aesthetic experience, akin to other forms (such as music) in which one perceives not narrative, but rather particular sense experiences" (Albrecht-Crane 2010: 248). Despite her call to consider the affective, aesthetic experience of adaptations, adaptation as a sensory experience does not seem to have been taken up by recent scholars. Nevertheless, adaptations continue to manifest as complex aesthetic and sensory experiences. Narrative capacity doesn't begin to account for the immersive experience of a Charles Dickens theme park (Dickens World opened in the county of Kent, England in 2007) or the recently revised choreography for Prokofiev's *Romeo and Juliet*. These experiences incorporate rhythm, movement, time, sound, smell, and spectacle in ways that exceed the questions that we are ready to ask of narrative. Rather than treating sensory experiences as means to the end of signification or narrative sense, it is perhaps time to grant that experience can be both an end in itself and a reflection on itself.

Technology and the senses

The capacity of our working axiom—that adaptation's primary expressive act is to convey not narrative but experience to the senses of an audience—is perhaps best conveyed when we consider its ability to reflexively address technological change. Our encounters with adaptation are inevitably shaped by—and indeed, often triggered by—technological change. From the birth of the moving image to the more recent possibility of immersive alternative realities, new media developments have dramatically altered the sensory experiences open to adaptors and their audiences. And indeed, the birth of a new medium seems to catalyze the need to reflect on those new experiences. This relatively obvious point comes with more significant—and eloquent—repercussions if we approach adaptation as experience expressing experience. If adaptations express experiences, then adaptations have the potential to become a particularly fertile locus for the reflexive navigation of media revolutions, which perhaps accounts for the profound amounts of self-reflexivity we witness in adaptations made at times of technological change. This potentiality has, for instance, been beautifully borne out by the history of film. Many of the earliest film experiments were re-workings of familiar scenes from the stage or the pages of a novel, which themselves reflect on or contemplate what it means to experience a new media. For example, *The Death of Nancy Sykes* in 1897 and *The Kiss* in 1896 were both adaptations of moments familiar from the stage (Bordwell and Thompson 2003: 20). Indeed, we might say that these new-media experiments specifically relied on the familiarity of the narrative content to allow a greater amount of reflection on the experience of the medium itself. While some definitions of adaptation might exclude these short scenes—they are, after all, direct transpositions of stage scenes onto the screen—it would be foolish to ignore the experiential significance they had for viewers, who could see/feel/hear past the narrative development to experience the film as a self-aware witness to a media transformation.

To illustrate this reflexivity more clearly, I wish to turn to a particular historical example—two, in fact—that not only demonstrate the ability of adaptation to shuttle between sensory registers, but that also demonstrate the peculiar aptness of adaptation for expressing complex experiences. The first films of Shakespeare, silent films that split their allegiances between the screen and the stage, will demonstrate adaptation's ability to articulate experience not only of a pre-existent text (in this case, the plays of Shakespeare), but also—with a double voice—to narrate an ambivalent experience of the emergent film media. As early iterations of both the medium of film and the practice of adaptation, they show quite readily their investment in both producing an experience and reflecting on that experience.

Reflecting on experience in Shakespearean silent film

The very first film adaptation of Shakespeare had a fascinatingly intermedial birth. In 1899, William K.L. Dickson filmed three scenes of *King John* (Figure 21.1). The scenes were shot in a studio on the Embankment, but featured sets, costumes, and actors from Herbert Beerbohm Tree's concurrent production at Her Majesty's Theatre. Both the film and the play premiered on the same September night. That temporal overlap—the fact that the same story was repeated on the same night in two different mediums—suggests that the narrative itself was not at all the motivating factor of the filmed performance. Rather, the experience of the film medium—the visual, spectacular simulacrum of real objects located not far away—seems itself to be an object of the endeavor.

The nature of Dickson's *King John* seems to bear out this suspicion. The film was actually films; three fragments of cinema depicting three important scenes from Beerbohm Tree's production of *King John*. Only recently have any of those fragments resurfaced. When Robert Hamilton Ball wrote the first history of Shakespeare on Silent Film in 1968, he could only guess as to which scenes had been photographed. He surmised wrongly. Ball logically suggested that the scenes photographed for the first Shakespeare film might have been visual tableaux, and Tree's non-Shakespearean added scene of the signing of the Magna Carta seemed like an obvious candidate, as it "needed no words, only pantomime for its effect" (Ball 1968: 23). When images from each scene did turn up, though, none were of strictly *visual* moments in the play;

rather, each of the three depict verbal exchanges important to the plot. The only viewable remaining scene depicts a death speech given by King John in the garden at Swinstead Abbey. Slumped in a chair in front of a painted screen, the King gestures to the attending men—men who are turned away from the camera to look, along with us, at the King. As his fervor rises, so does his body, and he struggles to sit forward in his chair. Repeatedly, the king pounds his chest: at first with an open clutching hand—perhaps gesturing toward the burning pain of the poison in his chest—but by the end of the scene his fist is closed, presumably as his speech ends on a reference to his body as "but a clod and module of confounded royalty" (*King John* 5.7, 57–8).

For a viewer profoundly familiar with Shakespeare's text, it might be possible to trace, through gesture, the entire progress of King John's last speech. Beerbohm Tree's gestures seem to encode Shakespeare's references to burning heat, John's appeals for cool water, and his rejection of the comforts offered by Prince Henry. But somehow, tracing the *linguistic* import of King John's gestures seems to be beyond the point of the entire exercise. The remaining film crystalizes instead around two non-linguistic and non-representational elements of the scene: the King stares directly at the camera, and the film ends before his death. The final, lingering image of this fragment of *King John* is not that of a dead king or a speech from Shakespeare. In the final moments, the surrounding figures lean toward the king as he pulls himself forward in his chair and, clutching fist to chest, appeals to an audience across the boundaries of the fiction. The whole point of the scene seems not to be the dying of the king in the diegetic world of the play, but rather the leaning in, the appealing gaze, the moment of sensible pathos. Screening *King John* to scores of undergraduate students has demonstrated that even to an audience viewing a century late, the film sensibly communicates an affective experience in the look across the screen, the haptic pounding of the king's chest, and the rising tension in each of the film's seventy-seven seconds.

In registering a palpable fascination with the *event* of King John's speech rather than the words of King John's speech, this first fragment of the Shakespeare film renders a space for pleasure and display that lies beyond the words of the Bard. Though the senses invoked by the film might seem to be *fewer* than those one might experience if they witnessed the stage play

Figure 21.1 Herbert Beerbohm Tree as King John in the first Shakespeare film by William K.L. Dickson, 1899

on that same night, they do not in any way seem *lesser*, and, in fact, the experiences as a whole seem to run parallel to one another. A viewer of the play would see the actor, hear his speech, and experience their affect rise with that of the audience members around them, with the added benefit of narrative context. Dickson, however, seems to have captured much of that affect in his fragmentary film, refracting it through the novel experience of the film medium. In the cinema audience, the specific sensory appeal is transformed, but the affective appeal would arguably be equally strong, and amplified by the new magic of the screen. In one sense, Dickson offers an expression of his own experience of *theatrical* pleasure, and in another, he offers pleasure to a new audience in a particularly *filmic* manner. Indeed, Dickson's film bears all the marks of a cinema enamored of its own medium. In Tom Gunning's formulation, the years before 1906 were a crucial period in which "the act of showing and exhibition" and the "harnessing of visibility" were the primary aims of cinema (1986: 64). This "cinema of attractions" exhibits a fascination not with story but with the "magical possibilities of the cinema," that is, with the possibility of subjecting the spectator to "aggressive ... sensual or psychological impact" (Gunning 1986: 64–65). As attractions, these films find their *telos* not in the delivery of a diegetic narrative, but in the ability to capture a viewer's sensations, to call upon feeling beyond the 'merely' visual, invoking a sense of space, of time, of movement, of bodily immersion. Put another way, the aim of the cinema of attractions is to produce sensory experience as an end. With *King John*, however, the particular intervention of adaptation comes into play. For Dickson's film not only explores the experience of cinema, but uses the cinematic encounter to relay his own theatrical one: *King John* adapts an experience of *King John*.

Dickson's first Shakespeare film celebrates both the affect of the theatre and that of the cinema. With this double allegiance, *King John* underscores the reflexive power of adaptations that is particularly potent at the birth of new media: a power to comment on new media experiences even while offering them to others. While Dickson's film seems relatively unexamined with regard to which scenes of the Shakespeare play it offers, other Shakespeare films reflect the capability of adaptation to perform both reflexive and analytic work in *how* they allow audiences to experience their theatrical source. By offering experiences which interrogate the practice of *viewing* (and not hearing) Shakespeare, these films turn their reflexive eye not only on the broadly articulated affect of the theatre or the cinema, but on the epistemological stakes of experiencing Shakespeare by these new means. Percy Stow's 1908 film of *The Tempest* enters in as a keen example here.

Judith Buchanan's recent reading of the 1908 *Tempest* has emphasized the strong ties between the Stow film and the 1904 stage production of the play by Herbert Beerbohm Tree. For Buchanan, the film shows significant tensions between modes of production, flip-flopping between cinematic autonomy and textual fidelity, filmic idiom, and stage convention, thus marking the film with "medium impurities" (2009: 316). In this sense, it echoes the dual investments of the Dickson *King John* by attempting to navigate multiple kinds of experience at once. While I think that Buchanan is entirely correct in pointing to moments of crossover between stage convention and filmic idiom as definitive of this piece, I want also to emphasize the appropriateness of that hybridity. As an experience, the picture reflects on the cultural emergence of the film audience, contemplating the kinds of visual experience—cinematic, 'realist,' spectacular—that were becoming available to mass audiences. *The Tempest,* then, makes it possible to *experience the question* of how the spectator ought to engage visual media in the brave new world of moving pictures. Viewed as experience expressing experience, *The Tempest* offers Shakespeare's play in the middle of a media transformation, caught between theatrical and filmic structures of visual pleasure.

Stow's *Tempest* begins with the pre-history of the Shakespeare play: we see Prospero arriving on the island with a toddling Miranda, Prospero discovering Caliban, Prospero freeing Ariel from a tree, Caliban attempting to woo Miranda. While each of these scenes does provide narrative context for the later progress of the play, I want also to point out two important things. First, none of these scenes are properly 'Shakespearean' *as scenes*. Rather, they are verbally related at later points in Shakespeare's play. The desire that produces them in the film, then, seems to be just as much a desire *to have them seen* as a desire to make narrative sense. Secondly, in addition to showing things that are unseen in Shakespeare's play, the early scenes of the film emphasize moments in which the senses can be cinematically surprised. Prospero frees Ariel from a tree using a familiar trick shot, and Ariel transforms from a girl to a monkey in rapid succession in a scene that leaves an audience as surprised as Caliban. The trick shots are obvious to the modern eye, but surprisingly sophisticated and still somewhat compelling. These surprises, along with the basic impulse to visualize what Shakespeare does not visualize, are, I suggest, the motivating factor of these early, imposed scenes. Moreover, they cue the audience into recognizing that cinematic visuality is itself a subject of interest in the film, a fact notably heightened by the absence of diegetic sound or speech that the medium necessitates.

This interest reaches its climax in the storm scene, a cunning cinematic endeavor, filmed on a studio set. The scene shows Miranda begging her father to desist as he conjures the storm. His wild gestures send explosions of doves out of a dais before him in a style powerfully evocative of vaudeville-style magicians. Quickly, though, the stage illusions give way to a celebration of the possibilities of the new medium. The film flashes to a few abstract, cloud-covered frames with bright, violent gashes signifying lightning. When the scene returns to Prospero, the wall behind him has opened to reveal actual footage of a storming sea, upon which a boat flounders and sinks. The illusion depends upon at least three layers of film cutting. It is powerfully cinematic, and also powerfully self-aware.

Indeed, we can read *The Tempest*'s storm scene as a lived reflection on cinematic audience behavior. Not only does the scene relish the enchantment of technology—drawing blatant parallels between the illusions of the screen and a more familiar kind of stage magic—but it also projects the roles of both viewer and filmmaker onto Shakespeare's characters. Throughout the scene, Miranda cowers in awe of the magic, while Prospero acts as the occult technician, a stand-in for the filmmaker rather than the traditional stand-in for the bard. Torn between the horror of the spectacle and her obedient awe towards her father, Miranda becomes an iconic citizen of a society of spectacle, an urban viewer of filmic representation. In her ability to survey the scene through the proscenium of the cave, she is granted a new kind of visual power. And yet, she has a false sense of the knowledge her vision entails. Remember, Miranda believes that all on the ship are lost, when in fact the men are spared, their clothing more fresh than when they first put it on. To Miranda, the spectacle registers as a simple representation of reality, even though the vision it screens is markedly distinct from actuality. Miranda's feeling of visual knowledge obstructs the illusionism taking place. By offering its audiences an experience of Miranda's visual confusion even while aweing them with a display of cinematic prowess, *The Tempest* captures the power of adaptation to engage in sensory meta-commentary. Not only does the film convey the Shakespearean Miranda's visual enchantment, it allows its audience to experience (and question) such enchantment too.

The Tempest, however, does not see this visual enchantment as purely pleasurable. It also underscores the deceptive power of a visual media that exceeds the understanding of its audience. If Miranda looks through a proscenium and sees a sinking ship (Figure 21.2), what is it that *The Tempest*'s audiences witness on the silver screen? In his modernist aphoristic treatise, Guy Debord describes a society of spectacle in terms that remind me of Miranda here. "In societies

Figure 21.2   Miranda views Prospero's illusion through the proscenium of the cave in Percy Stow's 1908
Tempest

where modern conditions of production prevail," he writes, "all of life presents itself as an
immense accumulation of *spectacles*. Everything that was directly lived has moved away into a
representation ... The spectacle presents itself as something ... indisputable and inaccessible
... The attitude which it demands is passive acceptance" (Debord 1967: 1.1, 1.12). As a play,
Shakespeare's *Tempest* plays upon the ambivalence between appearance and reality. As a film in
1908, however, that ambivalence has taken on a modern, and perhaps unsettling register. In a
world in which visuality claims to proffer both knowledge and pleasure, and offers the *makers* of
images a ready and submissive audience, how much more is at stake in visual deception?

In a sense, this is a question that the 1908 *Tempest* asks directly by playing up the relation
between illusion, affect, and visual consumption. The response it offers, however, is ultimately
rather flippant   pointing for the rest of the film to the pleasure of illusionistic experience, rather
than to the deception it involves. Later in Stow's film, the prince Ferdinand becomes another
stand-in for the audience perspective, contrasting the princess's terror for the ship with the
prince's obvious delight. One of the films longest scenes, and the scene right at the center, follows Ariel as she comes to lead Ferdinand to Miranda. Starting with a long shot across a field, we
see Ariel coaxing Ferdinand with wild gestures. Just as Ferdinand leaps forward to grab her, she
disappears. The two play the same game, weaving forward toward the camera, until Ferdinand is
standing right before us, arms extended, caught in a moment of both shock and chagrin as Ariel
disappears for a third time (Figure 21.3). The film's bewildered audience is, I think, figured precisely in the play between Ferdinand's grasping arms and his obvious hint of pleasure at Ariel's
vanishing. As much as they want to affirm her physical presence, they are delighted to realize
that their senses have been deceived, that the thing that has been attracting them, almost even
touching them, is an illusion.

This turn away from visuality as pat (and possibly deceptive) representation and toward the
technological production of visceral enchantment fills up the last minutes of the film. Indeed,
Ferdinand is not the only character to exhibit wonder at visual surprise. Time after time, the film
pauses on the expressions of characters that stare at the camera with unutterable awe (Figure 21.4).
Their hands, inevitably, are raised in wonder, signaling the potent (even haptic) experience of
what we can only call magic. Even a small amount of imagination would allow us to picture

Figure 21.3   Ferdinand attempts to capture Ariel in Stow's *Tempest*

the cinema audience mirroring back the awed expressions of these characters, as I myself still sometimes do when viewing the film for the twentieth time or more. The visually enchanting experience that the film is, on the one hand, interrogating, becomes at the same time a point of shared experience for the audience. *The Tempest*, then, brings to the fore the experiential and epistemological tension that marks adaptation as experience expressing experience. On the one hand, it offers a lived experience to an audience, and on the other, it invites them to question that experience as a product of an emergent culture of spectacle. Will the inhabitants of modern society submit (like Ferdinand) to a milieu of sensory wonder that yet remains self-aware, or will they (as Miranda suggests) be reconstructed as obedient and believing spectators? The film's profound ambivalence between these two positions—underscored by the fact that even these two characters flip-flop between approaches—indicates that the ongoing transformation of visual experience in the early twentieth century was itself a subject of wonder. By offering that transformation to the audience, not by means of narrative construction but as a thing to be experienced, *The Tempest* demonstrates the way that adaptation can locate its signifying power on and in the body of the receiving audience.

Figure 21.4   Expressions of wonder in Stow's 1908 *Tempest*

Both *King John* and *The Tempest* locate their reflective power in the experience they offer their viewers. The films' pleasures and their critical edge meet in the moment of sensory experience. This confluence of pleasure and critique—so familiar in the history of adaptation—also gestures to the intersubjective, community-making power of adaptive behavior. Here, the experience of the recipient and the experience the adaptor conveys reach a co-terminus in the seeable, sometimes hearable, and always sensible adaptation. In the act of seeing, hearing, touching, playing, receivers and adaptors meet, connecting across time and space by means of the body. This meeting, and the power it has to consider, critique, and comprehend the mediums by which it is accomplished, points out the deeper stakes of experiencing adaptation, stakes which our scholarship has yet to significantly explore.

To return to some of the terms with which I began, it seems that this enchantment—the incomprehension mingled with wonder which Alfred Gell attributes to our experience of a new medium—might affectively point to a particular but unexplored power of adaptation. Adaptation produces experiences that layer recognition with novelty, and that allow communities to congeal around moments of perception. What would it mean for our scholarship to invest in the critical and political power of those moments? Perhaps, as we saw with Miranda and Ferdinand, we would recognize the ways that new mediums position us as subjects? Perhaps we would uncover more ways in which experience can produce critique or accrue meaning in ways that exceed those of narrative. To re-center experience—with its attendant wonder and sensory frankness—might open up new avenues for understanding the cultural weight of adaptive behavior.

Works cited

Albrecht-Crane, Christa. (2010) "Lost highway as fugue: Adaptation of musicality as film," In Dennis Cutchins and Christa Albrecht Crane (eds.) *Adaptation Studies: New Approaches*, Madison and Teaneck, NJ: Farleigh Dickinson University Press, 244–262.

Ball, Robert H. (1968) *Shakespeare on Silent Film: A Strange Eventful History*, New York: Theatre Arts Books.

Bordwell, David and Kristin Thompson. (2003) *Film History: An Introduction*. Second edition. New York: McGraw Hill.

Buchanan, Judith. (2009) *Shakespeare on Silent Film: An Excellent Dumb Discourse*, Cambridge, UK: Cambridge University Press.

Cartmell, Deborah and Imelda Whelehan. (2010) *Screen Adaptation: Impure Cinema*, New York and London: Palgrave Macmillan.

Catrysse, Patrick. (2014) *Descriptive Adaptation Studies: Epistemological and Methodological Issues*, Antwerp-Apeldorn, Belgium: Garant.

Debord, Guy. (1967) *Society of the Spectacle*, Detroit, MI: Black & Red.

Gell, Alfred. (1992) *The Anthropology of Time*, London: Bloomsbury Publishing.

Gunning, Tom. (1986) "The cinema of attraction: Early film, its spectator and the avant-garde," *Wide Angle* 3(4): 63–70.

Heller-Roazen, Daniel. (2009) *The Inner Touch: Archaeology of A Sensation*, Brooklyn, NY: Zone Books.

Hutcheon, Linda. (2013) *A Theory of Adaptation*. Second edition. London and New York: Routledge.

King John (1899), dir. William K.L. Dickson. British Mutoscope and Biograph Company.

Merleau-Ponty, Maurice. (1945) *Phenomenology of Perception*. Donald Landes (trans.) London and New York: Routledge.

Sobchack, Vivian. (1992) *The Address of the Eye: A Phenomenology of Film Experience*, Princeton, NJ: Princeton University Press.

Stam, Robert. (2000) "The dialogics of adaptation," In James Naremore (ed.) *Film Adaptation*, New Brunswick, NJ: Princeton University Press.

Stam, Robert. (2005) *Literature Through Film: Realism, Magic, and the Art of Adaptation*, Oxford: Blackwell.

The Tempest (1908), dir. Percy Stow. Clarendon Film Company.

22

GREAT VOICES SPEAK ALIKE

Orson Welles's radio adaptation of Victor Hugo's *Les Misérables*

Bradley Stephens

While the field of adaptation studies has rightly challenged the primacy of fidelity as an analytical instrument, recognising an adaptation's relationship to its source remains a necessary step if we hope to understand a work's reception. This is true largely because adaptation is always an interpretive act that conveys at least one of the ways in which a source text can be read, and such acts of 'reading' are of course the focus of reception studies. By considering the relationship between adaptation and source in hermeneutic rather than simply communicative terms, a better understanding of a work's appeal within the dynamics of appropriation becomes possible. Orson Welles's adaptation of Victor Hugo's nineteenth-century novel *Les Misérables* for American radio in the summer of 1937 is an illustrative case in point. This radio drama performs what Lawrence Venuti might call an "interpretive operation" (2007: 33) upon *Les Misérables* that draws attention to the formal, thematic, and biographical contexts not only of Welles's own practice but also of Hugo's source novel. In its creative relationship with that colossal book, the 1937 miniseries stresses both Welles's imagination as what Paul Heyer has called an "auteur by means of adaptation" (2005: 213) and the pliancy of Hugo's prose as a work with universal reach.

The cultural history of a socially minded epic like *Les Misérables*, which has become embedded in a global popular consciousness (see Stephens and Grossman 2015), would be conspicuously narrow in focus if it were not able to explore exactly how such a story has travelled across different countries, forms, and periods. Just as it would be short-sighted not to recognise the critical benefits of broadening the perspectives of adaptation studies beyond the ubiquitous source-to-adaptation coupling, so would it be similarly rash not to consider how the comparison of an adaptation or a selection of adaptations with the relevant sources continues to be of use within this broader and more ambitious array of methodologies. Furthermore, and as numerous commentators have noted (Andrew 2011; Hermansson 2015), the need to engage with the likenesses between an adaptation and its source is enhanced by the extent to which fidelity still matters to audiences who have expectations of how adaptations reproduce their sources.

The much-maligned spectre of fidelity criticism that such case studies threaten to summon is kept at a distance by denying fidelity any privilege and instead thinking about the source-adaptation relationship in more objective fashion, as I will demonstrate in this essay. By acknowledging the diverse means of production and consumption through which adaptations are created, the various ways in which "adaptations can be understood without the crucial emphasis on literary origin" (Geraghty 2008: 194) have come into clearer focus in recent discussion.

Researchers and students alike are increasingly encouraged to look across the field of adaptation studies rather than just at fixed points within it, for example, using the kind of quantitative corpus-based research (Cattrysse 2014) that should become yet more prevalent in the digital age of macroanalysis or 'distant' reading. In turn, the intertextual spaces through which adaptation occurs have helped to unsettle the unhelpful hierarchies of aesthetic value onto which adaptations were once pinned. Such shifts in the critical orthodoxy allow for a return to fidelity as less a matter of aesthetic determination than as an issue of cultural function (Dicecco 2015: 173), a tool of analysis in hand as opposed to a definitive criterion, through which one may designate the faithfulness an adapter like Welles shows to a source such as Hugo's.

But it is not only consumers and critics who display an interest in fidelity, as Welles's adaptation of *Les Misérables* confirms. Consciousness of the source material can be equally important to the creative communities responsible for producing the adaptation in the first place. Their methods and motivations are as integral to a work's evolving reception as the audience's own practices. Welles's personal interest in Hugo's story as a work of both moral conscience and social justice produced an adaptation that was at once attentive to the source novel's composition and inventive in its representation of that material on the airwaves. The cultural capital of *Les Misérables* had proven itself to be a shrewd investment for adapters worldwide since the novel's extraordinary publication and global triumph in 1862, as demonstrated just two years before Welles's adaptation aired by the success of Richard Boleslawski's Academy Award-nominated Hollywood version. Welles himself unarguably bought into the novel's renown as an admired allegory for human rights, hence the aesthetic and thematic correlations between his radio adaptation and Hugo's source novel enrich rather than devalue an analysis of this work.

Welles was just 22 when the opportunity to develop this adaptation came about in 1937. He had already made his mark on Broadway as a rare creative talent, and at the same time had become a highly paid radio actor, raising his stock in the eyes of possible collaborators. In a concerted attempt to challenge rival radio stations with a quality dramatic production, the Mutual Network would offer him complete creative control over a mini-series of his choosing. The proposed summer scheduling gave Mutual some insurance against both potentially low audience figures and Welles's characteristic ability to court controversy, since the consequences of such outcomes would have been much more problematic during the prime-time season of September. Welles subsequently wrote, directed, and starred in a serialisation of *Les Misérables* that closely followed the fortunes of the novel's hero Jean Valjean, from his prison release to his death. Exercising the versatility of his baritone voice, Welles played the parts of both this central character and the narrator. The miniseries was broadcast in seven episodes, each around thirty minutes in length, beginning on 23 July 1937 at 10 p.m. ET.

Welles's adaptation of *Les Misérables* has easily been lost in the mix of a highly active and audacious period of his career, and even substantial studies of American radio drama during this period (Hilmes 1997; Deforest 2008) make only passing or anecdotal reference to it. The series is nestled between two particularly muscular moments in Welles's body of work, both of which were produced for the stage that year: *The Cradle Will Rock* in June, whose depiction of union activism forced a temporary shut-down of the production; and later, in November, his modern-dress adaptation of Shakespeare's *Julius Caesar*. These high-profile productions displayed an acute consciousness of working class exploitation on the one hand and the rise of fascism in Western Europe on the other. In contrast, Welles's radio dramatisation of *Les Misérables* maintained the novel's nineteenth-century setting, thereby appearing to be less contemporary and more formulaic. Welles's co-founding of the Mercury Theatre with John Houseman that same summer, named in homage to the iconoclastic *American Mercury* magazine, casts this adaptation into yet deeper shadow. The broadcast of *Les Misérables* predated the official formation of this independent repertory and so it risks being footnoted in its history, especially given the notoriety of Mercury's *The War of the Worlds* the following year – the famous broadcast in which simulated news bulletins about a Martian invasion caused public panic on America's East Coast.

The relative obscurity of Welles's *Les Miserables* is deepened by a methodological inequity within research on Welles, in which his radio productions are noticeably overshadowed by his work on screen. In 2015, for example, over 60 events worldwide were advertised on wellesnet.com, the leading web resource for Welles's life and career, to celebrate his centenary. The vast majority of these events concentrated wholly on his film and television work. Such a focal point reflects a wider trend in the arts and humanities, whereby radio is more often discussed in sociological and technological terms rather than artistic ones (Griffiths 2013: 17). Within adaptation studies – and somewhat ironically – radio remains rather invisible, in part due to this field's growth out of film and literature departments.

Yet these are precisely the reasons why Welles's radio version of *Les Misérables* warrants more extensive attention than it has received. Indeed, the question of how and why the future creator of *Citizen Kane* adapted one of the previous century's most globally recognised bestsellers yields insight into not only Welles's creativity at a formative time in his storied career but also the reception of Hugo's novel as one of Western literature's most adapted and most enduringly popular works. Certain motivations are immediately apparent. The trend of adapting successful literary works to other media in order to broaden audience figures was well established by the 1930s, and Welles's attraction to Hugo's social conscience as nineteenth-century France's most iconic writer is unsurprising, in light of his own belief that art had a civic role to play in questioning the social status quo. As Heyer (2005: xiii–xvi) notes: "Although never a socialist per se, [Welles] was pro-labor, a strong supporter of Roosevelt's New Deal, and a critic of monopoly capitalism" who received life-threatening hate mail and whose political sympathies prompted the FBI to open a file on him in 1942. Such a liberal spirit found obvious solace in *Les Misérables*, whose narrator insists that "Society must look these issues in the face since it is society that produces them" (Hugo 2008: 75). The biographical parallels between Welles and Hugo alone make for intriguing reflection: both men showed precocious talent at a young age that was cultivated by dutiful mothers and eye-opening travels, and both worked across different media in a desire to marry artistic innovation with widespread popularity and social reform (see: Bazin 1998; Callow 1995; and Robb 1997).

In keeping with Hugo's notoriously untranslatable French title, the concept of 'misery' consequently highlights a telling common ground between the novel and Welles's serialisation. In a literal sense, abjection and filth abound in this radio drama thanks to Hugo's imagination, from the downtrodden factory worker Fantine's plight in Montreuil-sur-Mer to the ex-convict Valjean's flight through the Parisian sewers. For Hugo and Welles alike, these instances serve an essential narrative purpose that relies on more figurative definitions of 'misery,' playing on the ambiguity of *misère* in French as a signifier of both material and immaterial wretchedness. In the socio-economic sense of degradation, and in the moral sense of corruption, *Les Misérables* offered Welles an allegory for the charged climate of 1930s America. He shared Hugo's drive as an artist to expose and expunge social ills through the edifying power of art, and to do so using a similarly popular sensitivity that would reach out to mass audiences rather than defer to elitist traditions.

Given that Hugo's story of humanity's resilience in the face of poverty's dehumanising effects had captivated audiences since the second half of the previous century, its significance to Depression-era America was certainly not lost on Welles. Echoing Hugo's own humanitarian causes as a writer, Welles would tell the Theatre Education League in New York in 1939 that "my right to having more than enough is cancelled if I don't use that more to help those who have less" (cited in Schwenn 2015: 63). As part of the Works Progress Administration that employed millions of Americans, he gladly accepted the invitation to join Hallie Flanagan's Federal Theatre Project in 1935, because he supported the FTP's ethos of "make work, make culture" as a means of rebuilding the country after the Wall Street Crash. As Simon Callow (1995: 218) explains, Welles understood that the WPA was a vehicle of widespread social change rather than just a machine to create jobs: "Flannigan saw herself and the FTP as being one with the workers building roads and the artists creating Post Office murals. America […] was being rebuilt by ordinary Americans" (218). Pumping his sizeable earnings as a radio actor into his ventures with the FTP, his support for the *misérables* of Manhattan was particularly evident in his stewardship of the FTP's Negro Theatre Unit with Houseman and in their popular *Voodoo Macbeth* (1936), which re-imagined Shakespeare's story during the Haitian Revolution. Dismayed by the squalor of the Great Depression and by the sordid self-interest of reactionary forces in America and beyond during the 1930s, Welles could strongly identify with Hugo's compassion for society's stigmatised souls and his impatience with the baseness of bourgeois materialism in nineteenth-century France.

This moral common ground between Welles's adaptation and Hugo's epic can, however, reveal yet greater depth when Welles's artistic affinity with the narrative he had chosen to retell on radio is recognised. Heyer's informative reading of Welles's *Les Misérables* (2005: 34–37) touches upon the use of 'all-enveloping' sound effects and emotive voices to draw the audience in, for example, but his otherwise sharp observations risk becoming descriptive, if not sugges- tive, in the absence of a critical sensitivity towards how the source novel itself thrives on such immersive and stirring techniques for its dramatic effect. Welles, like Hugo, had an interest in the power of oral storytelling and in particular in the medium of voice, highlighting a significant continuity between the two that helps to scrutinise how the former developed his adaptation. As a poet and a dramatist, Hugo created a discernible narrator for his fiction who directly addresses the reader throughout with invitations, imperatives, and most famously his lengthy digressions, which occupy well over a quarter of the entire text and reflect on subjects as varied as the Battle of Waterloo and Parisian slang. Hugo believed that writers should treat their readers not as passive consumers but as thoughtful individuals who needed to be emotionally and morally engaged by the drama. The structural similarities between his view of writing and how Welles aimed to use radio point towards a shared conception of narrative as being able to permeate into an audience's private space. This understanding probes Welles's interest in *Les Misérables* by highlighting the capacity of radio to function as what his long-time assistant Richard Wilson described as his friend's "theatre of the imagination" (cited in Heyer, 214). Upholding the idea of radio drama as an intimate medium in which consciousness is paramount, Welles's adaptation of Hugo's story on this inner stage infiltrates the listener's mind, stirring their thoughts and emo- tions in ways that tellingly resonate with Hugo's appeal to a thoughtful reader or *lecteur pensif* (see Roche 2007: 33–51).

Various theorisations of radio's ability to insinuate itself into its audience's inner space help- fully describe this effect, in no small part thanks to Martin Esslin's seminal 1971 essay, which conceives of radio drama unfolding in an internal world. In terms not dissimilar to Wilson's, Neil Verma describes radio as a "theatre of the mind," "as a voice echoed in an imperfect acoustic mirror, an utterance that begins inside and becomes a living voice" (2012: 11–12). Labelling this theatre a "blind" medium, as many do (Crisell 1994: 3–5), risks underestimating what Tim Crook calls its "visual force in the psychological dimension," but it does imply radio's unique status as an auditory mode of storytelling (1999: 9). With no visual stimuli for the eyes to focus upon – no images on a screen, no words on a page, no figure on a stage or at a lectern – "the sights and sounds of radio [dramas] are created within us, and can have greater impact and involvement" (McLeish 2005: 3). Our attention is directed inward as the mind becomes stimulated by the voices and sounds that reach into and occupy our space.

It is this incursive character of radio that speaks most audibly to Hugo's methods as a writer and, in turn, to the ways in which the similarity of an adaptation like Welles's to its source need not be automatically categorised as an expression of that origin's supposedly commanding influence. In essays composed at the time that *Les Misérables* was published, Hugo described the hold that a book exerts, working into and upon its readers like a conscience:

> You feel yourself seized, your thoughts seem no longer your own, and you become entirely focused, even consumed and then subjected. You are no longer able to get up and walk away, but who, you ask, is holding you back? That person is the book. […] It will only let you go once it has changed something inside you.
>
> *2002: 560*[1]

Hugo's conceptualisation of an artwork as arresting its audience and triggering their self-reflection in part anticipates the discourse around radio's incursion into the listener's mind. Radio possessed an undeniable cultural and commercial appeal alongside this invasive potential, of course, reaching directly into over 27 million American homes by the dawn of its golden age in the late 1930s and costing much less to produce than theatre or film. But Welles's version of *Les Misérables* confirms that radio's specificity as an aural medium offered compelling formal possibilities that could exert a potent psychological effect on its audiences, and that in turn set a standard for his future radio productions. Strongly believing in radio's capacity for a dramatic narrative experience, for this first broadcast Welles wanted to avoid the genres of comedy and melodrama that had dominated radio programming in favour of more poignant material that could insinuate itself into his audience's minds. That desire for a personal, intimate connection with the audience found ample gratification in both the source novel's humanist tones and the adaptation's aural medium. Similarly to Hugo, Welles aims for wretchedness on air to become a wretchedness inside us, an unease with human misery, in an attempt to dispel its contamination of the world and to inspire a better future that reflected radio's desire for "uplifting" material during the gloom of this period (Hilmes 1997: 33).

Whereas *Les Misérables* had already been adapted numerous times for stage and screen, no adaptation had gone out on American radio before. Yet Welles's originality lay not just in his choice of source. His approach to how radio adapted material from literature and other media was itself relatively fresh. He disliked the way popular radio dramas tended to reproduce the atmosphere of watching a spectacle in person, as if recorded in a theatre or cinema hall, such as NBC's anthology programme *The First Nighter* or CBS's *Lux Radio Theatre*. He made no secret of how he likened radio drama to prose fiction rather than a theatrical play, in that it embraces the act of telling rather than that of showing: physical presence and visual performance give way to aural effect and verbal communication. He thus chose not to have a studio audience in order to emphasise this distinction, both to the listeners at home and to the actors themselves, who could perform at ease with no spectators. At the same time, he was aware that radio drama "is never simply a literary work orally told", to borrow Elke Huwiler's astute identification of radio's intrinsic features as an auditive medium (2010: 139). As Huwiler outlines: "an adapted radio piece […] is an artistic work in its own right, working with much more varied medial features than only language and creating a story world with the intrinsic features of the auditive medium." Radio's key aural features of voice, music, noise, and silence – in conjunction with the editing, mixing, and stereophonic positioning of these features – could be used specifically to suit the medium's storytelling potential rather than simply record a reading or performance

As a result, Welles did not adapt a stage version of *Les Misérables*, as had been the case with a number of local British radio stations in the late 1920s that had broadcast a version of Norman McKinnel's 1901 play *The Bishop's Candlesticks*. Instead, Welles returned to Hugo's novel itself, more than likely referring to either the first English translation by Charles Edwin Wilbour (1862) or Isabel Hapgood's more recent version (1887).[2] He was clear in his artistic and moral purpose from the opening seconds. Before any formal introduction by the show's presenter, Welles's voice paraphrases Hugo's preface. Where Hugo's ten-line declaration culminates in his assertion that "as long as ignorance and misery exist in this world, books like the one you are about to read are, perhaps, not entirely useless" (2008: xlv), Welles succinctly states that "So long as these problems are not solved, so long as ignorance and poverty remain on earth, these words cannot be useless." A Master of Ceremonies then gives the following overview:

> These words set forth the soul and spirit of one of the world's great literary master-pieces, *Les Misérables*. Out of the depths of his pity for suffering mankind, Victor Hugo drew a compelling story, one that will live for so long as bewildered humanity shall continue to grope toward the light.

By ventriloquising Hugo even before the presenter invokes the author's spirit, Welles simultaneously aligns his artistry with that of his celebrated source and signs this radio play as his own. This double posture of veneration and invention is reinforced by the presenter's summary that the opening words channel Hugo's 'great' novel before noting that Welles is at once director, writer, and actor for a series he has made specifically for the radio. This voice then explains that Welles has necessarily condensed the sprawling source text in order to retell Hugo's story: "Each episode will depict some vital development in the epic of Jean Valjean … and those sections of the book itself, which in running narrative bind together the dramatic episodes, will also be read by [Orson Welles]." In choosing this structure, in which a narrator remained present and Valjean's fate as the symbolic heart of *Les Misérables* became magnified, Welles signalled his intention to mimic Hugo's approach, but less out of unquestioning reverence to his source than in order to respect the logic of his own creation.

Both the formal demands of radio as an 'invisible' stage and Welles's preoccupations as an artist stress the aesthetic and historical contexts in which adaptations as enacted receptions necessarily operate. Welles knew that his broadcast's action needed to remain tightly focused within each half-hour episode and across the seven-week schedule, so a narrative voice provided an effective means of directing listeners through the non-visual dramatisation. The novel's sizeable digressions had to be removed, however, due to these constraints. Too many character voices risked confusing the audience and losing their attention, so further edits were needed as well. Numerous characters and plot lines were either cut entirely, like the narrative arc of the forlorn Éponine, or considerably scaled back, such as the courtship between the young lovers Marius and Cosette. Especially noticeable in their relative absence were the Friends of the ABC, the group of revolutionary Parisian students who help to organise the June Rebellion of 1832. This alteration forces the insurrection to fall into the background of the drama's sixth episode as a seemingly spontaneous outburst of public discontent rather than as a more ideologically resolute uprising. The salutation of 'comrade' that is heard twice on the barricade as Valjean arrives to save Marius from the National Guard implies that Welles was hesitant to make more of that group at a time when revolution was associated in the American popular consciousness with Soviet Communism. This hesitancy is not the only biographical factor to come to light. If, as Callow believes, Welles's production of *Caesar* later that year revealed his concern with the "anguish of the liberal" in an unforgiving age of authoritarianism (1995: 322–323), then it follows that Valjean's quest for redemption against the backdrop of an uncaring monarchical society would have been of special interest to him. Such attention to one man's quest to establish his individual value can be stressed further still by citing Welles's reputation (which so fascinated the influential film critic André Bazin) as an auteur, whose ambition to convey a personal vision increasingly brought him into conflict with co-workers and producers.

This analysis of a 'blind' adaptation in 1930s America by a rising star would nonetheless be incomplete without illustrating how the narrator's role and Valjean's centrality aptly fulfilled Welles's conviction that radio could get into his audience's minds – 'inside the skull' – to play on Hugo's title of the third chapter in the seventh book of Part I, where Valjean frets over whether to confess his true identity to save the innocent man Champmathieu who has been mistaken for "the convict Jean Valjean." The character of Hugo's narrative voice – conversant and emphatic, compassionate and poetic, digressive and grandiose – looms large in Welles's adaptation. Hugo, after all, opened his novel by affirming its socio-political purpose, and his narrator is forever talking to the reader, as seen on the opening page when he addresses the rumours surrounding the Bishop Myriel's past but insists that these have no bearing whatsoever on the story he will be telling. But where Hugo carefully paints his scenes, Welles must minimise visual description and maximise the verbal interactions between narrator and reader, and between the characters themselves. He eschews the numerous particularities that Hugo's narrator picks out from Valjean's first appearance, such as his coarse, yellow, twill shirt, the game cooking away on a long spit in front of the kitchen fireplace in Jacquin Labarre's *La Croix de Colbas* inn at Digne, and the enclosing effect of the twilight clouds that concentrate the moon's light onto the earth. Instead, Welles provides briefer, more generalised descriptions. All the audience is told is that "He looked ragged and mean. He must have come far that day, for he looked weary." The trembling gruffness of Valjean's voice completes the characterisation of a pitiful soul who is coldly cast out by a nameless innkeeper into a thundery night.

This narrowing of the visual to its most basic components focuses the listener's mind increasingly on what is happening, rather than what it all looks like. The effectiveness for Welles's 'theatre of the imagination' is undeniable when the narrator recounts Valjean's backstory in the first episode, after the newly-released convict arrives in Digne in 1815. Like a memory that surges to the forefront of the mind, Valjean's court sentencing nineteen years earlier for stealing a loaf of bread to feed his sister's starving children is suddenly acted out. The magistrates' patronising laughter and haughty tone contrast with Valjean's bewilderment, followed by an eerie banging that at once evokes the falling of a judge's wooden gavel and the hard labour of the prison *bagne*. The immediacy of the expressive voices and the pitiless regularity of the bracing pounding oblige the audience's consciousness to undergo this recall of the past with a sense of both proximity and anxiety. The audience are immersed in Valjean's plight, and this immersion is stressed by the interlacing of the narrator's voice with those sounds, ensuring that he feels like a companion to Valjean's woes rather than a distant observer (Welles would even record the later sewer scenes in the men's restrooms so that his voice would echo as if the narrator were alongside Valjean). A lingering pause after Welles's narrator asks whether anyone had cared to ask about the fate of Valjean's family, using prose that is identical to the English translations, momentarily leaves the audience alone with this awkward question.

Welles's style is especially moving when he dramatises the novel's two most memorable forays into the tormented minds of its characters, namely Valjean's moral crisis during the Champmathieu affair in the third episode, and the Police Inspector Javert's mental derailment before his suicide in the sixth. In the first example and as in the novel, Valjean finds himself wracked with uncertainty as he contemplates the arrest of an innocent man and the fate of the town he is now responsible for as mayor, with the voices of the kindly Bishop Myriel and the fierce Javert replaying on the soundtrack. Myriel's vow that he has purchased Valjean's soul for moral virtue repeats with no warning, followed immediately by a replay of Javert's report that Jean Valjean has finally been detained. These voices echo in our minds as they do in Valjean's before his own monologue begins: "Why interfere? It doesn't concern me. I've brought happiness to this place. I've done some good. It is God's will that I do more." The relentless ticking of a clock in his chambers evokes both the inevitability of guilt "returning like a tide to the shore," to use the narrator's words, and Valjean's sense of time running out. Noble-sounding string instruments then herald the spectral presence of Myriel's guiding moral light towards Valjean's impossible decision, having earlier become a signature or motif for the Bishop's compassion. For Javert's breakdown Welles chose a different approach. Hugo focused on the scene's darkness, in which sound, not sight, is prominent through the sounds of the whirlpool in the River Seine beneath the Inspector and "the tragic whispering of waves" (1,088). Aware of how Hugo was using descriptions of sound to evoke a feeling of despair (and equally conscious that noise such as the "sound of foam" may have been technically challenging to reproduce in a studio), as Javert contemplates suicide Welles relies instead on a score of melancholy violins, foreboding cellos, and whistling flutes that wind and circle like the swirling water "which knots and unknots itself like an endless screw." Javert leaves the river temporarily for the local police station, where he is heard reading out his recommendations for police and prison reform, after which his voice is silenced as the narrator recounts the Inspector's return to the river and final moments atop the bridge: "Javert was standing exactly over the rapids of the Seine, perpendicularly, over that formidable whirlpool [...]. What was beneath was not water: it was chasm." The reference to Javert's rigid posture recalls the novel's likening of this crisis as the moment a locomotive is derailed from its fixed route, with Welles's voice itself rehearsing the twisting monologue of Javert's conscience. The lack of sound effects and the absence of Javert's own voice at the very end chillingly convey the character's sense of a world that has been emptied of meaning and agency.

In conclusion, Welles's radio version of *Les Misérables* enables a number of key methodological observations to be emphasised. Crucially, the notion of adaptation as an act of reception encourages the rehabilitation of fidelity as a key analytical tool, and such analysis implies two particular lines of enquiry that underscore the benefits of this approach. First, the frequency with which nineteenth-century European fiction, in its cultivation of the visual, is seen to be predominantly pre-cinematographic may be questioned. Such an association of page and screen implies a natural, predetermined affinity between the two that risks keeping radio in the dark, as it were, in spite of its status as a widely-used medium in literary adaptation, and that may also overlook how writers like Hugo can complicate narrative sight while appealing to the descriptive power of the other senses. Lest we forget that his narrator often stresses the shortcomings of his vantage points, such as the inability of any eyes to fix the exact shape of a battle like Waterloo (288–289), or that Valjean is rendered at once blind and deaf as he enters the sewers, where at first only his senses of touch and smell are able to engage (1,045). By extension, radio as an agent of trans-media adaptation warrants greater attention than has been the case, not least in a digital age where the medium is ubiquitous and where audiences have ever-growing access to past recordings (Mollgaard 2012: viii). Consequently, radio's intimacy as a 'theatre of the mind' requires theoretical strategies that respect its specificity as a narrative mode and therefore do not unquestioningly borrow the grammar of literary or screen analysis. Such methods facilitate a more dutiful reading of how radio has contributed to the reception of literary works by teasing out the power of fiction to infiltrate our minds.

Notes

1 The translation from the original French is my own.
2 Welles reproduces a noticeable mistranslation from both editions when presenting Valjean as a "galley slave … at the oar," confirming that he did not use the original French text. This image pervades many Anglophone adaptations, including many early productions of the beloved Boublil and Schönberg stage musical, but it is historically inaccurate: galley duty had been abolished in France in the mid-eighteenth century. The noun *galérien* remained in French usage, with the convicts at Toulon only sleeping on anchored decommissioned warships in between their days of hard labour ashore.

Works cited

Andrew, Dudley (2011) "The economies of adaptation," In Colin McCabe, Kathleen Murray, and Rick Warner (eds.), *True to the Spirit: Film Adaptation and the Question of Fidelity*, Oxford: Oxford University Press, 27–39.

Bazin, André (1998) *Orson Welles*, preceded by *Welles et Bazin par François Truffaut*, Paris: Cahiers du cinéma.

Callow, Simon (1995) *Orson Welles: The Road to Xanadu*, London: Penguin.

Cattrysse, Patrick (2014) *Descriptive Adaptation Studies: Epistemological and Methodological Issues*, Antwerp, Belgium: Garant.

Crisell, Andrew (1994), *Understanding Radio* (2nd ed), London and New York: Routledge.

Crook, Tim (1999) *Radio Drama: Theory and Practice*, London and New York: Routledge.

DeCecco, Nico (2015) 'State of the conversation: The obscene underside of fidelity.' *Adaptation* 8 (2): 161–75.

Deforest, Tim (2008) *Radio by the Book: Adaptations of Literature and Fiction on the Airwaves*, Jefferson, NC: McFarland.

Esslin, Martin (1971) "The mind as stage," *Theatre Quarterly* 1 (3): 5–11.

Geraghty, Christine (2008) *Now a Major Motion Picture: Film Adaptations of Literature and Drama*, New York and Plymouth: Rowman and Littlefield.

Griffiths, Kate (2013) "Labyrinths of voices: Émile Zola, *Germinal*, and radio," In Kate Griffiths and Andrew Watts, *Adapting Nineteenth-Century France: Literature in Film, Theatre, Television, Radio and Print*, Cardiff: University of Wales Press, 17–46.

Hermansson, Cassie (2015) 'Flogging fidelity: In defense of the (un)dead horse.' *Adaptation* 8 (2): 147–60.

Heyer, Paul (2005) *The Medium and the Magician: Orson Welles, the Radio Years 1934–52*, Oxford: Rowman and Littlefield.

Hilmes, Michelle (1997) *Radio Voices: American Broadcasting 1922–1952*, Minneapolis, MN and London: University of Minnesota Press.

Hugo, Victor (2002) *Œuvres complètes: Critique*, Paris: Laffont.

Hugo, Victor (2008) *Les Misérables*, trans. Julie Rose, London: Vintage Classics.

Huwiler, Elke (2010) "Radio drama adaptations: An approach towards an analytical methodology," *Journal of Adaptation in Film & Performance* 3 (2): 129–40.

McLeish, Robert (2005) *Radio Production* (5th ed), Oxford and Burlington: Elsevier.

Mollgaard, Matt (2012) *Radio and Society: New Thinking for an Old Medium*, Newcastle, UK: Cambridge Scholars Press.

Robb, Graham (1997) *Victor Hugo*, London: Picador.

Roche, Isabel (2007) *Character and Meaning in the Novels of Victor Hugo*, West Lafayette, IN: Purdue University Press.

Schwenn, Thomas A. (2015) "Prodigal, wayward: Orson Welles and adapting Richard Wright's *Native Son*," *Journal of Adaptation in Film and Performance* 8 (1): 61–69.

Stephens, Bradley and Kathryn M. Grossman (2015) "*Les Misérables*: A prodigious legacy," In Grossman and Stephens (eds), *Les Misérables and its Afterlives: Between Page, Stage and Screen*, London: Routledge, 1–16.

Venuti, Lawrence (2007) "Adaptation, translation, critique," *Journal of Visual Culture* 6 25–43.

Verma, Neil (2012) *Theatre of the Mind: Imagination, Aesthetics, and American Radio Drama*, Chicago, IL and London: University of Chicago Press.

23

LUX PRESENTS HOLLYWOOD

Films on the radio during the 'golden age' of broadcasting

Suzanne Speidel

While adaptation critics, such as Deborah Cartmell, Linda Hutcheon and Brian McFarlane, are accustomed to citing commercial interests as a key motivation for the act of adapting (Cartmell 2014: 160, Hutcheon 2006: 85, McFarlane 1996: 7), this has rarely been so conspicuously declared than in the adaptations of Hollywood movies, which were a staple element of broadcasting schedules during US radio's 'golden age.' From the mid-1930s to the early 1950s, networks such as CBS andd NBC Blue Network aired a variety of anthology series featuring hour-long and half-hour-long adaptations of recent Hollywood films. These productions were sponsored by commercial advertisers, whose products featured prominently in the series' titles and in weekly broadcast intermissions.

This chapter will consider two such series, *Lux Radio Theater* (1934–1952) sponsored by Lever Brothers, and the *Screen Guild Theater* (1939–1952), a series conceived by the Screen Actors' Guild as a means of raising money for the charity the Motion Picture Relief Fund, and sponsored variously by Gulf Oil, Lady Esther cosmetics and Camel Cigarettes. In common with many radio programmes of the era, these series switched networks several times during their existence, their production and creative personnel being overseen not by the networks themselves, but by advertising agencies employed by the shows' sponsors. The resulting programmes blended adaptations and commercials into single-entertainment packages for listeners and buyers, with Hollywood narratives segueing into advertisements for washing detergents, make-up, cigarettes and soap.

Within adaptation scholarship, the fact that adaptation offers what McFarlane terms "the lure of the pre-sold title" (1996: 7) has led critics such as Simone Murray, Christine Geraghty and Deborah Cartmell to identify the advertising of adaptations themselves as important adaptation paratexts. This is because adaptations are, as Cartmell puts it, "entertainment that strategically positions the consumer as the primary target" (2014: 163), and the marketing of adaptations sheds light on key creative decisions that have been made with consumers in mind. These include the degree to which films "exploit or undermine literary pedigree, how they translate characters into stars, how they tease us with the promise of our favourite film genres ... how they speak to the tastes of a contemporary audience, and how they locate themselves within a particular consumer group" (2014: 165).

For the makers of *Lux Radio Theater* (hereafter referred to as *LRT*) and *Screen Guild Theater* (hereafter referred to as *Screen Guild*), the importance of the consumer is clear, and the choice of adapting Hollywood films to radio pairs up the audience targeting embedded in adapting with the targeting of buyers of sponsors' products. The advertising within these shows is a key paratext of the adaptations, since it not only signals with unusual clarity the commercial aims of adaptation but also tells us much about the audiences addressed. As I shall demonstrate, the adaptations highlight the commercial contexts of the golden age of radio, and the radio industry's commercial contexts also elucidate how creative decisions in adapting are matched to consumer–listeners.

In this chapter I will consider how advertising paratexts illuminate their adaptations, through an examination of two anthology series, and through close analysis of two different radio versions of Disney's *Snow White and the Seven Dwarfs* (David Hand et al., 1937). The requirements of radio sponsors during this era have resonance for the study of adaptation, since, as Cynthia B. Meyers makes clear, advertisers sought to market products by employing "standard, recognizable appeals," while also needing to "innovate to build new markets" (2014: 150). Within adaptation scholarship, the pleasure of adaptation for audiences has been theorized by Linda Hutcheon who, taking her lead from genre studies, defines it as "repetition with variation," or "the comfort of ritual with the piquancy of surprise" (2006: 4). Thus, the act of adaptation itself meets the criteria of advertising agencies, since it marries and imbues commercial breaks and product endorsements with the familiarity and the novelty embedded in adaptation. In considering how patterns of repetition and variation play out in two versions of *Snow White*, I will show how audiences are simultaneously targeted as consumers of both products and stories. In examining the creative decisions underpinning these adaptations, I will therefore trace adaptation approaches which resemble the marketing strategies of the specific advertising agencies that produced them.

It is worth noting that such parallels at once mirror and also remodel the arguments put forward by Cartmell on the marketing techniques employed for classical Hollywood film adaptations. Cartmell proposes that the posters and press-packs of Hollywood adaptations traditionally drew on literary pedigree as a selling point, yet at the same time offered popular, commercial entertainment as a means of 'sweetening the pill' of canonical, literary authors such as Shakespeare and Dickens. Classic radio adaptations combine the cultural associations of their respective media in similar ways to those outlined by Cartell, yet in this instance film entertainment is offered as a counterbalance to the newer medium of radio's blatant commercialism. In other words, rather than sweetening the pill of high culture, film narratives make palatable the strong flavour of consumerism brought about by sponsorship in broadcast radio. Critics such as Robert Stam have previously noted the "class-based dichotomy" embedded in literature-to-film adaptation, in which "literature pays indirect, and begrudging homage to film's popularity, while film pays homage to literature's prestige" (Stam and Raengo 2005: 7). Golden-age radio adaptations of films make clear that such symbiotic plays on cultural capital are at once a staple of cross-media adaptation, and are also enacted in many more combinations and permutations of high-brow, low-brow, entertainment, consumerism, prestige et cetera than have been appreciated through adaptation studies' predominant focus on literature-to-film adaptation.

In her groundbreaking study of the history of commercial sponsorship in US radio, Cynthia B. Meyers dates the involvement of advertising agencies in radio production back to the early 1920s. Meyers records the rapid ascendancy of advertising in radio, with advertising revenues rising from $3.9 million in 1927 to $19.2 million in 1929. The 'golden age' of radio, from the 1930s to the early 1950s, saw radio increase its audience at a time when cinema was experiencing narrowing profit margins due to the Depression and the rising production costs which followed the introduction of sound. Whilst cinema's overall revenues remained greater than those of radio, the newer medium was able to attract a larger audience with considerably less expense (Meyers, 2014: 203), a situation that led to both competition and symbiosis between the two industries.

The radio industry became reliant on the film industry for its ready supply of performance talent, a commodity prized by its commercial sponsors, who exerted pressure on advertising agencies to increase its reliance on audience-generating film stars. Until 1935 there was little incentive for radio networks themselves – whose geographical hub was, like the advertising industry, New York – to broadcast from Hollywood, since the American Telephone and Telegraph Company insisted that programmes be transmitted along its telephone lines to New York before it would transmit them across the country to local stations. However, in 1935 a federal regulatory investigation forced a change in policy, and this meant that broadcast costs from outside New York were reduced (Meyers 2014: 202). Responding to the demands of advertisers for stars, and making use of newly established radio facilities, leading advertising agencies began in the late 1930s to set up radio-production headquarters in Hollywood.

The agency that produced *LRT* was J. Walter Thompson (JWT), which had a previously established history of using film stars in advertising, through its print-media campaign for Lux Toilet Soap, in which stars' pictures appeared alongside their endorsement of the product. Thus *LRT*, in which film stars performed and also praised Lux products during intermissions, grew out of JWT's pre-existing print campaign for Lever Brothers. *Screen Guild* was produced by Rubicam and Young (Y&R; Christman,18), an agency that was generating 30% of all its revenues from its Hollywood-based programming by 1938 (Meyers 2014: 205). While radio copywriters, who took the responsibility for the creative content of programmes as well as for advertisements, remained based in New York until the second half of the 1940s, directors and other technical staff were employed in Hollywood by the advertising agencies, which leased studio space and airtime from radio networks.

The film industry's collaboration with this already complex symbiosis was motivated by the fact that Hollywood film studios could see the marketing potential of radio, and its commercial breaks, just as readily as could the manufacturers of cigarettes and face powders. Initially there was reluctance on the part of studios, both to antagonize film exhibitors (for whom radio represented direct competition) and to undermine their control over stars, and this led to, from 1932 to 1933, a ban by studios on contracted performers participating in radio (Meyers 2014: 204). However, this proved difficult to enforce, and cooperation between film and radio was encouraged by instances of integration between the two industries. Notably, in 1932 the Radio Corporation of America, which had established RKO Pictures in 1928, appointed Merlin Aylesworth, the president of NBC, to be president of RKO as well, and his six-year tenure in both roles set a precedent for cross-promotion between film and radio (*Advertising Hall of Fame*). Thus, the number of star-led, Hollywood-inspired programmes grew in the late 1930s, with examples including *Kraft Music Hall*, *Hollywood Hotel* and *Old Gold Hollywood Screen Scoops*.[1]

'Package' agreements between film studios and radio sponsors were rare (although in 1937 MGM signed a deal giving coffee manufacturer Maxwell House access to all its stars, writers and producers), because they tended to result in conflicts of interest (for example, MGM star Clark Gable had an endorsement deal with rival coffee company, Chase and Sanborn [Meyers 2014: 204]). Such complicated commercial entanglements help explain the appeal for sponsors of adaptation anthology series, in which a different film was adapted every week. This format gave access to studio property, in the form of both actors and screenplays, which could be negotiated on an episode-by-episode, studio-by-studio basis. Whilst this was time-consuming for advertising agencies, it also broadened choice in terms of adaptation sources and the stars – whose availability was limited and costs high.[2]

The specific format of the adaptation anthology series also met the paradoxical advertising criteria of "standard and recognizable appeals" coupled with innovation "to build new markets." The series provided both familiarity (every week offered an adaptation, and so every week was the same) as well as novelty (every week featured a new film and new stars, and so every week was different). Episodes of anthology series usually finished with the series host telling listeners what is in store next week, as did *LRT*'s "Casablanca" (broadcast in 1944), whose host was no less than Cecil B. DeMille: "One of the big, dramatic prizes of the past year – the Metro-Goldwyn-Meyer hit, 'Random Harvest.'" Here DeMille entices audiences to return through a combination of popular entertainment and prestige – a romantic melodrama and an Academy-Award-nominated picture – neither of which needed to be directly stated, since both were known to cinema audiences.

Such concision, drawing on pre-existing audience knowledge, was precisely how advertisers were trained to present products to radio audiences, since the transience of the medium led to fears that listeners might miss broadcast content. Staff at JWT were advised to consider radio commercials as analogous to billboards rather than newspaper advertising, and to "'avoid quick transitions … or concepts which listeners cannot grasp as the words fly by'" (Meyers 2014: 74). Radio scholarship has since echoed this early, medium-specific analysis by advertisers, with critics such as Frances Gray, Richard Hand and Mary Traynor theorizing radio-listening as at once a "liberation" and "a secondary experience" (Hand and Traynor 2011: 35), since, even before the advent of transistor radios in the 1960s, it was often undertaken alongside other, household activities. Hand and Traynor posit that this led radio drama to "compensate" for surrounding distractions: "narrative structure tends to be uncomplicated, involving few characters … simplicity is the key to the most successful radio drama" (2011: 36–37).

LRT's "Casablanca" demonstrates how the repetition embedded in adaptation assisted in these compensatory tactics, since it enabled such complex scenarios as the European refugee-trail to the Americas via Casablanca and Lisbon to be summarized easily, audiences being already familiar with the geo-political context (which the film explicates with the aid of globes and maps). At the end of the play we hear strategies of repetition employed to advertise the series' next episode: "Random Harvest," with its story of post-war lost love, echoes the appeal of "Casablanca," yet at the same time its fresh appeal is signalled to listeners by the response of the studio audience, who give two rapturous gasps, once when the title is announced, and once for the film's and adaptation's stars, Ronald Coleman and Greer Garson. The gasps of pleasure were a customary, weekly response (although the audience performed their role with particular conviction on this occasion) helping to ensure that any inattentive listeners would then be eagerly awaiting the later recap of who and what was to come next week – which DeMille in this episode gives after a propagandistic appeal to listeners to buy more war bonds.

The remit of novelty and familiarity could also be met by adaptation anthologies through the variety of films and genres on offer. Women were transparently the principal target audience of US commercial radio, since they were the chief purchasers of consumer goods: as one JWT advertising manager put it when describing "Mrs Consumer," the target listener for the show *Kraft Music Hall*, "'She's your wife! She is first rate and high class and Kraft advertising is fashioned to meet those attributes'" (Meyer 2014: 220). Not surprisingly, genres such as the melodrama, the musical and the romantic comedy thus predominated adaptation anthologies schedules, with, for example, *LRT* dramatizing "Stella Dallas" (1937), "Wuthering Heights" (1939 and 1954), "Mrs Miniver" (1943) "Rebecca" (1941 and 1950), "All About Eve" (1951), "Show Boat" (1940 and 1952), "A Star is Born" (1939 and 1942), "His Girl Friday" (1940) and "The African Queen" (1952).

Within the broad market of the female listener, consumer advertisers also sought to target different demographics. Mothers were encouraged to listen with young children to adaptations of children's films such as *The Wizard of Oz* (Victor Fleming, 1939), and intermission advertising was shaped accordingly. Thus, regular guest "Hollywood reporter, Libby Collins" tells listeners of "The Wizard of Oz" (1950) that star Joan Bennett, the "glamorous mother of four lovely daughters," uses Lux Flakes on "everything from dainty party dresses to two-year-old Shelley's gay cotton play clothes." The Second World War brought changes both to the lives of target listeners and to the duties of advertising agencies. Women's roles in the war effort were recognized through commercials focusing on young, working women – for example, through the endorsement of "hard-working, Fox starlet," Susan Blanchard (*LRT*, "It's a Wonderful Life" [1947]), as well as through advertisements in which wives who have been "working all hours" and "neglecting beauty care" are advised by friends to use Lux soap before their husbands return home on war leave (*LRT*, "Casablanca" [1944]).

During wartime, the advertising strategy of using entertainment to 'sweeten the pill' of radio's address to consumers took on greater prominence and urgency because it was extended to propaganda. Advertising agencies were answerable to the Office of War Information (for whom copywriters volunteered) as well as to sponsors, which meant that commercial sponsors, as a means of promoting their own wartime contribution, effectively footed the bill for the US government's radio propaganda (Meyers 2014: 242). This also meant that series needed to simultaneously promote soap and gasoline while encouraging frugal product use and carpooling. Adaptations such as "Casablanca" met the demand for palatable, pleasurable propaganda, with its hero's noble, self-sacrificing protection of refugees ideally suited to the difficult task of persuading a war-weary US public to invest yet more money in the government's "Fourth War Loan Drive."

The balance between repetition and variation could sometimes be a precarious one. In a crowded market with many sponsors chasing similar target demographics, it was not uncommon for series to remake adaptations. Different anthology series also adapted the same films as one another (for example, *Screen Guild* also recorded "Mrs Miniver" [1942], "Rebecca" [1943 and 1948], "A Star is Born" [1937 and 1942], "Casablanca" [1943], "His Girl Friday" [1941] and "Wuthering Heights" [1946]). This meant that the offer of novelty and variety could be undercut through a multiplicity of radio adaptations, and across the major networks of NBC and CBS there was also a host of different adaptation anthology series on offer, such as *The Dreft Star Playhouse, Hollywood Star Time, Hollywood Playhouse* and *The Screen Directors' Playhouse*. The ubiquity of anthology series, as well as the need to find a new film every week, meant that such series had to adapt films and genres without an obvious, "safe" appeal to the female market. As a result, *LRT* also made versions of noir films, biopics, war dramas and Westerns.[3] Whilst some films were clearly more likely to attract radio audiences than others, the series' inherent diversity offered the potential for them to pick up new listeners from week to week, through the different appeal offered by such variety in films, genres and stars.

The combination of familiarity and novelty was particularly evident when it came to the presence of stars, whereby radio adaptations offered the pleasures of repetition, the pleasures of variation and the pleasures of variation combined with traces of repetition (or vice-versa). Thus, where the adaptation featured the film's original stars (as in *Screen Guild*'s "Casablanca," sponsored by Lady Esther), this was proudly announced when the stars were introduced, promoting the familiarity and authenticity of film stars in their (own) film roles. At the same time, such repetition was itself offered as novelty, since much of the appeal of the adaptations lay in the (then) ephemeral nature of films, which could not be re-experienced after their theatrical run unless they were subsequently reissued. Some of the series' complicated plays on repetition, newness and rarity can be heard in *LRT*'s "Wizard of Oz," which was produced in 1950, eleven years after the film's original release, but only one year after MGM reissued the film in cinemas for a new generation of children (Fricke 2013: 93). Host William Keighley's introduction of Judy Garland – "audiences have asked for her to be brought back again and again" – effectively uses the film's re-release to remind audiences that it will always be worth revisiting, while also offering up the specific radio performance (with a twenty-eight-year-old Garland providing a rather more throaty rendition of Dorothy) as a new, special, not-to-be-missed experience.

Conversely, where a film's original cast was not used in the radio adaptation, broadcasts provided the novelty of hearing familiar stars playing roles made famous by other actors – for example *LRT*'s version of "Casablanca" featured Alan Ladd, Hedy Lamarr and John Loder, instead of Humphrey Bogart, Ingrid Bergman and Paul Henreid. The unique ability of radio to bring audiences new combinations of well-known stories and stars was used to promote broadcasts, as DeMille's introduction to *LRT*'s "Casablanca" makes clear:

> Getting these players together in one drama is a talent scoop of the first magnitude in Hollywood and it's unlikely that it would ever happen in a picture, because our stars are under contract at different studios and are kept so busy on their home lots that they aren't "loaned out".

In this instance the film industry's ownership of stars is effectively used against it, with radio subtly presented as a newer, freer industry, able to circumvent the bureaucracy of powerful studios for the benefit of the audience's entertainment at home.

The fact that radio was indeed listened to in the home constitutes an important ingredient of its "medium specificity" (see Carroll 1996: 26), and part of the novelty, or 'variation,' offered by adaptation anthology series. This novelty was derived from the way these series exploited the inherent intimacy of radio. Again this was evident in the use of stars, since not only did actors share 'their' shopping and beauty tips through product endorsement, they also did so during 'behind-the-scenes' chats during programme intermissions. Thus, at the end of *LRT*'s "Casablanca," DeMille points out that Hedy Lamarr is also "Mrs John Loder," and Loder, Lamarr and DeMille exchange (slightly awkward) banter on whether DeMille advises married co-workers to "do a little rehearsing at home." The consumerism which brought about such listening pleasures also had the potential to undermine them, since the conversations' stiltedness arose from the fact that they were scripted by copywriters. This potentially undercut the impression of stars speaking intimately to listeners, although at the same time these segments still provided the novelty of stars "staying behind after the show" to talk to the audience. Radio also offered an unfiltered experience not available in the cinema, in that broadcasts were live, which meant that mistakes and off-script additions could not be removed. (Thus, Judy Garland shamelessly gives the game away at the end of "The Wizard of Oz," while reassuring DeMille that she is not away from her family for the Christmas-holiday broadcast: "I brought my three-year-old – my *four*-year-old – daughter, Liza, with me. It says three in the script, but really she's four.") The studio audience's reactions to both scripted jokes and unscripted contributions suggests a mixture of innocence and knowingness – a willingness to suspend disbelief in the spontaneity of star chat, and a concomitant delight when the illusion is broken and they are 'let in on the act.'

The seeming informality of these star segments was helped by the absence of any writers' credits. Neither the writers nor their employers, the advertising agencies who produced the series, are mentioned anywhere in the episodes, a policy Meyers presents as typical of commercial broadcasting at the time, ensuring that programmes appeared to originate directly from their sponsors (2014: 215). This downplays the status of the adaptations (and their chosen medium), since they are devoid of authorial markers that would distinguish them from sources (and promote radio as the product of creative talent). Cartmell argues that a similar practice occurred with classical Hollywood adaptations of literature: marketing stressed literary authors, film stars and also the words of talkies (which had the potential to offer fidelity to literary sources).

In both instances the adaptation promotes elements of the source medium with a view to evoking its (marketable) prestige, though the precise ways in which cultural capital is at play differ in each case.

The radio adaptations' authors are not just given a low profile, but are entirely uncredited. Authors are present, but they are cinematic, with directors such as DeMille (and later William Keighley) hosting (and billed as "producer"), in order to give the series cultural weight. Thus, film authors are evoked for radio during a period before the rise of auteurism gave them the profile they enjoy today (and when they were downplayed in film adaptations which promoted *literary* authorship instead). At the same time, film directors and stars promote sponsorship products, so that their presence in covering up radio authorship advances commercial goals as well as prestige; effectively, directors and stars assist in the project of audience misdirection away from the advertising-agency employees who are the adaptation authors.

There is a certain irony in the role that film occupies here: Meyers suggests that within the advertising industry there was reluctance to embrace radio advertising because it lacked the authority of the printed word, evoking instead a tradition of oral fakery associated with "travelling medicine shows" (2014: 61). Thus, in classical Hollywood film adaptations, spoken words are used to evoke literary, written words; these adaptations respond to what Stam identifies as a "class-based dichotomy" by linking film to literature through its "talkie" (rather than "movie") status. In golden-age radio adaptations, meanwhile, a key dichotomy emerges, not between words and images, but between spoken and written language; cinema's glamorous, moneyed voices offer industrial respectability and prestige to recorded, spoken words, despite the fact that when cinema adapts it is more usually understood to occupy the low-brow, populist, visual (non-linguistic) side of the adaptation coin.

Although radio adaptation anthologies effectively promote their own subordination to film, it is nevertheless possible to detect traces of authorship within individual adaptations. Specifically, adaptation approaches at times mirror the methods used for marketing sponsorship products, and the combined broadcast of the narrative and its sponsorship paratexts allows us to trace how approaches to adaptation are shaped by the targeting of audiences by advertising agencies as simultaneous consumers of stories and products. Frequently these creative decisions involve the manipulation of Hutcheon's definition of adaptation pleasure, that is "repetition with variation ... the comfort of ritual with the piquancy of surprise" (2006: 4). Two versions of Disney's *Snow White and the Seven Dwarfs* (David Hand et al., 1937) demonstrate this, one by *LRT* in 1938 and the other by *Screen Guild* (then named *Lady Esther Screen Guild Players*) in 1944. The Online Archive of California at the University of Santa Barbara lists the adaptor for seasons 1943–1947 of *Screen Guild* as Harry Kronman, but does not have the same information available about writers for *LRT*.

Meyers presents the J. Walter Thompson agency, makers of *LRT*, as "the foremost producer of star-studded popular entertainment derived from the Hollywood film industry" (2014: 224), and theirs was the only anthology series to record hour-long rather than half-hour adaptations. Meyers also offers their policy towards sponsors as standing in direct contrast to that of Young and Rubicam (producers of *Screen Guild*), citing JWT's radio-department head, John Reber, as disapproving of programs which "kid the product and make fun of advertising on radio" (2014: 162). Y&R, meanwhile, is described by Meyers as "a pioneer of disarming audience resistance to commercials by mocking their commercialism" (2014: 169), with one agency executive, Pat Weaver, even featuring in sketches making fun of advertising and broadcasting, in Fred Allen's *Town Hall Tonight*. Meyers considers Y&R to be leading exponents of "soft sell," which depended on "positive associations between products and consumers' emotions' (2014: 169), whilst JWT, though it also employed "soft sell," sometimes used "hard sell", or "repetitious hectoring, and direct, rational appeals" (2014: 8). On *LRT* stars were used to endorse the product directly (women stars often being required to gush about the facial-beauty benefits of Lux soap), whilst *Screen Guild* employed stars for the adaptation but used radio announcers to voice the commercials.[4]

LRT's "Snow White" was a Christmas-time broadcast, which took place just over a year after the film's theatrical release. Host DeMille introduces the film by stressing Walt Disney's trepidation and daring the year before – the film being Disney's first feature-length animation – and by recounting its success in extravagant terms:

> In the tiny hands of a little lady named Snow White lay the reputation and the future of Walt Disney. How this picture was received is history. It brought laughter and tears from the children and grown-ups of every nation. Praise came from the pulpit, from statesmen, from the press for this unassuming man, who even his switchboard operator calls 'Walt.'

The adaptation is clearly capitalizing on the prestige of the film and its maker, so much so that DeMille's bombastic tone is allowed to compete with humanizing assertions of Disney's humility, and more so with the episode's evocation of radio as an intimate medium telling listeners a bedtime story. (DeMille does not really modulate his voice during the broadcast, incongruously proclaiming the line, "Let's dim the lights a little. Let's sit down and shut our eyes. Forget the world and just imagine ….") The film poses an obvious difficulty for radio in that its groundbreaking technical achievements rest with the animated visuals. It is precisely here that both the film's appeal and prestige lie, which means that a key adaptation technique necessarily becomes talking about what the radio episode cannot show (the "one half million drawings" by "more than five hundred artists"). This build-up promotes the broadcast as a special event within the series, marking the end of *LRT*'s 1938 season.

The narrative itself is treated faithfully, with much of the screenplay reproduced in the radio play. This was *LRT*'s custom, and a major appeal of the adaptation anthology format was obviously the fact that large portions of its radio scripts had already been written by screenwriters. Some changes that are introduced directly respond to the needs of the sponsor, as when story events are reordered (for example, the Queen finds out that Snow White is still alive only after Snow White's second rendition of "Someday my prince will come,") in order to create three defined acts that will accommodate commercial breaks. Other changes accommodate the broadcast schedule of the radio network: the drama is shortened to under one hour by having DeMille's narration summarize minor scenes, by curtailing occasional songs, and by cutting the length of sequences that rely on visual humour. Thus, Snow White sings a shorter version of "Whistle while you work" as she cleans the dwarfs' house, since we cannot watch the forest animals comically helping her with chores.

Elsewhere we find additions designed to compensate for the absence of visuals: dialogue is added to impart information (as in Snow White's "there are eyes in the dark … staring at me … monstrous shapes and mouths agape"), while the episode also uses music to create "stylized sound effects" (Crisell 1994: 52) – for example, the name of each dwarf is accompanied by a different musical instrument playing a different note, so that musical pitch and timbre comically convey the relative size and sweetness of the characters. DeMille's narration occasionally also adds musicality (as in "We hope she'll be happy as a young girl should,/ With seven little men so kind and good"), which echoes the playfulness of the visuals it replaces and also evokes the film's fairytale source through child-friendly alliteration and rhyme. While *LRT* cannot reproduce the film's animation, it has the advantage that the film's performers are vocal artists, and the episode includes from the original cast Roy Atwell as Doc, Moroni Olsen as the Mirror,

Stuart Buchanan as the Huntsman and Billy Gilbert (the comedian famous for his noisy sneeze routines) as Sneezy. Most obviously, the radio version is able to reproduce the film's songs, and it plays out famous romantic and comic numbers – such as "Someday my prince will come" and "It's off to work we go" – in full.

The character of Dopey, who is famously silent in Disney's film, is included by the sound of a bell round his neck. The popularity of round-eyed Dopey with audiences was signaled by the fact that he featured on the front cover of the *Radio Guide* issue promoting the adaptation, and his inclusion via sound effect demonstrates the adaptation's determination to replicate the film's appeal. On one hand, such sound-effect additions draw our attention to visuals which the radio episode has lost in adaptation, yet it is worth noting that they also draw on medium-specific properties of radio which adaptations are particularly well-placed to exploit. Theories of radio repeatedly counter the notion of radio as a medium defined by lack ("a TV which gives us nothing to look at" [Hand and Traynor 2011: 3]) with the notion that radio provokes the visual imagination; Hand and Traynor specifically employ the notion of *anamnesis*, defined by Augoyard and Torgue as "an effect of reminiscence in which a past situation or atmosphere is brought back to the listener's consciousness, provoked by a particular signal or sonic context" (2011: 4). The radio episode's playful sound effects, such as Dopey's bell, make use of this capacity of radio sound, with the listener's visual memory of the source film being evoked through sonic substitutions and approximations. This technique manages to offer variation, which actually suggests repetition, and makes clear that the aim of the adaptation is fidelity to its source. Thus, the film is faithfully reproduced in a manner that reflects JWT's reverence towards product sponsorship, the hour-long broadcasts allowing time for the replication of much of the film's dialogue and songs, homages to the film through sound-evocations of its visual style and achievements, as well as product endorsement by stars.

This episode's guest star is Walt Disney himself, and he and DeMille engage in intermission chats which in themselves make the episode a noteworthy piece of film history. In these (scripted) interludes the two men compliment each other's work using the motif of adaptation, with each speculating how their films would be if adapted into the genre favoured by the other.[5] JWT's copywriters also put their advertising skills to work summarizing the Disney brand, with Walt Disney allowed a prolonged, 'folksy' speech endorsing the studio's wholesome ethos:

> Over at our place we're sure of just one thing. Everybody in the world was once a child. We grow up, our personalities change, but in every one of us something remains of our childhood … And it just seems to me that if your picture hits that spot in one person, it's going to hit that same spot in almost everybody. So in planning a new picture we don't think of grown ups, and we don't think of children, but just of that fine, clean, unspoilt spot down deep in every one of us, that maybe the world has made us forget, and that maybe our picture can help recall.

This speech encapsulates the symbiosis at work, with advertisers paid by Lever Brothers using radio in order to promulgate the 'cleanliness' of Disney pictures as a means of boosting sales of Lux washing detergent. Here the sponsor, the commercial paratexts and the adaptation seem perfectly matched, with all clearly addressing the family market and the mothers who were presumed to assess the suitability of washing powders, children's films and evening radio broadcasts.

LRT clearly utilizes Disney's *Snow White* as a means of self-promotion, its prestige and uniqueness stressed and mimicked in order to make this Christmas broadcast festive and special. However, towards the end of "Snow White" *LRT* also surreptitiously 'one-ups' the film, adding in extra lines as Snow White comforts the dwarfs before leaving with her prince: "Please don't be sad, little men, I'll be coming back … Once every year we'll meet." This partly addresses the medium-specific difficulty of the film's rapid, visually resolved ending (almost no words are spoken after Snow White wakes in her glass coffin and is led away into the sunset by her prince), but it also subtly promotes *LRT* as the means by which Snow White can be returned to us, replacing finite film releases with the comfort of radio seriality. Radio adaptation is thus associated with Christmas itself, bringing (the same) stories to our firesides year after year.

LRT's manipulation of both its source and its medium mirrors the selling tactics of JWT: the star endorsements, with their wholesome positivity are clearly soft-sell, but the cumulative repetition of endorsements for Lux products pushes the overall approach towards hard-sell techniques. Similarly, the fidelity of the adaptation, and DeMille's praise for Disney, speak of reverence for the cinematic hypotext and cinema itself; at the same time the hyperbole and bombast of DeMille's narration, coupled with the implication that broadcast radio can offer pleasures in repetition in ways that cinema cannot, reveals the precisely targeted, self-serving appeal to consumer–listeners which is at the heart of sponsored commercial radio.

Five and a half years later, *Screen Guild* and Y&R did indeed bring Snow White back to radio, thereby belatedly fulfilling (or perhaps breaking) *LRT*'s promise of annual fairytale outings. This was a much smaller production, and its appeal was its stars – not of film, but of radio – since the episode featured ventriloquist Edgar Bergen and his dummy side-kick, Charlie McCarthy. Bergen and McCarthy were much-loved radio stalwarts (despite the fact that Bergen's act of ventriloquism could not actually be seen on the radio), and were headline acts on *The Chase and Sanborn Hour* (a programme produced by JWT). *Screen Guild*'s "Snow White" has 'straight-man' Edgar Bergen tell the fairytale to his mischievous dummy and comic foil, Charlie McCarthy. Bergen explains that he has been contacted, he says, by *The Lady Esther Screen Guild Players*, who want him to play the prince in their upcoming new production. The episode is thus played out as if it is not part of the series at all, but rather an off-stage retelling prompted by a future 'real' *Screen Guild* production, in which Bergen hopes to star.

Just as Y&R made use of mockery to disarm audience resistance to sponsorship, in this version of "Snow White" the agency takes a humorous approach towards adaptation and the medium of radio. This strategy is effective given the high number of adaptations and adaptation series that were being broadcast at the time – the episode's comic elements clearly provide novelty, and in contrast to *LRT*'s proclaiming tone, the adaptation and its stars suggest self-deprecation. The adaptation's mise-en-abyme technique promotes, but also parodies, the notion of radio as the intimate teller of fireside tales, with McCarthy providing risqué, unchildlike asides (such as the suggestion that Snow White has had "a snoot-full" when she "stumbles on a cottage") during Bergen's "parental" recount. McCarthy's contributions ridicule Bergen, but they also poke fun at fairytales and their retelling through adaptation (he follows Bergen's "Once upon a time, long, long ago" with "Funny, how nothing ever happens *nowadays*"). The imaginative power of radio, so prized by medium-specific theorists, is also gently ridiculed: when Bergen announces a scene change, McCarthy mocks radio's low-budget, non-visual story-telling with "that's what I like about radio – it's so *flexible*". The episode features topical, wartime jokes ("the manpower shortage isn't that bad," responds a sceptical McCarthy to the notion of Bergen cast as prince) which occasionally poke fun at war propaganda. When Bergen invites us to "journey back to the time of Snow White," McCarthy responds "Yes, but before we go, is this trip really necessary?"

Although Disney's *Snow White* is not itself mocked (McCarthy gets caught up in the story, and its romance is told 'straight' through the inclusion of the film's songs), the McCarthy/Bergen "Snow White" makes clear how the advertising technique of "disarming audience resistance" was utilized in radio adaptation. Here potential wartime disillusion with both escapist fairytales and radio's propaganda role are anticipated and salved through self-aware, humourous entertainment. The episode indicates Y&R's consciousness of possible changes in consumer–listener tastes since the broadcast of *LRT*'s adaptation. It therefore seeks to sell the film anew to (women) listeners – whose experiences of war and work may have made them slightly cynical towards this wholesome tale of how Snow White replaces drudgery for the queen with drudgery for "seven little men," to be whisked off with barely a word by a returning prince to his fairytale castle in the sky. The popularity of this "Snow White" was such that *Screen Guild* reworked and rerecorded it four more times (in 1946, twice in 1948 and finally in 1950). Two were Christmas broadcasts, echoing *LRT*'s earlier intimation of radio as the means by which fairytales and films could be retold (and refreshed) year after year.

Radio, however, did not have a monopoly on the satire market. Hollywood was to offer a satirical commentary on commercial radio in 1947 through an adaptation of Frederic Wakeman's 1946 novel *The Hucksters* (Jack Conway). Both film and novel depict radio as enslaved to the advertising industry, with a programme output that insulted listeners with infantile unoriginality, dictated by the whims of despotic sponsors, such as the film's "Old Man Evans," president of "Beautee Soap." These were accusations often leveled at commercial radio, though they were refuted by those working for it, such as JWT writer Carroll Carroll, who argued that "If advertisers sometimes butt into the jobs of the writers and directors, so, too, does the money on Broadway and in Hollywood dictate to the creative echelon" (Meyers 2014: 215). This idea is echoed within contemporary adaptation studies, most obviously in Simone Murray's *The Adaptation Industry*, which refutes previously held dichotomies of cinema as commercial where literature is not, by tracing the complex networks of agents and publishing houses that make up the book industry. What Murray's work demonstrates is the need to understand all adaptations as the products of their industrial conditions and as targeted towards consumers of stories.

In 1949, director Robert Siodmak, when featured as a guest on the *Screen Director's Playhouse* radio adaptation of *The Killers* (Siodmak, 1946), was asked by star Burt Lancaster why he, a German, was assigned to direct such a "typical America" story. Siodmak, responding that as an outsider he has insights not obvious to Americans, elucidates as follows:

> When I first traveled across the United States I came to a small town with a big sign. The sign read 'Coca Cola.' So I said to myself, 'Ah-hah! I am now in Coca Cola, New Mexico.' And then I came through Burma Shave, Arizona!

As Siodmak's anecdote demonstrates, advertising is ubiquitous in the United States, just as vested financial interests, and the pursuit of profit, are inseparable from the production of narratives in any capitalist culture. What golden-age radio adaptations offer us is conspicuous sign-posting that this is so – that adaptations are always sold as well as told. In the parallels between the radio plays and their sponsorship paratexts, we can also see the techniques by which consumerism is rendered not just acceptable but pleasurable and entertaining within the adaptation's narrative landscape.

Notes

1 *Kraft Music Hall* was a variety show produced for the dairy manufacturer Kraft by JWT and hosted from 1936 to 1946 by Bing Crosby. *Hollywood Hotel*, produced by the agency Ward Wheelock for Campbell Soup, and *Old Gold Hollywood Screen Scoops*, produced by Lennen and Mitchell for the P. Lorillard Company, the makers of Old Gold cigarettes, were both gossip shows.

2 *LRT*, which was first broadcast on Sunday afternoons from New York and featured Broadway actors in stage-to-radio adaptations, moved to Hollywood – and to film-to-radio adaptations – in 1936, after a primetime broadcast slot opened on CBS on Monday nights. Its talent costs rose accordingly, from

$10,500 per episode in 1935 to $25,300 per episode in 1937 (Meyers 2014: 216), though clearly this increase was considered justifiable for primetime broadcasting, since what it purchased was the audience appeal of big-screen stories and stars.

3   Examples include thrillers such as "Angels with Dirty Faces" (1939) and "This Gun for Hire" (1943); biopics such as "The Life of Emile Zola" (1939); war dramas such as "Wings of the Navy" (1940); Westerns such as "My Darling Clementine" (1947) and "Red River" (1949).

4   Tom Lewis, Vice President of Y&R (whose marriage to Loretta Young is presented by Young's biographer as being partly motivated by his pursuit of stars for radio; Dick 2011: 231), was able to recruit actors for *Screen Guild* because the series raised money for charity: respective sponsors Gulf, Lady Esther and Camel paid Y&R and the host radio networks, but actors and studios waived their fees, which the Screen Actors' Guild – who commissioned the series – used to build and fund its Country Home and Hospital for retired, ill or destitute industry artists (Christman 1991: 18). JWT enticed stars to *LRT* through lucrative salaries, although its Hollywood Office president, Danny Danker also had a reputation for ruthlessness, and was rumoured to use blackmail to persuade reluctant stars to participate (Meyers 2014: 217).

5   DeMille's *Union Pacific* (1939) would feature a steam-engine hero falling in love with a beautiful coal car, and wooing her with the line "Baby, we gotta make tracks!"; De Mille also self-consciously tells Disney of the importance of Lux Flakes in keeping costumes clean during the making of epics.

Works cited

Carroll, Noel (1996) *Theorizing the Moving Image*, Cambridge, UK: Cambridge University Press.

Cartmell, Deborah (2014) "Teaching adaptations through marketing: Adaptations and the language of advertising in the 1930s," in *Teaching Adaptations*, D. Cartmell and I. Whelehan (eds.), Basingstoke, Hants, UK: Palgrave Macmillan.

Christman, Trent (1991) *Brass Button Broadcasters*, Paducah, KY: Turner.

Crisell, Andrew (1994) *Understanding Radio*, London: Routledge.

Dick, Bernard F. (2011) *Hollywood Madonna: Loretta Young*, Jackson, MS: University of Mississippi Press.

Fricke, John (2013) *The Wonderful World of Oz: An Illustrated History of the American Classic*, Camden, ME: Down East.

Geraghty, Christine (2008) *Now a Major Motion Picture*, Plymouth, MA: Rowman and Littlefield.

Gray, Frances (1981) "The nature of radio drama," in *Radio Drama*, P. Lewis (ed.) London: Longman.

Hand, Richard and Mary Traynor (2011) *Radio Drama Handbook*, New York: Continuum.

Hutcheon, Linda (2006) *A Theory of Adaptation*, London: Routledge.

McFarlane, Brian (1996) *Novel to Film: An Introduction to the Theory of Adaptation*, Oxford: Clarendon.

Meyers, Cynthia B. (2014) *A Word from our Sponsor: Admen, Advertising, and the Golden Age of Radio*, New York: Fordham University Press.

Murray, Simone (2011) *The Adaptation Industry: The Cultural Economy of Contemporary Literary Adaptation*, London: Routledge.

Radio Guide (1938) "'Snow White and the Seven Dwarfs' visit Lux Theater Monday," week ending December 31.

Stam, Robert and Alesandra Raengo (2005) "Introduction," in *Literature and Film: A Guide to the Theory and Practice of Film Adaptation*, R. Stam and A. Raengo (eds.), Oxford: Blackwell.

Radio series

Lux Radio Theater, National Broadcasting Company, Blue Network, (1934–1935); the Columbia Broadcasting System (1935–1954); the National Broadcasting Company (1954–55): "Stella Dallas" (10.11.37, CBS), "Snow White and the Seven Dwarfs" (26.12.38, CBS), "A Star is Born" (9.13.37 and 28.12.42), "Angels with Dirty Faces" (22.5.39, CBS), "The Life of Emile Zola" (5.8.39, CBS), "Wuthering Heights" (18.9.39, CBS and 14.9.54, NBC), "Show Boat" (24.6.40 and 11.2.52, CBS), "His Girl Friday" (30.9.40, CBS), "Wings of the Navy" (7.10.40, CBS), "Rebecca" (3.2.41 and 6.11.50, CBS), "This Gun for Hire" (25.1.43), "Mrs Miniver" (12.6.43, CBS), 'In Which We Serve (21.6.43, CBS), "Casablanca" (23.1.44, CBS), "Random Harvest" (31.1.44, CBS), "My Darling Clementine" (4.4.47, CBS), "It's a Wonderful Life" (10.3.47, CBS), "Red River" (7.3.49, CBS), "The Wizard of Oz" (25.12.50, CBS), "All About Eve" (1.10.51, CBS), "The African Queen" (15.12.52, CBS)

Screen Guild Theater (known variously as *The Gulf Screen Guild Show*, *The Gulf Screen Guild Theater*, *The Lady Esther Screen Guild Theater* and *The Camel Screen Guild Theater*), Columbia Broadcasting System (1939–1948); the National Broadcasting Company (1948–1950); the American Broadcasting Company 1950–1951; the Columbia Broadcasting System (1950–1952): "His Girl Friday" (30.03.41, CBS), "A Star is Born" (17.11.40, CBS), "Mrs Miniver" (6.12.42, CBS), "Casablanca" (26.04.43, CBS), "Rebecca" (31.05.43 and 18.11.48, CBS), "Snow White and the Seven Dwarfs" (24.04.44, 25.02.46 and 23.12.46, CBS; 07.06.48 NBC; 28.12.50, ABC), "Wuthering Heights" (25.02.46, NBC)

Kraft Music Hall (1933–1971, NBC), *Hollywood Hotel* (1934–1938, CBS), *Hollywood Playhouse* (1937–1940, NBC), *Old Gold Hollywood Screen Scoops* (1937–1938, CBS); *The Dreft Star Playhouse* (1943–1945, CBS); *Hollywood Star Time* (1946–1947, CBS); *The Screen Directors' Playhouse* (1949–1951, NBC).

Films

Snow White and the Seven Dwarfs (David Hand et al., 1937; Walt Disney Productions)
The Wizard of Oz (Victor Fleming, 1939; Metro-Goldwyn-Mayer)
Union Pacific (Cecil B. DeMille, 1939; Paramount Pictures)
Casablanca (Michael Curtiz, 1942; Warner Brothers)
The Killers (Robert Siodmak, 1946; Mark Hellinger Productions and Universal Pictures)
The Lion King (Roger Allers, Rob Minkoff, 1994; Walt Disney Pictures)

Websites

Advertising Hall of Fame: http://advertisinghall.org/members/member_bio.php?memid=528. Accessed 25/10/15.
Online Archive of California at the University of Santa Barbara http://www.oac.cdlib.org/findaid/ark:/13030/tf5779p13c/entire_text/. Accessed on 29/10/15.

24

RECONFIGURING THE NORDIC NOIR BRAND

Nordic Noir TV crime drama as remake

Yvonne Griggs

Nordic Noir TV crime drama foregrounds its Scandinavian identity, its particular noir-like mode of audiovisual expression, its complex long-form narrative, and its in-depth character studies, and though its national markers are an intrinsic part of its identity, it offers a branding template that has the capacity for cultural and geographical makeover on a global scale. With its roots in Scandinavian crime fiction dating back to the early twentieth century, the term *Nordic Noir* has become synonymous in contemporary times with quality television. First coined, according to Gunhild Agger, by the Scandinavian Department at University College London, and given mainstream exposure in a BBC documentary titled *Nordic Noir: The Story of Scandinavian Crime Fiction* (2016: 138), the label Nordic Noir has since been adopted by reviewers,[1] audiences, and production companies alike to classify film and television dramas that share a certain generic DNA, most readily aligned with crime drama and invariably employing a noir-style aesthetic. Steven Peacock sees the positive reception of screen adaptations of Stieg Larsson's *Millennium Trilogy*[2] and of Henning Mankell's *Wallander*[3] as instrumental in generating growing international interest in Nordic Noir products; such narratives, he argues, are particularly receptive to "repositioning as global texts" (2013: 98–99), as are recent TV crime dramas from other Scandinavian countries. Nordic Noir's global reach is evident across a body of television series that share its generic codes and its distinctive visual aesthetic. Through a process of *appropriation* rather than the more direct route of *adaptation*,[4] audiovisual markers of Nordic Noir crime dramas have become part of an embedded style signature that lends kudos to various British, French, Irish, American, and Welsh crime dramas,[5] but whether functioning as adaptation or appropriation, Nordic Noir TV crime series translate to other national, geographical, and cultural frameworks with ease. A more definitive adaptive relationship is established in a number of English-language remakes of Scandinavian TV series. Produced for an American audience, *The Killing* (2011–2014), a remake of Danish TV crime series *Forbrydelsen* (2007–2012), and *The Bridge* (2013–2014), a remake of Danish–Swedish coproduction *Bron/Broen*[6] (2011–), followed the release of the Scandinavian source texts, as did *The Tunnel* (2013–), an Anglo–French remake of *Bron*. Through analysis of the adaptive processes that inform the production of these Nordic Noir remakes, this paper explores what Linda Hutcheon terms the ongoing "dialogue" between not only source and remake but also the "dialogue between the society in which the works, both the adapted text and adaptation, are produced" and that in which they are "received" (2013: 149).

The status of the remake

Despite the commercial nature of both film and television production, the attendant prejudices that haunt the remake in cinema do not extend to its televisual counterpart. Tried, tested, and familiar products make for relatively safe investments within the worlds of film *and* television, but while discourse surrounding the cinematic remake highlights its often negative perception as an inevitably imitative and thus inferior form (Grindstaff 2001: 134; Braudy 1998: 327), the same prejudices do not attach to the TV remake of existing drama formats. Defined as "openly acknowledged and extended reworkings of particular other texts" (Hutcheon 2013: 16) that perform "a self-conscious balancing act" between "the familiar and the new, or the familiar and the 'transformed'" (Horton 1998: 174), the remake, within the realms of the TV industry, provides a viable commercial model. Negative notions of colonization attached to the film remake have limited currency within the context of TV production, where the sharing of successful formats is a recognized part of good industry practice, and the recycling of narrative forms an established production mode (Moran 2013: 1–5). In a digital age characterized by transmedia storytelling, the single story is becoming more of an anomaly for both financial and cultural reasons as producers harness the stability offered by pre-loved formats and audiences seek the pleasures of repeated consumption of narratives in various forms across numerous media platforms. Flaunting its mass media identity and its populist appeal, the TV remake, unlike the film remake, is able, according to Linda Grindstaff, to "realize most fully [André] Bazin's vision of the adaptation as capturing, not the *letter* of the original, which can be emulated in mechanical fashion, but the *spirit* of the original—its tone, values, and rhythms" (2001: 157). TV remakes that adapt or appropriate Nordic Noir drama embrace its existing style signature—its "tone, values, and rhythms"—as part of a profitable brand identity that lends artistic weight to its remakes, a desirable affiliation with its brand, and an operational mode that functions outside the prejudicial confines of fidelity-driven debates that traditionally circulate adaptation studies.

The Nordic Noir brand: defining a style signature

As with film noir, a term posthumously applied to a body of work that emerged during the golden age of the Hollywood studio system in the 1940s, contemporary Scandinavian TV crime dramas have garnered their Nordic Noir classification *post*-production. The term has become synonymous with a specific cultural moment within the evolution of the TV crime genre, exploring new territory and creating new generic markers now emulated in productions appearing in various global locales that cross spatial and cultural borders. Its very distinctive yet transportable style aesthetic is indebted to both the TV crime genre and cinema's film noir. It employs film noir's muted colour palette, its low-key lighting, and its dangerous cityscapes, all of which have become signifiers of a certain type of detective story, characterized by complex plotlines. Unlike the genre *film*, which according to Thomas Leitch "positions its audience to expect the conventions of the genre" while also "withhold[ing] any explicit knowledge that it is borrowing those conventions" (1990: 148), the TV remake in general and Nordic Noir crime drama remakes in particular highlight their "borrowed" conventions. Nordic Noir crime dramas foreground their dense noir iconography, playing out in rain-soaked urban landscapes and dimly lit claustrophobic settings that fuel, in both *Forbrydelsen* and *Bron,* a "mise of political and sexual intrigue" (McCabe 2013: 121). Series creator and chief writer on *Forbrydelsen* Søren Sveistrup notes the Danish Broadcasting Corporation's (DR) desire to adopt a "cinematic mode of expression" with a "Noir edge" from the outset, instigating a move away from the "talking heads" of the TV crime drama to a more "visual" cinematic style (Sveistrup in Redvall 2013: 177).

However, while film noir explores male anxiety in the aftermath of the Second World War, the anxieties that dominate its successor, Nordic Noir, are centred on the violence and the political machinations of our contemporary postfeminist age. The lone male detective of film noir becomes a female hybrid, part investigative loner, part dangerous femme fatale without the accompanying sexual potency. Jason Mittell notes the predominantly masculine appeal of contemporary crime dramas, many of which feature male antiheroes like Tony Soprano (*The Sopranos*, 1997–2007) and Vic Mackey (*The Shield*, 2002–2008) (2015: 151); similarly, antiheroic males serve as lead detectives in current British crime dramas like *Luther* (2010–2015) and *Good Cop* (2012). The "maleness" of the terrain, argues Mittell, is symptomatic of "cultural norms of particular genres" (2015: 150), but Nordic Noir crime drama and its remakes subvert the gendered norm, presenting antiheroic female detectives who are not defined by the "cultural norms" of the crime genre. Sarah Lund (*Forbrydelsen*) and Saga Norén (*Bron*) serve as prototypes for the construction of a different kind of female detective. They are "feminized versions of the traditional 'noir' detective" (Turnbull 2014: 182), and like their male counterparts in film noir they are first and foremost defined by their capacity to solve mysteries. Janet McCabe (2013: 126) cites Larsson's Lisbeth Salander and Peter Hoeg's Smilla Jasperson as Scandinavian forerunners of both Lund and Norén and notes their debt to earlier female detectives like Christine Cagney, Mary Beth Lacey (*Cagney and Lacey*, 1981–1988), and *Prime Suspect's* (1991–2012) Jane Tennison. But while Cagney, Lacey, and Tennison are constructed as women who must battle to retain their positions within a male-dominated work place, there is an assumption in Nordic Noir crime dramas that gender is irrelevant, equality the norm (McCabe 2013: 120).

The reconfiguration of the archetypal noir detective as dysfunctional and morally ambiguous protagonist is also maintained in Nordic Noir crime dramas and their remakes. Lund and Norén are intriguing constructs: they combine the characteristics of the highly intelligent and tenacious hard-boiled detective outsider and the dangerous (yet here decidedly unglamorous) 'otherness' of film noir's femme fatale. Female leads in Nordic Noir crime dramas and their remakes are dysfunctional within the personal realm. Familial instability is an established norm: Lund is a bad mother, daughter, and partner, while Norén is a socially inept and emotionally disengaged 'single'. The pattern is emulated in their remakes. *The Killing*'s Sarah Linden is presented as an even more troubled and dysfunctional mother, the product of a foster home upbringing and failed relationships. *The Tunnel*'s female detective, Elise Wassermann, is a convincing counterpart to *Bron*'s Norén, yet her emotional distance becomes part of her cultural reconstruction: within the context of a French–Anglo pairing, she is more moody, arrogant, sexually uninhibited French woman than autistic loner. When translated to the US remake, *The Bridge*, lead female Sonya Cross emerges as a tic-addled loner whose borderline autism teeters on the precipice of parody. The Nordic Noir crime drama remains, however, at the forefront of a wave of new forms of serialized storytelling that pushes the boundaries of gender representation within the genre. Synergies between Nordic Noir and other modern-day crime narratives featuring strong yet emotionally dysfunctional female detectives mark a similar reconfiguration of women in crime dramas: Stella Gibson (*The Fall* 2013–), Robin Griffin (*Top of the Lake* 2013–), and Marcella Backland (*Marcella* 2016–) have the same generic makeup as Lund and Norén. Their emergence denotes a contemporary fascination with the representation of women in crime drama as empowered yet unsettling; antiheroic, noir-like detectives rather than women whose gender defines them.

When moving into an "importing culture", the "retention" of the source text's "strangeness" and its "high value" markers is, argues Yuri Lotman, of vital importance (1990: 146). The Nordic Noir narrative's shift to a different geographical landscape, and a different cultural backdrop, is invariably characterized by retention of its "strangeness" and its "high value" markers: its affiliation with film noir plotlines, archetypes, and style aesthetic, and its complex, multistrand storylines are invariably recalibrated to work within the metalingual structure of the importing culture. Remakes like *The Killing*, *The Bridge*, and *The Tunnel*, for instance, retain "high value" markers of the Nordic Noir brand yet are characterized by their capacity to translate the narrative into a new cultural and geographical locale. Nordic Noir crime dramas and their remakes foreground the significance of their urban locale within their specific geographical location, the urban locale becoming a loaded signifier freighted with the cultural markers of nationhood. In *Forbrydelsen*, Copenhagen serves as backdrop to the unfolding of a crime story that imbibes national politics alongside the complexities of detection and the unravelling of a family embroiled in personal grief. In its remake, *The Killing*, the locale shifts to Seattle, but in both of these crime dramas the murder is enacted in a space outside the metropolis, in a forested area and the more open expanse of Native American land, respectively. In *The Killing*, an integral plotline plays out on Native American land, serving as a cultural referent that places the narrative firmly on American soil. It also creates a border territory redolent with tension—a space beyond the confines of state law that becomes intrinsically linked with the solving of the crime at the centre of the narrative. Here, as in *Bron* and its numerous remakes, it is this kind of border territory that creates a geographical and cultural demarcation of place. While the urban locales in which the detective work unfolds remain important signifiers of place and of identity, the foregrounded site of the initial crime becomes the dominant signifier of difference. These liminal spaces—marshland outside Copenhagen (*Forbrydelsen*), the reservation (*The Killing*), the Oresund Bridge (*Bron*), the Channel Tunnel (*The Tunnel*), the vast border between the United States and Mexico (*The Bridge*) are constructed as iconic signifiers of not only a physical divide but as sites synonymous with tension related to more than the crime that is being investigated. Tensions related to national identity and notions of invasion associated with border country pervade all five texts, creating much more than a visual aesthetics of difference.

Narratives that engage in what Mittell terms "complex serial poetics" present multiple plotlines and ongoing story arcs (2015: 18); they offer "thematically and tonally connected episodes" that go beyond the scope of episodic and serial forms (2015: 30). Nordic noir crime dramas employ this kind of structural "serial poetics". *Forbrydelsen* (Series One) consists of interconnecting narrative strands that create layers of complexity across a 20-episode season, though narrative momentum is ensured through the unravelling of a police investigation into the disappearance and murder of teenager Nanna Birk Larsen. The inclusion of a political backdrop that explores the lives of mayoral candidate Troels Hartmann and his electoral team provides another level of intrigue and a very specific sociocultural context to this Danish story. The series employs an innovative approach to the dramatization of the impact of crime upon the lives of its victims, the family's ongoing response to the murder being a prominent narrative thread. The private life of the female detective at the centre of the investigation is similarly foregrounded, creating a multilayered story that goes beyond the generic norms of the crime drama. Like its TV series predecessors, *Twin Peaks* (1990–1991) and *Deadwood* (2004–2006), each episode spans the course of one day, ensuring a "clear rhythm to its serialized narrative flow" (Mittell 2015: 28), but *Forbrydelsen* adds to this kind of rhythmic plotting by linking steps along the police procedural path to its daily structure and its multiple plotlines, using intertitles that foreground the moment as "Day One", "Day Two", and so on at the start of each episode, and adding further intertitled detail relating to specific times and locations within that day as a means to ensuring the drama's procedural clarity. Episodic closure throughout the series is replaced by a montage that acts as both summary and teaser, noting what progress has been made during the course of that day's investigation, its impact on the family and the political situation, and where the investigation will take us in the next episode. The montage creates what Mittell would term "window frames" that remind us what has happened, but potentially "distort [our] vision of the unfolding action" (2015: 53); these closing sequences provide the viewer with a complex variation on the cliff-hanger narrative across a 20-episode series aligned to a 20-day timeframe during which the hunt for Nanna Birk Larsen's murderer unfolds.

Constantine Verevis notes that remakes are "highly particular in their repetition of narrative units" (2006: 21); remakes of *Bron* are characterized by an attention to this kind of repetition. *Bron*, *The Tunnel*, and *The Bridge* present themselves, first and foremost, as police procedurals; their plotlines are primarily concerned with the clearly delineated solving of a series of crimes. All three employ the standard tropes of the crime drama within a backdrop of social politics and employ the requisite red herrings. *Bron* focuses on concepts of justice (in the first series) and political activism (in the second series), but like other Nordic Noir crime dramas it also foregrounds the familial, the social, and the political as part of its narrative fabric. The crimes committed in remakes of *Bron* are altered to fit the values and the cultural climate of each new geographical and national platform, but the nature of their detection remains the same, building on rather than deconstructing the story template provided by *Bron*. In the first series of *Bron*, the familial drama is linked to the solving of the crimes as the personal life of Norén's colleague, Martin Rohde, is later shown to be the prime motivation for the crimes. Here, source and remakes fulfil audience desire to follow the quest and to solve the crimes to a point of final resolution in the closing episode of each series, and any familial threads are linked intrinsically to the central plot, providing a neat police procedural framework. However, the narrative units employed in the remake of *Forbrydelsen* differ on several levels. Though the closing montage employed in each episode of *Forbrydelsen* is emulated in striking detail by its American remake, *The Killing*, the structure of the latter is altered significantly. *Forbrydelsen* consists of 20 episodes, each covering a day of the investigation and drawing to a point of series closure; *The Killing* plays out across not one but two seasons, spanning 26 episodes in total, its final reveal being delayed until the finale of season two. Narratives that engage in "complex poetics" invariably reject the need for definitive plot closure within every episode (Mittell 2015: 18), but Veena Sud, as series creator and head writer, takes this premise a step further: her decision to deny the audience any real sense of closure at the end of *The Killing* (season one), adds yet another layer of complexity to an already dense plotline. Where *Forbrydelsen* engages its audience in the kind of satisfying "forensic fandom" seen by Mittell as a marker of the successful crime drama (2015: 52), *The Killing* impedes viewer participation in the solving of the crime due to its convoluted plot machinations and its withheld point of closure. Season two functions here as a sequel that provides the final pay-off.

Arguing that, in an American context, "if a girl went missing ... no-one would care," Sud chooses to recalibrate the narrative into an even denser and more multilayered story, peppered with additional subplots and red herrings that are purposely left unresolved. *Forbrydelsen*'s multiple narrative strands are maintained yet probed in even greater depth. Instead of developing a clearly delineated remake of *Forbrydelsen*, Sud creates her own "jazz variation" of this Nordic Noir series, presenting what she terms an "anti-cop show" (Sud in Hughes 2011) that purposely denies her audience anticipated narrative closure at the end of season one. Some strands, like the culturally relevant terrorist subplot in *The Killing* (season one), are frustratingly underdeveloped, but contrary to Sud's claim that this is an "anti-cop show" the inclusion of a much more pivotal relationship between Sarah Linden and her colleague Steven Holder places the narrative into familiar genre territory within an American TV context. It becomes in part a buddy cop show of the type which has long-standing precedence in the US TV market, lending the series an affinity with successful American TV shows like *Miami Vice*, *Starsky and Hutch*, and *Cagney and Lacey*. Linden's backstory and that of recovering drug-addict Holder, her hoodie-wearing hipster partner, are developed across seasons one and two. Episode 10, season one, dedicated to their search for Linden's missing son, interrupts the series' narrative momentum by focusing on character development and backstory to the exclusion of anticipated police procedural progress. Though initially well received as a series, audience and reviewer dismay at its unresolved plot detours and its lack of satisfying closure followed the airing of the final episode. Some viewers saw the final 60 seconds of that last episode, "Orpheus Descending", as a "contemptuous psych-out of its audience" (Poniewozik 2011) that contravenes the logic of the police procedural and steps outside the structural parameters of the crime genre.

Appropriating Nordic Noir

The extent to which adaptations and appropriations identify their intertextual connections varies; in appropriations the "intertextual relationship" is deemed by Sanders to be "less explicit, more embedded" and yet appropriation remains an act of "re-interpret[ation]" (Sanders 2006: 2–3). The nature of the relationship between Nordic Noir crime dramas and TV series like *River*, *Hinterland*, *Braquo*, and *Marcella* remains open for debate, but it is a relationship that is invariably foregrounded during the marketing and/or reception of these various TV series. Disparate TV crime dramas, both remakes and dramas like *Hinterland* and *Braquo*, *Marcella* and *River* are marketed as part of the Nordic Noir brand; they work their connections with that brand as a means of garnering an audience whose thirst for all things Nordic Noir has yet to be satiated. Its brand identity is engineered through websites that foreground product similarity, generic expectation, and a desire for more of the same. Nordicnoirtv.com is one such site dedicated to the propagation and continuation of fan communities via blogs, Facebook, Twitter feeds, and Q & A events.[7] All of these series engage in what Christine Geraghty terms a process of "shadowing" or "doubling"; the "ghostly presence" (2008: 195) of the Nordic Noir crime drama haunts them on some level, whether through assimilation of its style signature, its noir-like, emotionally dysfunctional female detective, its multistrand approach to storytelling, or through its conscription of Scandinavian writers, actors, and production companies. Though none can be classified as remakes, they do nonetheless establish their connection with the Nordic Noir brand. All four series emulate the audiovisual codes of Nordic Noir crime drama and explore similarly dark content related to the solving of a crime or crimes. And like their Nordic Noir counterparts, all, with the exception of *River*, written by Abi Jordan, are produced by writing teams and utilize various directors across a series. By employing male lead detectives, *Hinterland*, *Braquo*, and *River* may align themselves with the crime genre's masculinist traditions, but Marcella Backland is modelled on the antiheroic, nonconformist female detective lead of Nordic Noir crime drama. The Nordic Noir influences at work at the level of plot and style aesthetic in *Marcella* can in part be attributed to Hans Rosenfeldt, series creator/head writer and former member of *Bron*'s writing team. Through its casting of Stellan Skarsgård in the lead detective role, *River* also intertextualizes its affinity with quality Scandinavian film and television, and is deemed part of the Nordic Noir brand of crime drama. *The Guardian*'s Sam Wollaston aligns it with *Forbrydelsen* as a similar study in loss and grief as well as an investigative quest (2015). If, as Regina Schober argues, adaptations are engaged in "a process of forming connections" which can be "explicit or implicit, total or partial" (2016: 89–91) then texts of this nature, which forge "implicit" and "partial" connections with Nordic Noir crime drama at the level of plot, style, and archetype can be read as appropriations of those dramas.

Conclusion: maintaining the brand

For many, the pleasure of viewing Nordic Noir narratives lies in the experience of immersion into another national space, both culturally and geographically, subtitles providing the only mediating signifier. *Forbrydelsen* was given an international platform in various overseas TV markets; delivered by the BBC as part of its strategy to create a more highbrow viewing experience on one of its channels (BBC Four), it was at the vanguard of a growing body of Scandinavian TV dramas imported to the UK. Though there is a limited audience for subtitled TV drama, for those viewers who are familiar with Nordic Noir in its original format, there remains the added pleasure of revisiting it as remake, the "new text" offering its audience an "intercultural engagement with an ever-expanding narrative network" (Evans 2014: 311). The remake as adaptation involves what Hutcheon terms "its *knowing* audience" in a pleasurable process of "conceptual flipping back and forth" between a prior text and its adaptation (2006: 139, my italics). However, though part of Nordic Noir's image as 'high-end' TV comes from its capacity to appeal to a niche audience of the BBC Four variety, as with subtitled cinema, demand for subtitled TV has its limitations. Whereas the BBC's successful transmission of *Forbrydelsen* and later *Bron* was dependent on its audience's desire for immersion in the otherness of a different culture, the successful remake of such dramas in the international arena is more often dependent upon the adapter's capacity to resituate the story and its characters within the cultural framework of its host nation, adding cultural referents that speak to the viewer's experience, that present familiar territory and familiar cultural preoccupations while retaining key narrative elements. Cultural shift is all-important. Speaking of format TV[8] and its global transferability, Moran adheres to the notion that "a good creative idea in one place can be successfully established elsewhere"; given a makeover that both realigns the screen text with its new audience and tells stories that are connected to that new audience's cultural space, TV programmes are ripe for remake (2008: 462). The same edict can be applied to the making of Nordic Noir crime drama. The creation (and longevity) of the successful TV series is part of a branding strategy employed by all TV broadcasters, whether publicly funded or commercially motivated, and Nordic Noir drama has emerged as a valuable brand identity for a number of Scandinavian broadcasters and production companies who foreground the significance of "brand identity" (Redvall 2013: 5). From the outset, the production team[9] behind the inception of *Forbrydelsen* sought to create "a new trend and brand for a certain kind of Nordic TV drama" that could "compete in the international domain" (Redvall 2013: 161, 179). Yellow Bird, the Swedish production company behind the *Wallander* series (Swedish and English–Swedish co-productions) and the original screen adaptation of *The Girl with the Dragon Tattoo* (2009) employs a similar strategy to ensure the global expansion of its products (Tourmarkine 2011: 52).

These Nordic Noir crime dramas are what Hutcheon would term "travelling stories", able to "adapt to local cultures" and "local environments" with relative ease (2006: 177). They offer a fertile site for further mediation of populist stories within a complex transmedial narrative network that transcends geographical and cultural borders. In a TV production context that proves conducive to the recycling of successful narrative formats, Nordic Noir crime series present as a brand of TV drama of growing transnational significance, characterized by a distinctive and highly adaptable style signature. Given this kind of production climate, both remakes and crime series that appropriate the style signature of Nordic Noir crime drama are set to flourish rather than fail. In this instance, the dialogue between source and remake results in the ongoing evolution of complex, quality television.

Notes

1 Though a term readily employed by print and online TV and film critics, academic interrogation of the term remains limited. With the exception of journalist and broadcaster Barry Foreshaw's populist publication (*Nordic Noir: The Pocket Guide to Scandinavian Fiction, Film and TV*) and a slim collection of essays edited by Steven Peacock (*Stieg Larsson's Millennium Trilogy: Interdisciplinary Approaches to Nordic Noir on Page and Screen*) there has, to date, been little academic work in the field of Nordic Noir TV drama.

2 Swedish–Danish screen adaptations of Larsson's trilogy, initially produced for television, were given cinematic release in 2009 (*The Girl with the Dragon Tattoo*; *The Girl Who Played with Fire*; *The Girl Who Kicked the Hornet's Nest*). An American remake (*The Girl with the Dragon Tattoo*) swiftly followed in 2011.

3 Mankell's *Wallander* has been adapted for Swedish TV (2005–2013) and UK TV (2008–).

4 Julie Sanders differentiates between the two terms: adaptations are said to "openly declare" their relationship with a precursor text, whereas appropriations have a "less explicit, more embedded" intertextual relationship to source (2006: 2).

5 *Braquo* (2009); *Hinterland* (2013–); *Lily Hammer* (2012–2104); *Amber* (2104); *River* (2015); *Marcella* (2016–).

6 All future references will employ the Danish term *Bron*.

7 As part of its marketing strategy, distributor Arrow Films (arrowfilms.co.uk/nordic-noir/) maintains brand identity post-TV broadcast by the release of clever online marketing materials that group together series that have any kind of affiliation with Nordic Noir product, presenting it as a body of work that falls within a carefully constructed brand identity.

8 Moran defines format TV as a formulaic guide to "programme adaptation" in other global locales (2008: 461).

9 *Forbrydelsen* was developed by Danish government broadcaster, DR.

Works cited

Agger, Gunhild (2016) "Nordic Noir: Location, identity, emotion," in Alberto Garcia (ed.) *Emotions in Contemporary TV Series*, Houndmills, UK: Palgrave Macmillan, 134–155.

Amber (2014–), dir. Thaddeus O'Sullivan. Perf. Eva Birthistle, David Murray, and Gary Whelan. Screenworks. DVD.

Archer, Neil (2012/2013) "*The Girl with the Dragon Tattoo* (2009/11) and the New 'European Cinema,'" *Film Criticism* 37(2): 2–20.

Arrow films (2104), "Nordic Noir and beyond showreel," YouTube 9 June: n. pag. Web [14 Feb. 2015].

Braquo (2009–), dir. Xavier Palud et al. Perf. Jean Hugues-Anglade, Joseph Malerba, Karole Rocher, and Nicholas Duvauchelle. Canal+ et al. DVD.

Braudy, Leo (1998) "Afterword: Rethinking remakes," in A. Andrew Horton, and Stuart Y. McDougal (eds) *Play it Again Sam: Retakes on Remakes*, Berkeley, CA: University of California Press, 1998, 328–333.

Bron (2011–), dir. Henrik Georgsson et al. Perf. Sofia Helin, and Kim Bodnia,. Filmlance International et al. DVD.

Cagney and Lacey (2004–), dir. Alexander Singer et al. Perf. Sharon Gless, and Tyne Daly. CBS (Columbia Broadcasting System) et al. DVD.

Deadwood (2004–), dir. Ed Bianchi et al. Perf. Timothy Oliphant, Ian McShane, and Molly Parker. Paramount Network Television et al. DVD.

Evans, Jonathan (2014) "Film remakes, the black sheep of translation," *Translation Studies* 7(3): 300–314.

Forbrydelsen (2007–), dir. Kristoffer Nyholm et al. Perf. Sophie Gråbøl, and Søren Malling. Danmarks Radio (DR) et al. DVD.

Forshaw, Barry (2013) *Nordic Noir: The Pocket Essential Guide to Scandinavian Crime Fiction, Film & TV*, Harpenden, UK: Oldcastle Books Ltd.

Geraghty, Christine (2008) *Now a Major Motion Picture: Film Adaptation of Literature and Drama*, Maryland and Plymouth, MA: Rowman & Littlefield.

Good Cop (2012–), dir. Sam Miller, and Susan Tully. Perf. Warren Brown. BBC Drama Productions et al. DVD.

Grindstaff, Laura (2001) "A Pygmalion tale retold: Remaking *La Femme Nikita*," *Camera Obscura* 47 16(2): 133–175.

Hinterland (2013–), dir. Ed Thomas et al. Perf. Richard Harrington, and Malli Harries. All3Media International et al. DVD.

Hughes, Sarah (2011) "It's a crime to remake a cult hit," *The Independent* 30 June: n. pag. Web [13 Feb. 2015].

Hutcheon, Linda (2006) *A Theory of Adaptation*, New York: Routledge.

"I'll let you know when I get there" (Ep. 10) *The Killing* (2011–), dir. Veena Sud, et al. Perf. Mireille Enos and Joel Kinnaman. KMF Films et al. DVD.

Leitch, Thomas (1990) "Twice told tales: The rhetoric of remake," *Literature Film Quarterly* 18(3): 138–149.

Lilyhammer (2012–), dir. Geir Henning et al. Perf. Steven Van Zandt, and Trond Fausa. Rubicon TV AS. DVD.

Lotman, Yuri (1990) *Universe of the Mind: A Semiotic Theory of Culture*, London: I. B. Tauris.

Luther (2010–), dir. Sam Miller et al. Perf. Warren Brown, Idris Elba, and Ruth Wilson. British Broadcasting Corporation (BBC). DVD.

McCabe, Janet (2013) "The girl in the Faroese jumper: Sarah Lund, sexual politics and the precariousness of power and difference," in Steven Peacock (ed.) *Stieg Larsson's Millenium Trilogy: Interdisciplinary Approaches to Nordic Noir on Page and Screen,* Houndmills, UK: Palgrave Macmillan, 118–130.

Marcella (April 4, 2016), dir. Charles Martin et al. Perf. Anna Friel and Nicholas Pinnock. Buccaneer Media et al. ITV broadcast.

Miami Vice (1984–), dir. John Nicolella et al. Perf. Don Johnson, and Philip Michael Thomas. Michael Mann Productions. DVD.

Mittell, Jason (2015) *Complex TV: The Poetics of Contemporary Television Storytelling*, New York and London: New York University Press.

Moran, Albert (2008) "Makeover on the move: Global TV and programme formats," *Journal of Media and Cultural Studies* 22(4): 459–469.

Moran, Albert (2013) "Global TV Formats: Genesis and Growth," *Critical Studies in TV* 8(2): 1–19.

Nordic Noir & Beyond. "Welcome to the UK home of Nordic Noir drama and beyond," nordicnoir. tv.com n.pag. Web [14 Feb 2015].

"Orpheus descending" (Ep. 13) *The Killing* (2011–) dir. Veena Sud et al., perf. Mireille Enos and Joel Kinnaman. KMF Films et al. DVD.

Peacock, Steven (2013) "Crossing the line: *Millennium* and *Wallander* on screen and the global stage," in Steven Peacock (ed.) *Stieg Larsson's Millennium Trilogy: Interdisciplinary Approaches to Nordic Noir on Page and Screen*, Houndmills, UK: Palgrave Macmillan, 98–117.

Poniewozik, James (2011) "The Killing Watch: Bloody Murder!" *Tuned In* 20 June: n. pag. Web [13 Feb 2015].

Prime Suspect (1991–), dir. Christopher Menaul. Perf. Helen Mirren, John Banfield, and Tom Bell. Granada Television. DVD.

Redvall, Eva Novrup (2013) *Writing and Producing TV Drama in Denmark: from The Kingdom to The Killing*, Houndmills, UK: Palgrave Macmillan.

River (2015–), dir. Tim Fywell et al. Perf. Stellan Skarsgård and Nicola Walker. Kudos Film and Television et al. DVD.

Sanders, Julie (2006) *Adaptation and Appropriation*, New York: Routledge.

Schober, Regina (2016) "Adaptation as connection—transmediality reconsidered," in Jorgen Bruhn, Anne Gjelsvik, Eirik Frisvold Hanssen (eds) *Adaptation Studies: New Challenges, New Directions*, London: Bloomsbury, 89–112.

Starsky and Hutch (1975–), dir. George McCowan et al. Perf. Paul Michael Glaser and David Soul. Spelling-Goldberg Productions. DVD.

The Bridge (2013–), dir. John Dahl et al. Perf. Sonya Cross, Demián Bichir and Thomas M. Wright. FX Productions et al. DVD.

The Fall (2013–), dir. Alan Cubit et al. Perf. Gillian Anderson and Jamie Dornan. Artists Studio et al. DVD.

The Killing (2011–), dir. Veena Sud et al. Perf. Mireille Enos and Joel Kinnaman. KMF Films et al. DVD.

The Shield (2002–), dir. Guy Ferland et al. Perf. Michael Chiklis, Walton Goggins, Kenny Johnson, and David Rees Snell. Fox Television network et al. DVD.

The Sopranos (1999–), dir. Timothy Van Patten et al. Perf. James Gandolfini, and Eddie Falco. Home Box Office (HBO). DVD.

The Tunnel (2013–), dir. Thomas Vincent et al. Perf. Stephen Dillane and Clémence Poésy. Canal+ et al. DVD.

The Girl with the Dragon Tattoo (2009–), dir. Neils Arden Oplev. Perf. Michael Nyqvist and Noomi Rapace. Yellow Bird et al. DVD.

Top of the Lake (2013–), dir. Jane Campion et al. Perf. Elisabeth Moss, Peter Mullan, and Thomas M. Wright. See-Saw Films et al. DVD.

Tourmarkine, Doris (2011) "Yellow Bird takes flight: Worldwide success of Millennium Trilogy propels Swedish production company," *Film Journal International* 114(7): 52.

Turnbull, Sue (2014) *The TV Crime Drama: TV Genres*, Edinburgh, UK: Edinburgh University Press.

Twin Peaks (1990–), dir. David Lynch et al. Perf. Kyle MacLachlan. Lynch/Frost Productions et al. DVD.

Verevis, Constantine (2006) *Film Remakes*, Edinburgh, UK: Edinburgh University Press.

Wallander (2005–), dir. Stephan Apelgre et al. Perf. Krister Henriksson. Yellow Bird et al. DVD.

Wallander (2008–), dir. Benjamin Caron et al. Perf. Kenneth Branagh. Zodiak Entertainment et al. DVD.

Wollaston, Sam (2015), "River review: Pairing personal demons with a peculiar partnership," *The Guardian* 14 Oct: n. pag. Web [31 Oct. 2015].

25

TWEETING FROM THE GRAVE

Shakespeare, adaptation, and social media

Anna Blackwell

In *Spreadable Media* Henry Jenkins, Sam Ford, and Joshua Green provide a revolutionary model of contemporary cultural practices, suggesting that older, top-down models of distribution are no longer viable. They argue that in contemporary culture "a mix of top-down and bottom-up forces determine how material is shared across and among cultures in far more participatory (and messier) ways" (2013: 1). Founded on the simultaneity of circulation and reception, the explosion of participatory media forms marks a shift in which the public are not "simply consumers of preconstructed messages" but are "shaping, sharing, reframing and remixing media content" (2). This quality (examined by Jenkins, Ford, and Green in a variety of "spreadable" media forms including film, television, advertising, and gaming) is, the authors argue, a key characteristic of contemporary culture, with the unique mode of public engagement that invites "reshaping the media landscape itself" (2). It is this potential, "both technical and cultural," for *audiences* to share content according to their own purposes, which is the focus of this paper and its discussion of Shakespeare's continuing adaptive legacy on social media and, in particular, on Twitter (3).

The key example I will be analysing in this chapter is the varied output of a community of Shakespeare and television fans called The Hollow Crown Fans (HCF). A multiplatform group founded in 2012, the Fans were initially motivated by their shared admiration for the BBC's adaptation of Shakespeare's Henriad plays of the same name. Their (largely inactive) WordPress site provides set reports alongside interviews with minor cast members and promotion of their more significant Twitter activity. Their Facebook page, too, typically displays information already posted on Twitter and, in particular, the community's weekly event: #ShakespeareSunday. Regularly appearing in Twitter's trending list with upwards of two thousand participants, #ShakespeareSunday is the Fans" most prolific and popular contribution to Shakespeare's digital presence. It presents an opportunity to rethink communality, participation, creativity, and cultural value through the study of one iteration of what Jenkins et al. describe as the many and varied "affordances of digital media" (2013: 3).[1] Through metadata tags like the #ShakespeareSunday hashtag, Twitter facilitates the coalescing of a high volume of short-form creative responses under one agentive aegis. The ubiquity of any one hashtag extends both laterally and vertically, used by individuals, groups, and companies; consequently, not only can the reception and circulation of #ShakespeareSunday be varied, but its content is also quite diverse. In addition to 140 characters, Twitter permits the simultaneous pairing of text with images and animated GIFs of up to 5MB, or videos of 30 seconds in length. The short but plentiful bursts of information that have characterised both Twitter's content and its delivery system since its invention thereby create a uniquely serendipitous mode of engagement. Submissions to #ShakespeareSunday or, indeed, to Twitter in general may appear through the Fans' main account or by chance according to the make-up of an individual's followed accounts, appearing as original entries, re-tweets, sporadically, unobserved, or not at all. Fans' content, furthermore, can exist in any combination of an infinite variety of intertextual relationships, functioning as independent information, linked through shared hashtags or common discussion of an event or phenomena, or in a more collaborative mode through replies to authors, quoted retweets, or 'mentions' (a subset of Twitter's replying function, in which interested parties can be directed to a post or invited to comment).

Jenkins, Ford, and Green acknowledge that the model of participatory culture they present in their conceptualisation of a "spreadable" media is not unique, existing beyond the "life span of specific technologies or commercial platforms" (2013: 160). Nor is its content or modes of expression restricted to native digital forms such as Internet memes. What a study of individual social media platforms such as Twitter does enable, however, is a framework through which to view instances of everyday engagements with Shakespeare's creative legacy as well as his broader cultural capital. Indeed, with a far greater quantity of tweets produced by the Fans in comparison to much larger and more-followed Shakespeare-centred accounts, these texts—which I will continue to argue constitute a micro-adaptive form—give voice to the producers and consumers who are rendered otherwise anonymous by traditional modes of cultural reception. Hollow Crown Fans are typically neither scholars nor theatre folk. They are, in a word, fans. A recognition of the creative facility made available by social media has the potential, therefore, to alter our understanding of the traditional means by which Shakespearean meaning is received and circulated in contemporary culture: the Fans, as an example of digital adaptors, should be understood as a community of participants, not merely observers or even audience members.

It is this quality of not only the Fans but the behaviour of fans online more generally that provides the unique mode of adaptation proposed in this chapter. As Paul Booth writes, fans "typically utilize their technological capacities, their communal intelligence, their individual knowledge base, and their social interaction skills to investigate and explore media" (Booth 2010: 20). They are, therefore, a "crucial analytic tool" (20) in the study of digital culture. Despite their popular representation as maladjusted social misfits, bearing only the most arcane knowledge, fans exist at the frontier of media behaviour; writing back to and about media texts before more traditional interactivity, like scholarly publication, even begins. They are sophisticated textual users who, as Henry Jenkins writes in his seminal work on the subject, *Textual Poachers*, are frequently subject to castigation because they are seen as squandering the cultural and social capital they typically possess as educated middle-class individuals, who should "know better" than to participate in fandom's "elaborate interpretations" (1992: 18). Fan culture, Jenkins argues, thereby "muddies boundaries" by employing reading practices "traditionally reserved for works of serious merit and applies them to the "disposable" texts of mass culture" (17). This unique level of attention and critical depth is apparent not only in the pre-digital tradition of fan fiction or fan art, however, but in fans' creative interactions on social media. I propose that these moments of engagement constitute instances of adaptation on a microscale.

Although adaptation typically occurs on a much larger scale, the examples analysed throughout this paper correspond with the nature of both fan labour and fan texts, as identified by figures such as Jenkins, Matt Hills, Jonathan Gray, and Cornell Sandvoss. Fans participate in critical and creative practices which engage comprehensively with their chosen subject, whether in terms of repeated and sustained consumption of the text and its commercial paratexts, extensive engagement with the minutiae of production, dialogue or characterisation, or in the form of creative rewritings. Theirs is an engagement with chosen texts that works from the level of word or sentence upwards; a mode of reading and interpretation that is surely recognisable to textual scholars across the humanities. This kind of engagement is of particular pertinence to giants within the literary canon such as Shakespeare. Indeed, fan practice offers a level of scrutiny with immediate parallels to the creative and critical work done in Shakespeare studies, in which the editing of any given play has, as Gabriel Egan argues, untold and—as of yet—unexamined impact "beyond" the page (2016: 52). Pascale Aebischer and Kathryn Price have similarly commented upon not only Shakespeare's historic focus as a site of scrutiny, but the practical effects of this in comparison to other early modern playwrights. They write that the sparser performance history of non-Shakespearean plays has resulted in "considerable latitude" in the ways that "the notion of a performance is applied to what would, in the case of Shakespeare, be classified as an adaptation" (2012: 3). Aebischer and Price's argument reveals an important point here: Shakespeare's ability to outpace many other cultural figures as the subject of widespread performance and criticism has, paradoxically, created a starker delineation between what is 'authentically' Shakespearean and what counts as an adaptation.

But because the labour analysed in this context is fan labour that takes place in social media's inherently intermedial space, it presents an opportunity to reevaluate the location and scale of adaptation, and perhaps come to Shakespeare's work with a little less cultural baggage. As W. B. Worthen suggests, digital culture has drastically altered our understanding of what constitutes a 'text' as well as the means by which we interact with it. In relation to Shakespeare, Worthen thus argues that our understanding of the Shakespearean "no longer oscillates dualistically between page and stage, page and screen, screen and stage" (Worthen 2008: 228). Instead, the digital screen "blurs" drama's traditional delivery system by representing "text as *image*," with both now "part of the same network … realizable wherever I can get a connection on my laptop" (228). A recognition of digital culture's ability to reorder the relationship between Shakespeare, text, and image would thereby seem to facilitate James Naremore's much quoted call for our understanding of adaptation to include the study of "recycling, remaking and every other form of retelling in the age of mechanical reproduction and electronic communication" (2000: 15). Indeed, the examples of #ShakespeareSunday analysed within this chapter (and which I have described as adaptive in nature) do more than just quote Shakespeare. Operating within the blurred space of the digital screen, they pair Shakespearean text with images that body forth their authors'' interpretation of not only Shakespearean meaning but his wider cultural capital and cachet. These images, in short, are not simply illustrations—they are a creative interpretation of the text. The wider cultural network which they frequently invoke asks us to consider Shakespeare's adaptability. Their framing within this chapter as *micro*-adaptive sites, meanwhile, acknowledges the scale of these adaptive endeavours as well as their uniquely ephemeral nature. Social media's micro-adaptive texts, alongside more general instances of digital representation, provide new territory for adaptation criticism. Social media's immediacy and participatory functions contain the potential for adaptation in collaboration, as well as a mode of adaptive production that is meta-adaptational and which illuminates an exegetical reception practice that takes place outside of academia.

Indeed, as Pascale Aebischer and Nigel Wheale have noted, adaptation studies tend towards a perception of popular cultural practices by critics as a "mode of production that imposes simplifying and popularising structures … often in conflict with the Shakespearean texts it attempts to remake" (2003: 8). But as I will continue to demonstrate, it is not always productive to view the adaptive relationship between Shakespeare and popular culture as an essentially combative one. The instances of micro-adaptation charted in this paper represent an adaptive practice which is collaborative, responsive, and critical, providing tangible instances of audiences engaging not only with popular culture but with Shakespeare in a manner that emphasises a plurality of meaning and which presents the most recent example of what Stephen O'Neill describes as the "continuing vitality of our contemporary 'Shakespeares'" (2015: 274). Certainly, for those in either Shakespeare or adaptation studies eager to deepen their understanding of Shakespeare's reception and circulation in contemporary culture, the playwright's digital presence is of increasing relevance. As O'Neill proposes in his summation of the critical field, "[p]erhaps, for Shakespeare scholars, social media platforms are compelling precisely because they constitute the latest locations where the phenomenon we call 'Shakespeare' ... finds iteration" (2015: 274). The work of Christy Desmet (2008, 2009), Maurizio Calbi (2013), and Peter Kirwan (Carson and Kirwan 2014), amongst others, similarly testifies to the role technological innovation plays in our interactions with media texts, determining the nature of these encounters and their (potentially) creative consequences.

Digital Shakespeare, however, is also the cause of potential anxiety in the field of literary studies. Jenni Ramone's criticism of digital texts, for instance, belies a fear that "the [online] text is shaped by its users and authorship is only an optional field of attention" (2011: 8). This apprehension that Shakespeare may suffer a "loss of authority" (2) online is, to my mind, indicative of a concern that the "spectral and imprecise" (8) bodies of digital culture are not appropriate "retell[ers]" of Shakespearean texts. In relation to the study of adaptation, though, an acknowledgement of such intangibility is not only fruitful but often inevitable, given Shakespeare's increasingly multilayered and multimediatised cultural afterlife. Indeed, the idea of haunting frames Maurizio Calbi's *Spectral Shakespeares: Media Adaptations in the Twenty-First Century* productively with its study of Shakespeare's mediation in experimental adaptations on film, television, and the Internet. Similarly, Judith Buchanan employs Stephen Berkoff's description of assuming a dramatic role already performed by theatre greats as "boxing with ghosts" (2009: 208) in her history of *Shakespeare on Film*. Buchanan uses this term in her chapter on Kenneth Branagh's Shakespearean films, and the image is a potent one when analysing an adaptive oeuvre *and* adaptor who repeatedly engages with his own theatrical and cinematic history. The image of the ghost or the idea of ghosting usefully conceptualises the insubstantial but present intertextual links that exist between productions. When applied in relation to digital culture, though, the same language of haunting has the potential to undermine not only the labour performed but the daily cultural and economic exchanges which are the occasion for the adaptation and evolution of Shakespeare's legacy. The Hollow Crown Fans' appeal for support in the Shorty Awards (a celebration of the best real-time short form content producers across social media) reveals precisely this anxiety.

Promoting the community's digital activity as a whole (including their significant Twitter presence), the blog's "keynote speech" (Admin 2015) explains that the HCF deserves representation in the Shorty Awards for its creative authenticity: "first and foremost ... [it is] not just a machine that generates un-curated content that simply matches key words." Despite the formal rigour of the statement, the authors repeatedly align the Fans on one side of a binary between automated slickness and self-generated industry; 'high' and popular culture; clickbait and journalistic integrity; academic Shakespeare and Shakespeare 'for everyone.' At the same time, the speech works to anchor the community's appropriation of Shakespeare. Theirs is not an abstract endeavour, it suggests, but one of 'work' of 'incredible effort and commitment,' requiring 'behind the scenes' attention, 18-hour days, as well as the 'balls' to ask guest Shakespeareans (including Kenneth Branagh, Samuel West, and Tom Hiddleston) to propose a weekly theme. The community's claim to 'authenticity' is thus also present in the authors' desire to convey the potential graft needed in running an online community: a reflection of their need to validate digital industry and industriousness to a wider society.

The Fans exist, nevertheless, as a strand within an interconnected cultural network—transmitting out and receiving information, and representing a diverse community of users whose purposes include the recreational, educational, academic, and commercial. Their actions, along with our own, witness the same intersections between commercial and non-profit or fan-based enterprise that structure our day-to-day experience of digital media. It is these exchanges, despite the intangibility of digital culture, which form its substance. Indeed, despite Barbara Hodgdon's argument that YouTube, as an example of a content-sharing platform, functions in a "top-down capacity" (2010: 313) and that its users exist "outside" of its value systems, audiences *do* partake in social media's "commercial creative practice" (2010: 314). Display and paid search adverts are its common currency, and the systems that facilitate an everyday user's digital experience are invested in (sometimes) discrete commercial exchanges. A cursory glance at Tumblr's privacy policy, for instance, avows their dedication to the "private nature of your information" while also detailing their targeted advertisements and provision of information to third parties. The Fans thus both use their Shakespeare capital and are used for it, whether by actively promoting the Shakespeare300 app, designed by James Reese, on their blog, or having the #ShakespeareSunday hashtag co-opted by organisations or companies seeking to capitalise on a unique weekly focusing of Shakespearean interest.

Other adaptive intersections can be seen online between fan-curated content and the framework of exchange that structures everyday life, both online and offline. As Simone Murray argues, adaptation "constitutes not discreet *sui generis* artefacts but outcomes of an encompassing economic system" (2012: 122). A parody account for King Richard III on Twitter provides a comic and—in the author's own words—"leicesterian" [*sic*] perspective on Richard's legacy, including his representation by Shakespeare, the debate regarding reinternment, his commemoration, and the East Midlands city that is his resting place. One characteristically self-conscious encounter sees Richard discuss cars with local dealership, the Sturgess Motor Group, after having requested a Porsche ('Got any Porsches? I've matured from my horse days! #aporsche,aporsche,mykingdomforaporsche'). Elsewhere opposing the potential sale of local museum, Newarke House, the account offers a satire on and criticism of the nature of the economy that has sprung up around Richard's death. 'Richard' highlights the moments of transparent economic gain for the council and private companies by directing followers to incongruous tie-ins, such as Richard III lolly pops, for example. Similarly, 'Richard' retweets an advertisement of a "Battle of Bosworth tour from just £279pp," stating: 'Unofficial tour. I take no part in this production. #wheresmycut??'

The Leicester 'Richard' occupies banal local traditions and infrastructure and dramatises what was felt by some to be an inappropriate response from a largely secular Britain to the discovery of a king, long dead and long maligned. 'Richard's' subversive voice works to highlight the intersections of literature and history where his identity has been founded and, in doing so, also undermines the narrative of redemption which was invoked by the city, the Cathedral, and even the Hollow Crown Fans at his burial.[2] @richard_third has none of the wistful philosophy attributed to him in the elegy written for the event by Carol Ann Duffy, nor the gravitas of Benedict Cumberbatch who performed the poem and who, in a pleasing adaptive connection, is not only a descendent of the king but plays him in *The Hollow Crown* (TV, dir. Dominic Cooke, 2016). His body and soul is not, as Duffy's Richard envisages, "emptied of history" or "incense, voting, vanishing" but is, instead, imagined driving a Jaguar F Type (Duffy 2015).

Shakespeare's presence in spreadable media presents further potential grounds for complication. Although participatory culture and fan labour theoretically levels the cultural playground—paying equal attention to more obscure cultural texts alongside those that dominate the mainstream—digital Shakespeare brings with it the same potential for traditional cultural hierarchies and judgments of taste that continue to persist in relation to Shakespeare "in real life." This is because, as critics such as John Fiske (1992) and Roberta Pearson (2007) have identified, fandom originates in a place of cultural lack. It is, Fiske argues, "a form of cultural labour to fill the gaps left by legitimate culture" and which works to "provides the social prestige and self-esteem that go with cultural capital" (1992: 33). Fiske continues, "fan cultural knowledge differs from official cultural knowledge in that it is used to enhance the fan's power over, and participation in, the original, industrial text" (43). In his example, a *Rocky Horror Picture Show* fan's knowledge will allow them to engage with and potentially rewrite the text, while a Shakespeare buff's understanding would not allow them to participate in the performance but to "discriminate critically between it and other performances" (43). The complicated (and potentially contradictory) nature of this relationship between Shakespeare and fandom is apparent in the framing of the work done by individuals or communities such as the Hollow Crown Fans or *Shakespeare* magazine, who present their digital activity as a discovery, or as a recovery of something only previously accessed by those with socially or culturally privileged knowledge of Shakespeare. The *Shakespeare* magazine website's tag line, for instance, "*At last!* A magazine with all the Will in the world" (emphasis added) echoes the editor's note that he hopes to "give a new voice to Shakespeare fans everywhere," as it also does the Fans' "keynote speech" in which they express their desire to reclaim Shakespeare for the masses. Shakespeare's canonisation thereby continues apace online, with the democratic ideals of Web 2.0 called upon to circulate an image of Shakespeare as a previously sidelined presence within popular culture.

As with all such communities, however, these groups are potentially exclusive as well as inclusive. An image of Shakespeare that is predicated upon his marginalisation is one which is also ironically used to enforce the omission of lesser-known contemporaries from the canon of early modern playwrights. Vimala C. Pasupathi (@Exhaust_Fumes) is castigated by *Shakespeare* (@UKShakespeare) for using the hashtag #ShakespeareWeek in order to draw attention to her own research on Shakespeare's almost forgotten collaborator, John Fletcher, for instance. *Shakespeare* challenges Pasupathi, "While we're on the subject of staying classy, @Exhaust_Fumes, maybe you could stop trolling Shakespeare to get attention for your research?"[3] The research to which *Shakespeare* refers is Pasupathi's #NotShaxButFletch bot: an adaptive, automated script which searches out tweets about Shakespeare that mention his name or his plays and retweets them with Fletcher-related information. A tweet on *The Taming of the Shrew*, for example, retweets with a reference to Fletcher's *The Tamer Tamed*, a tweet on Shakespeare's Kate to Fletcher's Maria, and Shakespeare to Fletcher. As an intellectual project, Pasupathi (an associate professor at Hofstra University, United States) seeks not only to draw attention to the works of a truly marginalised literary figure but to examine the differing cachet owned by Shakespeare and Fletcher, as well as the different sounds of their writing. Although the provocative nature of these authorial interventions (on the part of both Pasupathi and the Fletcher bot) may warrant its description as trolling, *Shakespeare*'s combative response articulates its distaste for an academic perspective in which Shakespeare's greater cultural capital is sidelined for a lesser-known (and apparently inferior) author. Pasupathi's perceived failure to be "classy" is thereby aligned with her lack of deference to Shakespeare's cachet, both generally and on the occasion of a dedicated #ShakespeareWeek in which he should have been assured uncompromised publicity. One response to *Shakespeare*'s criticism from Emily WeNNceslas (@battielove) articulated the incompatibility of this logic, however, stating: "Shakespeare's fine. The popular kids don't need your help" (2015).

Although *Shakespeare*'s attitude towards academic interventions in Shakespeare's ongoing social media life might seem counterintuitive (dismissing as it does a resource that draws attention to the connectedness of Shakespeare to his contemporaries), it expresses a feeling also shared by the Hollow Crown Fans. In the same "keynote speech," the Fans also explain their rationale for the community. The authors argue that the Shakespeare familiar to most of us is the one from school: a "dry, boring and tedious experience that belongs only to academia or those with a considerably advanced education. We choose to reject that notion. … Shakespeare is not the property of academia. He belongs to you, to all of us, together."

This refutation of academic Shakespeare continues with a specific creative outcome in mind. The Fans posit that the Shorty Awards present an opportunity to demonstrate to the media industry that there is a mainstream 'hungry' for more Shakespeare and that the playwright deserves a "place at the head of the table with other pop culture icons" (note that in the category of best fansite the HCF lost to one of these 'icons': Taylor Swift). And, just as the authors argue for their popular Shakespeare, they invoke the inherently popular and connected nature of the Internet platforms they have chosen, calling on their followers to "click the link or make the tweet to vote," because, "we stay silent and no one hears us." The physical gesture of clicking a link is thus aligned with catching the attention of "key players in the world of media and advertising" in order to highlight the contemporary relevance of Shakespeare.

It is by detailing this practical purpose that the Fans thereby distinguish the community's adaptations from what they regard as academia's monopoly on Shakespearean interpretation and its prohibitive requirement of "considerably advanced education." Both the Hollow Crown Fans" and *Shakespeare* magazine's engagement with the playwright mobilises the tools of mass culture in order to recover Shakespeare's inherent (but apparently under-appreciated) cultural capital by separating it from the institutionally and socially legitimated realm of academia. Roberta Pearson's account of the behaviour of fans of "high" cultural figures such as Shakespeare and J. S. Bach demonstrates that this suspicion cuts both ways. Pearson argues that while fan studies have extensively focused on the enthusiasts of popular and middle-brow materials, it has "almost entirely refused to engage with the high." She recounts the suspicion her younger colleagues expressed when she mentioned wanting to study Shakespeare as well as *Star Trek*. These colleagues viewed the study of academically ensconced Shakespeare as a kind of "dangerous apostasy" that threatened to "reinstate ideologically invidious cultural hierarchies" (2007: 99–100). In *Shakespeare*, this position is articulated through an acknowledgement of Shakespeare's high cultural hegemony, but the Fans' call for outspokenness from the Shakespeare community challenges traditional modes of reception assigned to the playwright, aligning him instead with popular cultural practices. Compared to Swift's supporters, who are not only able to but are actually expected to vociferously display their fan identities (shouting, screaming, buying merchandise emblazoned with their star's name and image), a more muted response is associated with theatre or cinema-goers. The HCF thus invite their community to vocalise their Shakespearean fannishness: first, through participating in a competition that could recognise the Fans' (and thus Shakespeare's) compatibility with mainstream culture, and second, through creative engagements with Shakespeare that prioritise both text *and* image.

Cristina Harper (@MCrisHarper), for instance, posts *King Lear*'s statement, "Ay, every inch a king," with an edited still from *The Hollow Crown*'s *Henry IV*. The pairing is, on the surface, an appropriate one, with the text summarising the image of the newly crowned King Henry V. But the post invites more complex thought; not least because the dramatic context within which Lear's line originated is one in which his affirmation is evidently false: Lear's subjects no longer quake; he is not king (1997: 4.6.108). Nonetheless, his statement rings well alongside an image of the Shakespearean monarch who, on a number of occasions, is called to enact his father's lesson on the artificiality of kingship. Harper's cropping of the original still to enhance an already tight close-up on Hiddleston, and use of chiaroscuro to convey introspection, works to further express the potential compatibility of text and image. By making explicit the emotional nature of the still, even to those unfamiliar with the play, the juxtaposition dramatises the contrast between inward emotion and outward show—the irony at the crux of Lear's statement and the complicated performativity of kingship for Henry.

In comparison to Harper's more straightforward response to the #ShakespeareSunday brief, a submission by C. S. Sinclaire (@CSessee) demonstrates not only the potential creative facility of fan practices but their adaptive autonomy and their freedom to engage with different multimedia forms.

Figure 25.1 sees Lucas Cranach the Elder's images of Judith with Holofernes and Salome with the Head of St John the Baptist edited onto a grey, black, and red background, accompanied by the Ambassador's lines to Horatio and Prince Fortinbras at the end of *Hamlet*. The severed heads held by the two women do not represent the only act of cutting that has taken place in the post, however; adaptation here is also transposition, with Judith and Salome severed from their original context. The effect of this recontextualisation and its juxtaposition with the text from *Hamlet* is transformative: the heads displayed become the hapless Rosencrantz and Guildenstern. The bloody handprints on the scroll underline this act of adaptation, working to connect and intensify the violence explicit in the severed heads but which is so vaguely described by the Ambassador. The image presents further nuance to the text, moreover. The portraits exist in an artistic intertext with other representations of biblical women; indeed, unlike the more complex emotions rendered in Caravaggio, Gentileschi, or del Piombo, Cranach's (anti)heroines regard the world outside the painting coolly, with detachment. Cranach's aristocratic women thus readily align with the English Ambassador who, both figuratively and narratively, bluntly delivers Rosencrantz's and Guildenstern's bodies. That Twitter's short-form structure lends potential abstruseness to such creative submissions—resisting definitive interpretation—is in keeping with

Figure 25.1   C. S. Sinclaire's creative interpretation of the First Ambassador's lines from, *Hamlet*, Act 5, Scene 2: "The ears are senseless that should give us hearing,/ To tell him his commandment is fulfill'd,/ That Rosencrantz and Guildenstern are dead./ Where should we have our thanks?"

the community's purpose of returning Shakespeare to the mainstream. An unwillingness to dictate a single path for the reader is congruent with the fan's desire to disrupt traditional cultural hierarchies and challenge an authoritative interpretation.

At the same time as this, however, and in a potential counteraction to its capacity for ambiguity, Twitter's immediacy as a social media platform enables one to reply to specific posts or individuals. With the exception of locked profiles (a rare occurrence in which a user closes their profile to public view), one is encouraged to engage with the platform by following its model of plentiful and spontaneous information by chasing links or profiles and using hashtags. In contrast to platforms such as Facebook, which are instead typically predicted upon the assumption of familiarity, interactions on Twitter can (and frequently are) made with virtual strangers. Tellingly, its 'Who To Follow' function extrapolates information from your browsing history and followers in order to recommend profiles that might match your interests, rather than the social dynamics Facebook maps, which suggests potential acquaintances. Individuals are thereby permitted to engage with profiles that they follow, as well as those that they don't and, unlike the conversational threads common in chatrooms, interested parties can be drawn to conversations or posts that they might otherwise not have seen. The serendipitous chance of viewing a particular tweet at a particular time by a profile that you have chosen to follow can thus be shared and become, instead, a collaborative experience of reading and engagement. The parody account @PopShakespeare, for instance, posted a Shakespearean translation of Meghan Trainor's song, "All About That Bass" ("For I am solely about thy bass, bout thy bass, no mischief"). The tweet was replied to by hazzzzz♡ (@harrietbwhite) who, evidently amused by the post, copied in her friends' Twitter handles in order to ask if it was an appropriate example for their schoolwork. Once alerted, her friends indicated their agreement and general enjoyment of the tweet.

A more comprehensive example of Twitter's potential for collaborative acts of engagement and reading is supplied by Cynthia Sykes (@cynsykes) in a submission to #ShakespeareSunday that also reveals a potential failure of the platform's abstruse format. Sykes posted a GIF of Marvel's *Avengers* alongside the Countess of Rousillon's counsel to her son from act one, scene one of *All's Well That Ends Well*: "be able for thine enemy rather in power than use, & keep thy friend Under thy own life's key" [*sic*]. The quotation, though sharing some of the dramatic context and aphoristic sentiment of Polonius' famous (and often misquoted) instructions to Laertes in *Hamlet*, demonstrates #ShakespeareSunday's frequent sharing of lesser-known Shakespeare plays. Indeed, the play and the quotation's relative obscurity was questioned by Citizen of Whoville (@Mamabear0772) who, directing her reply in a "*whisper*," asked what it meant. Citizen of Whoville's confession of incomprehension and later admission of the difficulty of reading Shakespeare as a dyslexic was not only met with a suggested reading of the line by Sykes but also her recommendation of Open Source Shakespeare, as well as referrals to Cliffs Notes from DirtyGirlLucille (@TWDTwerp) and the No Fear Shakespeare series from Adam's_Vamp (@adamsvamp). Meanwhile, a further admission from Citizen of Whoville in the same conversational thread that she had only understood *Coriolanus* "when someone made it a weekly cartoon here" was met with Sykes questioning, "even after watching that BRILLIANT production?" and posting a meme of Tom Hiddleston as Loki from the same Marvel Cinematic Universe. This reference to Hiddleston, who performed the role in 2014 after having gained fame in Marvel's superhero blockbusters, articulates the nature of Shakespearean capital for this group of Fans as culturally nonspecific and essentially interlinked. Hiddleston is Loki just as much as he is Coriolanus, and the evocation of one aspect of his acting career does not preclude the possibility of the other.

The community's micro-adaptations are thereby not limited to juxtapositions between Shakespeare and other "high" cultural texts, but often indicate their authors' other communal ties, as well the wider creative fan practices that online communities participate in. One #ShakespeareSunday prompt with the joint themes of "demons & darkness" and "spirits & magic," for instance, is accompanied by images of Matt Ryan (Fluellen from Michael Grandage's 2014 *Henry V*) from the NBC comic book adaptation, *Constantine*. The choice of Ryan and producer Christine Boylan to provide the prompt offers some Shakespearean and adaptive links; however, their inclusion sits at odds in comparison to other, more overtly Shakespearean guests such as Kenneth Branagh. But their presence here works to underline the HCF's argument for a mainstream context for Shakespeare by indicating the authors' conversance with and own participation within other pop cultural fan practices. Indeed, both the themes issued by the HCF and the reception of these invitations within the community demonstrate the sensitivity of social media to shifts in the cultural zeitgeist, with submissions registering anticipated releases, popular resurgences, or news items.

This responsiveness of Twitter and other social media platforms demonstrates their uniqueness as adaptive sites: they reflect the larger cultural trends that will have already determined the adaptation of a particular text for screen, or the popularity of one actor over another, while also presenting individual instances of engagement with said adapted texts or trends. It is this immediacy which often explains the origin of any non-Shakespearean Fans content and which ensures that, far from acting as a visual adaptation of the user's chosen quotation, the selection of image frequently precedes the text. The fact that any intertextual connections between Shakespeare and NBC's *Constantine* are potentially tangential or fleeting to some individuals reading the Hollow Crown Fans' Twitter timeline reinforces the nature of the adaptive work which the Fans perform: that is, they are clearly fans rather than individuals claiming any traditional cultural power, right, or authority to Shakespearean interpretation. The transferable, changeable, and adaptable Shakespeare of contemporary culture—the "open signifier," as Lanier describes it—is thus expressed through the culturally omnivorous practices of fan creativity (2007: 94). The disposable texts of mass culture which the fan engages with are, crucially, also the methods of circulation: the mainstream, user-generated content of social media.

A demonstration of this responsiveness can be seen in the Fans' hashtag, #BardBOND. Though not a #ShakespeareSunday theme, the hashtag was first utilised by the authors of the HCF and then adopted by the community, using the same formal principles. It was inspired by the news that that James Bond star Daniel Craig would be starring in *Othello* on Broadway, with the HCF posting an appropriate pairing of an image from *Casino Royale* of Bond and love interest, Vesper Lynd (Eva Green), alongside Cassio's warnings to Othello (with a wry insertion): "Look to her, Bond, if thou hast eyes to see: She has deceived her father, & may thee." Other contributions to #BardBOND demonstrated a similar willingness to adapt the text itself. Rose (@thelifeof_rose) asked, "Is this a Walther P99 which I see before me, The handle towards my hand?" while the HCF account paired the poster for *You Only Live Twice* (1967) with "'A man can die but once … or twice.'—Henry IV pt2." The fruitfulness of this pairing is striking. Beyond referencing the title of the Bond film, the sentiment expressed by Francis Feeble (and its interpolation of "or twice") is a useful encapsulation of a film franchise in which Bond has died numerous times and been reborn to different actors, directors, times, and styles. Indeed, while also gesturing to the titular Henry, who seemingly dies twice on stage, the ironic subversion of Feeble's statement encapsulates the deathless, endlessly circular and repetitive world of adaptation.

The example of #BardBOND provides one final qualification of the collaborative adaptive practice in which social media users such as the Fans engage. This process is, crucially, not simply creative but frequently meta-adaptational in its deliberate framing of the adaptive act. As indicated in the previous paragraph, the alignment of Bard with Bond is certainly a productive one, with adaptations of both figures characterised by their relationship to Britishness (culturally and commercially). Whether as a result of this, or potentially because of the frequent and recip-rocal exchanges between the action film genre and Shakespearean performance (whether in terms of cast or tone more generally), #BardBOND exemplifies the same productive "whirl of intertextual reference and transformation, … recycling, transformation and mutation," identified by Robert Stam in contemporary film adaptation (2000: 66). Contributions to #BardBOND thereby work to reveal the potential visibility of the adaptive process. Julie Bausman (@JuiyCakes), for instance, posts a still from *Skyfall* in which Bond and Q (Ben Whishaw) inspect a new gadget. It is accompanied by the line from *Coriolanus*: "You are never without your tricks: you may, you may." Delivered in the play by the Second Citizen, its re-contextualisation here refers to Q's technical expertise. Its adapted context does, nonetheless, retain some of the antipathy of the original statement for a scene in which a surly Bond informs his youthful advisor, "youth is no guarantee of innovation." And as Graham Holderness (Holderness 2014: 90–1; Blackwell 2014: 344–52) and myself have noted, *Coriolanus* is a productive intertext to the contemporary action film. The argument Holderness makes—comparing the play to *Skyfall* and other action films in its representation of military masculinity—is even echoed by the HCF. Kirsten (@Kirsten_STR) pairs Coriolanus' admission "Look, sir, my wounds! I got them in my country's service" with an image of Bond, bruised by his latest endeavour for MI6. Paul Booth's (2010: 12, 18) summation of the fan as an individual who does "more than passively view media" illuminates the irony of this point: "Fans make explicit what we all do implicitly: That is, we actively read and engage with media texts on a daily basis." Despite their disavowal of an academic Shakespeare, the fan community created by the HCF engage in a similar practice of exegetical reading; their insights, however, are expressed in 140 characters.

The creators detailed in this chapter thus present valuable case studies in the adaptation, understanding, and popular dissemination of Shakespeare's work within a solely digital environ-ment. They produce artefacts which articulate the curious intersections between Early Modern theatre, Internet, and fan culture, and which, at times, echo our own critical work as adaptation scholars. Ben Whishaw's presence in the above tweet by Kirsten underlines the inherently inter-textual and multiplicitous adaptive process at work in the HCF community. As Judith Buchanan states, in cinema "we read character through the determining filter of the specificity of the star" (2009: 224). Buchanan continues, "both screen actor *and* character have a prior existence." Marvin Carlson's extended metaphor of the stage as a haunted space argues similarly for the ghosts of past performances sitting alongside the present in a process of continual "recycling and recollection" (2003: 8). This is not a point limited to film or even theatre, however. Whishaw is to this group of culturally literate Shakespeare fans also Richard II, Ariel, and Hamlet. He is an actor whose presence readily conjures other literary and adaptive intertexts, including *Enduring Love* (2004), *Perfume: The Story of a Murderer* (2006), *Brideshead Revisited* (2008), *Bright Star* (2009), *Cloud Atlas* (2012), and *In the Heart of the Sea* (2015). Intertextuality—implicit in any cultural text—is, therefore, more explicit on social media, with its structure facilitating an immediacy of juxtaposition for its short-form adaptations. A viewer of *Skyfall* may or may not recollect the Shakespearean turns of Whishaw or Ralph Fiennes, but scrolling through the Hollow Crown Fans timeline, one is instantly reminded of these connections. The task of #ShakespeareSunday courts intertextuality, inviting us to think through the visibility of connections between texts and to their significance for our understanding. Its participants actively refute Fiske's imagining of the "Shakespeare buff," moreover. For the most part, their textual knowledge is not used for "discrimination in the dominant habitus" but, as he argues of the traditionally counter-cultural fan practitioners, to participate in the popular (1992: 43).

The Shakespeare evoked in social media adaptations is one that accords with Douglas Lanier's configuration of the playwright, therefore, "[as] one of the very few literary figures who have a double life in contemporary culture," serving "important iconic functions in both canonical and popular culture" (2002: 18). Lanier notes, for those critics "willing to embrace the post-modern tiger" our conception of Shakespeare should be, rather, "*Shakespeares*, a series of culturally specific, multiply-mediated historical events to which any given Shakespearean text is an incomplete and certainly not a regulatory guide" (1996: 188, my emphasis). #ShakespeareSunday or other such instances of Shakespeare in social media is precisely this: an acknowledgement of Shakespeare's mutability through an engagement with the winding, complex, sometimes fruitful, and sometimes fruitless connections between his works and other cultural texts. Social media iterations of Shakespeare thereby present a new framework within which to view and analyse the continuance of his cultural legacy as the latest instance of Shakespeare's presence within culture: the prevalence of some aspects of his representation, including the wielding of his cultural cachet to exert dominance over his dramatic contemporaries, while in other aspects the gradual democratisation of his work. This latter process is one which, I would argue, is facilitated not only by the work of fans but because of the position which they occupy, as individuals engaging with the mainstream in creative *and* critical ways.

Commenting upon the changing nature of selfhood in the digital age, Facebook founder Mark Zuckerberg famously stated, "You have one identity" (van Dijck 2013: 199). Although this statement has been contested by academics in the field of media studies and technology alike, it recognises the inescapability of technological proliferation and the diminishing space between our 'offline' and 'online lives.' Indeed, social media offers performative spaces in which individuals' identities, along with brands, adaptive texts, or authors like Shakespeare are renegotiated and redefined on macro- and microscales. Existing as both units of popular culture and adaptive sites, social media texts such as the micro-adaptations evidenced in #ShakespeareSunday thus represent doubly effective guides to those values associated with Shakespeare's contemporary legacy by users. This is, of course, not to say that Shakespeare's digital presence is without complication or conflict. But social media texts illuminate the events and economic exchanges which underpin Shakespeare's circulation in an increasingly digital world and, as readily accessible, open-access, and short-form texts, they highlight the role technological innovation plays in the process, reception, and understanding of Shakespeare in adaptation. #ShakespeareSunday, after all, is a task that the Fans are invited to undertake both creatively and critically. Its dual function is uniquely permitted by Twitter as a specific digital platform and by the changing landscape of spreadable media more generally, with its alteration of the way in which the public receive, engage with, and circulate cultural texts, companies, or individuals. Not only does spreadable media thus widen our purview as adaptation scholars, presenting us with the possibility of new, macro- and micro-sites of adaptation to study, but it encourages and facilitates engagement with Shakespeare for non-specialists or theatre practitioners on a platform that is immediate and connective. If users choose, they can remain observers, but the nature of spreadable media invites collaboration and participation, calling us to be active readers and providing us with the means to participate in what Booth describes as a "de Certeauan view of *productive* consumption" (2010: 18). We can retweet, reply, follow the example of others, and *create* our own interpretation of the theme.

Notes

1 In the week of 21–28 September, for instance, #ShakespeareSunday was referenced in 2,219 posts, with 9,885 in the month before (www.topsy.com). Arguably the platform best suited to the HCF's circulated format of images and text, Twitter ascribes the HCF 78,900 tweets and 12,600 followers as of October 2015. By contrast, Shakespeare's Globe has 129,000 followers but produced only 29,200 tweets since joining the platform in 2008.

2 In commemoration of the king's reburial in Leicester, the HCF dedicated two #ShakespeareSunday 'special' themes (Admin 2015). The first, "flowers," was chosen by the King Richard III Visitor Centre and the second, 'disgrace and redemption,' by Leicester Cathedral. The HCF's involvement of these institutions in #ShakespeareSunday represents an engagement with those bodies most pivotal in Richard's continuing cultural legacy at a time when the king's narrative was being rewritten.

3 This conversation was sparked by *Shakespeare* tweeting its disappointment that the Fletcher bot changed information about *Taming of the Shrew* into *The Tamer Tamed*, acknowledging that they ."wouldn't have minded, but Fletcher was talentless." Pasunpathi's rejoining comment for *Shakespeare* to ."Stay classy" was then met with the quoted tweet. It is interesting to note that *Shakespeare's* comments were met with amusement from both academic and non-academic Twitter users, who observed variously that *The Tamer Tamed* was a better telling of the Shrew story and that the Fletcher bot had inspired some "interesting conversations" abou the canon (@SLevelt, 2015).

Works cited

Aebischer, Pascal and Nigel Wheale (2003) "Introduction," in Pascale Aebischer, Edward J. Esche, and Nigel Wheale (eds) *Remaking Shakespeare: Performance Across Media, Genres and Culture*, Basingstoke, UK: Palgrave Macmillan.

Aebischer, Pascal and Kathryn Price (2012) "Introduction," in Pascale Aebischer and Kathryn Price (eds) *Performing Early Modern Drama*, Cambridge, UK: Cambridge University Press.

Blackwell, Anna (2014) "Adapting *Coriolanus:* Tom Hiddleston's body and action cinema," *Adaptation* (7)3: 344–52.

Booth, Paul (2010) *Digital Fandom: New Media Studies*, New York: Peter Lang Publishing.

Buchanan, Judith (2009) *Shakespeare on Silent Film*, Cambridge, UK: Cambridge University Press.

Calbi, Maurizio (2013) *Spectral Shakespeares: Media Adaptations in the Twenty-First Century*. New York: Palgrave Macmillan.

Carlson, Marvin (2003) *The Haunted Stage: The Theatre as Memory Machine*, Ann Arbor, MI: University of Michigan Press.

Carson, Christie and Peter Kirwan (2014) *Shakespeare and the Digital World: Redefining Scholarship and Practice*, Cambridge, UK: Cambridge University Press.

Desmet, Christy (2008) 'Paying attention in Shakespeare parody: From Tom Stoppard to YouTube,' *Shakespeare Survey* 61 (2008): 227–38.

Desmet, Christy (2009) "Teaching Shakespeare with YouTube," *The English Journal* (99)1: 65–70.

Egan, Gabriel (2016) "Shakespeare and the impact of editing," in Dominic Shellard and Siobhan Keenan (eds) *Shakespeare's Cultural Capital: His Economic Impact from the Sixteenth to the Twenty-First Century*, Basingstoke, UK: Palgrave Macmillan.

Fiske, John (1989) *Reading the Popular*, New York: Routledge.

Fiske, John (1992) "The cultural economy of fandom," in Lisa A. Lewis (ed) *The Adoring Audience: Fan Culture and Popular Media*, London: Routledge.

Hodgdon, Barbara (2010) "(You)Tube travel: The 9:59 to Dover Beach, stopping at Fair Verona and Elsinore," *Shakespeare Bulletin* 28(3): 313–30.

Holderness, Graham (2014) *Tales from Shakespeare: Creative Collisions*, Cambridge, UK: Cambridge University Press.

Jenkins, Henry (1992) *Textual Poachers: Television Fans and Participatory Culture*, London and New York: Routledge.

Jenkins, Henry, Sam Ford, and Joshua Green (2013) *Spreadable Media: Creating Value and Meaning in a Networked Culture*, New York: New York University Press.

Kidnie, Margaret J. (2009) *Shakespeare and the Problem of Adaptation*, London: Routledge.

Lanier, Douglas (1996) "Drowning the book: Prospero's books and the textual Shakespeare," in James C. Bulman (ed.) *Shakespeare, Theory, and Performance*, London and New York: Routledge.

Lanier, Douglas (2002) *Shakespeare and Modern Popular Culture*, Oxford: Oxford University Press.

Lanier, Douglas (2007) "Shakespeare™: Myth and biographical fiction," in Robert Shaughnessy (ed.) *The Cambridge Companion to Shakespeare and Popular Culture*, Cambridge, UK: Cambridge University Press.

Murray, Simone (2012) "The business of adaptation: Reading the market," in Deborah Cartmell (ed.) *A Companion to Literature, Film and Adaptation*, Malden, MA: Blackwell Publishing.

Naremore, James (2000) "Introduction: Film and the reign of adaptation," in James Naremore (ed.), *Film Adaptation*, London: Athlone Press.

O'Neill, Stephen (2015) "Shakespeare and social media," *Literature Compass* (12/6): 274–85.

Pearson, Roberta (2007) "Bachies, bardies, trekkies and Sherlockians," in Jonathan Grat, Cornel Sandvoss, and C. Lee Harringdton (eds) *Fandom: Identities and Communities in a Mediated World*, New York: New York University Press.

Ramone, Jenni (2011) "Online appropriations: Collaborative technologies, digital texts, and Shakespeare's authority," *Authorship* 1.1 (Fall 2011): 1–15. http://www.authorship.ugent.be.

Reader, Keith A. (1990) "Literature/cinema/television: Intertextuality in Jean Renoir's *Le testament du Docteur Cordelier*," in Michael Worton and Judith Still (eds) *Intertextuality: Theories and Practices*, New York: Manchester University Press.

Richard III (2016), dir. Dominic Cooke. Perf. Benedict Cumberbatch. BBC Two, 25 December.

Shakespeare, William (2013) *Coriolanus* (3rd edition), Peter Holland (ed.), London: Arden.

Shakespeare, William (1967) *Henry IV: Part Two* (2nd edition), Arthur Raleigh (ed.), London: Arden.

Shakespeare, William (1997) *King Lear (The Arden Shakespeare, Third Series)*, R.A. Foakes (ed.), London: Bloomsbury.

Stam, Robert (2000) "Beyond fidelity: The dialogics of adaptation," in James Naremore (ed.) *Film Adaptation*, New Brunswick, NJ: Rutgers University Press.

Van Dijck, Jose (2013) "'You have one identity': Performing the self on Facebook and LinkedIn," *Media Culture Society* (35): 199–217.

Way, Geoff (2011) "Social Shakespeare: Romeo and Juliet, social media and performance," *Journal of Narrative Theory* (41)3: 401–20.

Worthen, W. B. (2008) "Performing Shakespeare in digital culture," in Robert Shaughnessy (ed.) *The Cambridge Companion to Shakespeare and Popular Culture*, Cambridge, UK: Cambridge University Press.

Online resources

Admin., 7 January 2015 "EXCLUSIVE to Hollow Crown Fans…" *Hollow Crown Fans* [online]. Available from: http://www.hollowcrownfans.com/the-hollow-crown/exclusive-to-hollow-crown-fans-the-hollow-crown-succession/ [1 October 2015].

Admin., 11 February 2015 "Shorty Campaign 2015," *Hollow Crown Fans* [online]. Available from: http://www.hollowcrownfans.com/fans/shorty-campaign/ [3 October 2015].

Duffy, C. A. (2015) "Richard III," *The Guardian* [online]. Available from: http://www.theguardian.com/books/2015/mar/26/richard-iii-by-carol-ann-duffy [1 October 2015].

Hollow Crown Fans, 25 October 2014 "Join us on Twitter tomorrow for …" *Tumblr* [online]. Available from: http://hollowcrownfans.tumblr.com/post/100914280543/join-us-on-twitter-tomorrow-for [1 October 2015].

Twitter

17th Earl of Oxford (@EdeVere17).
Adam's_Vamp (@adamsvamp).
British Shakespeare Association (@BSAShakespeare).
Citizen of Whoville (@Mamabear0772)
Cristina Harper (@MCrisHarper).
C. S. Sinclaire (@CSessee).
Cynthia Sykes (@cynsykes)
DirtyGirlLucille (@TWDTwerp)
Emily WeNNceslas (@battielove)
hazzzzz♡ (@harrietbwhite)
Hollow Crown Fans (@HollowCrownFans).
Julie Bausman (@JuiyCakes).
Kirsten (@Kirsten_STR).
Richard III (@richard_third).
Rose (@thelifeof_rose).
Shakespeare (@PopShakespeare)
Sturgess Motor Group (@SturgessCarsLtd).

26

ADAPTATION, FIDELITY AND RECEPTION

Dennis Cutchins and Kathryn Meeks

Defining 'adaptation' has been an obligatory exercise in adaptation studies since the field was created. George Bluestone tried it in *Novels Into Film* in 1957, and he has been followed by dozens of scholars over the last sixty years. Despite the best efforts of adaptation scholars to understand adaptations as artifacts with more or less objective elements, however, texts occasionally touch readers in ways that are idiosyncratic, emotional, and highly personal. In short, sometimes texts are received in ways that defy definition. This is the bare fact of reception, and neither scholarly tradition nor the lack of sufficiently objective scholarly language should bar those who write about adaptation from acknowledging the receptor experience. We urge those who write about adaptations to shift the conversation in such a way that the reader/audience experience may be seen as a legitimate element in any understanding of adaptation. The process of making that shift, we believe, will reveal new aspects of adaptation studies.

We cheerfully acknowledge that this proposal is not revolutionary. In "The Limits of the Novel and the Limits of the Film," George Bluestone's introduction to *Novels Into Film*, he writes, "Differences in the raw material of novel and film cannot fully explain differences in content. For each medium presupposes a special, though often heterogeneous and overlapping, audience whose demands condition and shape artistic content" (1957: 31). Bluestone goes on to cite Sartre's observation that it is not a character's "behavior which excites my indignation or esteem, but my indignation and esteem which give consistency and objectivity to his behavior" (31). Bluestone's brief nod to what was then the nascent field of reception studies is easily lost, however, in his seminal treatise on adaptation. Fifty years later, in *A Theory of Adaptation*, which for many scholars is still the keynote book in the field, Linda Hutcheon seems to return to a reception-based definition of adaptation when she suggests that "we experience adaptations (*as adaptations*) as palimpsests through our memory of other works" (2013: 8). This statement appears to center the perception, and perhaps the definition, of adaptation not on the texts in question, but squarely with the one having the *experience*, the one in whose memory at least two texts exist simultaneously. An adaptation can only be an adaptation, after all, if at least one person perceives it as part of a palimpsest. But a few lines later Hutcheon seems to hedge this teleological position when she describes adaptations with three more structural bullet points. An adaptation, she writes, is:

- "An acknowledged transposition of a recognizable other work or works"
- "A creative *and* an interpretive act of appropriation/salvaging"
- "An extended intertextual engagement with the adapted work".

8

While we would not argue with any of these points, we might observe that the individual who recognizes an adaptation only because of his or her "memory of other works" appears to have slipped into the background with this more operational definition. Who, for instance, is acknowledging or recognizing other works? In whose eyes is the text in question 'creative' or 'interpretive'? And, finally, where does the intertextual engagement mentioned in the last bullet point actually take place? It may be true that the answers to each of these questions lead one inevitably to acknowledge the perceiver, the viewer, the reader, or the experiencer, but Hutcheon seems at least to want a definition that may be applied objectively.

Nico Dicecco reaches a similar moment[1] in his essay "State of the Conversation: The Obscene Underside of Fidelity." There he points out that

> There is a broad consensus among contemporary critics that adaptations require "repetition without replication" (Hutcheon xviii) and depend on an "inherent sense of similarity and difference between the texts being invoked" (Sanders 25). But problems emerge when we attempt to systematically account for where this 'sense' actually inheres, or to detail the theoretical boundary-lines that repetitions must remain within in order to count as adaptation.
>
> *2015: 162*

He then suggests that a useful definition of adaptation might shift "away from what adaptation *is* and towards what adaptation *does*" (162 emphasis in original). An adaptation, Dicecco suggests, is anything that functions as an adaptation for those who perceive it. Just before launching a wonderful case study of Tony Scott's *Enemy of the State* (1998) Dicecco concludes "that adaptation is not essentially a form at all, but rather a potential response to form" (166). The emphasis on reception is unmistakable in this statement, but a few lines later Dicecco seems to join Hutcheon in backing away from a fully-fledged reception-based definition of adaptation. Referencing *Enemy of the State*, he writes, "the possibility arises that audiences might *misidentify* a text as an adaptation and nonetheless experience the same enjoyment" (166 emphasis in original). If adaptation really is "a potential response to form," rather than a form itself, then there could be no such thing as a misidentified adaptation, since anything that functions as an adaptation would be, for all practical purposes, an adaptation. To be fair, Dicecco has reiterated his reception-based definition of 'adaptation' more recently in his theoretically rich essay in Thomas Leitch's *Oxford Handbook of Adaptation Studies*. There he addresses the question, "what would it mean if the 'adaptation' in adaptation studies did not refer to what certain cultural objects *are*, but rather to what certain audiences *do* (in a live and embodied sense) when they engage with the relationship between a work and its precursor(s)" (2017: 609 emphasis in original). For the next few pages we hope to address that question by looking at two very specific audiences—the authors of this chapter.

To be clear, the need for an understanding of adaptation that acknowledges reception is for us less a theoretical matter and more a result of practical experience. In the fall of 2014, Dennis and his wife attended a new adaptation of Jane Austen's *Sense and Sensibility* at the Utah Shakespeare Festival in Cedar City. Dennis writes,

> As we sat together we leaned toward each other and quietly whispered things we had noticed: ways the play differed from other adaptations we had seen or from Austen's

novel. We also observed the aspects of familiar texts, Austen's novel as well at films we had seen, that this production seemed to emphasize. I had interviewed the playwright earlier in the year, and was thoroughly enjoying the adaptation. During the intermission, however, we could not help but overhear the conversation of two couples who were seated at the other end of our row. One of the women mentioned that she was quite interested in watching the second half of the play to learn if Marianne would marry Willoughby or Colonel Brandon. I remember thinking at the time that these folks were in the same theater, and seemed to be enjoying the play as much as we were, yet their experience was completely different. They were not, indeed, witnessing an adaptation.

The experience of Dennis and his wife offers a clear example of the problem with any structural definition of adaptation. The play was clearly an acknowledged and extended intertextual engagement with another text, but without the recognition on the part of the audience it was not an adaptation. Hutcheon acknowledges this situation when she writes, "If we do now know that what we are experiencing actually *is* an adaptation or if we are not familiar with the particular work it adapts, we simply experience the adaptation as we would any other work" (2013: 120, emphasis in original). Perhaps this indicates the potential stability that a reception-based definition may bring to adaptation studies. Recognizing the central role of the receptor reminds adaptation scholars, as Dicecco and Hutcheon suggest, that adaptation is always an experience with a text, rather than a particular kind of text.

Although we feel that a reception-based definition of adaptation is necessary, we are forced to acknowledge that there are practical reason for caution. How, for instance, do scholars writing about particular texts acknowledge a personal dimension in their scholarship and still retain something like objectivity? Even the language of scholarship, with its reluctant use of first person pronouns, complicates this approach. Rita Felski, who stresses the general need for a more personal approach to literary scholarship in *The Limits of Critique*, admits that there is in scholarly circles "an understandable wariness of being tarred with the brush of subjective or emotional response" (2015: 4). Regardless of the potential problems with a reception-based approach to adaptation studies, however, we insist that the failure to acknowledge the element of reception seems to doom any approach to adaptation by miring it in the concrete of structuralism. Until adaptation studies finds a way to understand adaptations in terms of reception, it will continue to chase its own tail in regards to ontological questions.

Indeed, personal considerations inevitably play a role in most scholarly endeavors. As scholars, we often choose to study texts not for logical or empirical reasons, but because those particular texts have touched us or moved us. That moment of artistic connection is central, rather than coincident, both to artistic as well as to scholarly endeavors. Near the beginning of *Art and Answerability*, Mikhail Bakhtin's treatise on the role of art in everyday life, the author observes the uniqueness of every human perception, the isolation faced by all humans, and the role of art in mitigating this isolation. He imagines himself and another person standing near each other and notes, "at each given moment, regardless of the position and proximity to me of this other human being whom I am contemplating, I shall always see and know something that he, from his place outside and over against me, cannot see himself" (1990: 23). This includes "his head, his face and its expression, the world behind his back" (23). Bakhtin labels these things which the other cannot see, "excess" (23). Of course, these hidden visual elements are something of a metaphor for Bakhtin. The excess also includes any perception that is unique or personal, and for Bakhtin the isolation he identifies is not necessarily a cause to lament. At least one of the jobs of art, Bakhtin argues in the course of the book, is to make this excess, this material that is uniquely our own, from the simplest vision to the most complex point of view, available to those around us.

Kenneth Burke reaches a similar conclusion in "Art—and the First Rough Draft of Living." Like Bakhtin, he observes that there is a certain amount of alienation among human beings, and this alienation makes it impossible for us to communicate any experience with exactness. We must, Burke notes, resort to symbols to explain experiences or at least to communicate our feelings about those experiences, and thus reconcile our alienation. Burke explains, "whereas the sciences generalize, the arts particularize" (1964: 163). And here Burke puts his finger on a kind of paradox. The arts, for him, are simultaneously individualized and communal. Through art our human experiences are particularized, and yet that art potentially draws us together. As we identify in families, groups, and communities, what Stanley Fish in "Interpreting the Variorum" labeled "interpretive communities," we find that we can recognize expressive symbols that are common, and we discover similarities between our experiences (1998: 989). Burke writes, "the arts are continually coming up with 'universal' motives, in the sense that people in all times and places manifest the same range of emotions, though necessarily in widely varying situations" (163).

For Burke, then, we seek to express our particular experiences, our "excess," through artistic presentational symbols that potentially connect us to those around us. More to the point, those artistic moments seem closely related to the idea of adaptation. Philosopher of the artistic experience Susanne Langer, in fact, believes that all art is, itself, a kind of adaptation. "Images," she writes, are "our readiest instruments for abstracting concepts from the tumbling stream of actual impressions" (2009: 145). She goes so far as to suggest that "we must adapt all our biological activities. The mind, like all other organs, can draw its sustenance only from the surrounding world; our metaphysical symbols must spring from reality. Such adaptation always requires time, habit, tradition, and intimate knowledge of a way of life" (291). Langer is describing here a process whereby we adapt experiences to art, often narrative art, through a process of abstraction. The central role adaptation plays in this process should not be lost to those who study adaptation, but Langer is quick to insist that the artistic adaptation always marks those who experience it. This may be exactly what Laurence Raw and Tony Gurr point out when they write of adaptation as a kind of "learning" (2013: 1–3).

This learning process, this self-adaptation, if you will, may be the most important reason for a reception-based understanding of adaptation. Marielle Macé argues that, at least in some cases, we do more than learn from the texts we encounter. She believes that we may actually discover the building blocks of self in the stories we receive. "Individuals," she writes, "are not only composed of a body and its non-fungible allocation of time and space, but also of the images we project and receive, the scenes we occupy or reject, the mediations that we appropriate" (2013: 220). The stories we share, along with the stories we read or experience, in other words, may form part of our corpus, our bodies. Images, scenes, and mediations occasionally "bring profound rearrangements, such as, to take one example, the books that precede us, through which we simultaneously invent, recognize, or lose ourselves" (220). Paul Ricoeur suggested more than thirty years ago in his *Time and Narrative* series that we create ourselves with the stories that we tell (1984: 3), but Macé seems to be taking Ricoeur a step further, and arguing that we actually create ourselves in part by the stories we hear, read, or view. Fish may have recognized the transformative power of reading when he suggested that the reader's job was "to discern and therefore to realize (in the sense of becoming) an author's intention" (1998: 983).

Rita Felski makes a similar case in *The Limits of Critique*. She proposes that "We make ourselves out of the models we encounter; we give ourselves a form through the different ways we inhabit other forms. And we bring these differences to the event of reading, even as we are reoriented—sometimes subtly, sometimes significantly—by the sum of what we read" (2015: 172).

"Works of art," she suggests, "do not only subvert but also convert; they do not only inform but also transform—a transformation that is not just a matter of intellectual readjustment but one of affective realignment as well" (17). She argues that this dynamic relationship helps explain the profound attachments we often feel for the literature we read, the stories we encounter. We may have no objective way to judge the effect a text has on us, but that does not mean the effect is not real. In what may be the most delightful application of actor–network theory we have read, Felski writes, "The 'actor' in actor–network theory is not a solitary self-governing subject who summons up actions and orchestrates events. Rather, actors only become actors via their relations with other phenomena, as mediators and translators linked in extended constellations of cause and effect" (164). And for Felski, actors include readers and viewers: "A work of art is a potential source of knowledge rather than just an object of knowledge—one whose cognitive impact and implications are tied up with its affective reach. We are intertwined and entangled with texts, in ways that require further consideration" (84).[2]

The notion that those who listen to stories, read books, or watch films or television are actually in the process (at least potentially) of self-creation is potentially game-changing for adaptation studies. Perhaps the best way to grasp this idea is through the story of an experience. Below Katy describes how a particular story has become "intertwined and entangled" in her life.

In 2013 I returned from 18 months living in the slums of São Paulo, Brazil where I had been serving as a missionary. I flew home to live with my family in the Middle East for several months before returning to school. I had anticipated that the release from the pressures of missionary work and being reunited with my family would make me feel relaxed and relieved. Instead, I found myself profoundly depressed and feeling an immense amount of self-inflicted guilt.

During that dark time, a close friend gave me Norman Maclean's *A River Runs Through It* which he said was the "finest piece of American literature" he had ever read. I couldn't imagine how a story of men and fishing could mean anything to me—a young woman living far from the forested lakes and rivers of Montana. But Maclean begins his novella by describing his father holding his fly-fishing rod right "where it trembled with the beating of his heart" (2001: 2). For reasons that, to this day, largely defy explanation, I was touched by this sentence, and by the sentences that followed—they had an almost numinous quality for me.

In *A River Runs Through It* Maclean created something his father would call "beautiful." The story, written as a reminiscence, follows brothers Norman and Paul to Paul's untimely death. Maclean's minister father taught, "all good things—trout as well as eternal salvation—come by grace and grace comes by art and art does not come easy" (4). My moment of grace in a dark time came through reading Maclean's story, his written art. And to this day, my memories of the novella are painful because they are tied to a time of dark depression and the loss of someone I loved deeply. At the same time, Maclean's story was a source of light and healing—it was a beautiful text that helped me understand my own relationships.

In 2016 I watched Robert Redford's *A River Runs Through It* (1992) for the first time. I was nervous because I had actually tried to watch the film a year earlier but I had turned it off since it reminded me of the friend who had recommended Maclean's novella—a friend who I had lost. But when I returned to Redford's film I was pleasantly surprised. The film's Academy Award for Best Cinematography and nominations for Best Music, Original Score, and Best Adapted Screenplay were well-deserved. I enjoyed it, but as good as it was, this gorgeous film simply did not replicate the experience I had with the novella three years earlier. The film, of course, differed from what Maclean had crafted in his novella. I recognized that the creators of the film deserved space for interpretation and so these differences were intellectually acceptable to me. What was less acceptable, however, was that my *experience* with Maclean's novella was different from my *experience* with Redford's film. My depressed and lost condition at the time I picked up the novella set the stage for me to find release and an escape in Maclean's novella. And, as I was in the Middle East, very far away from my friends, this novella was a connection to a friend I love deeply. This context and setting is in contrast to my experience with Redford's film. Much in my life had changed by the time I viewed Redford's film on a warm, sunny, afternoon as a school assignment for course in adaptation studies. Redford did an excellent job using cinematic tools in an effort to create a wonderful film. Yet, Redford could not possibly recreate what Maclean's narration and descriptions did *for me* when I read them. Even now I still feel that the beautiful prose of the novella actually healed me. Sometimes Maclean's descriptions of the land, fly fishing and the people he loved did not even completely sink in, sometimes they just allowed me to feel something other than darkness. Maclean took me with him on a long afternoon fishing with his brother Paul. I watched them wade through the river and call out to each other. I watched Maclean stop and clean his Eastern Brook Trout and lay them in his "basket between layers of wild hay and mint where they were more beautiful than those painted on platters" (2001: 39). When I closed my eyes I could see the shiny scales of the fish, and I could smell the hay and mint.

No matter how carefully it was filmed, a camera could not capture this serene and sensory moment I experienced. At the end of an afternoon fishing, Maclean's narrator notes, "I glanced at the sky which I had forgotten about since the world had become no higher than a bush" (45). As I read I knew exactly how it felt to be so immersed in a task that everything else faded away. I was wrapped up in Maclean's world, and as much as I try, it is difficult to describe how these simple and sublime images and impressions made me feel whole then. Only now do I look back and understand that this release was exercising my heart at a time when I thought my heart had stopped. Wandering the land of western Montana and being integrally involved in Maclean's longing to be close to his lost brother was cathartic in the truest sense of that word. I certainly enjoyed the cinematography and the score of Redford's film, but the effect was simply not the same.

The different experiences Katy describes having with the novella and the film likely have less to do with the difference between film and literature or with the inherent qualities of the two texts and more to do with Katy, herself. How, indeed, could her two experiences have been the same? Novella aside, Redford could not possibly replicate on the screen what Katy had experienced while reading Maclean's novella at a particular moment in her life. In fact, even rereading Maclean's novella again would be unlikely to create the same experience. And this suggests two points, both of which are important to understanding adaptations. First, people actually build parts of their lives on experiences with texts, as Katy's story seems to indicate she did. This idea, as we discuss in the next section of this paper, has to be acknowledged when we study adaptations. When that happens, of course, understanding the text objectively becomes practically impossible. As particular textual experiences are embedded in our lives, the attachments we feel for those texts may be based on reasons that are quite personal and, at least sometimes, may even be inexpressible. The second point is suggested by the first. The fidelity response that most of us have felt at one time or another, may be based on experiences similar to Katy's, and may have little to do with fidelity to a text and everything to do with fidelity to an experience.

We would like to pause here to note that we are not arguing for the primacy of literature over any other medium. For years scholars have assumed that readers of books have felt frustration with adaptations because the reader/viewer has discovered differences between the primary text and the secondary text—typically expressed, "the book was better." As Frances Bonner and Jason Jacobs, among others, have pointed out, however, the "primary" text may not be the first one created, but is usually the first one experienced (2011: 38). For most of us, for instance, the primary text dealing with the land of Oz is not Frank Baum's *The Wonderful Wizard of Oz*, published in 1900, but rather Victor Fleming's *The Wizard of Oz* (1939).[3] Neither are we arguing for the quality or value of one text or one medium over another, but rather for the significance of one experience over another, and experiences are always idiosyncratic. The distinctive nature of experience with text and medium is illustrated above in Katy's experience and again here in an experience Dennis had. He writes,

I saw Phil Alden Robinson's *Field of Dreams* (1989) in the theater when it first came out. Having lived during high school in a small, Illinois town surrounded on four sides by cornfields I instantly recognized the rural landscapes of the film. Cornfields serve as the backdrop for half the shots in the film, and actually play an integral role in the story. But I was infinitely more moved by the vexed family relationships portrayed in the film. Ray, the movie's protagonist, played by Kevin Costner, is struggling with his marriage, his money, and his in-laws. Voice-over narration soon reveals that his relationship with his own father was also strained to the breaking point before his father's death. That death meant that apologies and reconciliation were utterly impossible for Ray.

Though I don't remember thinking about it at the time, looking back on the experience I believe I must have seen something of my own life in the film's characters and situations. My middle name is Ray, though I don't imagine that coincidence played much of a role in my experience watching the film. But in 1989 my wife and I had been married a few months, and like most newlyweds, we were struggling with money. I was learning how to get along with my new wife, who by another coincidence resembles Amy Madigan, the actress who plays Ray's wife in the film. I was also learning to get along with her large family. Unlike Ray, both of my parents were alive in 1989, but they had divorced just before I started junior high school, and the break up was difficult. My relationship with my father, never warm when I was growing up, was finally maturing into something closer as I was learning to see him as a person, not just as a parent. Perhaps all of this helps explain why, to this day, I am moved to tears during the film's final moments as Ray discovers that the "he" of the film's famous catch phrase, "If you build it he will come," is actually his father. I don't think I've ever watched the scene of Ray playing catch with his long-dead father except through tears. Friends and colleagues sometimes scoff at the film's unlikely and admittedly corny magical-realism. But it never fails to move me.

In graduate school I finally read the novel the film was based on, W. P. Kinsella's *Shoeless Joe* (1983). More complex and layered than the film, the novel is a good read. I've actually taught it in my classes and even published an essay about its relationship with J. D. Salinger's *Catcher in the Rye* (1951), but the novel has never touched me emotionally in the same way the film does. Both the film and the novel center on the theme of second chances, but somehow the film has always seemed more direct and concrete and perhaps more real to me.

Dennis's story suggests that attachments to characters or fictional situations may have less to do with medium or performance, and more to do with the way an audience member relates to those characters or situations. Dennis mentions the coincidences of his middle name and the fact that his wife resembles Amy Madigan, but even small and utterly incidental moments of recognition like these can make a world of difference. Linda Hutcheon may have acknowledged this when she argued that adaptation, "like classical imitation … is not slavish copying; it is a process of making the adapted material one's own" (2013: 20). And that sentiment applies to both artists and audience members. What Aristotle called imitation or "mimesis" does not occur exclusively on the stage, screen, or page. Mimesis also occurs inside the reader. Something about the plight of a character, the plot development, the mastery of literary language, or even something as small as a character's hair color speaks to the reader/viewer and allows them to feel a connection, a moment of recognition or ownership. Empathy, after all, may be considered another form of replication, since in order to experience it we have to put ourselves in 'their shoes,' as the common phrase goes. Mimesis, then, may be more than just the imitation of life on stage or in art; mimesis is also the replication of life, someone else's life, *within* the reader or viewer. Ricoeur makes a similar, though more complicated point in *Time and Narrative* vol 1 when he suggests, "the reader is that operator par excellence who takes up through doing something—the act of reading—the unity of the traversal from mimesis$_1$ to mimesis$_3$ by way of mimesis$_2$," (1984: 53).

This may be exactly what Bakhtin was getting at when he wrote that art in general, and stories in particular, make the receiver, the listener, the reader "answerable" or responsible in some ways to both the teller and the story. "I have to answer with my own life," he writes, "for what I have experienced and understood in art" (Bakhtin 1990: *Art and Answerability* 1). He goes on to argue that in order to truly apprehend a text, "one must enter as a creator into what is seen, heard, or pronounced, and in so doing overcome the material" (305). "I encompass [a text]," he writes, "give it a form, and consummate it" (306). This oddly intimate image is echoed by Pablo Picasso, who once expressed a similar idea in an interview. He noted that when a picture is "finished, it still goes on changing, according to the state of mind of whoever is looking at it. A picture lives a life like a living creature, undergoing the changes imposed on us by our life from day to day. This is natural enough, as the picture lives only through the man who is looking at it" (Cooke 1978: 60). Because we consummate some texts, make them a part of ourselves, interpretations or adaptations that do not match our own, often idiosyncratic, creative readings of those texts are not simply challenges to the text, but personal challenges—threats to our very selves. They tread, as a psychologist friend once said, on the sacred ground of self.

Without denying the importance of Stanley Fish's interpretive communities, the profoundly personal ways we interpret some texts, built as they are on the foundation of the self, may be the most important reason for understanding adaptation in the light of reception. In 2010 Christa Albrecht-Crane and Dennis Cutchins published a collection of essays on adaptation. In the introduction to that collection, they argued that "a stubborn insistence on fidelity certainly has kept adaptation theory from maturing" (2010: 12). They were not alone in their critique of fidelity criticism. Robert Stam had said something similar in 2005, Linda Hutcheon in 2006, and Tom Leitch in 2007.[4] Their argument was sound, and likely still applies when attempting to understand the relationship of one text to another, but it was also wrongheaded in some important ways. In their critique of fidelity, they considered the relationships of texts to one another, but failed to consider the relationships texts have with people. They failed, in other words, to consider the central importance of reception and the necessity of acknowledging the role audience plays in any theory of adaptation. One of the reasons the specter of fidelity continues to resurface in the adaptation conversation is because readers and audience members who experience adaptations respond, at least sometimes, based on a perception that the adaptation has either been true or untrue not to an 'original' text, but to their original experience with a text. This, then, is a brand of fidelity that is both inescapable and utterly invaluable to adaptation studies. Citing Roland Barthes, Marielle Macé believes that reading is "'rewriting the text of the work within the text of our lives,'" and this kind of connection, this fidelity, if you will, is one of the main reasons human beings seek art in the first place (2013: 217).

So what does our insistence on recognizing both the roles texts play in the lives of real human beings, and the roles human beings play in the lives of texts mean to the practice of adaptation studies? What is the practical application of what we have been discussing? Perhaps the first consequence of recognizing the personal ways texts often touch us is accepting the often-irrational nature of our relationship with texts. Unlike scholarly language, stories such as the ones we have recounted here *bear testimony* instead of provide data or create an argument. Readers might question a storyteller's motives, doubt their sincerity, or even find some truth in a recounted story, but stories are difficult to ignore. We would add that adaptations will also be judged by the creativity, engagement, and interpretive skill of the audience. This notion of adaptation as a radically personal experience seems to agree with our own experiences we have noted above. The third-person, objective language we typically use in scholarly discourse certainly does not lend itself to a discussion of this highly personal reception, but, more to the point, it ignores what are often the most important elements of a text. To remedy this incongruence, we encourage more personal experience and language in scholarly writing about adaptation. When scholars write about adaptation, they should allow themselves to be present. They should acknowledge and perhaps even embrace their own subjective responses to the texts in question. After all, at least part of what draws us, both as scholars and as people to these texts, are the personal connections we feel with characters, situations, music, places, events, media, and so on. As we suggest above, most of our scholarly endeavors are at least partially based on personal experience, and so our writings about a text should reflect at least a few of these personal elements.

A second consequence of defining adaptation in light of reception is recognizing that all texts are not created equal. Some texts are more important to us than others. That seems an obvious point, but in literary studies in general and adaptation studies in particular we tend to ignore affect and thus inadvertently flatten our perception of texts. We recognize that our experiences with texts change over time. As we mentioned, we may not have the same or even a similar experience when we re-encounter a text at a different time and in a different space. But the fact that our experiences with a text are bound to change does not negate the fact that we may have had an emotional and impactful experience with the text at one point that, to some degree, remains in our memory. Some texts remain with us, and are thus more important to us than others. The strong fidelity response we have to some adaptations, we believe, stems from this embedding.

We began this chapter by hinting that it may be time for adaptation scholars to reconsider their relationship with fidelity. And we're not talking here of textual fidelity. That was always illusory. Rather, we are talking about a fidelity of reception: a faithfulness to the experiences with texts and fragments of texts that are embedded in our lives, and which actually help structure our lives. The reason audiences sometimes choose to describe adaptations with words like 'faithful,' 'true,' or 'betrayal,' is because their personal experiences with texts are potentially very powerful. These emotionally loaded words are not too strong to express the feelings that one is being personally attacked by an adaptation. The fidelity response makes much more sense once it is viewed through the lens of the receptor and the powerful nature of stories and personal experience is acknowledged. We are not arguing for a re-crowning of fidelity criticism, but rather a reconsideration of what fidelity means in adaptation studies.

Notes

1 See also Cassie Hermansson's article, "Flogging fidelity: In defense of the (un)dead horse" (2015).
2 Kenneth Burke suggests an analogy that acknowledges this central role stories potentially play as the building blocks of our lives. He says that our lives are like the rough draft of an essay "hastily organized" (1964:161). Burke makes the point that we cannot revise our essay of life as we are unable to review our notes or rewrite. However, he believes that we can encounter art that allows us to gain a broader sense of the human experience. Art becomes our tool to "*personally* consider many more possibilities than we could otherwise" (162). We can learn from our responses, thoughts, and imaginations of art how certain events and possibilities would transpire in our own "rough draft" of living. Burke says, "You'll see all kinds of personal experiments worked out for you, projected at times even into their most perfectly conceivable developments" (161). Our lived experience, or our 'rough draft,' becomes much richer as we encounter art, and despite our inability to revise and re-do, we can, through art or stories, draft a life that builds and deepens as it is being written. Burke explains, "The great advantage I see in the arts is their ability to make us feel such shifts of attitude not merely from without but from within … only the arts can saturate themselves with such changes of attitude imaginatively, personally. And surely this is their great virtue, as regards their contribution to the arts of living" (158–59).
3 Thanks to Kate Newell for suggesting this example
4 See Stam's *Literature Through Film* (2005: 3), Hutcheon's *A Theory of Adaptation* (2013: 4), and Leitch's *Film Adaptation and its Discontents* (2007: 3).

Works cited

A River Runs Through It. (1992). [DVD] USA: Columbia Pictures.
Albrecht-Crane, C. and Cutchins, D. (2010). *Adaptation Studies: New Approaches.* Madison, NJ: Fairleigh Dickinson University Press.
Bakhtin, M. (1990). *Art and Answerability.* Austin, TX: University of Texas Press.
Bluestone, G. (1957). *Novels into Film.* Baltimore, MD: Johns Hopkins Press.
Bonner, F. and Jacobs J. (2011). "The first encounter: Observations on the chronology of encounter with some adaptations of Lewis Carroll's Alice books." *Convergence*, 17(1), pp. 37–48.
Burke, K. (1964). "Art—and the first rough draft of living," *Modern Age*, vol. 8, no. 2, pp. 158–63.
Cooke, H. (1978) *Painting Techniques of the Masters.* New York: Watson Guptill Publications.
Dicecco, N. (2015). "State of the conversation: The obscene underside of fidelity." *Adaptation*, 8(2), pp. 161–75.
Felski, R. (2015). *The Limits of Critique.* Chicago, IL: University of Chicago Press.
Field of Dreams. (1989). [film] Hollywood: Phil Alden Robinson.
Fish, S. (1998). "Interpreting the Variorum." *The Critical Tradition: Classic Texts and Contemporary Trends.* 2nd ed. Ed. David H. Richter. New York: Bedford St Martin's, pp. 977–90.
Hermansson, C. (2015). "Flogging fidelity: In defense of the (un)dead horse." *Adaptation*, 8(2), pp. 147–60.
Hutcheon, L. and O'Flynn, S. (2013). *A Theory of Adaptation.* 2nd ed. New York: Routledge.
Kinsella, W.P. (1983) *Shoeless Joe.* New York: Ballantine Books.
Langer, S. (2009). *Philosophy in a New Key: A Study in the Symbolism of Reason, Rite, and Art.* 3rd ed. Cambridge, MA: Harvard University Press.
Leitch, T. (2007). *Film Adaptation and its Discontents.* Baltimore, MD: Johns Hopkins University Press.
Leitch, T. (2017). *Oxford Handbook of Adaptation Studies.* New York: Oxford University Press.
Macé, M. (2013). "Ways of reading, modes of being." *New Literary History*, 44, pp. 213–29.
Maclean, N. (2001). *A River Runs Through It and Other Stories.* 25th Anniversary ed. Chicago, IL: University of Chicago Press.
Raw, L. and Gurr T. (2013). *Adaptation Studies and Learning.* Plymouth, UK: Scarecrow Press.
Ricoeur, P. (1984). *Time and Narrative,* Vol 1. Chicago, IL: University of Chicago Press.
Salinger, J.D. (1951). *The Catcher in the Rye.* New York: Little, Brown and Company.
Stam, R. (2005). *Literature Through Film.* Malden, MA: Blackwell Publishing.

PART V

Technology

Eckart Voigts

Technology is pivotal to any form of text processing and production no matter what the level of sophistication may be. Thus, adaptation must be, necessarily, intertwined with and embedded in uses and displays of technology. This can, of course, vary depending on the medium and its specific technology. When *The Lion King* is adapted from film to the stage, for instance, animation changes to technologies surrounding puppetry. But even within animation film, various technologies, from stop-motion animation to animatronics or 3D computer animation, shape its media history, as in the case of the BBC–Netflix remake of *Watership Down* in 2017. Frequently, the display and employment of technologies in adaptations is pivotal to their genesis, as is the case in, for example, game-based, web-based, and virtual adaptation. This section aims to investigate the relationship between adaptation and technology not only with regards to specific adaptations, but also asking whether, and to what extent, the nature of technology available shapes the process, the product, and the reception of adaptation. While many papers in this section eschew notions of technological determinism, many focus on moments of technological transition: they show how crucial moments such as the beginnings of film, the shift from silent film to talkies, or the advent of digital games are illustrative of changing modes and approaches in adaptation. Some chapters address the effects that the advances of digital, networked technologies have had on adaptation (Stobbart: video games, Cochrane: streaming, Voigts: remix and social media), but another focus lies on the remediation of earlier technologies (Cochrane and Gaudreault/Marion: theatre and early film; Goggin: literature, illustration, magic lantern shows; Grossman: monochrome silent film).

Joyce Goggin discusses the case of Charles Dickens – traditionally thought of as a key example of 'visual' writing transferred to a 'visual' medium. Her investigation has wider implications for the study of media representation of thought processes. Medium-inherent properties posited by the distinction between 'temporal' and 'spatial' arts are Goggin's starting point. What emerges in the case of Dickens is a complex net of media cross-pollination that shapes early Dickens adaptations, including conventions such as thought bubbles in illustrations or comics and magic lantern shows. Dickens adaptations thus become a key example of how complex and permeable networks of text and image are, contrary to linear notions of literature-into-film models.

The question of fidelity is the starting point in the essay by André Gaudreault and Philippe Marion. They propose the terms *adaptogénie* and *médiagénie* to account for the spectrum between a highly adaptable text and one that is intrinsically linked to its specific media setup. Gaudreault and Marion focus on the intermediality of early cinema and theatre and sketch the minute transitions which emerged when the 'kinematograph' assumed an awareness of its own mediality. In their 'ecological' perspective on adaptation, the authors show how, beginning as a mere auxiliary to pre-existing genres, early cinema adapted to the new media 'ecosystem' at the beginning of the twentieth century.

The next chapter shows the wide gulf between a discussion of theatre and film under the vastly different media conditions at the beginning of the twenty-first century. In the case of Bernadette Cochrane's contribution, the relationship between theatre and film is re-cast in view of contemporary satellite and streaming technologies and high-definition cameras. Cochrane debates the losses (immediacy, liveness, and interaction) and gains (accessibility) in adaptations of theatre to the event cinema of live relays. The chapter points out that live transmissions of theatre into cinemas do not simply follow the 'theatre into film' paradigm in adaptation, but enforce the textual authority of the theatre productions. Cochrane also illustrates the way in which live relays adapt theatre audiences as cinema audiences, which experience different texts and materialities.

In some ways, Julie Grossman also addresses the early film that is the focus of Goggin and Gaudreault/Marion – her example is a case of intentional eschewal of available technologies as a deliberate aesthetic strategy. Her example, Charles Lane's silent movie *Sidewalk Stories*, anachronistically breaks out of its technologically defined conventions in order to throw contemporary viewing practices into sharp relief. Remaking the supposedly obsolete Charlie Chaplin style, the film is shot in black and white, making viewers aware of technological connotations of 'old' and 'new'.

Malcolm Cook and Max Sexton note the surprising persistence of notions of medium specificity in adaptation studies, and they approach this question from a perspective of technology. Cook and Sexton respond to the historical contingency and ontological instability of medium specificity by restricting their discussion to a distinct period (television drama from the late 1940s to the early 1960s) and a specific region (the United Kingdom). Technology remains central to this discussion, but does not determine it. Just like Shannon Brownlee (see her contribution in this volume) they conclude that medium-specificity is a problematic concept, and they highlight the fact that the technological dimension of adaptation is always dependent on a historically determined cultural, social, and political situation, as well as on institutional frameworks and aesthetic preferences.

While discussions of visuality dominate Cook/Sexton's analysis of television drama, Richard Hand's chapter demonstrates how fruitful a foray into radio drama can be for adaptation studies. Both a purely visual medium (Grossman and Voigts, for instance, discuss the absence of sound) and a purely sonic one have special heuristic potential. Not only have practices of adaptations dominated radio drama from its inception but also its 'visionlessness', while ostensibly bringing it down in the sensory hierarchy of adaptations, can offer fascinating versatility, as Hand's case studies amply illustrate.

While several papers discuss how we see adaptations, Hand's chapter reminds us that we can also hear adaptations, and Dawn Stobbart's contribution points out how we can also play adaptations. Stobbart notes in her chapter that video-games studies have yet to fully engage with the adaptive functions of video games as remediations of other media. Stobbart, however, discusses many productive case studies, from Lewis Carroll's *Alice's Adventures in Wonderland* as a computer game, to the way the 2007 video game *BioShock* adapts Ayn Rand's 1957 novel *Atlas Shrugged*. Stobbart's contribution illustrates how video game adaptations have evolved from clumsy beginnings to remediate a variety of media (such as other games, text, cinematic scenes, music, and more) to create complex new forms of adaptation.

Finally, Eckart Voigts reviews the effects that the circulation of remixes, mashups, and samples in social media have on adaptation studies. His case study of animated GIFs, while clearly dependent on the 'recombinant appropriation' facilitated by digital technologies, is reminiscent of the early cinema of attractions, and thus might be viewed as yet another case of the inextricable linking of old and new technologies.

27

ADAPTATION FROM THE TEMPORAL TO THE SPATIAL

Materialising Dickens's imaginings

Joyce Goggin

Introduction

This chapter discusses a significant element in the adaptation of narrative content from one medium to another, namely the reproduction of subjective thought processes, along with the conventions that have developed around it, and why. Representing thought as, for example, through the novelistic convention of interior monologue, is a job that prose is often said to do particularly well, while film has developed other means of visualising and communicating characters' thoughts, such as the classic voice-over. In what follows, I will delve into this aspect of adaptation, taking Dickens as my source-oriented study, and focusing on various means that he and other artists have experimented with over time for the representation of interiority, from illustrations, to magic lantern shows, and film. In doing so, I am attempting to go beyond the parameters of the single text-to-film case study by looking at various modes of adaptation and how they have attempted to represent the inner musings of both Dickens and his fictional characters. Thus, rather than taking a singular case-study approach, in this chapter the emphasis will be placed on the history of adaptation in this one key area.

This essay will also address how adaptation – in this instance the adaptation of Dickens's narrative properties – is necessarily intertwined with and embedded in uses and displays of technology from page to lantern to screen. I will therefore investigate the relationship between adaptation and technology, not only with regards to specific adaptations, but also with regards to how the nature of available technology has shaped the process of adapting and portraying thought in the media just mentioned.

For some, the focus of this discussion as I have just outlined it will raise the red flag of medium specificity. However, while I would certainly argue that the medium in or on which narrative content reaches us (printed text consumed solipsistically, a shared viewing of magic lantern slides, cinema screenings, serial binge viewing, interacting with an avatar in a videogame world) is of the utmost importance, I would also argue that specificity is always blurred with or bears the trace of other media platforms.[1] Given this, I prefer to see the historical development that I am describing here as part of a continuum, rather than a matter of neatly cordoned off, discrete platforms that do not mingle and merge various aspects of how they tell stories. In what follows here, then, I want to focus on how story or narrative content moves from one platform to the next, taking with it some features of what came before or along with what is being remediated. Therefore, I will be concerned with one feature that seems regularly to come to the fore in the representation of thought across the media under discussion here, and that appears to remain more or less stable when thought is represented in various media, namely metaphor and metonymy. This, in turn, will entail a discussion of adaptation from so-called temporal media, such as the nineteenth-century realist novel, and how such texts are then adapted into spatial arts, such as film wherein the passage of time is necessarily condensed. As I will attempt to show, the representation of thought from text to various other media is accomplished by means of conventions, such as thought bubbles in illustrations and magic lantern shows, and how thought is later expressed in film as the image becomes animate.

Adapting thought

When Foucault asked, "What Is an Author?" back in 1969, he could scarcely have predicted what has happened to the figure of the author in the wake of postmodernism. In his time, Foucault (2010: 1481) argued that the name of the author amounts to a function that "serves as a means of classification," holding "together a number of texts … thus differentiat[ing] them from others." In that same essay, Foucault (2010: 1489) also argued that the romantic concept of authorship, based on "tiresome" notions of authenticity and originality, is a fairly recent one, and one which was well on its way to desuetude by the time he was writing.

More recently however, critics have begun tracing the emergence of a sharply defined and carefully managed author figure whose consciousness functions as the seat of textuality, hence Simone Murray's answer to Foucault's closing question – "what matter who's speaking?" (Foucault 2010: 1490) – in her 2012 book on the adaptation industry. Here, Murray analyses the figure of the contemporary author which, far from being deconstructed and dispersed into the discursive diaspora, has persisted and been reborn as the unique source of any given text. As she writes, Foucault, as well as Barthes, who also famously wrote about the death of the author in 1968, were both arguing from a "dematerialised view point" and understood authors "almost solely as sites of hermeneutic and aesthetic confrontation between literary critics" rather than as "creative professionals with artistic and commercial motivations of their own" (Murray 2012: 30). As she points out, the Barthesian and Foucauldian versions of authorship are increasingly under pressure in the face of the current importance of intellectual property rights and authorial image management in the literary market, which "fundamentally undercuts the pervasive anti-bourgeois, anti-capitalist rhetoric of both essays" (Murray 2012: 30).

This same point could be argued by looking at Genette's notion of paratext, that is, those bits of 'textuality,' such as titles, frontispieces, illustrations, authors' portraits, prefaces, and other addresses to readers, which somehow fall outside of the threshold of what would generally be considered the principle text (cf. Genette 1997). Were one to look at some of Dickens's work from the perspective of Genette's notion of paratext it would become evident that, far from being commensurate with a sort of generic author function, Dickens was very actively involved in branding his work through various forms of paratext, including the many products that his fiction spawned, and branding himself as author through portraits published in various newspapers as well as his own periodicals, *All the Year Round* and *Household Words*.[2] Given this, it is of particular interest to the present essay on the topic of technology and adaptation to note how often readers were encouraged to anchor Dickens's work in the author as personality, focused through images and photographic portraits of Dickens, engaged in thought.

Such images, produced both during and after Dickens's lifetime, provide us with a sort of voyeuristic view of the man, alone in his study, arrested in the act of bringing to the page those household words for which readers waited so impatiently. And, as Grahame Smith (2003: 7) wrote, it is this packaging of Dickens that "inspir[ed] all his constant readers … with a sense of habitual dependence on their contemporary, the man, Charles Dickens, for a continuous supply of entertainment which only he could furnish."

Dickens is certainly not the only author of popular serialised fiction in the nineteenth century who has been represented in the act of writing, and Kamilla Elliott (2012: 182–4), for example, has discussed representations of Thackeray composing *Vanity Fair* captured in early film adaptations of his work as a sort of guarantor of authenticity. Indeed, this kind of authorial endorsement or branding arguably got under way late in the eighteenth century, but what distinguishes Dickens here is how he and others staged representations of the great man as author. Dickens's singularity in this regard is the preponderance of images of the author not writing, precisely, but rather deeply engaged in thought, and this is interesting for a variety of reasons. The compelling suggestion here is that Dickens's thought process, his imagination, is somehow rendered transparent and immediately available to viewers and readers in the form of his narrative fiction, for which these images function as a mark of authenticity. One result of this practice is that many of Dickens's characters take on an afterlife as branded commodities or iconic cultural figures and continue to be linked, at least metonymically, back to the brain of the author whence they emerged, so that Dickens's characters and stories are forever connected to iconic images of the writer.[3]

Thought is, however, obviously and thankfully invisible, which is why we are able to entertain untoward thoughts in the company of others with impunity. Yet one of the many distinctive pleasures of novels of various genres is precisely their capacity to create the voyeuristic illusion of access to the intimate thoughts of others. In novels, this function is usually taken up in some form of omniscient narration, frequently involving free indirect discourse. As Kittay and Godzich argued in *The Emergence of Prose* (1987), prose fiction that relies on free indirect discourse arose out of various literary traditions, and out of multiple forms of text, including verse, letters, and so on. According to them, the particular strength of prose, and indeed the secret to its enormous, sustained success as a form of entertainment, is "the way [that] prose withholds itself from view" (Kittay and Godzich 1987: 197), so that any marks of the absent subject who wrote or performed the story are present in a "writing whose only marks, whose only 'styles,' are to be those discourses it contains and frames" (Kittay and Godzich 1987: 196). Point of view in prose "is considered omnipresent" and "is meant to have no place; prose does not happen. Prose is what assigns place," hence "[i]t will under-stand and under-write speech and verse" (Kittay and Godzich 1987: 197–8). The result, they argue, "is one of seamless flux, of constant displacement" that absorbs the "introspective dimension, what we call a subjectivity … the mental life of the subject," allowing "shifting deixis" on the one hand, and fleeting "deictic anchoring from which to gain a view" on the other (Kittay and Godzich 1987: 206–7). In other words, prose sets up the illusion of neutrality in sustained narratives, such as those of Dickens, and has the capacity to absorb almost anything in its path and to narrate virtually anything, including multiple characters' invisible thoughts, rendering them transparent to the reader and turning the text into a game of seemingly perfect information.

In images of Dickens then, we see the author in the act of transferring the products of his imagination to paper, where the omniscient, telepathic power of the author finds expression. Translated into pictorial form, the presentation of Dickens's thought in the process of becoming text is itself a highly creative enterprise. For example, in a cartoon that was one of Dickens's favourites, he is portrayed as having an enormous head to which he gesticulates, indicating that he is thinking and that his thought will be transposed directly onto the paper on which he is writing (Figure 27.1).

FROM WHOM WE HAVE
GREAT EXPECTATIONS

Figure 27.1   Dickens's thought transferred to paper. Drawing by Charles Lyall photographed by Herbert Watkins, 1861

The implication is that Dickens's literary output came directly and almost immediately out of his head – his imagination – so that the text is essentially thought rendered visible: thought made manifest. Or, as was noted in *The Publisher's Circular* in 1861, Dickens's serialised publications offered readers works that sprang "warm from the brain" of the author (qtd in Sutherland 1976: 21).[4] This branding strategy likewise spawned many nineteenth century cartoons that catered to readers' pleasure in imagining Dickens's creative process, pictured as a swarm of his characters, each of which is capable of conjuring up a dense package of associations in the mind of the viewer (Figure 27.2). Similarly, Dickens's thought process was sometimes represented by balloons, circulating around the author's head, encapsulating familiar characters as well as whole scenes from his fictional worlds, much in the way that we might imagine a magic lantern show of Dickens's work.

In this last image, the parallel drawn with how Dickens's imaginative process works and a magic lantern show is probably no coincidence. As D.W. Griffith (1875–1948) and Sergei Eisenstein (1898–1948) both claimed, Dickens work is highly "visual" or vividly imagined, and the prevailing notion that "Dickens's prose style somehow pre-figured later cinematic modes of narration" has, according to Michael Eaton (2012: 4), become something of a truism. This is not to say that visuality was not a sort of common purpose of nineteenth-century realist prose fiction more generally, and indeed Dickens's contemporary Flaubert wrote that he wanted his work to conjure up *tableaux* that the reader could see, while author Barbey d'Aurevilly (qtd in Tooke 1994: 155) described his fellow writer's fiction as "the projections of a '*lanterne magique*,'"

Figure 27.2   Robert William Buss, *Dickens's Dream*, 1875. Courtesy of the Charles Dickens Museum

while their countryman and contemporary Stendhal (qtd in Tooke 2000: 20) wrote, "my head is a magic lantern."[5] That said, moreover, many of Dickens's novels contain references to magic lanterns, including *David Copperfield*, *Little Dorrit*, and *Bleak House*, as well as *Nicholas Nickleby*, wherein the hero complains about dramatisers of his work who put together performances that appeared "sometimes faster than [the novels] had come out … within the magic circle of [their] dullness" and who "hastily and crudely vamp up ideas not yet worked out by their original projector" (Dickens 1995: 591). In this last passage, Dickens, like Stendhal, imagines his brain as the projector in a magic lantern show, and he would later refer to London as an immense magic lantern that "supplied something to [his] brain," such as a "'crowd of objects' which would suddenly appear and then just as quickly 'dissolve, like a view in a magic lantern'" (Dickens qtd. in Smith 2003: 10). When linked to magic lantern shows in passages such as these, images of Dickens as an "original projector" of thought, and representations of the invisible products of his imagination in emblematic balloons, take on added resonance and significance.

Across media: brain, image, text, film, and beyond

I want to move now from images of Dickens that invite us to imagine the great novelist cooking up new characters and adventures beyond the furrow of his knitted brow, to focus on the process of these thoughts, moving from their expression on paper in the form of text, frequently supported by images, into stand-alone images in magic lantern shows and subsequently images in motion in film. While it is not my purpose here to make definitive or conclusive statements about the mechanics behind the technology involved in representing thought in visual media, I want to suggest a few avenues for future exploration, particularly where the study of Dickens, both the man and his work, in various media is concerned. While the study of the complex and porous division between text and image that has occupied Mieke Bal in much of her work underwent a considerable vogue, not to mention a thorough deconstruction, in academic discourse in the

1990s, Dickens's French contemporaries Flaubert, Balzac, and Zola were also preoccupied with the respective capacities of words and images to communicate the products of the creative imagination. These authors wrote from a tradition that included French art which, according to Bal (1991: 29), had turned to representing "characters so absorbed in their own mental occupations that the visual representation of their states of mind was the ultimate challenge painters had to face." In narrative text this effect can be achieved through interior monologue, yet the combined properties of text and image were also explored in the many forms of illustrated text that amplified in number over the course of the nineteenth century, thanks to advances in printing that developed along with the mass audiences who feverishly bought serialised novels.[6]

Illustrations in serialised fiction serve the obvious purpose of adapting, in a different and stimulating form, some moment of the narrative in which they are embedded. As such, illustrations in text effectively collapse temporality, or insert a different temporality into the text, while activating various cognitive operations in the mind of the subject who perceives them. These operations have their parallel in thought and its expression in language, and in this regard Ricoeur's theory of metaphor is particularly enlightening. According to Ricoeur (1981), metaphors serve as rhetorical images and constitute a sort of micro-event that lifts us fleetingly out of the macro-discursive event constituted by the text in which they occur. As he explained, when we activate and re-enact a given text in our heads, metaphor effectively conflates two or more semantic fields (roses and women, men and wolves), so that the density of metaphor lifts us momentarily off of the written word and the mandatory exercise of reading from right to left. Metaphor, therefore, "emerges as a unique and fleeting" event, "a local event in the text [which] contributes to the interpretation of the work as a whole" (Ricoeur 1981: 168, 180).[7] But whereas Ricoeur (1981: 180) wants to overwrite the power of images with "emergent meanings in our own language" so that "imagination would be treated as a dimension of language" and a "new link would appear between imagination and metaphor," I would like to suggest that such images or illustrations physically embedded in text perform metaphor in a different medium by pictorially taking up the function of rhetorical tropes, thereby offering a fugacious lift-off from the rigid temporality of the act of reading. This, then, is the power of the image in its capacity to adapt complex events in the text, even in this fixed, non-cinematic form.

In light of these remarks, as well as the preceding observations concerning the representation of thought more generally, let us now consider the more specific case of Victorian thought balloons as illustrations in or about Dickens's work, which would apply equally to the work of many of his contemporaries. I want to argue that thought balloons – at least early thought balloons and certainly many contemporary ones – seem to work on two rhetorical principles. First, they suggest a panoply of individual characters involved in various typifying actions, springing out of Dickens's own head as in Figure 27.2, or externalised out of the heads of his readers, or those of characters in his novels.

A second kind of balloon, which may or may not be directly perceived as a thought balloon *per se*, and which often appeared in nineteenth-century frontispieces, represents characters imagining the narrative events contained in the volumes that they grace, often in the form of a tree that presents branching thought bubbles, each containing entire narrative segments. I would argue, therefore, that balloons or bubbles of the first variety seem to represent thought – for example, in the case of Dickens's thought – as a collection of characters marked by traits and sartorial trappings, layered to form one condensed, dense image which hopefully awakens impressions and associations in the mind of the viewer. On the other hand, images such as those represented in the form of a branching tree move from one scene to the next and thereby introduce a certain temporality into each frame and into the composition as a whole – hence, the frequent inclusion of an hourglass at the base of the tree in such images.

This last point was perhaps made first and best by Hogarth in *The Analysis of Beauty* (1753), in which he discusses his many techniques for "lead[ing] the eye a merry chase" (Hogarth 1997: 33). According to Hogarth (1997: 33), viewers derived enjoyment from his comic 'progresses' or panels as a function how they direct the movement of the eye, as it "must course ... to and fro with great celerity ... yet amazing ease and swiftness," because ocular pleasure "is still more lively when [the eye] is in motion." In an effort to explain how this works, Hogarth supplied two illustrative plates for *The Analysis of Beauty*, in the first of which the artist adapts a typical view of the collection of cheaply reproduced objects that was regularly on display at Henry Cheere's statuary yard, Hyde Park Corner, thereby inducing the eye to move rapidly from one object to the next, although the plate adapts one static moment frozen in time. On the other hand, in plate 2, Hogarth (1997: 105) explains how to read images *across* panels and *across* the particular time frame that any given panel may represent, as a succession of events forming a line that "cuts through the air; the equal continuation of which, is varied by ... curveting from side to side." Hogarth (1997: 104) mused, therefore, that action should be reducible to a sort of language, "which perhaps one time or other may come to be taught by a kind of grammar-rules" and a cataloguing of rhetorical figures.

Centuries later, Scott McCloud (1993: 17) forwarded a similar theory in *Understanding Comics*, when he wrote that comics art constitutes "a language all its own." And while he dated the drive to adapt or "capture motion" in still images as occurring well after Hogarth's time, McCloud (1993: 108) suggests that by 1880, "inventors the world over knew that 'moving pictures' were just around the corner," which observation prompts his analysis of serial progression, time, and motion from panel to panel as well as within panels in comic strips. To illustrate his argument, McCloud (1993: 95) composed a panel in which action is to be read across the frame just as in Hogarth's plate 2, but then McCloud goes Hogarth one better by providing another panel directly below which adapts the same scene, but includes a clock to indicate the possible duration of the temporal narrative segment depicted. As McCloud (1993: 7) puts it quite simply, when adapting motion from one medium to another, "space does for comics what time does for film!"

Following both Hogarth and McCloud then, I want to suggest that single panels representing one densely layered thought, for example, in the form of a character in a thought bubble indicating that s/he is in another character's thoughts, may be contrasted with panels that show movement and events travelling across the frame. However obvious or banal these observations may be, they form the basis of my deeper argument here, namely that these two standard modes of representation have their parallels in twentieth-century models of the unconscious mind at work – engaged in thought – that have become part of a more or less standardised 'language of the cinema.'

As I have been arguing, thought balloons, and indeed illustrations more generally, introduce another different temporal order into the written word, so that, as suggested by Ricoeur, they have the capacity to act as a metaphor that condenses possible meanings and lifts us off the temporal regiment of the printed text. More specifically, single condensed images, such as portraits of memorable characters, work on the order of what Freud referred to as "condensation" in the dream work, whereby images and characteristics are conflated to suggest a fullness of things at the same time.[8] Following this train of thought, individual panels that contain sequences to be read across the panel may then be said to work on the principle of "displacement," as our mind moves from one image/thought to the next along a specific trajectory.[9]

In "The Agency of the Letter in the Unconscious" (1957), Lacan (2010: 1169), building from Freudian theories of the unconscious, famously wrote that "the psychoanalytic experience is the whole structure of language," and he went on to translate Freud's model of how dreams work in the unconscious into language, whereby condensation is expressed as metaphor, and displacement as metonymy. Lacan's major addition to Freud then, was to generalise the application of these terms, and the mechanisms they describe, to the processes of thought, both conscious and unconscious, which he equates with language. If we then take both Hogarth and McCloud at their word, images like those in comics and illustrations are also a sort of language that expresses thought, so that my argument has now come full circle.

But these are not the only twentieth-century thinkers to whom such insights have occurred. In "Dickens, Griffith, and the Film Today" (1944), Eisenstein (1977: 205) remarked that some of Griffith's best ideas came from reading his favourite nineteenth-century author, and that the American filmmaker "was led to the idea of parallel action by – Dickens!" Eisenstein opens his famous essay by contrasting the dizzying tempo of D.W. Griffith's world with the context in which Dickens's wrote, suggesting that the marriage of the two gave form to the cinematic language that Griffith pioneered in his various film adaptations. As he explained, Griffith found a language in Dickens adequate to the "duality behind the dynamic face of America" and American cities, wherein on the one hand, "high-powered automobiles are so jammed together that they can't move much faster than snails … halting … for the counter-creeping of the cross traffic," and rural America where "you can fly along as fast as you wish … in the exact opposite of the metropolitan congestion [and] that frantic activity choked in the stone vises of the city" (Eisenstein 1977: 196–7). Clearly, the first image of grid-locked traffic corresponds to condensation or metaphor, while the second image of a rural freeway along which one speeds corresponds to displacement or metonymy, and the two have, of course, become standard conventions in the adaptation and exposition of thought in various media.

So, to follow a somewhat obvious, if very partial, chronological progression in the history of adapting narrative content from one medium to another, I now want to move from text to illustration, and briefly to magic lantern slides, which were, of course, already popular when Dickens was writing.[10] To take just two examples from sets of magic lantern slides depicting scenes from Dickens, the slides in Figure 27.3 display the standard mechanisms of metaphor and

Figure 27.3   Glass slides from author's collection showing (left) a character sketch from Dickens and (right) a scene from *Dotheboys Hall*

metonymy in operation, both of which adapt narrative segments from Dickens in the condensed form of a character from Dickens as represented by Cassell (left), and in a scene from *Dotheboys Hall* (right). However correct or incorrect it may be to see the magic lantern show as a precursor to the cinema, Lynda Nead (2007: 46, 52) has notably remarked that "the transition from an inanimate to an animate state" is a "process of immense cultural and psychological potency," and that "merely the slightest move" was required to advance from the magic lantern to "the projected images of film and the haunted gallery of cinema." Similarly, in his history of American film, Benjamin Hampton (1970: 1) writes that "many men for many years, searched for ways to make pictures appear to move … [and] the invention of the 'magic lantern' was a part of this search for pictorial motion."

Nead (2007: 234) has also conjectured that "[f]or a creative thinker in the 1870s it was not a huge imaginative leap to move from the still photograph to dissolving lantern slides to continuously moving images, or 'living pictures', as film was also first called," and this is precisely what the British Film Institute has attempted in their "collection of early adaptations of perhaps Britain's favourite (and after Shakespeare) most adapted author."[11] In *Dickens Before Sound* (2012), the BFI has produced a film adaptation of *The Story of Gabriel Grub, or the Goblins Who Stole a Sexton* from painted glass slides, thereby introducing motion into this magic lantern show which would have been presented with two projectors and fade outs.[12]

Interestingly enough, one image suggests a screen onto which the goblins from the story project 'pictures' from their own collection, including filmic annotations in some of the slides to 'alter the scene' and to cut to another slide in which we follow a narrative sequence that metonymically describes the passing of a sick child. In other words, the film is produced from slides that adapt *Gabriel Grub* as a series of thought balloons that work along the lines of metaphor and metonymy, which are then transferred from a static to a moving medium.

As film technology developed, Griffith, who claimed to have learned parallel editing from Dickens, experimented with techniques for exteriorising thought in *For Love of Gold* (Eisenstein 1977: 206). As Eisenstein (1977: 255; emphasis added) explained, Griffith created a scene whose effectiveness "depended upon the audience's awareness of what was going on in the minds" of the characters. This he accomplished through the "only known way to indicate a player's thoughts … by double-exposure 'dream *balloons*.'" This would later be supplanted by "insert[ing] a picture of the object of a player's thoughts, and then by cutting from one scene to another" (Eisenstein 1977: 225). Hence, Eisenstein (1977: 234) continues, "montage thinking is inseparable from the general content of thinking as a whole," and would eventually result in a multiple chain of images that, through parallel editing, reproduce the movement of metaphor and metonymy, incorporating the rhythm and syntax of natural language, which becomes the language of cinema. This notion is crystallised in Terry Ramsey's assessment of Griffith's *Intolerance* (1916) as "a giant metaphor," and Eisenstein's assertion at the end of his essay that film images become connected as a function of metonymy, as the verbal image "follows a kind of kinematical development," ultimately as a means of representing "inner monologue" as understood in film (Ramsey 1954: lxi, Eisenstein 1977: 250, 247).

Over the decades that followed Griffith's thought balloons, the problem of representing thought, particularly in adaptations, has continued to present a challenge to film makers. According to George Bluestone (1957: 48, 45), who wrote the first full-length study of film adaptation, in "contrasting [the] ability of film and novel to render conceptual consciousness … the film image, being externalized in space, cannot be similarly converted through the conceptual screen [since] the compacted luxuriance of the trope" is alien to the screen, hence dreams, memories, and thought more generally "cannot be adequately represented in spatial terms." Therefore, he continues, "[t]o show a memory or dream [or thought], one must *balloon* a separate image into the frame, or superimpose an image" (Bluestone 1957: 48; emphasis added). Since the advent of sound, one frequently used "portal to the brain," to quote Charlie Kaufman in *Adaptation.* (2003), is the voice-over, which is used to clue the viewer into the interior monologue of a character whose thoughts we would otherwise not know. In that same film, however, the character Robert McKee cries, "God help you if you use voice-over in your work … Any idiot can write a voice-over narration to explain the thoughts of a character" (Kaufman and Kaufman 1999: 87).

That said, however, Eckart Voigts-Virchow (2006: 256) has argued that, "for a long time, film has been thought to be particularly ill-suited to represent the mind, consciousness and interior processes. This is wrong." While I would agree at least partially with this statement, I would also hesitate to be quite as categorical as Voigts-Virchow on this score, and prefer to see the representation of interior processes as something with which film has indeed struggled, and for which film has found various, sometimes competing and overlapping strategies. Significantly, as he goes on to point out, Charlie Kaufman's vehement put-down of the voice-over should be understood in the context of a film that itself contains many voice-overs. Hence, we would probably do best to read this self-reflexive, self-conscious moment in *Adaptation.* as yet another postmodern, autoparodic gesture in a film adaptation that constantly folds back on itself, as characterised by the figure of the ouroboros which features prominently in the film.

Following on from this observation, I would like to conclude with a few brief insights from Bolter and Grusin's *Remediation*, namely that "all current media function as remediators" (Bolter and Grusin 2000: 55) so that one development does not entirely supersede or wipe out its predecessors. Rather, the authors offer "genealogy of affiliations, not a linear history," whereby newer media remediate and even maintain various aspects of older media, and "older media can also remediate newer ones" (ibid.). Hence, "[m]edia are continually commenting on, reproducing, and replacing each other" so that media need each other to function as media at all (ibid.). Given Bolter and Grusin's observations, I would like to close this discussion with a related and particularly evocative example from David Lean's adaptation of Dickens's *Great Expectations* (1946), in which the opening shot is quite literally a shot of the first pages of the novel accompanied by a voice-over that reproduces Pip's childhood memories (Figure 27.4).

Interestingly enough, with sound and other advanced technologies at his disposal, Lean opts to return us to the *text* as a means of visualising Pip's childhood memories, which then slowly

Figure 27.4   Still from the opening sequence of David Lean's *Great Expectations* (1946)

fades, suggesting a bridge into the visualising techniques of the adapted filmic narrative. This moment in Lean's film then provides an excellent example of how methods of representing thought merge and flow out of and into each other as one medium adapts another, so that one may trace a genealogy of representing thought in visual media as I have done here. Moreover, although this genealogy ends with film, it in no way seeks to exclude the possibility of extending the discussion into the area of video games or other entertainment platforms that channel Dickens's narratives.

Conclusion

In this essay, I have attempted to trace a loose genealogy of the representation of thought from text, to illustrated text, to magic lantern slides, silent film, and finally talkies, specifically in relation to images that congregate around Dickens and his fiction. In writing 'loose genealogy,' I again mean to stress the fact that, as film technology develops, older techniques for representing thought do not simply disappear – hence my discussion of Bolter and Grusin. Likewise, it is important to keep in mind that older representational conventions are re-mediated in more recent media, as the example of the opening 'pages' of David Lean's *Great Expectations* graphically illustrates, through an image of pages that flutter up and fade out, drawing us into the cinematic adaptation of Dickens's novel, accompanied by a voice-over that prepares us for the filmic representation of Pip's innermost thoughts.

Moreover, I would be the first to admit, in the words of Joseph Conrad (1981: 21), that I have succeeded mostly in presenting "a weary pilgrimage amongst hints for nightmares," which is to say that this essay was intended to lay some groundwork for future investigation through the presentation of a case study that raises the issue of the word/image opposition in the history of adaptation. However, to say anything conclusive or definitive about this issue would necessitate further rigourous empirical studies in the style, for example, of Forceville, Veale, and Feyaerts (2010) who have developed a 'science' of comics balloons, to which they refer as 'balloonics.' While they conclude that "thought balloons are often the equivalent of interior monologue," they have also constructed what they call an "ontology of the balloon" and spent several years counting many thousands of balloons and sorting them in terms of type, shape, colour, location, frequency, and so on (Forceville, Veale, and Veyaerts 2010: 67). Likewise, some would surely welcome more hair-splitting around the exact delineation of what I have identified as balloonic renditions of metaphor and metonymy, given that some images doubtless contain both.[13] However, while much more bean counting would be necessary in order to support my argument empirically, I do think it is safe to conclude that the representation of thought in various media follows developments in describing the cognitive operations involved in human thought, and that a more detailed, thorough study may well lead us to a deeper understanding of the representation of human thought in various modern processes through image, film, and right into cyberspace.

And finally, I want to concur with Grahame Smith (2003: 11) who raises "the question of consciousness as part of the progression towards film … along with the relationship between consciousness and social change," in other words, through the pressure of modernity and modernisation. Smith (2003: 7) insists on the notion that the effects of various aspects of urban modernity (street lights, train travel and railway time tables, the working day), along with the intensity and speed of movement and huge increases in the number of urban dwellers that have accompanied modernity, have had a powerful effect on human consciousness and more specifically on Dickens's consciousness as a major narrator of city life whose abundant serialised publications are replete with "proto-filmic elements." If film is the "space and time machine of the imagination," and if "thought [in film] is for the first time adequately projected," then those thoughts are shaped by modernity and technology and expressed in visual culture in some of the ways I have been outlining (Smith 2003: 47). In this regard, it is worth noting that Freud himself opened *Beyond the Pleasure Principle* (1920) by explaining that his goal was to introduce an economic point of view into his account of consciousness, that is, a modern, utilitarian account, the backbone of which is the metaphorical double-ledger that assists us in balancing pleasure against pain (Freud 1961: 1). So if "Dickens's manipulation of objects, gestures, body language … represents his mode of accessing the unconscious as well as the conscious springs of human behavior" (Smith 2003: 47) and is "a method peculiarly appropriate to the urban world which is his central theme," then I would like to suggest that the relationship between technology, modernity, and how we have come to adapt and represent thought is one of reciprocity, and that Dickens's engagement with external reality, as well as the technology through which the author and his work are adapted into various media, both represent and shape how we think of thought.

Notes

1 For a more detailed treatment of this issue, see Stam (2005: 18–24). See also Voigts-Virchow (2006: 249–56).

2 On the commodification of various Dickens characters and their transposition onto products such as fabrics, cups, and umbrellas, see Steinlight (2006: 135–6).

3 As Lynda Nead (2007: 111) has suggested, this connection also inheres between Dickens and 'his' Dickensian London, such that "writers in this period saw London through a filter of Dickens." In other words, 'Dickens' as a brand of fictional writing also explicitly involves his thought processes and the way he apprehended London for himself and then for others in writing.

4 This quotation comes from an anonymous publication of 1861 about Dickens's imaginative process.

5 "Ma tête est une lanterne magique; je m'amuse avec les images, folles ou tendres, que mon imagination me présente." Quoted in Tooke, this passage originally occurred in Stendhal's "L'Italie en 1818" (1973: 238). In a similar vein, according to Smith (2003: 12), Dickens's imagination is best described as what Baudelaire called "a kaleidoscope gifted with consciousness."

6 On the sheer volume of serialised fiction that circulated in England in the nineteenth century, see Turner (2014), particularly pages 16 to 20.

7 This same effect of metaphor was also noted by Nabokov (1969: 185) which he described as a "triumph over the ardis [or arrow] of time."

8 In "The Work of Condensation," Freud (1983: 399, 403) describes condensation in the dream state as resulting in "collective image[s]" containing "contradictory characteristics," or "collective figures and composite structures" whose meaning then becomes "over-determined."

9 Freud's notion of displacement, which he describes in "The Work of Displacement," amounts to a kind of dream-distortion, involving a shift of emphasis from important to unimportant elements, or the rapid replacement of something with an illusion or parallel element (Freud 1983: 414–9).

10 According to Nead (2007: 50), the magic lantern dates back to at least the seventeenth century. Smith (2003: 21–7) cites numerous early examples including the use of the *camera obscura* to project images in 1558, as real actors moved in front of projected scenes.

11 The second quotation here is taken from the jacket notes to the BFI's *Dickens Before Sound* (2012).

12 As Fred Guida (2000: 51–5) has explained, various standard filmic techniques were introduced into the magic lantern show, such as the panning shot, which was "easily achieved by pulling … a long horizontal slide through the projector" and "the use of two projectors made it possible to cut directly from one slide to the next in much the same way in which a film … cuts from one shot to another."

13 See, for example, J.R. Brown's "Dickens Surrounded by His Characters" (1889–90).

Works cited

Adaptation. (2003), dir. S. Jonze. Columbia. DVD.

Bal, M. (1991) *Reading Rembrandt: Beyond the Word–Image Opposition*, Cambridge: Cambridge University Press.

Bluestone, G. (1957) *Novels into Film: The Metamorphosis of Fiction into Cinema*, Berkeley, CA: University of California Press.

Bolter, D.J. and Grusin R. (2000) *Remediation: Understanding New Media*, Cambridge: MIT Press.

Brown, J.R. (1889–90) "Dickens surrounded by his characters," in *The Victorian Web*. Image. Online. Available HTTP: http://www.victorianweb.org/authors/dickens/gallery/24.htmll (accessed 11 July 2017).

Conrad, J. (1981) *Heart of Darkness*, 1899, London: Penguin.

Dickens Before Sound (2012), dir. various. BFI. DVD.

Dickens, C. (1995) *Nicholas Nickleby*, 1839, Hertfordshire: Wordsworth Editions.

Eaton, M. (2012) "Old curiosity shots," in *Dickens Before Sound*, dir. various. BFI. DVD.

Eisenstein, S. (1977) "Dickens, Griffith, and the film today," 1944, in *Film Form: Essays in Film Theory*, trans. and ed. J. Leyda, San Diego, CA: Harvest, 195–255.

Elliott, K. (2012) "Screened writers," in D. Cartmell (ed.) A *Companion to Literature, Film, and Adaptation*, Chichester, UK: Wiley-Blackwell, 179–97.

Forceville, C., Veale, T., and Feyaerts, K. (2010) "Balloonics: The visuals of balloons in comics," in J. Goggin and D. Hassler-Forest (eds) *The Rise and Reason of Comics and Graphic Literature: Critical Essays on the Form*, Jefferson, NC: McFarland, 56–73.

Foucault, M. (2010) "What is an author?" 1969, trans. D.F. Bouchard and S. Simon, in V.B. Leitch (ed.) *The Norton Anthology of Theory and Criticism*, 2nd edn, New York: Norton, 1475–90.

Freud, S. (1961) *Beyond the Pleasure Principle*, 1920, trans. J. Strachey, London: Penguin.

Freud, S. (1983) *The Interpretation of Dreams*, vol. 4, 1899, trans. J. Strachey, London: Penguin.

Genette, G. (1997) *Paratexts: Thresholds of Interpretation*, Cambridge, UK: Cambridge University Press.

Godzich, V. and Kittay, J. (1987) *The Emergence of Prose: An Essay in Prosaics*, Minneapolis, Minnesota: University of Minnesota Press.

Great Expectations (1946), dir. D. Lean. Universal Pictures/Cineguild. DVD.

Guida, F. (2000) *A Christmas Carol and Its Adaptions: A Critical Examination of Dickens's Story and Its Productions on Screen and Television*, Jefferson, NC: McFarland.

Hampton, B. (1970) *History of the American Film Industry from Its Beginnings to 1931*, New York: Dover.

Hogarth, W. (1997) *The Analysis of Beauty*, 1753, ed. R. Paulson, New Haven, CT: Yale University Press.

Intolerance (1916), dir. D.W. Griffith. Triangle Film Corporation, DVD.

Kaufman, C. and Kaufman, D. (1999) "Adaptation.," in *The Daily Script*. Film script. Online. Available HTTP: http://www.dailyscript.com/scripts/adaptation.pdf (accessed 11 July 2017).

Lacan, J. (2010) "The agency of the letter in the unconscious," 1966 trans. A. Sheridan, in V.B. Leitch (ed.) *The Norton Anthology of Theory and Criticism*, 2nd edn, New York: Norton, 1169–81.

McCloud, S. (1993) *Understanding Comics: The Invisible Art*, New York: HarperCollins.

Murray, S. (2012) *The Adaptation Industry: The Cultural Economy of Contemporary Literary Adaptation*, New York: Routledge.

Nabokov, V. (1969) *Ada or Ardor: A Family Chronicle*, New York: McGraw-Hill.

Nead, L. (2007) *The Haunted Gallery: Painting, Photography, Film c. 1900*, New Haven, CT: Yale University Press.

Ramsey, T. (1926, 1954) *A Million and One Nights: A History of the Motion Picture*. Abington, Oxon, UK: Frank Cass & Co.

Ricoeur, P. (1981) *Hermeneutics and the Human Sciences: Essays on Language, Action and Interpretation*, in J.B. Thompson (ed.), Cambridge, UK: Cambridge University Press.

Smith, G. (2003) *Dickens and the Dream of Cinema*, Manchester, UK: Manchester University Press.

Stam, R. (2005) "Introduction: The theory and practice of adaptation," in R. Stam and A. Raengo (eds), *Literature and Film: A Guide to the Theory and Practice of Film Adaptation*, Oxford: Blackwell, 1–52.

Steinlight, E. (2006) "'Anti-Bleak House': Advertising and the Victorian novel," *Narrative*, 14(2): 132–62.

Stendhal (1973) *Voyages en Italie*, Paris: Gallimard.

Sutherland, J. (1976) *Victorian Novelists and Publishers*, London: University of London Press.

Tooke, A. (1994) "Flaubert on painting: The Italian notes (1851)," *French Studies*, 68(2): 155–73.

Tooke, A. (2000) *Flaubert and the Pictorial Arts: From Image to Text*, Oxford: Oxford University Press.

Turner, M.W. (2014) "The unruliness of serials in the nineteenth century (and in the digital age)," in R. Allen and T. van den Berg (eds), *Serialization and Popular Culture*, London: Routledge, 11–33.

Voigts-Virchow, E. (2006) "Adaptation, *Adaptation* and drosophilology, or Hollywood, bio-poetics and literary Darwinism," in C. Houswitschka, G. Knappe, and A Müller (eds), *Proceedings of the Conference of the German Association of University Teachers of English*, vol. XXVII, Trier: Wissenschaftlicher Verlag Trier, 247–65.

28

AN ART OF BORROWING

The intermedial sources of adaptation[1]

André Gaudreault and Philippe Marion

Adaptation raises frequently recurring general questions, one of which is, as we know, the great motif of 'fidelity.' We have chosen this motif to begin our discussion because it is in a sense the 'blind spot' of every approach to adaptation. We will see, however, that this question of fidelity deserves to be downplayed and neutralised by being much less subjective about it than usual. We know that the way in which the question of fidelity was posed until recently was simply not well suited to a theoretical discussion of the problem of adaptation. Hence the temptation, to which several scholars have succumbed since the 1980s, to kick it out of the field of adaptation studies at any price.[2]

We have also chosen this motif to begin our discussion because, despite these numerous attempts, the question of fidelity, like any repressed worthy of the name, has always taken a malicious pleasure in returning to haunt adaptation studies. To such an extent that today it is impossible to keep track of the texts which seek in some way to rehabilitate the notion – taking care, nevertheless, to strip it of the normative and 'moralising' cloak in which it had, more often than not, been confined.[3] Hence the relatively new positions which view, as Nico Dicecco (2015: 174) proposes here, "not ... the enemy of scholarly conversation, but a key tool in studying the cultural stakes of media interpretation," or those which advance, in the words of Casie Hermansson (2015: 156), that the notion of fidelity is to be included "in the intertextual toolbox of adaptation criticism."

All this supposes that the concept of adaptation opens up a broad spectrum of possible approaches: It can, of course, open the door to questions of intertextual writing across media, but it can also be seen, more narrowly, as a mere translation or even simply, more narrowly still, as a limited and conventional 'copy' of the source work.

Identity in adaptation's very fibre

By examining the meanings and definitions contained within the word 'adaptation' along with all their connotations, a constant theme soon becomes evident: adaptation is always associated with transformation. This simple observation is revealing, because it enables us to view adaptation from the perspective of the identity of the source and target works. For every transformation supposes the co-existence of the 'same' and the 'other,' the 'similar' and the 'different.' Transformation thus involves a degree of alterity and identity. For a phenomenon to be transformed, some element of it must be perpetuated, extended. At the same time, this phenomenon must become part of a process of change, must join the flow of an evolution, a differentiation. This dual dynamic, between constancy and change, is the stuff of every phenomenon every time that the question of identity is concerned.

Two simple and related questions nevertheless persist with respect to the conception of adaptation as a transformation of the 'identity' of a text:

1   What, in the course of the adaptation, changes?
2   What remains the same?

This two-fold interrogation also exists in certain biological conceptions of adaptation. Thus, in classical theories of evolution (Darwinian in particular), mimicry is seen as a form of adaptation. This term describes a living being as seeking to become a part of its environment in an optimum manner. For our part, we believe that this mimetic quality deserves to be preserved on the semio-medial terrain of the word: to adapt is to respond to new situations and to fit in with them 'mimetically.' Note that, from this perspective, the mimicry in question concerns the environment of the target text (the hypertext, text B) more than it does any imitation of the source text (the hypotext, text A). It is as if adaptation were also the ability to adapt to what is 'already there' in the surroundings, and will be a contextual retention, an unavoidable residue of the hypotext.

It should be noted in passing that this 'ecological' concept of adaptation has been explored fairly recently by scholars such as Gary Bortolotti and Linda Hutcheon, who adopt a critique – which we share – of fidelity reduced to a narrow deterministic process:

> Our hope is that biological thinking may help move us beyond the theoretical impasse in narrative adaptation studies represented by the continuing dominance of what is usually referred to as 'fidelity discourse.' This common determination to judge an adaptation's 'success' only in relation to its faithfulness or closeness to the 'original' or 'source' text threatens to reinforce the current low estimation (in terms of cultural capital) of what is, in fact, a common and persistent way humans have always told and retold stories.
>
> *Bortolotti and Hutcheon 2007: 444*

From this perspective, to adapt thus consists in accustoming the text (or certain elements of it) to another context. Thereupon another batch of questions is raised. For what is meant exactly by this 'other context'? The answer to this question will determine the sense that is given to the notion of adaptation. This other context may be a different audience or cultural community of reception (in a geographical, socio-cultural, or historical sense), but it may also be a change of genre or especially a change of medium: in this respect, another medium is a 'new situation,' a new 'ecosystem' to which the text one wishes to adapt must be accustomed.

Now is undoubtedly the time to address a more contemporary dimension of this process of accustoming to 'another context': today, adaptation must not only absorb the shock of a change of medium; often it must also negotiate with a complex pluri-media environment. In the age of the digital and of hypermedia, adaptation must be able to manage a mimicry whose breadth is openly intermedial. It is no longer just a change of 'medium' in the singular, but also a change of 'media' in the plural, which is to say organising the passage from text A to several media regimes in text B. In this case, the notion of hypertext opens onto every possible intermedial sense of the term that the digital world has multiplied exponentially (see Gaudreault and Marion 2015).

The very mainstream but nevertheless emblematic case of the 2015 episode of *Star Wars* (*The Force Awakens*), in the multitude of these media crystallisations offered alongside the film's release, gives an idea of this potential.

More precisely, we are in the presence of a kind of adaptation that is typical of our media culture. Through a kind of metonymic continuity, today the principal work can sometimes be accompanied by 'spin-off media products.' This is a kind of 'adaptation marketing,' an opportunistic adaptation which sets out to multiply, in transmedial fashion, the 'principal' work, thereby demonstrating its vitality and reach. In this respect, the case of that 'cultural machine' *Star Wars,* now under the wing of the Walt Disney Company, is a fine example: 'new' productions are released in what we might call a pre-programmed adaptation package orchestrated by a strategy of media roll-out.

In the field of stories told in images, we could also take the more elitist example of the work of the graphic novelist Marc-Antoine Mathieu. His *3 secondes* is a work available straightaway via two concurrent media systems (or at the very least two reading systems): a classic book version (on paper) and a digital version. In the latter, the drawn images come to life in the real-time continuity of an uninterrupted zoom, as if governed by a persistent and dizzying morphing (Mathieu 2012).

In our digital and multimedia era, many artistic productions (live entertainment, performances, concerts, etc.) are conceived and put together from the outset for simultaneous intermedial distribution (video, stage, screens, etc.) in a way we might call 'performative adaptation.' Such a form of adaptation is no longer a process that postdates the work, but is apparent from the very moment of its conception, making this conception multimodal in nature. The digital could thus be said to promote, in a manner of speaking, real-time adaptation. Contemporary hypermedia productions can thus be seen as a kind of simultaneous 'hyper-adaptation.'

The remake: a substandard adaptation?

In the opposite sense to the opening begun above, we can also try to better understand adaptation by limiting the concept to mere accommodation, intended for another audience but without changing the medium. Here adaptation merges, at least in the case of the cinema, with the remake – if we accept, of course, the definition which holds that a remake is a "reworking by another filmmaker of a previously made film, using the same script" (Serceau 1989: 6) in order to bring it up to date or make it suit the tastes of a new audience.

Let us linger for a moment on the path that the remake opens up for us. With a remake, we are clearly in the presence of a text that has undergone a process of adaptation, but we are running on the spot in a sense, because in passing from one text to the next we remain in the realm of the same medium. While we can readily agree that adaptation is based on a binary articulation and that it supposes the existence of at least two texts, we also tend to think that it equally supposes the involvement of two media. This is not necessarily the case, at least if we accept the idea that producing a remake is a process of adaptation. Nor would this be the case if we consider the history of adaptation itself, which initially meant a textual variation within the same medium or, at the very least, a textual variation within the sphere of influence of the same institutional domain (in the event, the literary domain).

This idea of intra-generic (intra-medial, intra-institutional) variation is in fact quite present in some of the dictionary acceptations of the word 'adaptation.' The French *Robert* dictionary gives the following definition, which dates from 1885: "Very free rendering of a stage play with numerous modifications to bring it up to date."[4] This dictionary adds the following more familiar definition, which is meant to be more up to date but, like the first, finesses the intermediality supposedly involved in an adaptation: "Transposition to the stage or screen of a work of a different literary genre (particularly a novel)." Here we see the extent to which such a conception tends implicitly to maintain the 'adaptive process' under the thumb of one and the same institution: the literary, or '*Literature*.'

This appears as a kind of super-medium, within which are found genres: the novelistic, the 'theatrical,' and the 'cinematic.' These ideas resonate with those of Bortolotti and Hutcheon mentioned above. Indeed, the authors remark that "Shakespeare transferred his culture's narratives from page to stage and made them available to a whole new audience; we did not begrudge him his creative borrowing" (Bortolotti and Hutcheon 2007: 444).

Here there is no intermedial concern: the work of adaptation thus occurs within each media institution, between the 'genres' that each accepts and authorises. This reference to the history of the concept demonstrates that at one time, adaptation could very well have been seen as being limited to a given expressive domain. Take, for example, the first definition of adaptation offered by the linguist Jean Giraud (1958: 41) in his French film lexicon, in which he takes up the definition of the *Grand Dictionnaire Universel du XIXème*: "Literary work by means of which a writer, taking as his text the work of another author, transforms it into a similar production, yet differing in some points from the first."

Here we find a mono-medial conception of adaptation, so to speak, for we remain in the 'purely' literary world. Giraud, probably conscious of the deficiencies of this initial definition of adaptation inherited from a major dictionary, hastens to assert the need to adapt it to cinema:

> This definition applies only in part to what are called film adaptations. A novel need only be arranged to make it into a play; putting it on screen requires a complete reworking by different means of expression. For it is no longer simply a matter of changing the literary category, but rather of rewriting for the cinema.
>
> *Giraud 1958: 41*

For Giraud, turning a novel into a play does not merit the status of adaptation, because the transfer consists simply in changing the literary category. This consists in the end of nothing more than 'little arrangements between friends.' When on the other hand cinema takes hold of a novel, we are in the presence of a true rewriting ("a complete reworking," Giraud remarks) which takes into account the semio-medial singularities of the target medium. The family atmosphere that was supposed to join the theatre stage and the movie screen is thus shattered. Giraud's conception appears to be based on the self-awareness that the very early film milieu developed in the 1910s. This self-awareness enabled it to distinguish itself from the theatre by claiming the role and status of a true autonomous medium (whereas the theatre, despite its singular status as a form of live entertainment, would for its part always remain more or less integrated into the literary system).

Five aspects of adaptation

To better frame the problem of adaptation, we propose to set out five parameters which should enable us to grasp the phenomenon from a new perspective. We thus believe it important to pose the question of the intensity of the adaptation process, the extent and cumbersomeness of this process, then the intention behind it and finally its media orientation.

A methodological caution: these parameters are in no sense mutually exclusive categories. They are, rather, bearings, prototypical qualifications, and 'paradigms' which will make possible, as we shall see, a great many crossovers and intersections.

We will now present these parameters briefly, beginning with the intensity of the adaptation process.

A topic, theme, motif, or narrative can demonstrate a greater or lesser degree of reticence towards lending itself to adaptation. A text can thus be adaptable to a greater or lesser extent (we will return to the reasons for the greater or lesser adaptability of texts below). The greater the reticence towards adaptation, the more intense the labour of the adapter is presumed to be. This is the intensity of the adaptation process, and it is not unrelated to Giraud's general remark, quoted above: "putting [a text] on screen requires a complete reworking by different means of expression."

Alongside a kind of general adaptability, which will vary according to the divergences between the source medium and the target medium, we can also express an opinion as to the intensity required by the adaptation of a particular text. In this respect, we can imagine a spectrum stretching between two poles: on the one hand *adaptogénie*, and on the other what one of the present authors has proposed we call *médiagénie*.[5]

Adaptogénie ◄—————————► Médiagénie

The *adaptogénie* of a text is also its greater or lesser ability to lend itself to adaptation, to migrate easily to another medium and thus to demonstrate a great transmedial vocation. We might suppose that the adventures of James Bond, for example, meet this criterion; in any event their on-screen success appears to be a practical demonstration of this. In the case of *Star Wars*, as we mentioned, *adaptogénie* even appears to constitute a feature of texts well rooted in popular culture, which in turn is intimately connected to media culture. We might thus remark that, since the nineteenth century, great popular narratives have shown a keen propensity for intermediality. Think, for example, of the flexible *adaptogénie* of the *Zorro* saga. And we need only mention the exemplary case of *Marvel*, with its gallery of heroes ready to leave their original frame of comics to live out other media adventures. Here, too, we must acknowledge that digital culture has accelerated this *adaptogénie* dynamic, in particular with respect to iconic expressions that can be modulated ceaselessly.

At the other end of our continuum lies the *médiagénie* pole, or a text's status as being literally cast in the medium in which it found its initial expressive form. *Médiagénie* can thus be defined and measured as the intensity of the interpenetration between a medium's possibilities and the expressive project in question. It is a reaction, in an almost chemical sense of the term, between expressive project and medium configuration. *Médiagénie* proceeds not from what the medium makes possible, but rather what it encourages; it concerns, in a sense, this medium's elective affinities, meaning what it does better than its peers. In a narratological context, the most mediagenic stories appear capable of occurring in an optimal manner by choosing the media partner best suited to them, and by negotiating intensely their 'plotting' with all this medium's internal apparatuses. Moreover, it was long believed that Tintin's adventures were practically impossible to adapt without irremediable alterations. One might wonder if the recent adaptation by Spielberg and Jackson (*The Adventures of Tintin: The Secret of the Unicorn*) has finally made it possible to banish this weak *adaptogénie* (or this strong *médiagénie*) from Hergé's work (opinion is divided on the matter).[6]

The more mediagenic a text is, the more difficult its adaptation to another medium will be. Hence the presence of a strong intensity of the adaptive process in such a case, so as to 'disembody' the text from the original medium with which one might say it had fused.

The extent of the adaptation concerns the scale of the adaptation project. Is the adapter content, for example, to adapt only the plot of the text, or does his or her net catch other elements relative to media expression or narrative? Note that it is difficult to pin down with precision the criterion of extent, in that it varies according to the point of view adopted. The extent of the labour is greater if one seeks to render, transpose, or translate every element of the source text. In such a case, one must adapt not only the plot of the text, but also a share of its media specificities.

If we were to look at the question from another point of view, we could say that there is a great extent of 'creative' labour in the case of an adapter taking up only a slim proportion of the elements which make up the source text. In a situation such as this, there will be many gaps to be filled. We might thus posit two kinds of adaptive extent, marked by two different concerns: on the one hand recreating and on the other filling in. Here the spectre of fidelity appears once again, because at first sight we might think that in any attempt at adaptation this is what most often motivates the concern for recreating.

Hence the next concept, directly related to the previous one: that of the cumbersomeness of the adaptation. Beyond the content alone, beyond the plot or storyline, an adaptation can also take into consideration those aspects of the hypotext which have to do with its 'form' (the choice of décor, the characters, the style, the tone, etc.), along with those aspects which have to do with the putting into signs of a given medium. In sum, the adaptation can also take into account the way in which the text's storyline or argument was adapted to the singularity of the source medium and attempt to transpose it to the target medium. In cases where all these elements coincide (the adapter tries to respect, maintain, and preserve the storyline, the semio-expressive tone, and the role of the source medium), we can state that the adaptation process may be very cumbersome. This cumbersomeness thus depends on a complete evaluation of the 'givens' of the source text which the target text reformats, or tries to reformat.

'Tries to reformat?' Here is a good segue to our next parameter, more pragmatic in nature: that of the intention of the adaptation. When this intention is manifest, it enables us to distinguish between adaptation and various other acts of textual appropriation such as piracy, plagiarism, quoting, and even the remake and novelisation. This criterion is connected to the wish, shown more or less explicitly, to invite (to convene, to 'summon to appear') a new re-presentation of an already existing text. Such a re-presentation is accompanied by a strong desire (which leaves traces in the hypertext) to modify the hypotext, if only by virtue of the simple fact that it is implanted in a different media context.

Put differently, there exists in the adaptation – the object – a greater or lesser trace of the adaptation (as intentional act or discourse). This parameter thus consists in taking into consideration the indicators of such an adaptive intentionality. In sum, it is a matter of identifying the traces of the adaptation as discourse.

This brings us back to the two principles outlined above: on the one hand, the principle of permanence – a degree of constancy must be maintained in the new text so that a more or less recognisable trace of the source text remains therein – and the principle of alterity, on the other hand, for to adapt is inevitably to transform.

There remains a fifth aspect to introduce: that of the adaptation's media orientation. Generally speaking, an adaptation leans in one of two opposite directions: towards the target medium or towards the source medium. When it leans to the source medium, or hypotext, the adapter must find in the target medium the means to evoke the expressive qualities and singularities of the source medium. The adapter must thus create a medium effect to invoke the absent medium. In this way, for example, the cinema can endeavour to create a graphic novel effect (as in Ang Lee's *Hulk* or Steven Spielberg's *The Adventures of Tintin*). Nevertheless, one may also take the opposite tack and overlook the source medium, by putting the potential of the target medium to full use. To the extent that one is working with a narrative configuration, the *fabula* must then lend itself to the specificity of the new medium. It is in this that there may exist, paradoxically, a mediagenic adaptation. In the former case, the *fabula* is not just the *fabula*: it bears traces of the source medium. In the latter case, little attention is paid to the source medium, and one tries instead to employ the *fabula* within the target medium.

The genealogy of cinema and adaptation

Another question arises at this stage of our discussion. What might be the connection between adaptation and media genealogy? In order to better understand what adaptation is exactly, and to better grasp the issues it raises, we must look to the past. Our demonstration will thus focus on cinema's early years. The second question, a corollary of the first: did this idea of adaptation (in the strong and thus 'limited' sense we discussed above and which supposes a change in medium between the source text, which we call the hypotext, and the target text, which we call the hypertext) arise with the advent of cinema's base technology in the late nineteenth century?

Our conception of film history draws a strict dividing line between two great paradigms: on the one hand, "kine-attractography," and on the other, "institutional cinema" (see Gaudreault and Marion 2007; see also Gaudreault 2011). Each of these two paradigms has occupied, successively, the forefront of film history, and they are in a sense the high points of a diachronic sequence, even though they are not necessarily mutually exclusive. In fact, their relative succession made possible a degree of overlapping. The first paradigm was dominant until around 1908–10, and continued to be present in a fairly significant way until around 1914; while the second paradigm began to appear on the horizon of film history a little before 1908, precisely, becoming dominant with the new decade a couple of years later before reigning uncontested, or almost, in the new socio-cultural institution that cinema was in the process of becoming.

Whereas institutional cinema, the cinema of the institution, is a regulated, rule-governed, and officially agreed-upon practice (with organisms as well for regulating, governing, and agreeing upon), kine-attractography was a relatively little-overseen practice subjected to no rule, unique to the medium apart from the strictly technical constraints that came with it. Kine-attractography was in a sense a practice without rules and regulations – or at least without intrinsic rules and regulations, for the rules and regulations running through kine-attractography were those of practices exogenous to the kinematograph, and on which it was modelled.

Kine-attractography brought together a hodge-podge of institutions, a hodge-podge of all those institutions from which it took not its first steps but its first, shall we say, borrowed ideas. 'Borrowed?' Is this not where adaptation pokes its nose over the horizon? Is adaptation not, precisely, an art of borrowing? Even though, on a strictly technical level, the protagonists of the two paradigms, 'kine-attractography' and 'institutional cinema,' use the same 'base apparatus,' in kine-attractography the kinematograph is mostly a reproduction device (capturing is the order of the day), while in institutional cinema the kinematograph is more a production device.

This device's ability to reproduce continued on (and was even refined, on a purely technological level), but it became subordinated to the production spirit typical of institutional cinema. Conceived as a simple 'reproduction' device, in order purely and simply to 'record' what was placed in front of it (or before which it was placed), the kinematograph of kine-attractography was limited to 'capturing' what passed before its lens at the precise moment the operator set the device in motion.

Note also – and this is important, because it goes against conventional wisdom – that the paradigm institutional cinema is also that of the consecration of narrative cinema: the idea of a story told by this singular medium, the cinema, really only took hold under this paradigm (something attested to by a concomitant form of institutionalisation, the consecration of scriptwriting in the 1910s). The story, whose importance for adaptation is clear, was not at all a priority under the kine-attractography paradigm. What was dominant under attraction was not the story, but the attraction itself: the novelty effect of this fascinating technology, which made it possible to capture and restore slices of real time and space.

Adoption or adaptation

Let us try to connect more systematically these genealogical considerations with the question of adaptation. We will first take up a position based on the way in which moving pictures were made at the time of kine-attractography. This position can be expressed in a provisional typology taking the form of three gradients. First, depictions of actions whose arrangement and unfolding seemingly owe nothing, or very little, to the camera operator, who only captured, in a relatively neutral manner, beings and actions for whom 'God' was the sole director. Something like a sur-veillance camera.[7] If we were to express this in terms of our concept of borrowing – adaptation as borrowing – what is being borrowed is reality itself in the form of the capturing–restoring taking place. Here we find the kind of zero degree of adaptation discussed above: a minimal adaptation, in the very weak sense of the word, in the form of a 'mere' re-presentation. Or, more precisely, as a re-presentation of real space–time. Put differently, the adapted text merges with reality-as-text. To employ the categories proposed by the filmologist Étienne Souriau (1953: 7), here the profilmic, or rather the 'afilmic,'[8] operates as the adapted text. This is the case, for example, with the film *Sortie des usines Lumière* (*Workers Leaving the Lumière Factory*, 1895), in that it was shot with a hidden camera without any prior understanding with the workers we see exiting the factory and whose movements were not the result of any agreement between filmer and filmed.

We might note in passing that if we were to agree to view the reality being captured as a 'text,' it would be situated less to the side of the afilmic than it is to the side of the profilmic (with the connotation associated with this term of a degree of 'arrangement' of the reality being captured by the camera), as the notion 'text' also conveys the quite similar quality of an 'arrange-ment of signs.'

Then there are scenes whose arrangement and unfolding are, on the contrary, the work of the camera operator or one of his associates (director, actor, assistant, production manager, helper, etc.). *Partie d'écarté* (*Card Game*, Lumière, 1896) is a good example of this. The picture recounts a card game whose progress is clearly the result of collaboration between filmer and filmed. In this case, it is still a matter of borrowing from reality, but this reality has been arranged to create an 'intentionalised' profilmic prepared with a view to filming this adaptation–borrowing.

What interests us most, however, with respect to the question of adaptation, is the inter-mediary stage between the two poles we outlined above. Between the two, we find numerous examples in which the camera operator chooses to record a scene composed before the putting onto film in a different sphere than that of kinematography, and which the filmer must of course adapt to the various limitations of his filming apparatus. We have chosen as our example here the film *Sandow* (Edison 1894) (see Figure 28.1). In it, the arrangement and unfolding of the char-acter's body-building movements are determined first and foremost by a "cultural series" (see Gaudreault 2011) other than kinematography. Sandow's impressive flexing of his muscles comes out of the sphere of fairground performance and brings with it the memory of this other series. It is the same situation as that of Méliès when he served up stage routines, which he adapted for the screen. Here adaptation consists in adapting or, better yet, if the reader will allow us this play on words, in adopting, by means of a new technology, one or more pre-existing cultural series. It is thus a case of borrowing an ensemble which has been pre-determined by the cultural series 'pre-constrained package' to which it belongs.

Figure 28.1 Photogram taken from the Kinetoscope film *Sandow*, recorded at the Edison studio on 6 March 1894

In the beginning, then, one did not adapt, at least in the strong sense of the term: one reworked or adopted the know-how and practices of surrounding cultural series. It was only with the arrival of institutional cinema and the rise of narrative that the question of adaptation could take on its strong meaning. It was only then that the cinema would systematise its quest for 'scripts,' with other narrative media (theatre, literature, and even the printed press and its sensational little news items) serving it as a vast pool of hypotexts or 'template texts.'

Adaptation and piracy: theatre versus cinema

Once we view adaptation from an intermedial perspective (see Rippl 2015), as writing across media, a number of observations on the relations between theatre and kine-attractography will enable us to advance our argument. In fact, there exist revealing traps, directly tied to our criterion of the intention of the adaptation and to the dimension of transformation through transferring in which the idea itself of adaptation can be found. A relatively naïve view of the matter would conclude that the risk of betrayal when adapting from one of these two media to the other should be relatively small, given the superficial resemblance between these two systems of expression: do not both media 'act out' a story by embodying it in flesh-and-blood characters?

We will develop this idea by commenting on the relations between theatre and cinema at the moment the kinematograph appeared. One of the first questions that crank turners at the turn of the twentieth century were confronted with was how to find subjects to film. And, of course, one of the solutions they came up with right from the start was to draw on the theatre. Indeed, the cinema seemed to involve the power of mimetic illusion as much as the theatre. Each showed living characters, in action, whose story unfolded in the space–time of a scene.

Thus, Méliès saw himself as a stage artist, even a man of the theatre, focusing on magic sketches, fairy plays, and fantastic scenes. But people soon saw that this filmed theatre, this filmic adaptation of another medium – or, if one prefers, this adoption of a well-rooted cultural series, this serial adoption – was not ideal. And the films made in this context (photoplays, *ciné-drames*) had only fleeting success.

In fact, this situation was the result of instability, of the period of turbulence around a new medium's definition and identity. The reason cinema sought to lay its hands on theatre at a certain point in time is that these practitioners had not yet become conscious of its singularity as a medium. They saw no reason for dissociating it from the cultural series live theatre. Because the new capturing–restoring apparatus had not yet attained the institutional visibility of a medium, those working in kine-attractography were content to reproduce shows, convinced that they were becoming a part of a traditional cultural series in their own manner, not truly perceiving the adaptation dimension – in the strong sense of a change of medium – of their work.

Piracy and … adaptation

In its own way, the question of copyright is a good indication of the spirit of this age, when the kinematograph did not dream of distinguishing itself from other attractional cultural series such as the theatre, fairy plays, and itinerant fairground shows. Here the idea of piracy encounters that of adaptation like a sort of shot–reverse shot, raising the question: is it piracy to simply 'borrow' the idea behind a literary text? Implicitly, the question being asked here is whether the act of borrowing this idea while using radically different signs in another media context is well and truly a form of adaptation. Here we use the term in the strong sense: that of a change of medium, in tandem with the intention of re-arranging the source work (as a template text) for a new medium. We might note in passing that this also means that the use of different mise-en-scène and putting into signs does not rule out piracy, because the definition of adaptation is not limited to reworking the storyline, or the *fabula*, to use the terminology of the Russian formalists. Here we return to the problem of evaluating the adaptational cumbersomeness.

With this idea of pirating and the gradual awareness of the need to establish a legal framework for this new means of expression, the question of media specificity and singularity gradually arose. And with it, the gradual awareness that an intentional and defined form of adaptation was necessary to supply the cinema with stories and scripts. This narrative aspect was, moreover, concomitant, we repeat, with the paradigm of institutional cinema and with the assertion of the cinematic identity characterising it.

In this respect, the reticence expressed by Edmond Sée in the early twentieth century demonstrates in its own manner the rise of a discourse on the specificities of, and thus the major differences separating, theatre and cinema. Speaking of playwrights, he remarked (in 1908):

> Nevertheless, they are well aware that their field is a different one, in which they have a duty to remain and reign *strictly* … For the beauty of theatre, of the art of theatre, is that each night it gives the impression of creating its own life for that night only! And the nobility of the actor is that he conveys this life always somewhat by chance, gropingly, and that he is thus a true creator also! But trepidation has immobilized (if I may be permitted the expression) cinema scenes "once and for all," and the way actors nonchalantly deposit their voices in the phonograph, the way they deposit their gestures of a single night – fixed, dead-revived every night here on a screen … how shameful! How sad! And what an indictment!
>
> *Sée qtd in Carou 2002: 97; emphasis in original*

Here we can see the extent to which there existed, even in an explicit value judgement such as this, a false resemblance between the two systems of expression: because of its recording system, its 'putting on record,' cinema is in no sense a live performance!

Another question of the same sort was posed at the time: was cinema's reworking (so as not to say adaptation) of a literary work a form of theatrical show or an illustrated publication? In 1909, the publisher Calmann-Lévy brought a suit against the heirs of Alexandre Dumas, who wished to profit from the Dumas repertoire, a true milk cow. A court judgement that same year described the publisher's tell-tale pretensions: "the publication in the form of a cinematographic film of a script taken from a literary work does not constitute theatrical exploitation of this work but rather only an illustrated edition of a particular kind" (Carou 2002: 244). In other words, while "theatrical exploitation" appears here to mean an adaptation in the strong sense of the term (one takes the 'script' of a novel and intentionally transposes it into another medium and expressive system), putting it on film is just a 'little' editorial variation and thus an adaptation in the weak sense of the term.

The response of Dumas's heirs was just as significant, but in this case in the opposite sense, that of taking into account cinema's singularity as a new adapting medium: "unlike illustration, the cinematic projection of a dramatic work gives rise to the representation of an animated scene" (Carou 2002: 245). They add that adaptations "compete with theatrical exploitation of a dramatic work and not with the distribution of published editions of that work" (Carou 2002: 245). In terms of financial compensation at least, cinema thus found itself on the side of theatre. This proximity of the two kinds of 'show,' and even their interchangeable nature, despite a degree of awareness of a difference, is also visible in the following comment by Nozière (in *Gil Blas*) in 1908:

> Writers have given us little books that can be played in a theatre. They have not made use of the resources offered to them by the kinematograph. The tableaux depicting "The Assassination of the Duc de Guise" could be played on the stage of the Comédie-Française … It is an ordinary piece rather than a fantasy for the kinematograph.
>
> *Nozière qtd in Carou 2002: 106*

Others, however, while acknowledging a kind of equivalence between cinema and theatre, tipped the scales more on the side of cinema's singularity, as seen in this remark by Edmond Benoît-Lévy in 1908:

> Cinema depicts scenes that could have been played in the theatre … Yes, we note this with pleasure; but for presenting these works to the public, it is quick, cheap and goes everywhere … without a stage, without actors. This is an essential difference between the two modes of representation.
>
> *Benoît-Lévy qtd in Carou 2002: 107*

Conclusion

These hesitations around the status of kine-attractography with respect to theatre in particular, and even literature, make it impossible to clearly situate the status of adaptation as a change of medium.

By inheriting an apparatus at the crossroads of various already existing intermedial combinations, the medium remained nothing more than an auxiliary to pre-existing genres, tasked with facilitating access to those genres by improving their performance and offering them the opportunity to be more widely disseminated. The medium thus remained strictly 'in the service of.' On the level of adaptation, the capturers–restorers of kine-attractography had no consciousness of adapting cultural series (in the strong sense) because they had little or no sense of being a part of them. In sum, in this kind of ambient, diffuse intermediality, one did not adapt; one was content to adapt to the surrounding cultural series. Here we find the ecological conception of adaptation discussed above. Or, if one prefers, and to return to our play on words of a moment ago, one adopted the contours of other cultural series to which one felt one was indebted. There is no adaptation in the strong sense here, simply little borrowings among friends.[9] This phase of coalescence was thus characterised by a kind of 'spontaneous' intermediality.

When it became institutional cinema, the kinematograph acquired visibility as a medium. Clearly distinguished from other media, institutional narrative cinema grasps hold of storylines or scripts fully aware that it is adapting in the strong sense of the term. Meaning that, as a medium that has become responsible and autonomous, it manages its intermediality by adapting content from other media. In this very act, we can see a kind of paradoxical consciousness, one of singularity and complementariness alike: adaptation has come to enable cinema to feed its singular media imagination fully aware of what it is doing.

Notes

1 This text was translated by Timothy Barnard. The research on which the present text is based has benefited from the financial support of the Fonds de recherche du Québec – Société et culture (FRQSC), the Social Sciences and Humanities Research Council of Canada (SSHRC), and the Canada Research Chairs program, through the intermediary of three university infrastructures headed by André Gaudreault: GRAFICS, the research partnership TECHNÈS, and the Canada Research Chair in Cinema and Media Studies. The Belgian half of the authorship team's reflection were carried out under the aegis of the Observatoire du récit médiatique (ORM) of the Université Catholique de Louvain. The present text extensively reworks one previously published in French: Gaudreault, A. and Marion, P. (2006) "Un art de l'emprunt: Les sources intermédiales de l'adaptation," in C. Fratta (ed.) *Littérature et cinéma au Canada (1995–2005)*, Bologna: Edizioni Pendragon, 13–29.

2 This is what J.D. Connor (2007) calls the "fidelity reflex": "What I am calling the fidelity reflex, though, is not the persistence of the discourse, but the persistent call for it to end."

3 Already the very notion of 'fidelity' is not without a moral connotation, at least when one associates it with a certain conception of human love. Robert Stam takes pleasure in recalling the tendentiously moralising vocabulary one finds in every direction in discussions of adaptation: "Terms like 'infidelity,' 'betrayal,' 'deformation,' 'violation,' 'bastardization,' 'vulgarization,' and 'desecration' proliferate in adaptation discourse, each word carrying its specific charge of opprobrium. 'Infidelity' carries overtones of Victorian prudishness" (Stam and Raengo 2005: 3).

4 Translated by Timothy Barnard.

5 For an initial discussion of the neologism "*médiagénie*," see Marion (1997).

6 On these questions, see Marion (2003, 2013).

7 The metaphor of the surveillance camera may seem abusive. For the surveillance camera has a static and impersonal quality: it is neither more nor less than a 'blind eye' which records indiscriminately everything that is placed and takes place before it, without any kind of expressive intention. Despite the presence of an undeniable kind of intentionality on the part of the earliest crank turners, as minimal as it may have been and even if only in their selection of topics to film, the image of the surveillance camera serves well, we believe, to illustrate the dominant practices under the capturing–restoring paradigm.

8 On this question, readers may also consult our volume, *The End of Cinema?* (see in particular Chapter 4 – From shooting to filming: The *Aufhebung* effect). There we establish a link between this recording and the afilmic and the almost ontological "archiving" function associated with this perspective of "minimal borrowing" of space–time carried out by the capturing–restoring apparatus: "This is the ability to archive that is proper to the medium cinema, which is based on the recording capabilities of the film stock (the silver gelatin base that gathers or collects what will *make* the document) and defines the *allographic* quality of the kind of *ontological archiving* involved in capturing–restoring what Souriau called the *afilmic*" (Gaudreault and Marion 2015: 95; emphasis in original).

9 Note, however, that in certain situations and despite the fact that it has been institutionalised for a long time, cinema today can recover this posture of respectful adoption of the features of the cultural series to which it feels indebted. This is the case in particular of the capturing–restoring of opera, when the film

medium makes it a point of honour to retransmit great operatic performances as if on the tips of its toes. For more on this topic, see chapter four of our volume *The End of Cinema?* (Gaudreault and Marion 2015).

Works cited

Bortolotti, G.R. and Hutcheon, L. (2007) "On the origin of adaptations: Rethinking fidelity discourse and 'success' – biologically," *New Literary History*, 38: 443–58.

Carou, A. (2002) *Le cinéma français et les écrivains: Histoire d'une rencontre (1906–1914)*, Paris: École nationale des Chartes/Association Française de Recherche sur l'Histoire du Cinéma.

Connor, J.D. (2007) "The persistence of fidelity: Adaptation theory today," *M/C Journal*, 10(2). Online. Available HTTP: http://journal.media-culture.org.au/0705/15-connor.php (accessed 13 March 2017).

Dicecco, N. (2015) "State of the conversation: The obscene underside of fidelity," *Adaptation*, 8(2): 161–75.

Gaudreault, A. (2011) *Film and attraction: From kinematography to cinema*, trans. T. Barnard, Champaign, IL: University of Illinois Press.

Gaudreault, A. and Marion, P. (2007) "Pour une nouvelle approche de la périodisation en histoire du cinéma," *Cinémas*, 17(2–3): 215–32.

Gaudreault, A. and Marion, P. (2015) *The End of Cinema? A Medium in Crisis in the Digital Age*, trans. T. Barnard, New York: Columbia University Press.

Giraud, J. (1958) *Le Lexique français du cinéma des origines à 1930*, Paris: CNRS.

Hermansson, C. (2015) "Flogging fidelity: In defense of the (un)dead horse," *Adaptation*, 8(2): 147–60.

Hulk (2003), dir. A. Lee. Universal. Film.

La Sortie des usines Lumière (1895), dir. L. Lumière. Société Lumière. Film.

Marion, P. (1997) "Narratologie médiatique et médiagénie des récits," *Recherches en communication*, 7: 61–87.

Marion, P. (2003) "Médiagénies de la polémique: Les images 'contre': de la caricature à la cybercontesta-tion," *Recherches en communication*, 20: 127–54.

Marion, P. (2013) "Spielberg au pays de la ligne claire: de la graphiation à la performance capture," in M. Grosoli and J.-B. Massuet (eds) *La capture de mouvement ou le modelage de l'invisible*, Rennes, France: Presses Universitaire de Rennes, 203–24.

Mathieu, Marc Antoine (2012) "3 secondes," on *YouTube*. Online. Available HTTP: https://www.youtube.com/watch?v=mhZbRLi0J2s (accessed 13 March 2017).

Partie d'écarté (1896), dir. L. Lumière. Société Lumière. Film.

Rippl, G. (ed.) (2015) *Handbook of intermediality: Literature – image – sound – music*, Berlin: de Gruyter.

Sandow (1894), dir. Dickson, W. K.-L. Edison. Film.

Serceau, M. (1989) "Un phénomène spécifiquement cinématographique," *CinémAction*, 53: 6–11.

Souriau, É. (1953) "Préface," in É. Souriau (ed.) *L'Univers filmique*, Paris: Flammarion, 5–10.

Stam, R. and Raengo, A. (2005) *Literature and Film: A Guide to the Theory and Practice of Film Adaptation*, Oxford: Blackwell.

Star Wars. Episode VII: The Force Awakens (2015), dir. J.-J. Abrams. Lucasfilm. Film.

The Adventures of Tintin (2011), dir. S. Spielberg. Paramount/Columbia. Film.

29

BLURRING THE LINES

Adaptation, transmediality, intermediality and screened performance

Bernadette Cochrane

This chapter considers the thriving practice of screening live theatre performances in cinemas with a two-fold purpose. In tracing out the problems of nomenclature arising from the incorporation of new technologies vis-à-vis audience reception into the live performance paradigm, and considering notions of liveness and mediatisation, I make the case that the live relay is a particularly sophisticated and complex adaptive instantiation in its own right.

Before shifting the chapter proper, the matter of nomenclature needs first be settled. As Margherita Laera (2014: 1) observes, "theatre returns, it always does. It returns to place where it has already been before." But if theatre always returns, so too does adapting theatrical productions for live screening. Since the 1938 screening of J.B. Priestley's *When We Are Married* from St. Martin's Theatre London (Wyver 2011), the live transmission of theatrical productions to screen – television or cinema – has had a rich and varied history. The National Theatre of Great Britain's (NT) "NT Live" (or "National Theatre Live" as it is sometimes known) and the Royal Shakespeare Company's (RSC) "Live from Stratford-upon-Avon", therefore, follow in this tradition.

With the development of both digital technologies and cinema screenings of live performance, the performative scope of companies such as NT and the RSC has expanded to reach not just national audiences but international ones. And "event cinema" (Ellingson 2014), as the broader cultural phenomenon is sometimes known, has "a huge and loyal worldwide audience" (Berwick 2014). Improvements in satellite and streaming technologies and high definition cameras mean, therefore, that "digitization is having a profound effect upon the accessibility, financing and business models of performing arts organizations, on their productions and on performers" (Towse 2013: 311). For large countries with widely dispersed populations, such as Australia, the combination of improvements in digital satellite technology and high definition cameras, as well as changes in audience expectations, is also having an impact on the activities of flagship performing arts companies. For example, and in the Australian context, the advent of such technology has meant that companies such as Black Swan State Theatre Company and Queensland Theatre can sometimes augment their respective touring programs to regional and remote areas with live relays of main-stage productions, albeit not so frequently as to "cause alarm" apropos the live relay being a "substitution for, rather than an addition to, live tours" (Fotheringham 2016: 16). The initial concerns, apropos audience cannibalisation, of the Met, the NT and the Major Performing Arts companies of Australia have been largely unrealised (Bakhshi and Throsby 2014; Fotheringham 2016; van Eeden 2011; cf. Tommasini 2013; Midgette 2016). Indeed, in the case of NT Live and vis-à-vis audience attendance, Hasan Bakhshi and David Throsby (2014: 7) found that live screenings are complementary rather than substitutional. The question of how to refer to the collective contemporary phenomenon, however, remains unsettled.

As Frances Bonner and I (2014: 121–2) have discussed previously, other scholarly work sometimes refers to live-streaming, suggesting an online activity, but, as we note, the distinguishing component is the presentation in the cinema. A repeated inclusion is the word "broadcast." Jaume Radigales (2013: 164) talks of "opera broadcasts," Daisy Abbot (2014: 8) uses "broadcast theatre" and "digital broadcasts," and Paul Steichen (2011: 443) refers to "HD broadcasts." "Broadcast" does carry the import of transmission. While the use of the word "broadcast" signals the both technological and digital status of the adaptation, for the most part, the events are, in fact, narrowcast. And, as Richard Fotheringham (2016: 10) points out,

> the transmitted material is also being recorded on an encrypted DCP (Digital Cinema Package; the 'digital equivalent of a 35mm film print'18) for subsequent licensed cinema screenings in (at the Met[ropolitan Opera], as of early 2015) some seventy other countries including Australia, usually several weeks after the live performance.

As Bonner and I note, Martin Barker, by way of contrast, uses the term "livecasting," thus avoiding the suggestion of broad dissemination, while stressing the liveness, which is crucial to the promotion of the practice (2014: 122). While "livecasting" is certainly sharper, nonetheless, to accommodate the later discussion of delayed reception (Fotheringham's "subsequent licenced screenings") the preferred term in this chapter for the phenomenon is "live relay."

With the terminology now settled, for this chapter at least, attention can now turn to the live relay and adaptation. Barker (2013: 81–93) in his conclusion to *Live to Your Local Cinema: The Remarkable Rise of Livecasting* points to a variety of lacunae when it comes to writing about and on the live relay of performances, one of these omissions being the link between adaptation studies and livecasts. For example, at the time of writing, *Live to Your Local Cinema* is the only monograph dealing with the live relay. Besides Barker, the two most sustained examinations of this relationship, to date, are the special issue "From Theatre to Screen – and Back Again" in the journal *Adaptation* (Cartmell and Parsons 2014) where three related articles focused on the live relay of performances and the ongoing work of John Wyver, with both Illuminations and the RSC. To return to both Barker (2013: 91) and adaptation, he notes in his conclusions that over the last fifteen to twenty years, adaptation studies has moved from being focused on medium theory, faithfulness and making "normative judgements" to being more broadly concerned with the "textual, contextual, paratextual, historical, institutional and reception aspects of adaptation." He observes, also, that nothing as yet "link[s] the study of livecasts with the big changes that have taken place in the past two decades in adaptation studies" (Barker 2013: 90). It is against these observations that Barker (2013: 91) asks, "when, and to whom, does a livecast become an 'adaptation'?" This chapter goes some way to answering Barker's question, locating the answers chiefly in matters of reception.

The reason for the shift in emphasis to reception is that medium theory (or questions of medium specificity) has been one of the dominant debates apropos the live relay thus far, and so has played a significant role in most discussions of these adaptations. The discussion here moves away from the considerations of medium specificity, for the most part. Too often, conversations around medium specificity carry hierarchal import, and concomitant value judgements can imply that the adaptation is inferior to the adapted work. The intention is not to engage in a form of *paragone*. The emphasis here is on how technological advancements open up theatrical productions to new and extended audiences, often remote in time and space, and on how the live relay broadens the field of adaptation studies. To this end, the live relay can be understood as "an encounter between two media – theatre and film – wherein the emphasis is on the tense and complicated, multifaceted nature of this meeting" (Béatrice Picon-Vallin qtd in Pribisic 2010: 149). Rather than examining and elucidating the live relay through a 'paragonic' investigation, I posit that a more enriching way of apprehending the phenomenon is to understand it as a complicated and multifaceted complex of lateral relationships. The multifaceted nature of the complex does mean that there are times where medium specificity does play a role in the various strands of the argument. Nonetheless, the default position here is that, understood laterally, each instantiation of the work – adapted text or adaptation – has its own validity and its own integrity. The recognition of these attributes, in turn, allows the multi-part relationship to be examined and explicated.

The discussion here of why this phenomenon can, and should, be considered an instance of lateral adaptation – both as process and product – takes up John Wyver's notion of these events as being doubled adaptations, given that each theatrical production is, inherently, an adaptation of the playtext (Wyver 2014: 104; cf. Krebs 2014), which are, in turn, further adapted through mediatisation, both transmedial and intermedial. In this way, the chapter both challenges and extends the arguments apropos the relationships between 'theatre film' and theatre and cinema (Pribisic 2010: 149), arguing that the live relay of is an adaptive instantiation in its own right. And, as such, the live relay extends the reach of adaptation studies.

The adaptive relationship between the theatrical production and the cinematic screening of that production would seem to be a given. The transferral of the content material from the theatrical medium to the cinematic medium is a straightforward transmedial shift in form. Where the transmedial shift becomes more complicated relates to Chiel Kattenbelt's observation that "at the level of the content the concept refers in particular to those media changes which become absent, for example, the way that the specific features of the source medium become lost in the process of transposition" (2008: 23). There are losses in the cinematic version of the live relay, such as the audience's right to direct their gaze where and when they will. The cinema audience is subject to the broadcaster director's decisions apropos shot sequencing, or whether they choose to go from wide shot to mid-shot to close up, or to start close and stay close, or start close and come out to wide (Caron qtd in Warner 2016: 4). This loss, while demonstrable, is, however, balanced by gains. The loss of autonomy of how and where a theatre audience member may choose to look is offset by the fact that a sensitive broadcast director can augment what the director wants to achieve on stage (Caron qtd in Warner 2016: 2). Often the cinema screenings will incorporate pre- or post-interviews and discussions with the creative team, something not usually available to the theatre audience. For audiences remote in time and space from the producing company, there is the opportunity to see productions that otherwise would simply not be available to them. The increase in accessibility implies a democratising process for this globalised phenomenon, or at least for those audience members who can access the live relay. This democratisation of access, moreover, goes in some ways to counteract the loss of immediacy, liveness and interaction. These attributes being the historical "defining and distinguishing concepts" of theatre, according to Christopher Balme (2008a: 81). The two experiences – theatrical and cinematic – will, without a doubt, be different. The live relay is, nonetheless, an inherently transmedial adaptation with the two components of that adaptation existing in a complementary and lateral relationship.

If the live relay is transmedial, so too is it intermedial. It is this intermedial quality that goes some way to militate against the loss of immediacy, liveness and interaction. If, as Balme (2008b: 206) argues, "intermediality is … understood to be: (1) the transposition of diegetic content from one medium to another, (2) a particular form of intertextuality, [and] (3) the attempt to realize in one medium the aesthetic conventions and/or patterns of seeing and hearing in another medium," then the live relay is, most definitely, an intermedial adaptation. As we saw in the previous sequence on transmediality, the live relay does not necessarily attempt to realise the aesthetic conventions of the theatre in the cinema. That said, there are some caveats to this statement to which I will return shortly. For the moment, however, following Balme's formulation above, the transposition of diegetic content and the overt intertextuality of the live relay renders it almost palpably intermedial.

In general terms, the live relay can be classified as belonging to the broader 'stage to screen' paradigm. It shares much in common with any stage play that has been made into a film. There is a significant difference, however, between the live relay differs and the more conventional stage-to-film model. Unlike the stage-to-screen model, the live relay exists in a smaller, and more accurate, domain. This domain being the live-to-digital. This difference is because, with the live relay, the shift between media is not merely a two-part transposition of playtext to film, with the cinema standing in for the theatre in this transposition. The live relay requires a three-part transposition: playtext to theatrical production to film. Rather than a cinematic version of Henrik Ibsen's *Hedda Gabler* being created, the live relay supposes that Ibsen's *Hedda Gabler* is realised in production by a theatre company such as the NT and that the material of that particular production is then transmitted to the cinema. The live relay is a guarantee by the producing companies that the diegetic content of the theatrical production will be transposed to the cinematic version of that production. While, as discussed, there are losses and gains in this transposition, for the most part, the cinema audience will expect to see a version of the theatrical production. The issue here is not one of fidelity but one of authority.

The relationship between the two versions of the production is a complementary one. There is, nonetheless, a two-fold authoritative tension in this relationship. In the instances of the NT and the RSC, both companies invoke the authority of the company and assure quality. For example, the strapline for National Theatre Live is "National Theatre Live: Experience the *best* of British theatre at a cinema near you" (National Theatre Live n.d.; emphasis added). Given that some 5.5 million people have attended NT Live since its inception in 2009 and that it is relayed to over 2,000 cinemas around the world, of which only 650 are in the UK, the cinema may well be near you. The strapline for the RSC's Live from Stratford-upon-Avon is similar: "Theatre at its *best*. Made in Stratford-upon-Avon. Shared around the World for Everyone" (Royal Shakespeare Theatre 2016; emphasis added). In both cases, the companies call upon brand recognition. In both cases, the companies are presenting *their* live relay as being superlative. In both cases, the companies are highlighting the inclusive nature of the experience. To the NT and the RSC can be added the Kenneth Branagh Theatre Company (KBTC). The strapline for the KBTC 2016 live relay project was: "The Kenneth Branagh Theatre Company, in partnership with Picturehouse Entertainment, will broadcast three productions of its year-long Plays at the Garrick season live to cinemas worldwide." (Branagh Theatre Live 2016). While the KBTC does not make the same assertions as the NT and the RSC do apropos quality, it does invoke the authority of the company in the figure of Branagh, and it does emphasise its inclusivity.

It is not just the producers of the live relay content that highlight the relationship between the theatrical and cinematic. So too do cinema chains, particularly those who cater to delayed and encore audiences. In Australia, for instance, under the drop-down menu of "films" on the Palace Cinema chain website, potential viewers are offered three choices: "now showing," "coming soon" and "stage to screen" (Palace Cinemas n.d.). Whether producer or presenter, in each case it is made clear that the cinematic experience originates in the theatrical event. To borrow from Linda Hutcheon (2013: xvi), with the live relay complex, the cinematic adaptation is both announced and presented as being deliberate by all parties, the purposive intention being something that feeds into the intermediality of the live relay.

Before discussing the extension of the intermedial nature of the live relay, however, attention should be given to the second point of authoritative tension; a tension which lies in the use of the word 'live.' In each of the three cases above, the companies are at pains to accentuate the live antecedents of the relay. In effect, each of the straplines stresses one of the historical defining features of the theatrical experience as being intrinsic to the cinematic experience. Notions of liveness, as it relates to the live relay, can be problematic (see Cochrane and Bonner 2014; cf. Wyver 2014). Nonetheless, what is important here is how these claims to liveness create the conditions for Balme's second point apropos intermediality, being "understood as an extension of the term 'intertextuality'" (Balme 2008b: 206). While not disagreeing with the contention that all texts are to some degree or other intertextual, it is Balme's narrowing of the term to mean "a specific strategy of explicit reference to particular pretexts" that is of interest (Balme 2008b: 206). The guarantee by the producing companies apropos the transposition of diegetic content is reinforced by their assurances of quality. This reinforcement is, in turn, further bolstered by the explicit reference to two given pretexts. The first pretext is the actual production, as per the argument that the producing companies are promising a version of a particular production above. The second pretext is the producing company itself; each of the straplines above stress that theatre is the core business of the company.

While I dismissed Balme's third point apropos intermediality (the attempt to realise in one medium the aesthetic conventions and/or patterns of seeing and hearing in another medium) as not being relevant to the production, it is pertinent to both the company as a pretext and to notions of liveness. One of the ways in which the producing companies flag the cachet associated with an event being live is to draw attention to the conventions of live performance. This drawing of attention to theatrical conventions is notable around the highlighting the role of the theatre audience in the preshow introduction that accompanies most cinematic screenings. It is all but obligatory that the cinema audience is shown the theatre audience taking their seats. This showing serves a three-fold purpose. First, in a harkening back to notions of inclusivity, showing the theatre audience settling in to watch the production is a reminder to the cinema audience that they are sharing the experience of the production. It is a prompt to the cinema audience that even though their experience may be a different experience, there is sufficient experiential overlap for the two audiences to have a common cause. Second, it is an affirmation of the antecedent text being a theatrical text. And finally, it is reminder of the core business of the producing company, something that we have already seen is bound up in notions of quality, all of which creates the conditions for "a particular form of intertextuality" (Balme 2008b: 206). The intertextual particularity of the live relay is created because, without the theatrical pretexts, the cinematic version just cannot exist. The live relay is predicated on specific companies producing specific productions. Remote audiences, connected via digital technology, transmediation and intermediation, can knowingly access a version of the original theatrical production.

Questions about what is known or not known by the audience relate directly to Barker's question of "when, and to whom, does a livecast become an 'adaptation'?" (2013: 91). The ontological and the epistemological implications of Barker's question are simultaneously involved yet straightforward. The live relay, in effect, conflates the two aspects of Barker's question as both considerations that resolve in the figure of the audience/s. It is at the level of reception that, ultimately, the live relay can be reconciled as adaptation – an adaptation that is technologically driven and enabled. There are two considerations here. First, there is the matter of how many audiences exist, where they exist and when they exist. Second, there is the matter of how these audiences understand what it is they are viewing.

We have already seen that discussion of the audience necessitates the conceptual separation of theatre audience and cinema audience. On the matter of the quantity, location and temporality of the cinema audience, the answer to the question depends on where and when the cinematic version of the stage production is viewed. In broad terms, there are three opportunities for cinema audiences to experience a live relay. There is the 'simultaneous' (cinema audiences watching the theatrical performance at the same time as the theatrical audience), the 'delayed' (cinema audiences watching the theatrical 'film' at a later date but on its first showing and often elsewhere in the world) and the 'encore' (repeat screenings of the original film). These variant positions, singular and plural, are created by the transmedial and intermedial 'object' that is the cinematic version of the live theatrical performance.

Vis-à-vis the simultaneous opportunity, if, as Kattenbelt (2008: 22) argues, "time and space are still the two main dimensions by which we distinguish media from each other and determine their specificity" and that "such a determination of [this] specificity … is usually related to their materiality," then with the live relay we have two distinct media operating at the same time, albeit sympathetically. The two events may be working in tandem temporally, but they are separate spatially. The two audiences are experiencing different materialities. There are two different ontologies at work: the theatrical and the cinematic. With the 'delayed' and 'encore' opportunities, the difference between these materialities becomes accentuated because the spatial separation is compounded by the temporal separation. The cinema audiences are experiencing the adapted work at one remove, which is, in and of itself, an innately adaptive relationship with that work.

This relationship is, in turn, both inflected and reinforced by considerations of *a priori* knowledge. Linda Hutcheon (2013: 120–1) in her discussion of the knowing and unknowing audience argues that for an audience to experience the adapted work as an adaptation, they need to both recognise the work as an adaptation and to know its adapted text. For the audience to experience the adaptation with what Hutcheon (2013: 120) refers to as a "palimpsestic doubleness," the "knowing" audience needs a conscious understanding of the underlying material. The notion of the "knowing audience" is predicated, therefore, on them having an *a priori* knowledge or experience of the adapted work. With regard to the "unknowing audience," Hutcheon (2013: 121) states, "without foreknowledge, we are more likely to greet a film version simply as a new film, not as an adaptation at all." The "unknowing" audience, lacking the apriorism of the knowing audience, consequently experiences the adaptation not as an adaptation but as an original singularity. Hutcheon's observations apropos the knowing audiences bring about a revisitation to the question asked by Barker vis-à-vis "when, and to whom." If the "when" is understood as encompassing the simultaneous, the delayed and the encore opportunities, then the "who" is all cinematic audiences, for, almost without exception, all cinema audiences are "knowing audiences" when it comes to the live relay.

Much of Hutcheon's work on the "knowing audience" is founded on this audience being aware of the correspondences between the adapted work and the adaptation. For Hutcheon, understanding, and often appreciating, the adaptation as an adaptation is reliant on the audience apriorism. The *a priori* situation with the live relay is somewhat more complicated. In the case of text-based theatre, the cinematic audience may well have a familiarity with the source material, that is, the actual playtext. In the instance of canonical texts such Shakespeare's *Romeo and Juliet* or Ibsen's *Hedda Gabler*, they may have seen multiple previous productions; all of which will inform their understanding of the live relay. The cinema audience may not be thinking of the screen version as an adaptation but, in the instances above, they will be experiencing it as an adaptation. And, in an oblique extension of Wyver's notion of doubled adaptations, where the production is, in and of itself, an adaptation of the playtext, in the instance of the simultaneous audience, they may already have seen the theatrical production thus creating a triple adaptation. To rework an example previously given by Bonner and myself, having attended an NT Live performance at the NT in London, on the night of the recording, and then having seen the cinema screening of the same performance in Brisbane, Australia, several weeks later, I can be confident that nothing confirms the adaptive experience quite so graphically as seeing the back of your own head sitting in an audience in front of you (on the screen) watching a performance that you saw some weeks earlier on the other side of the world (2014: 126). In this instance, the correspondences between the theatrical experience and the cinematic experience are manifold.

For the live relay, the primary correspondence may well be the theatrical production and the cinematic version of that production. There is, however, another correspondence or apriorism that is in play. This being an observation which brings about a return to the proposition of the live relay both being and marketed as a version of the production to a "knowing audience." Given the previously discussed overt signalling of the origins of the cinematic event by the producers and the distributors, there is little ambiguity about what is on offer. What is being sold by the producing company is the opportunity for them to have experience equivalent to that of the theatrical audience for that particular production. For the simultaneous audience, they *know* the theatrical performance is taking place at the same time. They *know* that they see a version or an adaptation of that performance. So too, the deferred or encore audiences also *know* that they see a version or an adaptation of the theatrical performance. The theatrical event is, after all, the *raison d'etre* of their attendance, as it is for the simultaneous audience.

While the above may seem a bold claim on behalf of the cinema audiences, in the first NT Live audience survey, "ninety-one percent of the cinema audience had been to a play that year, and only four percent said they had little or no knowledge of theatre" (NESTA 2011: 40). In a more recent survey of English theatre, "the most commonly identified motivation for people to attend [a live relay] is that it saves travel time," this being some sixty-seven percent of respondents (Karpf Reidy et al. 2016: 47). That said, some eighty-eight percent of the surveyed audiences did claim that "[b]roadcasting live theatre opens up new ways of seeing this artform [sic]" (Karpf Reidy et al. 2016: 58). And, being offered new ways of to see or understand a work or a medium is at the core of the adaptive experience. Live relay cinema audiences – simultaneous, deferred, or encore – bring with them the *a priori* knowledge of the theatrical condition. It is this apriorism that allows them the opportunity to engage with new ways of seeing and understanding the theatrical art form.

There is a final complication vis-à-vis *a priori* familiarity. Given that thirty-nine percent of cinema audiences choose to see the live relay because a production has sold out (Karpf Reidy et al. 2016: 48), this suggests a familiarity with the producing company at hand. To some degree, such apriorism is no more than to be expected from any regular theatre-goer. Nonetheless, combine this general familiarity with production specificity, and the situation changes. It is this very specificity of content with the live relay that allows all of these audiences to be positioned as "knowing audiences." The express content of each event changes with each production, but the live-to-digital paradigm is stable. And, in each and every instance of the live relay, the cinema audience is being offered an experience analogous to that of the theatre audience. Given that over 2,000 cinemas screen NT Live and 1,300 of these are located in "delayed territories" (NESTA 2011: 14), and that revenues can generate in excess of a million pounds, it is not surprising that producing organisations such as the NT would want to capitalise on these "knowing audiences" and their desire to share the theatrical experience. The broadcast technology of the twenty-first century extends the artistic reach of these theatre companies to audiences around the world via the adaptation of their core business.

To conclude, as can be seen from all of the above, that located in the live-to-digital paradigm, the live relay is a complex and sophisticated instance of adaptation. The primary focus of this chapter has been on audiences and reception as a way of determining when and for whom the live relay is an adaptation. It has also been implied throughout that there are further avenues of investigation vis-à-vis adaptation studies. For example, consideration could be given to how the technology that underpins the live relay interrupts the conventional temporal progression of the adaptive process. Does the live relay offer the opportunity to reassess notions of the first encounter? Or, bearing in mind Caron's comments on the role of the Broadcast Director in the live relay, aesthetic considerations would be fertile territory for future examinations. The live relay may be all but ubiquitous in manifestation, but there remains much about the live-to-digital paradigm to be discussed.

Works cited

Abbot, D. (2014) "Old plays, new narratives: Fan production of new media texts from broadcast theatre," in M. Hudson et al. (eds) *Proceedings of the Interactive Narratives, New Media and Social Engagement International Conference*, Toronto: University of Toronto, 8–18.

Bakhshi, D. and Throsby, D. (2014) "Digital complements or substitutes? A quasi-field experiment from the Royal National Theatre," *Journal of Cultural Economics*, 38(1): 1–8.

Balme, C. (2008a) "Surrogate stages: Theatre, performance and the challenge of new media," *Performance Research: A Journal of the Performing Arts*, 13(2): 80–91.

Balme, C. (2008b) *Cambridge Introduction to Theatre Studies*, Cambridge, UK: Cambridge University Press.

Barker, M. (2013) *Live to Your Local Cinema: The Remarkable Rise of Livecasting*, Basingstoke, UK: Palgrave Macmillan.

Berwick, I. (2014) "Hot ticket: Live-streaming theatre productions in cinemas," *Financial Times*, 20 August. Online. Available at: https://www.ft.com/content/d5571c1a-2dcf-11e4-8346-00144feabdc0 (accessed 3 March 2016).

Branagh Theatre Live (2016) Online. Available at: http://www.branaghtheatrelive.com/ (accessed 6 January 2017).

Cartmell, D. and Parsons, E., eds. (2014) "From theatre to screen – and back again!" Special issue of *Adaptation*, 7(2).

Cochrane, B. and Bonner, F. (2014) "Screen from the Met, the NT, or the House: What changes with the live relay," *Adaptation*, 7(2): 121–33.

Ellingson, A. (2014) "Event cinemas a growing business for movie theatres," *L.A. Biz*, 11 April. Online. Available at: http://www.bizjournals.com/losangeles/news/2014/04/11/event-cinema-a-growing-business-for-movie-theaters.html (accessed 1 May 2014).

Fotheringham, R. (2016) "Screening live performance: Australia's major theatre companies in the age of digital transmission," *Australasian Drama Studies*, 68: 3–33.

Hutcheon, L. (2013) *A Theory of Adaptation*. 2nd ed. Abingdon, UK: Routledge.

Karpf Reidy, B. Schutt, B., Abramson, D., Duraki, A., Casale, L., and Throsby, D. (2016) *From Live-to-Digital: Understanding the impact of digital developments of audiences, production and distribution*, AEA Consulting, October. Online. Available at: http://aeaconsulting.com/uploads/200002/1476388267228/AEA_-_From_Live_to_Digital_-_complete__FINAL.pdf (accessed 10 January 2017).

Kattenbelt, C. (2008) "Intermediality in theatre and performance: Definitions, perceptions and medial relationships," *Cultura, Lenguaje y Representación/Culture, Language and Representation*, 6: 19–29.

Krebs, K. (2014) "Ghosts we have seen before: Trends in adaptation in contemporary performance," *Theatre Journal*, 66(4): 581-90.

Laera, M. (2014) *Theatre and Adaptation: Return, Rewrite, Repeat*. London: Bloomsbury.

Midgette, A. (2016) "HD broadcasts, once the future of opera, are now seen by some as its demise," *The Washington Post*, 15 July. Online. Available at: https://www.washingtonpost.com/entertainment/music/hd-broadcasts-once-the-future-of-opera-are-now-seen-by-some-as-its-demise/2016/07/14/d576 3340-3406-11e6-8758-d58e76e11b12_story.html (accessed 20 July 2016).

National Theatre Live (n.d.) Online. Available at: http://ntlive.nationaltheatre.org.uk/ (accessed 7 January 2017).

NESTA (2011) *Digital Broadcast of Theatre: Learning from the pilot season – NT Live*. Online. Available at: https://www.nesta.org.uk/sites/default/files/nt_live.pdf (accessed 12 February 2013).

Palace Cinemas (n.d.) Online. Available at: https://www.palacecinemas.com.au/ (accessed 16 January 2017).

Pribisic. M. (2010) "The pleasures of 'theatre film': Stage to film adaptation," in D. Cutchins, L. Raw and J.M. Welsh (eds), *Redefining Adaptation Studies*, Lanham, MD: Scarecrow Press, 147–59.

Radigales, J. (2013) "Media literacy and the new entertainment venues: The case of opera in movie theatres," *Communication & Society/Comunicación y Sociedad*, 26(3): 160–70.

Royal Shakespeare Company (2016) "Winter 2017 Season Guide." Online. Available at: https://cdn2.rsc.org.uk/sitefinity/season-guides/rsc-winter-season-2017-guide.pdf> (accessed 6 January 2017).

Steichen, J. (2011) "HD opera: A love/hate story," *The Opera Quarterly*, 27(4): 443–59.

Tommasini, A. (2013) "A success in HD, but at what cost?" *The New York Times*, 14 March. Online. Available at: http://www.nytimes.com/2013/03/15/arts/music/mets-hd-broadcasts-success-but-at-what-cost.html (accessed 10 February 2016).

Towse, R. (2013) "Performing arts," in C. Handke and R. Towse (eds), *Handbook on the Digital Creative Economy*, Cheltenham, UK: Edward Elgar, 311–21.

Van Eeden, S. (2011) "The impact of *The Met: Live in HD* on local opera attendance," Master of Arts, Vancouver, BC: University of British Columbia.

Warner, C. (2016) "Interview with Ben Caron – broadcast director," in *Romeo and Juliet by William Shakespeare – Education Pack: Interviews with Cast & Creative Crew*. Kenneth Branagh Theatre Company. Online. Available at: https://drive.google.com/file/d/0BzNRGoo0co1WNGpYNk9nbmNoMHM/view (accessed 6 January 2017).

Wyver, J. (2011) "In the beginning: *When We Are Married* (BBC, 1938) 1.," *Screen Plays: Theatre Plays on British Television*, 22 September. Online. Available at: https://screenplaystv.wordpress.com/2011/09/22/in-the-beginning-when-we-are-married-bbc-1938-1/ (accessed 10 April 2015).

Wyver, J. (2014) "'All the trimmings?': The transfer of theatre to television in adaptations of Shakespeare stagings," *Adaptation*, 7(2): 104–20.

30

SIDEWALK STORIES
Re-sounding silent film

Julie Grossman

Long before the homage to silent film *The Artist* won the 2011 Academy Award for Best Picture, independent filmmaker Charles Lane adapted classic silent film to contemporary subject matter with his underappreciated *Sidewalk Stories,* a 1989 black-and-white silent-film reworking of Chaplin's *The Kid* that focuses on contemporary homelessness in New York City. Considering Verevis and Loock's distinction between remakes and adaptations, *Sidewalk Stories* is like a remake in that it offers a "version of another film," but also it is an adaptation in its "movement between *different semiotic registers*" (Verevis and Loock 2012: 6), adapting a form that is generally understood as tethered to an historical moment and thus has an altered semiotic effect in a contemporary context.

No longer defined by what is technologically possible, sound-era silent films raise questions about why filmmakers might adapt an obsolete cinematic form. In *The Artist*, Michael Hazanavicius paid tribute to early film, with *The New York Times* gushing that the film is "an irresistible reminder of nearly everything that makes the movies great" (Scott 2011). While the film allows audiences to luxuriate in a nostalgic view of early cinema, *The Artist* has little critical distance from its source. Instead, it relies upon the charm of clever references to silent film and the charisma of its lead actor, Jean Dujardin, to disarm viewers. Both of these features make most of the film very enjoyable to watch, but *The Artist* lacks the rich texture of other films adapting silent tropes in the sound era.

In contrast, Charles Lane's *Sidewalk Stories* denies nostalgia by jolting audiences into reflecting on a contemporary social problem. Rather than only delighting audiences, which no doubt it also does, *Sidewalk Stories* seeks to unsettle viewers. *Sidewalk Stories* exemplifies how technology may play a crucial role in adaptations without there being necessarily an advance or progression (i.e., 'new technology=updated text'). Further, the silent film form breaks out of its technologically defined historical era to make viewers aware of the assumptions we bring to our viewing practices: what is old, for example, and what is new. Deconstructing this binary, the newness of Lane's film relies on an old form but without seeking a nostalgic return to the past.

Sidewalk Stories exemplifies Peter Brooker's understanding of refashioning earlier works as postmodern Brechtian artmaking, whereby an adaptation "'re-functions' both the form and content of its source text so as critically to address the changed cultural and political circumstances of its own time" (2007: 114). Lane's film presents an unusual portrait of urban homelessness by employing silent-film form to display a "dialectical tension between its whimsically nostalgic formal approach and its bold representation of pressing contemporary issues" (Clark 2013). The film toggles between an appealing humour drawn from Charlie Chaplin, Buster Keaton, and Harold Lloyd (as well as Lane's own talent for pantomime) and a distinct presentation of homelessness, "a living reproach to the workaday world" (Hoberman 1989). *Sidewalk Stories* finds in the homeless street artist's adaptive responses to exigency an analogue to a work of art's creative adaptation of its sources. Like Chaplin's *Tramp*, Lane's Artist demonstrates imaginative improvisation, a psychosocial skill that helps to navigate the pressures and constraints of contemporary life.

As *The Village Voice* observed in a review of the film in 1989, Lane "[evokes] the compassion of the little man [Chaplin's *Tramp*] and his impotence in living in a sociologically defeatist environment." However, the honorific element of *Sidewalk Stories* cloaks its "quiet [radicalism]" (Clark 2013). The film disrupts viewers' expectations, questioning not only the boundaries that divide old forms from new content but also, within that content, the boundaries that typically separate urban spaces from one another in the cultural imagination. *Sidewalk Stories* accentuates the breakdown of social pacts, these fractures appearing more extreme when they are seen through the absurdist lens of silent comedy. As Noel Murray said in 2014, *Sidewalk Stories* creates "a new frame around familiar sights … forcing the audience to reconsider them."

Especially when contrasted with *The Artist* (2011), which announces its nostalgia in the blurred fonts of its opening credits, *Sidewalk Stories* illustrates not only the omnipresence of multiple sources for adaptations (cf. Leitch 2003; Klein and Palmer 2016), but a call to draw distinctions between critique-oriented adaptations, such as Lane's film, and the limits of *homage* in a film like *The Artist*. If adaptation has been seen as the poor cousin to its often (more) literary or (more) treasured sources, an assumption critiqued by Thomas Leitch, Kamilla Elliott (2003), and others, *Sidewalk Stories* calls into question the cliché that the progression of forms and ideas is necessarily an additive process. Instead, art forms may progress through diminution, by denying or withdrawing conventional traits. Such is the case in a film like Russell Rouse's *The Thief* (1952), which includes diegetic sound but no voiceover or dialogue. This postwar film-noir movie about stolen atomic secrets relies, instead, on expressive acting by Ray Milland, an engaging moving camera, and rich cinematography. *The Thief* experiments with a minimalist aesthetic, but in withdrawing dialogue directs attention to visual style, camera work, and performance.

Sidewalk Stories goes further to withdraw diegetic sound almost entirely, producing for contemporary viewers a relentless cognitive dissonance. The film is allied in this sense with John Cage's avant-garde musical composition *4′33″* (1952), which challenged its audiences by providing all the trappings of an old form, a concert replete with a baby grand piano, only to withhold conventional performance and its familiar sounds. The 'music' of this piece inhered in the silence that followed. *Sidewalk Stories* functions similarly to withhold. In an effort to emphasise 'ways of seeing,' to borrow John Berger's well-known title (2009), Lane introduces dissonance by upsetting expectations for sound and dialogue and merging a contemporary social issue with an antiquated form.

As adaptations provide multiple perspectives on known material, *Sidewalk Stories* presents different viewpoints on urban poverty, interpreting imperviousness to homeless individuals as noxious, but also portraying these outsiders as stealthy artists. The silent-film form allows Lane to stage physical comedy alongside a "flexible score" by Marc Marder (Clark 2013). Drawing from multiple music traditions, the score is a crucial voice in the film, very pointedly adapting the eclecticism of urban street life into varied classical and contemporary musical forms of expression.

Like Chaplin's first-directed feature film *The Kid*, *Sidewalk Stories* is a tale of two cities, documenting the upscale and squalid sides of urban modernity. If Chaplin's 1921 film was meant to evoke the late-Victorian London of the filmmaker's childhood (though it was filmed at Chaplin's studio and on location around Los Angeles), Lane's film begins with contrasting views of commercial New York City's Wall Street and the Greenwich Village street on which men and women with no resources live and perform for small means of subsistence. The film quickly establishes human creativity and institutional exploitation as parallel characteristics of urban space. While the silence of the film is a metaphor for indigent people having no voice, Lane's performance, cinematographer William Dill's lush black and white photography, and composer Marc Marder's score counter the depressing tone of a visual documentary of homelessness with vignettes combining humour, art, and social critique. In its dramatic reorientation on Chaplin's *The Kid*, *Sidewalk Stories* shows the potential for social commentary embedded in creative appropriations of known texts and genres.

As a return to sources that provides a different perspective on culturally familiar material, *Sidewalk Streets* exemplifies critique-oriented adaptation. Elsewhere, I argue that such adaptations can be seen as "monstrous," forcing viewers and readers to assimilate to reconsiderations of sources that are difficult, that lack at least initially the pleasure theorists have often associated with the adaptation's work to create "repetition with variation" (Hutcheon 2006: 4). "Hideous progeny" of their sources, adaptations that are aggressive in bringing a new perspective to light may jar viewers who are forced to grapple with a return to treasured sources made uncanny (Grossman 2015). The figure of the monster seems particularly useful when the text under consideration, Lane's *Sidewalk Stories*, takes on the subject of urban homelessness by adapting *The Kid* to Ronald Reagan's America, where street people were seen as monstrous blemishes on the splendor of modern New York City. Infrastructural failures combined with a ruling ideology of indifference, exemplified in Reagan's declaration, "I don't believe that there is anyone going hungry in America by reason of denial or lack of ability to feed them; it is by people not knowing where or how to get this help" (Anon., *New York Times* 1986). A decade after the release of *Sidewalk Stories*, New York City's Mayor Rudolph Giuliani began his initiative to sanitise New York by 'cracking down' on the homeless, but the work-for-shelter programs that followed were seen by many as very unforgiving, failing to address the larger class and mental-health issues at the heart of urban poverty (cf. Morse 1999).

Lane's film introduces the Artist figure, played by Lane himself, accompanied by his own Joplinesque musical refrain in the score, a version of which Marc Marder had composed twelve years earlier when the two collaborated on "A Place in Time," Lane's 1977 short precursor to the later feature film. *Sidewalk Stories* begins with the Artist sitting by his easel, biding his time, as the scene that surrounds him is full of expressive city dwellers, many of whom, like the Artist, have no home. His peers perform for dollars and cents, while, in the background, we see a banner that conveys one socio-economic context for these sidewalk stories: urban gentrification. In New York City in the late 1980s, Greenwich Village, like other parts of the city, was threatened, in Lane's words, by "a capitalist bulldozing movement which would besmirch much of the artistic, bohemian landscape with small to modest sized structures which celebrate the height and stature of each individual" (Lane 2015). Lane emphasises the lack of "height and stature" of his character in *Sidewalk Stories*, a street artist seen here vying against the competition. The scene introduces Lane's visual humour, as he spars with the much bigger man after trying to steal away his customer (Figure 30.1). J. Hoberman (1989) observes that the large man may also represent audience resistance to "little-guy" Charles Lane and his strange mashup of an outdated form and challenging content.

Lane uses strategies from silent film, visual gags that rely, as the musical score does as well, on repetition: the Artist is pushed by the taller man several times, going down to the ground and repeatedly springing back up. The final time he arises, he lies down before his opponent has a

Figure 30.1   The Artist and his competition in *Sidewalk Stories*

chance to push him again. The comedy is familiar from Chaplin and Keaton, but lacks the pacing of gags dictated by the technology of silent film, when undercranking (filming at 16 frames per second [fps] then projecting at 21–24 fps) served the cinema, since film stock threatened to catch fire if it was not projected at a higher speed, and projecting films at a higher fps also economised delivery of content, with more films able to be shown during theatrical exhibition slots (cf. Model 2015). In *Sidewalk Stories*, the first scene shows several business people pulling a guy out of a taxi on Wall Street because they think they hailed the cab first. Were it shot at 14 fps and projected at 21 or 24 fps, the gag would have Chaplin's silent era frenetic 'pop' associated with early film comedy.[1] Charles Lane's avoidance of undercranking during chase scenes and physical gags in *Sidewalk Stories* marks its hybrid tone and adaptation of silent comedy into a more serious, naturalistic portrait of an authentic contemporary reality. As Lane comments in an interview included in the DVD extras, he wanted to strip the "histrionics" from the silent film form, exemplifying Julie Sanders' understanding of appropriations whereby "a political or ethical commitment shapes a writer's, director's, or performer's decision to reinterpret a source text" (2016: 3).

In the condemned building where he sleeps, the Artist in *Sidewalk Stories* is similar to the character in Ralph Ellison's *The Invisible Man*, as Clark also notes in her review of the film in *Film Comment* (2013). Having "stolen electricity," he has jury-rigged the wires to "[tap] into and [siphon] from electrical sources" (Lane 2015), providing his makeshift household some light. Chaplin also engineered a domestic haven in *The Kid*, after this baby's mother had abandoned him in despair on the streets. In one scene, we see young baby John in his hammock cradle drinking from a tea kettle fashioned into a baby bottle (Figure 30.2).

Another film Lane credits with having influenced *Sidewalk Stories* is *Tiger Bay* (J. Lee Thompson, 1959). Shot on location by the Cardiff docks and hillsides, the film helped to usher in the British New Wave. *Tiger Bay* is about a Polish sailor Korchinsky (Horst Buchholz) on the run with the eleven-year old tomboy Gillie (Hayley Mills) who witnessed him kill his girlfriend in a fit of jealous passion. Korchinsky and Gillie's unlikely bond creates a space for expression on the margins of a social world from which they are both estranged – seen, for example, in the final two images below of their imaginative and energetic playacting in the Welsh countryside. For outcasts The Tramp, Korchinsky, Gillie, and Lane's Artist, improvisation and fabrication serve

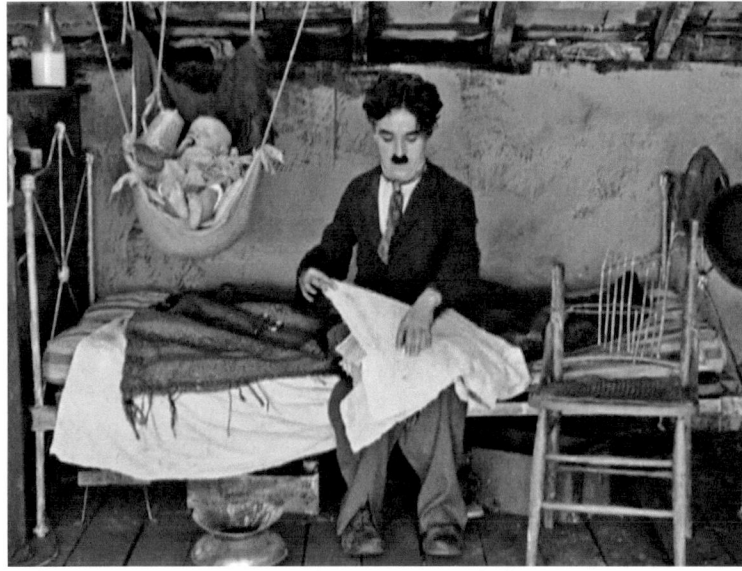

Figure 30.2 Charlie Chaplin takes care of Baby John in *The Kid* (1921)

Figure 30.3 Korchinsky observe the wedding dance in *Tiger Bay* (1959); Korchinsky and Gillie playact

as adaptive ways of surviving meaningfully. More importantly, however, in *Tiger Bay*, Lane was drawn to the depiction of a racially diverse and vibrant street culture, strikingly illustrated in the scene in Figure 30.3 of the black couple's wedding dance on the nighttime streets, which Korchinsky and Gillie observe. Earlier, Gillie had sung in church during the couple's wedding mass, further underscoring the blended nature of the community (Figure 30.3).

The main story of *Sidewalk Stories* is set in motion by a man knifed to death because of a gambling debt. The Artist has inadvertently observed the crime, squatting alongside the murder victim's young child, whom he had earlier sketched on his sidewalk easel when the family passed by. The Artist now attempts to take the knife out, in a shot that resonates with Hitchcock's *North by Northwest*: like Roger Thornhill (Cary Grant's character), the Artist is implicated in a crime he did not commit, and so is positioned as defensive throughout the narrative. However, he has also found himself with the murdered man's toddler, and takes her 'home.' Here, Lane parodies a domestic ritual (the child's breakfast), sound effects adapting a feeling – a desire to connect – into audio gestures. With its trumpeted reveille, the score introduces the challenges of domesticity without its appliances (the milk carton sits by the open window so that it stays cold). In his insistence that the girl enjoy their moments together, the Artist attempts to teach her to wink, accompanied in the score by pings on a xylophone. When he hangs the girl's laundry in his living quarters, the score adapts "This is the way we wash the clothes … so early in the morning." The film's score repeatedly emphasises an ironic disjunction between individual resilience and hostile living conditions. Later, as wealthy mothers in the playground inch away from a homeless man on a bench near them, Marder quietly weaves "My Country, 'Tis of Thee" into his score. In a scene taking place at a Burger King, the Artist pretends with the little girl that she is on a working carousel (he physically shakes a mini-merry-go-round to simulate the sensation of movement for the little girl), the score riffing on "Merrily We Roll Along." Music creates pleasure for the viewer not only in the repetition of musical themes for many of the characters, but also in its allusion to recognisable ditties, emphasising the film's commingling of human pleasure and deprivation. As the Artist considers approaching a police officer on the street to help him find the little girl's home, the score plays with "East Side, West Side (All Around the Town)." The reference functions ironically, since the workings of life in the city, 'east side, west side,' will fail the Artist. Here, he well understands that he will be seen as a criminal if he engages the police to help him find the girl's mother. Repeatedly, the score samples music to express characters' desires for 'normalcy,' a wish for comfort represented in the presence of familiar pieces of music, as in the score's invocation of "All around the mulberry bush / The monkey chased the weasel," when the Artist tries to bathe the little girl quickly during a dinner at the apartment of a woman (Sandye Wilson) who had earlier been a customer. In the context of extreme living conditions, the musical references exemplify stealth and adaptation as artistic acts keyed to human need and survival.

Charlie Chaplin was a master of adapting sentiment to score and sound effects, seen most deftly in *Modern Times*, to which *Sidewalk Stories* also refers. If Lane's film adapts a story and character patterns from *The Kid*, it borrows a fundamental *raison d'être* from the later Chaplin film, which was released in 1936, seven years after the silent film era essentially ended. As Jeffrey Vance observes, "Chaplin took a great risk in making an especially speechless film so many years removed from the end of the silent era" (2003: 219). If Chaplin was at first plagued with a "depressing fear of being old-fashioned" in making *Modern Times*, he channeled such anxiety into a brilliantly self-reflexive cinematic discourse. *Sidewalk Stories* shares this bold design, eschewing even the intertitles that would facilitate communication of the story. But it is dialogue rather than sound effects that is mocked in *Modern Times*, in which talking often signals cacophony and Chaplin's initial disdain for sound movies. From the factory owner's autocratic spoken announcements to work harder and faster in the opening sequence, to the spoken directions in the factory about how to use the feeding machine, to Chaplin's performance of a nonsense song at the end of the film, audible words distract from the pleasures of the visual. Like *Modern Times*, *Sidewalk Stories* presents sound as intrusion and violation, becoming an aural figure for the failure of society to *see* humans for their individual worth. The conclusion of *Sidewalk Stories* places the Artist in a park among the indigent, as the film gradually gives voice to these people at the margins, literally, by introducing sound in the final moments of the film to broadcast the pleas of the homeless.

Before its conclusion, *Sidewalk Stories* shows the Artist comically inhabiting public spaces while revealing their general unreceptiveness to figures of the Other. Lane shoots the interior of a children's clothing store from the point of view of a surveillance camera. Like other homeless men and women of color, the Artist is unseen when it matters, when common humanity might compel observation of the plight of homeless people. Here, however, the Artist steals clothes in plain sight, calling attention to the perspective from which he is seen: a panoptical viewpoint of institutional authorities. In this particular sense, the city's marginal characters are always watched.

In an interview in 2014, Lane explains how he came to make the film. He was on a subway in New York in November 1988, having just seen what would later be known as a legendary boxing match between Sugar Ray Leonard and Donnie Lalonde. A homeless person approached Lane, who exhibited the discomfort of guilt and sympathy, expecting a request for money. Instead, the person wished to participate in Lane's conversation about the boxing match. The film reflects closed habits of seeing within urban space that are motivated by viewers' assumptions and judgments, calling these biases into question and encouraging us to see a humanity in our environment rather than a reflection of our own sometimes rigid categories of understanding.

The vignettes or 'sidewalk stories' alternate in Chaplinesque fashion between showing the intransigence of public authorities and institutional exploiters on the one hand, and the potential of individual resourcefulness on the other. One scene shows the Artist teaching the little girl to dance. A few minutes later, Lane shows the girl's desperate mother in a police station portrayed as bureaucratic and unconcerned, as we see the cops glad-handing one another while the mother suffers alone.

Public institutions are antagonistic, pitted against individual imagination, resourcefulness, and pathos. In *The Kid*, the County Orphan Asylum representatives come to take young John away from his informally adoptive father, leading to the film's perhaps most famous image of Jackie Coogan pleading from the backside of a truck for his 'daddy' (Figure 30.4).

"It is," Jeffrey Vance observes, "a powerful, raw performance from a young child that has lost none of its emotion with time" (2003: 111). In *Sidewalk Stories*, the Artist knows well enough to avoid public authorities and problem-solves on his own, outside of the institutional structures which document only in order to restrain or damage him. When the Artist considers approaching a police officer for help, Lane cuts from a medium close-up of the contemplative artist to a fantasy sequence, in which he imagines how his story will be received (Figure 30.5): in the lineup, he drags on a cigarette, then is seen receiving his sentence. Though presented through the lens of comic irony, the shots foreground the racism that keeps the Artist and little girl on the

Figure 30.4 Jackie Coogan as 'The Kid'

Figure 30.5 The Artist's understanding of his bleak social status

Figure 30.6 The Artist sees himself; mirrors in *The Artist*

margins. For this reason, Ed Guerrero places Charles Lane alongside other African-American filmmakers – for example, Matty Rich, Julie Dash, Bill Duke, Carl Franklin, and others – who, on the eve of the 1990s, had "begun to take responsibility for *framing blackness* away from the dominant Hollywood apparatus" (1993: 1). The celebration in *Sidewalk Stories* of a racially diverse street culture, alongside the film's confrontations with racism and homelessness, achieves a unique affective tone that appropriates Chaplin's Tramp(s) for a different audience. With *Tiger Bay* and its representation of racial blending serving as an interpretant, to borrow Lawrence Venuti's term designating texts that mediate among other texts in translations or adaptations, *Sidewalk Stories* withdraws Victorian sentimentality from its hypotext *The Kid* (2012).

Adapting to his new paternal role, the Artist steals some clothes for the little girl, although the benevolent owner turns out to be a woman whose picture the Artist drew on his sidewalk easel. She has already established an affinity for him and, now, the little girl. They all meet repeatedly, culminating in the Artist and little girl coming to the woman's apartment for dinner. If the film invokes a fantasy that there might be a person able to transcend the biases of her socio-economic class, Lane nevertheless undercuts the happy domestic portrait by punctuating the dinner scene with visual reminders that the Artist is utterly out of place and must make use of the setting to advance his and his little girl's well-being, sneaking baths and stealing candelabras. The woman invites him to stay the night, and he indulges a fantasy of sexual escapades with her – but then is brought back to the reality of his circumstances with the little girl, especially when he discovers subsequently that the homeless shelter mission is full.

While he is at the woman's apartment, the Artist catches sight of himself in a mirror, a jolting experience (Figure 30.6). His humanity often unseen, he is nevertheless watched, but he is never seeing himself, the mirror being an icon of privilege.

Here, as Lane (2014) has said, "The Artist is startled by the immediate, unexpected image of a homeless stranger he does not instantly recognize." The scene invites comparison with

Hazanavicius's *The Artist*: Mirrors in *The Artist* are about the luxury and narcissism of seeing one's self and crafting a persona in a celebrity culture. The aptly named Peppy draws a mole on her face so that she can be 'different,' making her the same as other starlets; mirrors point to the illusions that undergird our sense of self – George Valentin's mugging, for example, a performative gimmick that won't survive the coming of sound.

Critics and theorists have increasingly relied upon the assumption that adaptations need not be nostalgic, that they can be forward-looking rather than mainly "archeological" (Leitch 2012: 99). Moreover, according to Leitch, adaptations can advance a literacy which carries the "inevitable element of being able to recombine elements, not just gobble them up" (Leitch 2008: 21). In contrast, *The Artist* falls into outdated theoretical categories for understanding the work of adaptation because it is an insular film that lacks the critical power to provide new perspectives on familiar material. Instead, *The Artist* refers to silent-movie conventions to titillate viewers with its cleverness and nostalgic A-Star-is-Born story about the rise and fall of celebrities. The fantasies portrayed in *Sidewalk Stories* are grounded instead in our knowledge of a larger context, the socio-economic position that oppresses its protagonist, despite his seeming imperviousness to trauma. The trauma in Lane's film challenges viewers because it characterises the socio-economic setting, not the individual characters in it, as is the case in *The Artist*.

Hazanivicius's film begins with a movie premiere featuring heartthrob George Valentin (Jean Dujardin) playing a dashing swashbuckler tortured by electrocution to "SPEAK!" as the intertitles tell us. George Valentin's character's sidekick dog Uggy saves him, and we see the audience's reaction to the scene – shock, pleasure – without seeing the dog's rescue of George's character, foregrounding the idea that the cinema exists for the pleasure of the audience (as producer Al Zimmer [John Goodman] later affirms, "the public is never wrong"). George's character's escape from the villains includes his donning his superhero mask with a mug to the camera. He is seen in a wide-canted angle so that the shot, which dramatically features expressionist chiaroscuro shadows, includes the grandeur of the theater, with orchestra and conductor. The film within the film ends with George saving 'the girl' and piloting a plane for an escape to safety, as the title "Long live free Georgia" reminds audiences that the film is one in a line of crowd-pleasing hero pictures (the next will be "A German Affair"). After the curtains fall, Valentin greets his adoring audience, doing repeated jigs and a tap dance on the stage. His arrogance is evident in his invitation to his dog, not his leading lady, to join him on stage.

The film is filled with clever gags and ironic references to sound and speaking (signs backstage at the premiere read "Please Be Silent Behind the Screen"; later, George's wife complains to him, "Why do you refuse to talk?"). As the film develops into an overlong and melodramatic sequence foregrounding George's inability to adapt to the sound era, a motif most famously pursued in *Sunset Boulevard* (1950) and *Singing in the Rain* (1953), Hazanavicius depicts George's experience in a nightmare in which objects around him make sounds, but he cannot speak. Just after this, the film introduces Bernard Hermann's famous *Vertigo* score, a choice that elicited an extreme reaction from actress Kim Novak, who starred in Hitchcock's film and who described the 2011 appropriation as a "rape" (cf. Lang 2012). Novak's unfortunate comparison misses the point made in some trade papers and magazines that the musical reference to *Vertigo* is arbitrary, overplaying George Valentin's loss and validating Hazanavicius's announced plan to pay tribute to great films from the past, including reference to the mealtime montage of failed marriage in *Citizen Kane*. Unmotivated by anything intrinsic to the story or form of *The Artist*, the reference to *Vertigo* may remind viewers that most of what is in *The Artist* is repetition *without* variation, to tweak Hutcheon's phrase quoted earlier.

In *Sidewalk Stories*, as is often the case in Chaplin, a perversely comic interaction between pleasure and socio-economic need is staged in public places, consumption sites, and social institutions (prisons, work houses, department stores), where psychological adaptation to privation and suffering involves an imaginative transformation of public spaces based on individual desire, modelling the extent to which adaptation as an artistic act is inseparable from a psychosocial response to the difficulty and challenge of source material.

Inverting expectations, the Artist teaches the little girl how to adapt to street life. She becomes an artist herself, as a line forms on the street to have her scribble a likeness onto the easel. Scenes like this show the adaptability of those who inhabit 'sidewalk stories,' their insistence on making use of their moments and opportunities and their ingenuity. Early in the film, the Artist makes pictures of Disney and Sesame Street characters to decorate the little girl's new 'home.' In one scene, the Artist does wheelies with the little girl in her carriage on the city streets, accompanied in the score by a minuet.

The film explores the spaces of the city and the different perspectives we bring to bear on them based on our predispositions and biases. Lane surveys public settings, such as the playground, where wealthy women assault the Artist when they look up at the wrong time and assume their rich little boy was bullied by the girl (when it was, of course, the other way around). Sound effects underscore the absurdity of these women's superior stance, a mother's alarm whistle registering sound as a hysterical intervention. In a way, this is much like Charlie's nonsense song in *Modern Times* (with which Marder's score in this scene particularly resonates). Chaplin decries cinema's move to sound, using absurd lyrics to make fun of the idea that the visual mode is insufficiently 'meaningful.' Sound here adds only incoherence, and 'modern' entertainment is portrayed as gibberish. Chaplin once described the virtues of silent film: "[P]antomime is far more poetic and it has a universal appeal that everyone would understand if it were well done. The spoken word reduces everybody to a certain glibness" (Vance 2003: 365). Lane's Chaplinesque reliance on pantomime, sound effects, and music affirms Chaplin's anxiety about 'glibness:' All the acculturated men and women in *Sidewalk Stories*, except the Artist's companion, are racist buffoons or authoritarians. In the playground in *Sidewalk Stories*, for example, Lane depicts bigoted and narcissistic moms and their spoiled children failing to recognise fellow humans in the Artist and little girl. The scene reveals some 'sidewalk stories' to be solipsistic narrative projections about urban squalor, self-satisfying tall tales about what it means to be homeless. The playground women's classist aversion to the Artist and the little girl is painful, in a physical space that should signify fun and play.

In another public space, the public library is portrayed as only concerned with rules. When the little girl raps on the table, the expressionist soundtrack smashes on the piano keys to emphasise the institution's inhospitable treatment of its creative or playful occupants. Within these unreceptive environments, the Artist is seen as constantly having to regroup. His unflappability is tested when he returns home to find the abandoned building in which he has dwelled demolished. The Artist is unfazed, but the pathos of the scene is expressed in the score's use of violin. He takes the little girl to the Bowery mission house for the homeless, where he cleverly ties her to himself so that he may sleep. Later in the film, when the mission is full, the Artist and little girl wander through Penn Station, looking for a place to rest. The next scene shows where they have slept, a shocking but strangely funny and charming shot of the little girl arising from a cardboard box on the street, holding her toothbrush (Figure 30.7).

It is at this point that Lane explicitly connects the two characters' survival to multiple silent film motifs. Neighbourhood thieves undertake to kidnap the girl for ransom. The Artist steals a horse and carriage, where a couple (featuring a young Edie Falco) kisses in the back, never noticing that the vehicle has been hijacked. The Artist's desperate chase recalls the moment in *The Kid* when young John is pulled from his father figure, Chaplin's Little Tramp, and thrown on the back of a truck. The Artist chases the thieves, who are in a taxi with the screaming little

Figure 30.7 The Artist and little girl waking up on the street

girl, whose crying annoys the taxi driver so much that she and her two kidnappers are kicked out, where they are overtaken by the Artist. Jackie Coogan was chosen by Chaplin for *The Kid* because of his uncanny method acting and expressiveness. In filming the interior of the taxi in *Sidewalk Stories*, Lane himself performs a stealthy maneuver: during filming, he hid under the seat, telling the little girl, played by his daughter Nicole, "You have to take a nap now"; a two-year-old method actress, she is in fact crying, "I don't want to take a nap!" The chase scene is strongly reminiscent of the improvisational Buster Keaton as much as Chaplin's rooftop chase in *The Kid*. The sequence reinforces how pointedly allusive Lane is, addressing cinema and how movies shape our vision of things.

Doing what adaptation does best – shifting audience's perspectives on familiar form and content – *Sidewalk Stories* uses strategies from silent film to force a reconsideration not only of what we see in the modern city but also our habits of viewing. Just as John Cage experimented with music and silence, filmed stories without spoken dialogue rejigger our modes of reception. An adaptive work need not use the bells-and-whistles technology of its film-industrial or cultural moment, pace Baz Luhrmann's *Romeo + Juliet* (1996), or *The Great Gatsby* (2013). *Sidewalk Stories*, by contrast, strips contemporary conventions down, avoiding the nostalgia that *The Artist* would later rely upon. Lane's film updates Chaplin's critique of mass-organised productivity, valuing instead an art of independent, resilient, and quietly expressive improvisation.

Note

1 In another silent film in the sound era, *Silent Movie* (1976), Mel Brooks uses undercranking to exaggerate action sequences (such as a surprising slapstick chase of Paul Newman in a wheelchair) and to enhance the physical comedy of his actors, particularly Marty Feldman. In the film, Brooks plays Mel Funn, trying to enlist famous performers (who portray themselves) to appear in a silent film to assist a studio chief (Sid Caesar) ward off a corporate takeover by Engulf (Harold Gould) and Devour (Ron Carey). Funn and his two sidekicks, Marty Eggs (Marty Feldman) and Dom Bell (Dom DeLuise), court Burt Reynolds, Liza Minnelli, James Caan, Paul Newman, and Anne Bancroft, as well as Marcel Marceau, who refuses Funn's request, uttering the film's only spoken word, "Non!" Brooks confected his silent film to entertain and to make audiences feel "emotionally satisfied," as the filmmaker commented when *Silent Movie* was released (Pressbook). The film expresses Brooks's comic sensibility, presenting its humour without manipulating audiences as *The Artist* does through sentimentality and nostalgia.

Works cited

Anon. (1986) "Opinion," *New York Times*, 25 May. Online. Available at: http://www.nytimes.com/1986/05/25/opinion/ignorant-about-hunger.html (accessed 15 August 2016).

The Artist (2011), dir. Michel Hazanavicius. Studio 37. DVD.

Berger, J. (2009) *Ways of Seeing*, New York: Penguin.

Brooker, P. (2007) "Postmodern adaptation: Pastiche, intertextuality, and re-functioning," in D. Cartmell and I. Whelehan (eds.) *The Cambridge Companion to Literature on Screen*, Cambridge, UK: Cambridge University Press, 107–20.

Cage, J. (1952) *4′33″*, composition first performed by pianist David Tudor on 29 August in Woodstock, New York.

Clark, A. (2013) "Rep diary: *Sidewalk Stories*," in *Film Comment*. Online. Available at: http://www.filmcomment.com/blog/rep-diary-sidewalk-stories-charles-lane/ (accessed 8 August 2016).

Elliot, K. (2003) *Rethinking the Novel to Film*, Cambridge, UK: Cambridge University Press.

Grossman, J. (2015) *Literature, Film, and Their Hideous Progeny: Adaptation and ElasTEXTity*, London: Palgrave Macmillan.

Guererro, E. (1993) *Framing Blackness: The African American Image in Film*, Philadelphia, PA, Temple University Press.

Hoberman, J. (1989) "Young, gifted, black," *The Village Voice* 67: 1.

Hutcheon, L. (2006) *A Theory of Adaptation*, New York: Routledge.

The Kid (1921), dir. Charlie Chaplin. Charles Chaplin Productions. DVD.

Klein, A. and Palmer, R.B., eds (2016) *Multiplicities: Cycles, Sequels, Spin-offs, Remakes, and Reboots: Multiplicities in Film and Television*, Austin, TX: University of Texas Press.

Lane, C. (2014) Interview with the author. 5 October. Syracuse, New York.

Lane, C. (2015) "Re: checking in," message to the author. 21 September 2015. Email.

Lang, D. (2012) "Kim Novak clarifies 'rape' comments on *The Artist*'s use of 'Vertigo' music," *The Hollywood Reporter*, 6 Mar: http://www.hollywoodreporter.com/news/kim-novak-rape-the-artist-vertigo-297151 (accessed 4 April 2017).

Leitch, T. (2003) "Twelve fallacies in contemporary adaptation theory," *Criticism*, 45(2): 149–71.

Leitch, T. (2008) "Panel presentations and discussion, the persistence of fidelity," in *In/Fidelity: Essays on Film Adaptation*, eds. David L. Kranz and Nancy C. Mellerski, Cambridge, UK: Cambridge Scholars Publishing.

Leitch, T. (2012) "Adaptation and intertextuality, or what isn't an appropriation, and what does it matter," in D. Cartmell (ed.) *A Companion to Literature, Film, and Adaptation*, West Sussex, UK: Wiley-Blackwell, 87–104.

Loock, K. and Verevis, C., eds. (2012) *Film Remakes, Adaptations and Fan Productions*, London: Palgrave Macmillan.

Model, B. (2015) "A study in under-cranking," *The Kid* (DVD Supplements). Criterion. DVD.

Morse, J. (1999) "Cracking down on the homeless." *Time*, 12 December: http://content.time.com/time/magazine/article/0,9171,35839-1,00.html (accessed 4 April 2017).

Murray, N. (2014) "Review of *Sidewalk Stories*," *The Dissolve*. Online. Available at: https://thedissolve.com/reviews/1156-sidewalk-stories/ (accessed 20 September 2016).

North by Northwest (1959), dir. Alfred Hitchcock. MGM. DVD.

Sanders, J. (2016) *Adaptation and Appropriation*, New York: Routledge.

Scott, A.O. (2011) "Sparkling, swooning and suffering wordlessly," *New York Times*, 24 November 2011: http://www.nytimes.com/2011/11/25/movies/the-artist-by-michel-hazanavicius-review.html (accessed 25 September 2016).

Sidewalk Stories (1989), dir. Charles Lane. Palm Pictures. DVD.

Silent Movie (1976), dir. Mel Brooks. Crossbow Productions. DVD.

Silent Movie Pressbook (1976), British Film Institute, Reuben Library.

Singin' in the Rain (1952), dir. Stanley Donen, Gene Kelly. MGM. DVD.

Sunset Boulevard (1950), dir. Billy Wilder. Paramount. DVD.

Tiger Bay (1959), dir. J. Lee Thompson. The Rank Organisation. *Amazon Prime* (accessed 1 August 2016).

Vance, J. (2003) *Chaplin: Genius of the Cinema*, New York: Henry M. Abrams.

Venuti, L. (2012) "Adaptation, translation, critique," in T. Corrigan (ed.) *Film and Literature: An Introduction and Reader*, London and New York: Routledge, 89–103.

31

ADAPTATION AS A FUNCTION OF TECHNOLOGY AND ITS ROLE IN THE DEFINITION OF MEDIUM SPECIFICITY

Malcolm Cook and Max Sexton

The process of adaptation necessarily engages the technologies of the chosen media and associated ideas of medium specificity. In adapting works between media, producers must negotiate between fidelity to the source and perceived specificities of the chosen technological medium. The degree to which this is explicitly acknowledged varies from instance to instance. It is the relationship between source and target media that makes a work an adaptation, distinct from other terms used to describe commensurate production processes: remake, reboot, production, performance, cover version, translation, reproduction, and edition. As Sarah Cardwell (2002: 21) states in her discussion of the adaptation of costume drama to television, "to call something an adaptation of another text is to highlight the conscious, complex process of implementing changes necessary to re-present the source text under new conditions (in a new medium)."

Critical engagement with adaptations must therefore necessarily adopt a position on the idea of medium specificity, whether theoretically rationalised or else implicitly and intuitively grasped. For example, Cardwell (2002: 45–8) discusses how George Bluestone, in his landmark *Novels into Film* (1957), aligns himself with a medium-specific approach by defining films and novels as ontologically discrete media, while his working methodology fails to adhere to such strict divisions. The comparative approach that has dominated adaptation studies since the 1970s (Cardwell 2002: 51), along with widespread journalistic discourse in periodicals and newspaper reviews, allows for greater shared characteristics between media, which are necessary for adaptation to be meaningful. Yet it commonly falls back to value judgements that reinforce distinctions between media, especially between literature and cinema. This is apparent in the praising of cinematic adaptations for their fidelity to a literary source, or conversely the denigration of film or television adaptations for being too literary or theatrical.

The recent adaptation for BBC television (2015) of the 2004 novel *Jonathan Strange and Mr Norrell* by Susanna Clarke reminds us of the growing complexity in the relations between what, until comparatively recently, have been understood and interpreted as the separate media of film and television, and their relation to the novel. *Jonathan Strange and Mr Norrell*, which was originally slated to be a film, instead became a television series. Its adaptation to the 'small screen' is an indication of television's rising status and the declining status of the cinema that can be, in part, attributed to television's increased technological prowess. According to John Caldwell (1995: 4, 261), since the 1980s, televisuality has pushed the formal and narrational possibilities of television by foregrounding a "visually based mythology, framework, and aesthetic based on an extreme self-consciousness of style" that also deploys "self-contained and volatile narrative and fantasy worlds." *Jonathan Strange and Mr Norrell* in part achieves this through the extensive utilisation of digital visual effects, but it should be recalled that as television has developed stylistically, its increased expressive opportunities did not depend exclusively on technology. Rather, the formal organisation of a drama such as *Jonathan Strange and Mr Norrell* relies on the serial structure of the long form on television, and the specificities of the image are predicated on the ontological relationship of the viewer to the medium. The extent to which the range of metaphysical meanings in *Jonathan Strange and Mr Norrell* are fully realised is therefore the result of a constantly shifting assemblage of temporary coherencies rather than a fixed form of electronic textuality. *Jonathan Strange and Mr Norrell* also flaunts other ontological distinctions between history/text and reality/fiction, as the programme variegates space and time to form a complex continuum, disrupting a clear sense of identity between the cinematic, the televisual, and the literary. In a related discussion of a similar adaptation, Eckart Voigts (2015) provides an account of the BBC's *Sherlock* (2010–) as an exemplar of intermedial transposition, transfer, or transcoding.

The technological basis of media and the idea of medium specificity may thus be seen as central both to adaptation as a process and to critical engagement with adaptations as texts. In his critique of medium specificity, Noël Carroll (1985: 6) defines it as the idea that "each art form has its own domain of expression and exploration … determined by the nature of the medium." These arguments conflate art form and technological medium and treat them ahistorically. From a philosophical perspective, Carroll has persuasively argued that the logical basis for these ideas is flawed. Carroll (1984: 135) shows how, in some cases, these arguments are trivial: if a medium is truly incapable of something, then it does not require admonitions against it. Conversely, while a technology may make certain practices very easy, this is not a basis for valuing those characteristics in an art work (Carroll 1984: 136). Crucially, Carroll notes that there is frequently disagreement in specifying the characteristics of a medium, and a medium may even hold contradictory characteristics that cannot be reconciled, something that is apparent in the specific moving image examples discussed later. In addition to these considerations internal to a medium, Carroll also comprehensively debunks the construction of medium specificity on the basis of comparisons between media. The things a medium does best in comparison to others is not necessarily the thing it does best in and of itself (Carroll 1984: 143). Equally, what a medium does uniquely in comparison to others, if it can even be identified and agreed upon, is not necessarily a basis for its valuation (Carroll 1984: 145). The suggestion raised by such arguments would ask us to "forgo potential excellence for the sake of purity," an untenable position for Carroll (1984: 144) and, one would expect, for most critics.

Such logical flaws evident in medium-specificity arguments anticipate the ontological uncertainty about a medium, such as television, due to digitalisation. The underlying technologies of different media have previously provided a common-sense rationale for their division, which concealed the contradictions Carroll exposes. For instance, in terms of moving image media, the association of television with the mode of liveness had led to the characteristics of television being understood as being determined by the technologies of electro-magnetic transmission. This was despite that association being historically unsound, as will be shown later in this chapter in relation to the use of celluloid film and the development of 'visual expansion.' The process of visual extension was itself displaced by the development of visual distinction in the 1980s. There was, furthermore, general disagreement about what essential qualities the technological characteristics of transmission actually promoted (Carroll 2003: 266–70).

Digital technologies have further rendered such assumptions indefensible. The convergence between platforms has seen the transmission of live events to cinemas, such as a Metropolitan Opera performance of Richard Wagner's *Tannhäuser*,[1] become commonplace, and a succession of home video formats offer transcription media that can be projected domestically. Moreover, both film and television are being superseded by both static and mobile forms of digital consumption, including the PC, tablets, and mobile phones. These devices, which are capable of interactive communication, represent a convergence, not only of moving image media, but also literary, musical, and other media. Furthermore, these devices contribute to the proliferation of screens in contemporary life, ranging from small, wearable versions, through handheld mobile devices and tablets, to large-scale public displays. This range of scale challenges the small-screen/ large-screen distinction commonly used to distinguish cinema and television, which is central to the examples discussed further in this chapter. Changes to the technological basis of artistic media through digital convergence and diversification become increasingly prevalent each year, but their implications in further undermining notions of medium specificity were already anticipated by Carroll (2003: 265–80) over a decade ago.

This convergence should not be considered at a purely technological level. Henry Jenkins (2006: 3), in his influential book on convergence, argues "against the idea that convergence should be understood primarily as a technological process … instead, convergence represents a cultural shift." This is clearly evident in the example of opera broadcasts to cinema. André Gaudreault and Philippe Marion (2015: 90–4) discuss the critical debates around these presentations, demonstrating that a medium's definition is determined by more than its technological invention. Likewise, John Wyver (2014: 118–9), in the discussion of live theatre broadcasts, indicates how changing aesthetic definitions and valuations of the theatrical, the televisual, and the cinematic have shaped the critical reception of these presentations, with recent stagings receiving increasing acceptance and acclaim.

Yet even in the face of this overwhelming change, the notion of medium specificity has remained surprisingly persistent. Some writers have looked to define digital or new media as having specific characteristics. Lev Manovich's *The Language of New Media* (2001: 10) and Jay David Bolter and Richard Grusin's *Remediation* (1999: 54n2) could both be seen in this light, although other readings of their nuanced discussion of new media are possible. In tandem, other writers have looked to refine existing technologically based medium specificities. Artist Tacita Dean (2015), for instance, has argued for distinguishing film as a medium distinct from a broader category of cinema and distancing it from digital technologies. While Dean makes valuable points about archival practice, her argument not only rehashes distinctions between 'film' and 'cinema' that arose from structural film in the 1970s (Wollen 1976), but also espouses an ill-conceived account of film's indexicality that misses the nuances other writers have identified in this topic (Doane 2007; Gunning 2007). Ultimately, attempts to either define a digital medium with its own inherent characteristics, or to refine existing media definitions on the basis of a technological specificity fall foul of Carroll's logical critique of medium-specific arguments.

With the idea of medium specificity undermined by both theoretical flaws and the historical development of digital technology, the very concept of adaptation might also seem in crisis, dependent as it is upon distinctions between media. Bolter and Grusin (1999: 44–5) certainly see adaptation as merely another example of new media's insatiable remediation of other forms. As they observe, adaptation is, if anything, growing in importance in the digital age, with the cross-fertilisation between comic books, video games, movies, television, online video, and other platforms interacting with the mainstream acknowledgement of transmedial franchises in the form of 'expanded universes.' However, we take these activities not as evidence of a convergence or collapse of medium boundaries, nor as markers of essentialist, ahistoric medium specificities, but as further indications that the idea and definitions of media are, and always have been, historically and culturally constructed. As a practice that directly engages with this process, adaptation has both served as a vital tool in the construction and maintenance of medium-specific boundaries, and is a valuable key for studying how this occurs.

In the field of literary theory, the study of intermedial or transmedial characteristics, such as narrative or metareference, has led to a growing awareness of medium specificity and the idea of medium as culturally constructed. Within narratology, for instance, the extension of the study of narrative into intermedial forms, beyond its traditional basis in the study of literature, has led to the idea of medium being addressed in these terms. Marie-Laure Ryan (2005) describes a cultural, as well as semiotic, dimension to media and denies that technology alone would determine their characteristics. Nevertheless, Ryan (2005: 16) sees some media as inherently having more "narrative potential" than others, implying an ahistoric and essential definition of individual media. Similarly, Werner Wolf (2011: 166), drawing on Ryan's work, proposes a definition of the idea of medium that clearly incorporates "conventionally and culturally" determined properties that are not only influenced by technology. However, Wolf (2011: 166) still suggests that "different media have different capabilities for transmitting as well as shaping narratives." He proceeds to create a categorisation of media based on their capacity for narrative, retaining an unchanging definition of 'film,' 'music,' 'painting,' and other media, rather than recognising these forms as both culturally and historically contingent.

A number of writers have acknowledged the contingent nature of definitions of media while looking to retain a concept that is both deeply ingrained and, as Carroll (1985: 6) and Doane (2007: 129) both observe, underpins the division of academic disciplines. Doane (2007: 129) remarks: "Despite its essentialist connotations, medium specificity is a resolutely historical notion, its definition incessantly mutating in various sociohistorical contexts." Cardwell (2014: 6–7) looks to refine the idea of medium, especially that of television, by distinguishing "medium," describing the technological base, and "art form." Cardwell addresses wider historically and culturally constructed practices that may be influenced by, or attributed to, associated technology ("medium"), without the "art form" ultimately being determined by it. This approach echoes Henry Jenkins's discussion of "delivery technologies" (Jenkins 2006: 13–4). John Belton (2014: 470) equally shows keen awareness of the way cinema technology has varied historically and culturally, writing that "the cinema is constantly changing in tandem with changes to its basic technology, constantly redefining itself" – Belton rejects simple technological determinism, while looking to retain cinema as a meaningful description rather than embracing "death of the medium" narratives.

Valuable as these contributions are, they are primarily directed at shaping the future discussion of the idea of medium, no doubt reflecting the disciplinary affiliations of both the authors and the journals their arguments appear in. Yet we must also acknowledge the historical role medium specificity arguments have played. Even if they are unsound in philosophical, technological, and historical ways, they have, in historiographical terms, shaped the ongoing development of what we understand as different media. Lisa Gitelman offers perhaps the most pertinent account of media for the present study in this respect. Gitelman not only acknowledges a cultural and historical dimension to media, but understands their very definition as contingent to those dimensions. She defines media as "socially realized structures of communication … as such, media are unique and complicated historical subjects [and] their histories must be social and cultural" (Gitelman 2006: 7) and proceeds to examine a historical subject (the phonograph) in these terms. Rather than dismissing medium specificity, we can acknowledge it as being culturally and historically formed and trace this process.

Adaptation, as a practice that actively engages with the media it crosses between, can be seen as a crucible for these wider concerns and thus a pertinent focus for this investigation. This serves not simply as a corrective to the historical record, valuable as this may be, but also offers to shed light on present day processes of redefining and negotiating media boundaries. In this we follow William Uricchio (2014: 266), who in discussing these issues, and himself drawing on earlier historiographical discussions, observes that "history's power derives from its ability to illuminate the present." Furthermore, Uricchio (2014: 275) points towards television as a particularly suitable ground for such work, arguing that "television's ability to remain conceptually intact in the face of unrelenting change may offer valuable lessons" in understanding the complex determinants that make up a medium.

By examining adaptation on the medium of television, within a distinct period (from the late 1940s to the early 1960s), primarily in one region (the United Kingdom), and focusing on the science-fiction genre, we can address the historically and culturally specific nature of this discussion while making observations that may be more generally applicable to other media, places, and times, especially the present day. Given the prior arguments that move the idea of medium beyond ahistoric and technologically determined definitions, our understanding of adaptation is also more expansive. The adaptation of technology from other media is seen as a crucial way in which medium specificity is both maintained and negotiated. Yet technology is not the only area in which medium specificity is observed to be adapted and determined, but rather takes in institutional, generic, and textual concerns.

Case study: defining British television through science fiction adaptation

The critical discourse constructed around the early function of television was concerned with television as an aesthetic medium, but which aesthetic was 'natural' to television? In 1970, Joan Bakewell and Nicholas Garnham (1970: 14) commented that "a whole theology was erected to justify television's existence as an autonomous art in terms of its liveness." When such debates were raised, it was television's artistic and ontological specificity, its relations to other media – particularly cinema – and its proper programme forms that were the core issues that formed the debate. For example, Ted Kotcheff, a Canadian, who had been working in British television as a director for ABC-TV with Sydney Newman on *Armchair Theatre* from 1958 to 1960, was frustrated by the technical difficulties of working in TV and aspired to direct for the cinema:

> [Y]ou have no absolute control of [television's] picture quality. Your lighting effects which you took great trouble over, may go out of the window once the show is transmitted, because someone has their brilliance up too high, or contrast down too low … So you're working in a pictorial medium where you run the risk of losing 50 percent of your picture quality! … [A]dding up all the points, you find that ultimately working in TV is like sculpting in snow, it's all gone in the morning sun and everyone's forgotten it was ever there. This is why every TV director wants to get out into films.
> *Kotcheff qtd in Porter and Wicking 1965: 230–2*

Such comments suggest not simply the limits of television at the time but also include the assumption that television was the poorer relation of the cinema. Kotcheff's complaint against television in this instance can be interpreted as the distinction between television as an audio-visual medium restricted to the technology of transmission and film as a medium of the moving image, which by extrapolation is a possible art form. In this way, the logical form for television was different from the cinema but also reflected attitudes to the cultural value and status of each medium.

However, the relationship between genre, specifically science fiction, and television technology offers an insight into how its modalities were capable of a visual expansion that belies the assumption that television was only capable of live transmission. Television's technological development, and especially the application of the recording medium of film, played an initial role in its visual expansion. The use of film in programme construction when television was considered to be live suggests a desire for a more aesthetic approach to genre, as well as a thematic one, in spite of the ephemeral nature of its programmes.

In 1949, fourteen years prior to the British science fiction show *Doctor Who* (1963) and four years before *The Quatermass Experiment* (1953), the BBC adapted for television, as a 'visual experiment,' a science fiction novel by the celebrated writer H.G. Wells. In *The Time Machine,* the Traveller departs the present to journey into the far future. As was typical of much of early television, no copy of the show exists to view or to be reviewed. From 1947, the ability to film a programme by pointing a camera at a television screen (telerecording in the UK or Kinescope in the US) was possible, but it was a primitive and cumbersome process and few programmes were recorded to be saved later (Morgan 1955: 88). Due to television being broadcast live, the possibility of the television producer altering the show as it was recorded, by reacting to the visual dramatisation of the script as it unfolded, was very limited.[2] This may have been an important consideration owing to the fact that, although the story is simply narrated in the novel of *The Time Machine,* the decision was taken to produce an adaptation that 'showed' rather than 'told' the narrative. The Wellsian fantasy included a relatively simple point-of-view and focalisation, but the adaptation also used recorded images on film, such as a possible giant land crab three million years in the future, for television.

One technique of early television was for film to be put into the live show to provide 'inserts' usually lasting no more than a few minutes. Robert Barr, who dramatised and produced the story as a one-hour play, used both inserts and film as back projection.[3] A new picture-script system was used to simplify the matching and the superimposition of pre-shot film sequences and futuristic models. Back projection was used to show the passage of sped-up time and to represent 800,000 years in two and a half minutes. As Nigel Kneale (qtd in Chapman 2006: 24), the creator of *The Quatermass Experiment* (1953), a seminal and important early science fiction television drama, was to point out a few years later, the use of film "adds both physical freedom and atmosphere," and could "provide a most useful extension of the story beyond the cramped studio sets." Barr's comment in the *Radio Times,* "faint heart never made good television," suggests that experimentation was not only allowable but encouraged so as to overcome television's technical constraints in order to 'enhance' the script by interpreting it visually.[4] Notwithstanding, this point of view was also conveyed using sound: the Traveller's thoughts are heard after they had been recorded onto a 78 rpm disk cut at the BBC's Broadcasting House.

Science fiction adaptation was a site for the negotiation of what was desirable or achievable, and the incorporation of other media (film, audio recording) was commonplace. The example of *The Time Machine* serves to remind us that science fiction is a genre whose narrative elements often require a high degree of visualisation and the accumulation of spectacle when adapted to the moving image. Sci-fi, unhampered by restrictions of realism, has been attractive to producers such as Barr because it offers viewers the opportunity for an enhanced use of the medium that eschews the belief that television operates a mirror-like or relay function.

At the same time, institutional beliefs in emphasising broadcast technology as television's unique property would become axiomatic of much of professional thinking about the medium throughout the 1950s and 1960s. According to this view, television was a medium that had a privileged relation to reality: "[T]elevision is not merely a substitute for the theatre or cinema which holds up a mirror to life: it is rather an open window to life" (Swinson 1955: 21). Because it was live, television's reliance on the time-based nature of its programmes was used to distinguish it from the movies, where "everything is recorded" (Metz 1982: 43).

Where technical innovation in the development of the medium was to be encouraged, its greater purpose in a British context was to serve the requirements of public enlightenment, not to challenge medium-specific boundaries. Thus, according to one source, "[t]he BBC ... aims to use television as an additional means of helping to bring about an informed democracy and generally raise public standards. A compromise with the film industry would gravely damage this work."[5] In both the UK and US, there was a marked tendency to produce drama that examined social problems, although this declined radically in the US after 1960.[6] As a consequence, television drama in the 1950s sought those techniques that helped in "their contribution to the sense of immediacy … [Audiences] feel that what they see and hear is happening in the present and is therefore more real than anything taken and cut and dried which has the feel of the past" (Seldes 1952: 32). It was also felt that, because television was to be watched in the home, unlike the cinema, it would have a different appeal and "a small group in the living-room will demand a different tempo and a different feeling in their television programmes from what they will welcome when they visit a theatre … it is hard to imagine the typical film trailer, all explosions and superlatives, raising anything but a laugh in the home" (Gorham 1949: 31).

Despite such emphasis on transmission as the basis for television's specificity, from the 1950s, the ability to record programmes either on film or, by the decade's end, on videotape[7] began a process in which certain types of images – for example, the close-up that showed a strict fealty to the actor – would be allowed to compete with images that were able to represent visual style and action. Moreover, a character-built script whose naturalistic style television drama had relied upon might also incorporate the promise and pleasure of watching action. Such action was possible because, as we have seen, film could be used to record additional scenes beyond the confines of the electronic studio. However, the ability to show physical action and movement in space did not necessarily become central to science fiction's generic identity. Instead, the genre was to seek other sorts of transformations, both at the level of narrative and viewpoint, as well as themes that incorporated ideas that had been more difficult to represent within the electronic studio.[8] For example, in an episode of the anthology series *Out of the Unknown*, which began broadcasting on the BBC in 1965, the original short story by the science fiction author John Wyndham was adapted and used an opening mood which is slow and lyrical but fraught with tension as the protagonist, Bert Foster, ponders his existence on Mars.[9]

> The episode calls on the vision mixer or "switcher" to create an extended montage of superimposed images. At first the camera tracks towards Foster for a close up, before there is a superimposition … Voice-over reveals his reservations about the offer to travel to Venus. This conveys not simply an interior monologue by Foster, but the ideas and thoughts of the writer: The montage is multi-layered and becomes a stream of consciousness … There is a complex soundtrack, consisting of the sound of lapping water and contemplative music, before a dissolve takes the audience to film of Max Berg's strikingly modernist Centenary Hall in Breslau, accompanied by the strains of the "Ode to Joy" from Beethoven's Ninth Symphony. … This impressive short montage of dissolves and superimposition conveys sensations of extreme emotional states, and this is clearly how Foster feels.
>
> *Cook and Sexton 2015: 77*

Within this complex sequence, the person whose job it had been to do the switching would be operating a battery of push buttons, levers, and dials, which controlled the transition from one camera's pictures to another's. In addition, some images would not have come from cameras in the studio but from projection units such as telecine to provide additional material such as Berg's Centenary Hall. In this example, television was neither ascetically limited to broadcast characteristics, nor merely aping or co-opting film, but rather developing a new aesthetic that used science fiction to redefine what the medium of television could express.

Here then is how questions about the role of technology and its adaptation of the text to television also begin to raise additional questions about practices at work, as well as the professional ideologies of programme-making within television. Technology is not the sole or primary determinant, but institutional discourses play a central role, even if couched in technological terms. Since the production of *The Time Machine* in the 1940s, several discourses had operated beyond the boundaries of the text, which had produced tangible effects on the process of adaptation to television. According to John Caughie (2000: 41), television was a technology for relay and adaptation, and "the demand for a 'creative' television which spoke with its own voice … struggled to be heard above the routines of production." However, the television text is part of an array of technologies that are constantly changing and being updated within various industrial practices. Science fiction is involved in these relations in particular ways that affected it in the 1960s.

Technical quality can function as the criterion for forming a category of value to be assigned to a particular programme. Yet a comparison that assigned value on this basis alone – in the past a programme either shot on film or recorded on video would be misleading. Although some professional ideologies regarded television as a distinct medium, others did not. By the 1960s and 1970s, the script, which may have been limited by the possibilities of set design and lighting in the electronic studio, could be far less aesthetically limited by the professional ideologies of both the technical crew as well as the producer if shooting on film. For example, shooting on film with an experienced crew was very different from recording on tape in the television studio. Television engineers in the studio would strongly disapprove of brilliant whites and solid non-reflective blacks, especially if they were in large quantities, because they were of the opinion that the electronic signal of the TV system would not cope with highlights and lowlights. Alan Hume, a lighting cameraman at the time, comments that:

> What, for example, looks more ludicrous than a person creeping about with a torch when there is so much light that you can see everything as clearly as in daylight? This sort of result is usually caused by the cameraman *thinking he must play safe, he probably being of the opinion that the T.V. system will not cope with dark, unlighted areas* … there are surely times … one is getting good pictures which are interesting, or even exciting; the lighting is adding to the atmosphere and the suspense, and plays a very important part in helping the director and scriptwriter to put over their story and to make it more interesting.
>
> *Hume 1968: 4–8; emphasis in original*

The collaborative nature of production can allow a solution to a technical problem that can lead unexpectedly to the creation of an artistically more interesting scene content that differs from the subject matter of the script. Of course, the degree of difference can vary from a minor alteration to a complete break from the script. Hume makes it clear that on another science fiction series for television, *The Avengers* (1961–9), the crew's overriding purpose was to maintain the integrity of the script and to serve it rather than alter or replace it. The desire to maintain the integrity of the script should not be understood as a slavish desire to keep to the exact nature and meaning of the written word, but a collaborative effort to use the available film technology to its maximum extent, including under very difficult lighting conditions. This approach complicates our understanding of adaptation because while the fidelity to the source script is still considered important, it is not necessarily primary. Rather, it incorporates a 'complex approach' capable of a multi-layered diegesis, consisting of scripted dialogue and the camera image, as well as the possible innovative use of post-production sound. Whereas the electronic technology of the studio would be made subordinate to the script and invisible in order to disguise the technical nature of television – additional lighting might, for example, be added inside the studio – a film crew's approach was, on the whole, more inventive than within the TV studio. For Hume, the experience of shooting on *The Avengers* was that lighting would be carefully chosen for each set-up, whereas, in the television studio, the set tended to be over-lit to avoid problems of continuity and to placate the television engineers as they sought to apply the strict technical standards laid down by the television regulator.[10]

The adaptation of genre on television and the readings of individual programmes rely on an awareness of specific televisual qualities that vary during television's history. These qualities can consist of how space and lighting will be manipulated to contribute to the narrative. Textual readings of television programmes within a genre can offer insight into understanding how theme and narrative have been developed and adapted to television. For example, how television was able to produce a sophisticated use of electronic montage in *Out of the Unknown* contributes to the understanding of how thematic transformations to the genre – in this case the subversion of older ideas about invading Martians and peaceful Earthlings – were made. At the same time, changing codes of visibility by the 1960s, such as the greater use of physical action and, by the 1980s, the use of digital special effects raises further questions about how programmes connect with the genre and its relation to broader sociocultural contexts. The professional ideologies within television, controlled by an institution such as the BBC, have often created a discourse about value and quality, which continues to play a role when adapting a text to television. Technological developments within television as methods of programme construction must be considered in terms of how the representation of technology on television – from the use of film in the 1940s to today's computer-generated imagery – has over time become known and understood by practitioners, but also by audiences.

More generally, while the idea of medium specificity can be seen as logically flawed, by understanding it as historically and culturally constructed and examining the particular arguments used in distinct times and places, we can see how those ideas have shaped the development and understanding of specific media. Technology remains central to this discussion, but does not determine it. Rather, it is implicated in institutional discourses and aesthetic practices, as well as wider cultural, social, and political debates. Adaptation, with its close alignment to and engagement with medium specificities, is not undermined by this complication. It can, instead, be seen as a critical site for the construction of medium definitions, offering a particularly valuable opportunity for examining that process, both historically and in the present day.

Notes

1 This was relayed from New York to the Curzon Mayfair in London, and other cinemas, on 31 October 2015.
2 The second time the programme was broadcast live and its producer described it as "not much as a second performance as a first full version." There would be more time to concentrate on continuity and a hope to create more polished effects. The script was also partly re-written. See Anon. *Radio Times*, 4 February 1949, 24.
3 Barr worked closely with the set designer, Barry Learoyd.
4 *Radio Times*, 21 January 1949, 25.
5 "Television and the Film Industry," Draft Paper for the Board of Governors, 11 July 1948, 14.
6 The Golden Age of Television in the US was a brief period, generally considered to be between 1949 and 1960, when several anthology dramas broadcast the work of such writers as Paddy Chayefsky, Sidney Lumet, Rod Serling, Reginald Rose, and Gore Vidal. By 1960, this had largely ceased (37 Wheen 1985).
7 The BBC introduced a system of video recording of their own called VERA, which stood for Vision Electronic Recording Apparatus. The American broadcast companies used the Ampex Videotape Recorder. After 1961, the BBC did eventually use the Ampex system.

8 BBC Television Centre, with its seven major studios and some smaller ones allowed the production of 3,000 hours of television programme material every year. See Sir Gerald Beadle, *Television in Britain: An Address before a Private Audience in New York on 26 January 1960* (London: British Broadcasting Corporation, 1960), 2.

9 Wyndham was also responsible for the novel *The Day of the Triffids* (1951), which was adapted for radio, film and television.

10 However, this is not to claim that stylised lighting was not possible in the studio, but it was more difficult generally. James Chapman (2002: 66) points out that some stylised lighting was used in series three of *The Avengers*.

Works cited

Anon. (1948) "Television and the film industry," *Draft Paper for the BBC Board of Governors*, 14.

Anon. (1949) "Television," *Radio Times*, 4 February, 24.

Anon. (1949) "To the world's end in sixty minutes," *Radio Times*, 21 January, 25.

Bakewell, J. and Garnham, N. (1970) *The New Priesthood: British Television Today*, London: Allen Lane.

Belton, J. (2014) "If film is dead, what is cinema?" *Screen*, 55(4): 460–70.

Bluestone, G. (1957) *Novels into Film*, Baltimore, MD: Johns Hopkins University Press.

Bolter, J.D. and Grusin, R. (1999) *Remediation: Understanding New Media*, Cambridge, MA: MIT Press.

Caldwell, J. (1995) *Televisuality: Style, Crisis, and Authority in American Television*, New Brunswick, NJ: Rutgers University Press.

Cardwell, S. (2002) *Adaptation Revisited: Television and the Classic Novel*, Manchester, UK: Manchester University Press.

Cardwell, S. (2014) "Television amongst friends: Medium, art, media," *Critical Studies in Television*, 9(3): 6–21.

Carroll, N. (1984) "Medium specificity arguments and self-consciously invented arts: Film, video, and photography," *Millennium Film Journal*, 14/15: 127–53.

Carroll, N. (1985) "The specificity of media in the arts," *Journal of Aesthetic Education*, 19(4): 5–20.

Carroll, N. (2003) *Engaging the Moving Image*, New Haven, CT: Yale University Press.

Caughie, J. (2000) *Television Drama: Realism, Modernism, and British Cinema*, Oxford: Oxford University Press.

Chapman, J. (2002) *Saints and Avengers: British Adventure Series in the 1960s*, London: I.B. Tauris.

Chapman, J. (2006) "Quatermass and the origins of British television SF," in J.R. Cook and P. Wright (eds), *British Science Fiction Television: A hitchhiker's guide*, London: I.B. Tauris, 21–51.

Clarke, S. (2004) *Jonathan Strange & Mr Norrell*, London: Bloomsbury.

Cook, M. and Sexton, M. (2015) *Adapting Science Fiction to Television: Small Screen, Expanded Universe*, Lanham, MD: Rowman & Littlefield.

Dean, T. (2015) "Anyone who works with film …" *Artforum*, October. Online. Available at: https://artforum.com/inprint/issue=201508&id=54974 (accessed 13 October 2015).

Doane, M.A. (2007) "The indexical and the concept of medium specificity," *Differences: A Journal of Feminist Cultural Studies*, 18(1): 128–52.

Doctor Who (1963), creator: Sydney Newman et al., BBC, "An Unearthly Child," tx. 23 November, 1963. *H.G. Wells' The Time Machine*, BBC, tx. 25 January, 1949.

Gaudreault, A. and Marion, P. (2015) *The End of Cinema?: A Medium in Crisis in the Digital Age*, New York: Columbia University Press.

Gitelman, L. (2006) *Always Already New: Media, History and the Data of Culture*, Cambridge, MA: MIT Press.

Gorham, M. (1949) *Television – Medium of the Future*, London: Percival Marshall.

Gunning, T. (2007) "Moving away from the index: Cinema and the impression of reality," *Differences: A Journal of Feminist Cultural Studies*, 18(1): 29–52.

Hume, A. (1968) "Filming for colour television series," *British Kinematography Sound and Television*, 50(1): 4–8.

Jenkins, H. (2006) *Convergence Culture: Where Old and New Media Collide*, New York: New York University Press

Manovich, L. (2001) *The Language of New Media*, Cambridge, MA: MIT Press.

Metz, C. (1982) *The Imaginary Signifier: Psychoanalysis and the Cinema*, London: Macmillan.

Morgan, T.J. (1955) *The True Book about Television and Radio*, London: Frederick Muller.

Porter, V. and Wicking, C. (1965) "The making of life at the top," *Film and Television Technician*, 31(249): 230–32.

Ryan, M. (2005) "On the theoretical foundations of transmedial narratology," in J.C. Meister (ed.) *Narratology beyond Literary Criticism: Mediality, Disciplinarity*, Berlin: de Gruyter, 1–23.

Seldes, G. (1952) *Writing for Television*, Garden City, NY: Doubleday.

Sherlock (2010–), creators: Mark Gatiss and Steven Moffat. BBC/ WGBH Boston. Television.

Swinson, A. (1955) *Writing for Television*, London: Adam and Charles Black.

The Avengers (1961–1969), ABC-TV, ITV, tx. 7 January 1961–21 May 1969.

The Quatermass Experiment (1953), creator: Nigel Kneale, BBC, "Contact Has Been Established," tx. 18 July, 1953.

Uricchio, W. (2014) "Film, cinema, television … media?" *New Review of Film and Television Studies*, 12(3): 266–79.

Voigts, E. (2015) "Literature and television (after TV)," in G. Rippl (ed.) *Handbook of Intermediality: Literature – Image – Sound – Music*, Berlin: de Gruyter, 306–24.

Wheen, F. (1985) *Television: A History*, London: Century Publishing.

Wolf, W. (2011) "Narratology and media(lity): The transmedial expansion of a literary discipline and possible consequences," in G. Olson (ed.) *Current Trends in Narratology*, Berlin: de Gruyter, 145–80.

Wollen, P. (1976) "'Ontology' and 'materialism' in film," *Screen*, 17(1): 7–25.

Wyver, J. (2014) "'All the trimmings?': The transfer of theatre to television in adaptations of Shakespeare stagings," *Adaptation*, 7(2): 104–20.

32

SOUND STORIES
Audio drama and adaptation

Richard J. Hand

From the first wax cylinders through to digital downloads, from network output captured on crystal sets through to Digital Audio Broadcasting, audio listeners have consumed culture entirely mediated by technology. These audiences have enjoyed music on vinyl records, transistor radios, or iPods, and in so doing have determined the evolution of music, sharing and propagating its popularity. Audiences, however, have also used their audio technology to listen to the spoken word. Early broadcasters realised the potential of radio technology to disseminate news and sports for its listeners to consume in immediacy and simultaneity. The pleasures of radio would subsequently encompass book readings, which would evolve into 'drama.' Although it is one of the most neglected fields of performance culture, throughout its history audio drama has been prolific and impactful. Early in the advent of broadcasting, radio drama proved itself to be enormously flexible, creating different formats of drama, from serialisations to standalone works, as well as inventing genres such as the soap opera and developing distinctive forums for science fiction, fantasy, whodunits, and other popular narratives. In the twentieth century, radio featured adaptations of fiction which were as (in)famous as *Mercury Theatre on the Air*'s "War of the Worlds" broadcast (1938) and as monumental as the BBC's adaptation of the complete *Sherlock Holmes* (1989–98). In the twenty-first century, the internet has created a new era of audio drama: there has never been a more fluid range of options through which we can consume network or independent radio, and, in addition to this, there are websites streaming archival materials as well as podcasts of experimental or amateur work. In short, with a plethora of available works from past and present, there has never been a richer time to be a 'listener.'

Adaptation has been a central practice since the beginning of radio drama: indeed, the creation of 'original' plays for radio is a trend that emerged sometime after plays began to appear on radio. Initially, radio drama was essentially the recitation of stage plays: in the US, Eugene Walter's popular stage melodrama *The Wolf* (1908) was arguably "the first 'on-air' drama" (Blue 2002: 1) when it was broadcast in August 1922; in the UK, short extracts from Edmond Rostand's *Cyrano de Bergerac* (1897) were broadcast in October 1922 (Crook 1999: 4), and in the following year the BBC broadcast extracts from Shakespeare before airing a full-length version of *Twelfth Night* in May 1923. After the success of broadcasting stage plays on the air, radio dramatists turned to fiction as a source for audio drama. For the BBC, this commenced with an audio dramatisation of Charles Kingsley's 1855 novel *Westward Ho!* in April 1925 (Briggs 1985: 63).

Audio drama has continued this close relationship with adaptation to the present day. As well as reworking stage plays and prose fiction for the airwaves (very often using works which are safely – and appealingly – out of copyright), radio drama will also present significant sub-genres of other adaptive processes, including a close relationship with cinema (in which audio versions of films are produced) and fact-based dramatisation (including distinctive examples of biographical drama and docudrama). In addition to the wide range of source materials selected, the format of these plays has been as diverse as the genres that have been chosen. From readings and audiobooks to the complexity of binaural and interactive productions, audio listeners have experienced one-off dramas and serialisations, differing in length and ambition. In terms of strategy, audio adaptation can be found to use the techniques of allusion or hybridisation as much as a more conventional or 'completist' approach. In this chapter, we will explore different types of adaptation in audio drama. A range of case studies will be used for analysis to ensure that the topic is explored in the most diverse way: in addition to classic works of radio drama and output from the major radio networks, the chapter also features analysis of independent podcast audio drama and examples of 'experimental' sonic culture.

This essay will focus on examples of audio drama from the UK and US. Radio drama has been comparatively disregarded in academic study, above all non-Anglophonic work, which (despite outstanding output in, for example, German and Italian contexts) represents an overlooked but rich critical mass worthy of exploration. Indeed, the academic neglect of radio drama as a field of performance culture is an interesting phenomenon. It cannot be for lack of material (the BBC has always produced vastly more radio drama than television drama) or lack of ambition (the BBC's 1981 *Lords of the Rings* totals thirteen hours in duration, compared to the eleven hours of Peter Jackson's combined 'extended versions' on screen (2001–3)). Nor is radio hard to access; indeed, the potential for radio to be consumed while doing something else makes it the perfect medium for contemporary, 'multitasking' life. Admittedly, however, the general perception of radio might be that it is a forum for music, news, sports broadcasting, and 'talk radio.' Perhaps most critical for audio drama is its 'visionlessness': It is an 'invisible' medium that might suggest to some that it somehow 'lacks' the richness or sophistication of the visual. This is exacerbated further when it comes to adaptation: as Sibylle Bolik (1998: 154) observes, despite the fact that it is the mode in which radio drama began, the "radio adaptation of literature is regarded as subordinate to the 'true,' the original acoustical play." Radio adaptation is thus impugned for both its lack of visuals and its lack of originality. Furthermore, in writing about radio adaptation, Linda Hutcheon (2006: 41) assesses that "most radio plays concentrate on primary characters alone and therefore simplify the story and time-line." This implies that, as an adaptive medium, radio 'simplifies' the comparative 'richness' of both the reading and the viewing experiences. In many ways, such debates reveal what Tim Crook (1999: 54) describes as the "sensory hierarchy" in contemporary culture that assumes that sound is always inferior to vision. In fact, auditory culture is not only extremely sophisticated – in many ways we need to 'learn' how to listen – it has an almost limitless potential (one that is, perhaps, not hampered by the four corners of a flat screen). Audio drama is ideally suited to the epic and the intimate. Therefore, the complex comic universe of *The Hitchhiker's Guide to the Galaxy* (1978–2005) is ideally realised as radio, as is the sophisticated interiority of Samuel Beckett's *All That Fall* (1956), as well as, in popular genre, Lucille Fletcher's thriller "Sorry, Wrong Number" (1943). Furthermore, although radio drama is not 'cheap,' it is economical. Radio writers have long appreciated being able to create narratives completely unrestricted by geographical location, historical period, or dramatic canvas.

If we return to radio's appropriateness for a culture of multitasking, the consideration of where and how we listen to audio is profound. Hugh Chignell coins the word "secondariness"

(2009: 70) to describe how audio tends to be an art form we uncontroversially consume while we are doing something else. Certainly, it is worth acknowledging that a radio listener is not enthralled by a screen and may well be travelling, exercising, trying to sleep, or even writing on a laptop while they experience an audio narrative. Although we may have an assumption (reinforced by countless marketing images) that early radio listeners crowded as a family around their wireless receiver by the fireside, it is worth noting that car radios became increasingly popular from the 1930s and, before that, the amateur receiver that revolutionised and popularised radio listening – the crystal set – was a strictly 'in-ear' (hence personal) device. In this regard, the potential 'spaces' of the audio listener are significant. Compared to the experience of listening to radio, the consumption of theatre, cinema, and even television can seem somewhat conventional. Arguably, many of the best examples of audio drama are aware of the challenge of the countless ways of listening. This is part of the reason that, despite the efficacy and versatility of audio drama as a form, it can be extremely difficult to create. As Hand and Traynor (2011: 103) write, there is "a peculiar dichotomy in audio drama between its *constraints* and its *limitlessness*" (emphasis in original): despite the scope of its potential, it can face a particular challenge in 'hooking' the (typically multitasking) listeners to make them 'see' and hold the unfurling story in their mind's eye. In this regard, Tim Crook (1999: 156) is uncompromising when he warns radio writers that "[b]oredom is not listed as one of the seven deadly sins, but … [when] this happens you do not exist as a dramatist." Audio drama needs to be engaging and lucid, a thoroughly co-creative process between artist and audience.

Radio drama at its best succeeds in an efficient assimilation of script, voice, and sound: it can deploy lucid and simple techniques to create profound and complex narrative experiences. To return to Hutcheon, perhaps it is more apposite to stress that radio drama 'streamlines' more than 'simplifies.' To illustrate this, let us consider examples from two particularly renowned shows we have already mentioned. In the *Mercury Theatre on the Air*'s "War of the Worlds" (1938), the perfect imitation of 'breaking news' broadcasting (a comparatively new form in the 1930s) mediates a carefully crafted script, so that by the time the cylinder of the alien craft begins to creak open (in fact, merely a manhole cover being dragged across the concrete floor of a CBS studio), the experience is compelling and (so the reception legend goes) believable. In this respect, H.G. Wells's Victorian work of scientific romance is adapted into the language of a contemporary medium. In other words, the novel is appropriated by radio drama and retold using conventions of another radio medium: news broadcasting. It is perhaps easier to retell the source 'radiophonically' rather than 'theatrically,' but the impact is vastly more potent. Similarly, in the realms of science fiction, halfway through the opening episode of *The Hitchhiker's Guide to the Galaxy* (1978), the listener hears the total destruction of planet Earth (in fact, a sustained sound effect created by the BBC Radiophonic Workshop). This is not a multimillion-dollar movie but an inventive example of radio: through a simply realised narrative moment, the audience holds the concept of an 'Earth-less' universe as we follow the playful pan-galactic adventures of the surviving earthling, Arthur Dent.

As we have already indicated, early radio drama developed distinct formats for drama, ranging from standalone to serialised works. These can vary enormously in duration. BBC Radio continues to feature slots for forty-five-minute (or longer) self-contained plays to serialised dramas in daily fifteen-minute or longer weekly instalments. Within these different formats, adaptation is recurrent. British radio has tackled a vast amount of nineteenth-century fiction on the air. The English novel (as well as numerous examples of French, Russian, and other literature in translation) offers rich source stories with well-developed characters and evocative periods and locations. Appropriately enough, in the case of the dramatisation of Victorian fiction, the popular Classic Serial format, which adapts the source texts into instalments, ties in appositely with the original context's serialisation of fiction.

Emily Brontë's novel *Wuthering Heights* (1847) has been consistently popular in radio adaptation, with versions as far back, at least, as *The Lux Radio Theater*'s post-Hollywood film-to-radio version in September 1939, and with new versions more or less in every decade since. The appeal of the novel is clear: it is an intense tale of love and revenge set evocatively in the Yorkshire moors. However, despite the fact that the novel spans several decades (principally two generations and the life of Heathcliff from young childhood to his death in his late thirties), adaptations have tended to focus on the dramatic core to the first half of the novel: the relationship between Cathy and Heathcliff, which is probably second only to Shakespeare's *Romeo and Juliet* as one of the greatest tales of tragic love in English literature. Notably, radio has attempted to rectify this adaptive decision in dramatising Brontë's novel. Lucy Gough's adaptation of *Wuthering Heights* for BBC Radio 4's *Woman's Hour* in 2003 is a comprehensive and inventive audio adaptation. The dramatisation is a serialisation, breaking the novel into fifteen instalments (all fifteen minutes in duration), played each weekday morning over a three-week period. In writing the adaptation, Gough (2013: 158) wanted to come further in adapting the fuller scope of the novel, while, at the same time, being determined to avoid it becoming the poorest type of adaptation: "a book on legs." *Wuthering Heights* is, however, a book of many challenges for reasons of style, story, and popular reception: "Apart from the obvious challenge of adapting such an unwieldy uncompromising novel with numerous characters with confusing names and a time scale which is tricky, it is everyone's favourite novel" (Gough 2013: 158).

Lucy Gough's solution is to consider the potential of radio. What can audio give its audience that other media cannot? She focuses on 'Wuthering Heights' itself, that is, the house. This anthropomorphist decision focuses all action in the central context and character of the house, resolving complicated issues of timespan and the many people who pass through the location. For those who love the novel, Gough provides a refreshing 'take' on this classic novel in a substantial total duration of nearly four hours. The adaptation opens with the house and the bleak, wintry surroundings:

> *(A wind and snowstorm are buffeting around the house. The house talks as if it is*
> *full face to the wind and being buffeted hard.)*
> HOUSE: High up.
>   I stand.
>   High up.
>   Weathered walls jut the storms
>   Narrow deep set eyes defend interior thoughts
>   *(Beat)*
>   In my hearth, my heart.
>   And in the space of each room.
>   My Soul.
>   *(A man is walking up the snow covered path, struggling against the wind.)*
>   A visitor.
>   *(The man stops at the door and knocks. The dogs inside start barking, the wind is*
>   *howling)*
>
>   *Gough 2013: 163*

Gough immediately creates for the listener an intimate relationship with the house and a sense of narrative interiority. The house is going lead us through this epic drama; it will be the constant focal point for the audience while generations of characters in the story come and go. This short example demonstrates the efficacy of radio drama: the soundscape of the winter storm creates a mood and narrative context, while the lyrical voice of the house locates an evocative, otherworldly perspective that will capture the emotional intensities of the novel. In so doing, Gough 'solves' the adaptive challenge of Brontë's novel, placing the listener, as it were, 'inside' the constancy of the house.

In her two-part adaptation of Virginia Woolf's *Mrs Dalloway* (1925) for BBC Radio 4 in 2012, Michelene Wandor also 'rethinks' a much-loved classic of English fiction. In many respects, Woolf's modernist narrative is an ideal source and style for radio drama: its use of the interior monologue is a perfect strategy for radio, and it is also a novel which is structured by and through sound (especially the chimes of Big Ben); while First World War veteran Septimus Smith's hallucinations (which melt reality into traumatic visions that only he can see) have the potential to be powerfully realised in a solely auditory medium. In fact, it is worth noting how some of the pioneering examples of radio drama have exploited these modernist advantages. For example, in Dalton Trumbo's antiwar novel *Johnny Got His Gun* (1939) the focal character is Joe Bonham, a severely injured soldier who lies in a hospital bed. Joe is unable to communicate, but the reader is placed within the mind of Joe and experiences his stream-of-consciousness existence as they are led through his memories, his fantasies, and the despair and agony of his miserable present condition. In 1940, Arch Oboler adapted the novel into a radio version on NBC featuring a virtuoso performance by James Cagney. The audio dramatisation was extremely successful, the medium placing the listener 'inside the head' of Joe and creating a powerfully intimate experience somewhat different to the 1971 film version (directed by Dalton Trumbo himself) and the 1982 stage adaptation written by Bradley Rand Smith. Although the film and stage play are effective works that capture Trumbo's passionately held antiwar sentiments and arguments, their respective media can nonetheless seem 'distanced' when compared to the intense interiority of the source novel and the radio dramatisation of it. The film and stage versions can evoke the pathos of the viewer, but the novel and radio play seem to thoroughly implicate the reader/listener in the narrative. In this regard, *Johnny Got His Gun* and its adaptations demonstrate that the interior monologue can reveal the close affinity that can exist between the media of prose fiction and audio drama.

To return to Wandor's Woolf adaptation, a central approach has been to rework the voices of the characters, above all the central figure of the protagonist, Clarissa Dalloway. Just a few paragraphs into Woolf's novel, we read:

> For having lived in Westminster – how many years now? over twenty, – one feels even in the midst of the traffic, or waking at night, Clarissa was positive, a particular hush, or solemnity; an indescribable pause; a suspense (but that might be her heart, affected, they said, by influenza) before Big Ben strikes. There! Out it boomed. First a warning, musical; then the hour, irrevocable. The leaden circles dissolved in the air. Such fools we are, she thought, crossing Victoria Street. For Heaven only knows why one loves it so, how one sees it so, making it up, building it round one, tumbling it, creating it every moment afresh ....
>
> *Woolf 1925: 8*

The writing is allusive and experiential, placing personal perspectives within a public forum. Wandor appropriates some of the material contained within this is to construct the opening of the radio play:

MORNING
(Music: Erik Satie: Gymnopedies)

SCENE ONE

CLARISSA (*over*) A June morning in London. Soft blue grey air. I build my London around me, creating it every moment afresh. I am at peace in the midst of carriages and motor cars, omnibuses and vans. I am at ease among the triumph and the jingle, the shops and parks. In this moment of summer, there is a solemnity, a suspense, a hushed moment just before Big Ben strikes.
(Music continues, and Big Ben begins to strike eight o'clock.)

CLARISSA (*over*) Big Ben strikes, on a June morning in London. The leaden circles dissolve in the air.
(Music fades, and Big Ben mixes into a silvery clock in dining room. Breakfast)

As we can see, Wandor takes key concepts such as 'building London' and the 'leaden circles,' as well as the setting of the city street within earshot of Big Ben. In addition, the possibilities of audio drama language are used: the appropriately impressionistic/modernistic music of Satie and the morphing of Big Ben into a dining room clock. What is most radical, however, is the way Wandor rewrites the voice of Clarissa. We encounter the character through an intimate 'I,' in contrast to Woolf's novel in which the narrator mediates the thoughts and feelings of Clarissa: Woolf undoubtedly gives the reader an 'interior' experience, albeit presented in a third-person style. Although the differences in style are radical, Wandor arguably captures the 'feel' of Woolf's narrative very authentically. While the construction of language and point-of-view are fundamentally rethought, this nonetheless demonstrates an effective translation of prose fiction into a wholly different medium. In short, Michelene Wandor completely 'rethinks' Virginia Woolf's story into the performance medium of radio.

Broadcast radio is not the only source for audio drama and adaptation. Since the advent of the World Wide Web in the 1990s and other advances in digital technology, we have entered into a realm of unprecedented access to audio. We can access websites that stream recordings of radio plays from the past or sell them to us as MP3s. We can subscribe to a host of production companies that are producing all-new audio drama that we can listen to on a regular basis or in an equivalent to televisual 'box-setting' consumption. Some contemporary companies such as Chatterbox Audio Theater and the Wireless Theatre Company recreate the classic practices of 'live' radio drama with Foley artists and an ensemble of voice actors. In contrast, others use the advanced technology afforded by complex editing and digital mixing. One noteworthy technological approach is in the area of binaural recording. Although this technology was pioneered in the earliest days of sound recording, it has increasingly come into its own in the digital age. Binaural audio creates an extremely high quality of stereophonic sound with a depth and gradation of recording detail that is, effectively, three-dimensional. This has opened up opportunities to take recording out of the studio and to 'capture' distinctive and remarkable spaces. For instance, ZBS Foundation's 1998 adaptation of Karl Edward Wagner's short story "Sticks" (1974) is a thirty-minute horror play set in an abandoned farmhouse and the ritual chamber discovered beneath it. Although the short story is a somewhat conventional 'post-H.P. Lovecraft' narrative, by using the binaural recording of suitable locations, the audio drama is an intense experience that can make the listener 'be present' among the doomed characters in the play. The ZBS Foundation has produced audio drama since 1970 and, along with *Sticks,* its adaptive binaural repertoire has also included a version of Carlos Fuentes's *Aura* (1962) in 2008. Fuentes's novel is eerie, lyrical, and even erotic, alluding to Charles Dickens's *Great Expectations* (1861) as well as the broader literary traditions of the Gothic and the ghost story. The novel is set inside a house of near total darkness, a challenge for the screen but ideal for translating into the immersive experience of binaural audio.

As part of its 2015 Halloween season, BBC Radio 4 included an adaptation of Nigel Kneale's television play *The Stone Tape* (1972) in both conventional and enhanced binaural versions. Adapted by Peter Strickland – writer/director of the film *Berberian Sound Studio* (2012), itself a vivid exploration of horror and sound – with Matthew Graham, the play retains the 1970s context of the source. The story is about a 'haunted' mansion and a group of sceptical scientists who strive to explain the phenomenon rationally. While the screen version can include visual phenomena, the radio adaptation can wholly emphasise the auditory. The scientists' 'torture' of the building in their attempts to unleash the sonic 'memories' they assume are captured within the literal stones of the walls – and their protracted analysis and manipulation of screeds of tape – turn Kneale's post-Lovecraftian, technophobic parable into a compelling experience about sound and horror. In a theatrical experience before the official broadcast of *The Stone Tape*, the audio drama organisation *In the Dark* arranged a site-specific airing of the play on 23 October 2015. A limited number of audience members entered the crypt of a church in Holborn in London, where, after walking past 1970s reel-to-reel tape machines and radios, the audience sat in darkness in a catacomb chamber and experienced the play through wireless binaural head-phones. The only consciously theatrical effect during the listening was at the end of the play, when eerie, green lights began to pulse against the walls of the crypt. However, the experience as a whole was highly theatrical inasmuch as the listener/spectator was acutely aware of the environment they had been placed in: the voices and reverberations, and the uncanny soundscape the scientists are trying to unlock, mapped seamlessly onto the walls and pitch-black corridors of the vault. A vivid dramatisation in its own right, the communal listening experience in such an evocative underground space enhanced the adaptation.

The radio adaptation (and theatrical shared listening) of *The Stone Tape* reveals a particu-lar aptitude of the medium: namely, that audio can work better than any other performance medium in realising the neo-Gothic and horror genre experiences of total darkness. It can also create uncanny environments and moods through sound design and narrative suggestion, both of which can work very effectively in combination with the imagination of the listener. While the light-based oblong of the television or cinema screen has a physical delineation, the experi-ence of audio drama can place us central to the experience (above all, if we listen on earphones), unfurling a narrative between our ears in darkness or mapping invisibly but potently onto our physical environment. In regard to making an environment uncanny and strange, the very corridors of the BBC itself have been a locus for experimentation. In a powerfully symbolic experiment during the impending closure of the BBC's Bush House (home of the BBC World Service from the 1940s) in 2012, sound producer Robin The Fog recorded the empty hallways, rooms, stairwells, and elevators of the building before editing and mixing them into soundscape compositions. The subsequent album – *The Ghosts of Bush* (2012) – is an eerie experience, its echoes and reverberations (occasionally combined with fragments of World Service broadcasts or automated elevator announcements) creating a sonic (re)construction of a location which was once in the business of producing 'sound.' *The Ghosts of Bush* is, in effect, an example of hauntological adaptation, a unique and literal physical space with a significant history that is captured, appropriated, and reworked into an uncanny audio experience.

It is an easy but mundane task to compare and contrast the poetics of audio with those of video. After all, in the simplest terms of reception, it has to be acknowledged that although the fact there is 'nothing to look at' with radio may be its greatest advantage (in terms of its con-comitant effect on the imagination), it is for many people its greatest deficiency. Regardless of this – or maybe because of it – there has always been a significant relationship between radio and cinema. In addition to *The Stone Tape*, the 2015 BBC Radio 4 Halloween featured Anita Sullivan's *The Ring*, an adaptation of Koji Suzuki's novel *Ringu* (1991). Similarly, the previous year's Halloween offerings on BBC Radio included Robert Forrest's adaptation of William Peter Blatty's novel *The Exorcist* (1971). Although there is no doubt that Sullivan and Forrest turned to the original novels as their adaptive sources, the radio plays undoubtedly drew on the popularity of the film adaptations (directed by Hideo Nakata in 1998 and William Friedkin in 1973 respectively), which are commonly regarded as classics of horror cinema. However, the radio/cinema relationship has a much longer history. One of the most prominent examples is *The Lux Radio Theater* (1934–55).

The Lux Radio Theater aired fully live one-hour adaptations of Hollywood movies in front of a studio audience. The broadcasts were contemporaneous with the release of the movies they were adapting and featured the film's cast reprising their roles for the airwaves. For example, in 1951 *Lux* presented *All About Eve*, an adaptation of Joseph L. Mankiewicz's 1950 film (itself based, albeit not credited, on Mary Orr's 1946 short story "The Wisdom of Eve"). The *Lux* version featured Bette Davis, Gary Merrill, and Anne Baxter reprising their roles (although Reginald Gardiner replaces George Sanders as the 'narrator,' Addison DeWitt). Similarly, on Christmas Day 1950, *Lux* listeners experienced a seasonal special with (an admittedly maturely voiced) Judy Garland recreating her role of Dorothy in *The Wizard of Oz* (1939). The *Lux* broadcasts functioned as an important part of movie publicity, equivalent to a trailer or magazine coverage. This was particularly evident as the broadcasts would sometimes feature an interview with members of the cast and, for nearly a decade of its run, *Lux* featured the epitome of the larger-than-life Hollywood director, Cecil B. DeMille, as host. In addition, the *Lux* plays were also a forerunner to video/DVD/Blu-Ray release, effectively offering audiences "a slice of Hollywood to be enjoyed in your own home" (Hand 2006: 44). In this regard, *Lux* represents a significant example of marketing and (for writers, actors, and other members of the production team) copyright-led income generation in the pre-digital culture of popular adaption. At one hour, the *Lux* radio adaptations are shorter than their source films, but they remain exemplary demonstrations of audio adaptation, rethinking their visual narratives into efficient audio drama.

Shorter lived, but in the same style as *Lux*, was *Academy Award Theater* (1946), which aired live adaptations of successful films in even shorter thirty-minute versions. This included Humphrey Bogart, Sydney Greenstreet, and Mary Astor reprising their roles from John Huston's 1941 film version of Dashiell Hammett's *The Maltese Falcon* (1929); and Cary Grant (with Ann Todd replacing Joan Fontaine) recreating for the airwaves his role in Alfred Hitchcock's 1941 film *Suspicion*, based on Francis Iles's novel *Before the Fact* (1932). The repertoire of *Lux* and *Academy Award Theater* reveal a popular subgenre of adaptation that existed on radio, drawing on the 'dream factory' of Hollywood to create – and retain – millions of radio listeners who wanted to relive (or had not had the opportunity to experience) popular motion pictures. However, the relationship between radio and cinema may be even more symbiotic. Focusing on Arch Oboler's radio series *The Adventures of Mark Twain* – commissioned by Warner Brothers to adapt and promote the contemporaneous film *The Adventures of Mark Twain* (Irving Rapper, 1944) – Matthew A. Killmeier demonstrates how intricate the marketing and publicity links between cinema and radio really were. The Twain biopic was implicitly "intertwined with the film's origination rather than simply a part of the post-production promotional campaign" in a way that reveals that "[c]ontemporary media convergence and synergies are not a simply a recent practice tied to industry consolidation, conglomeration and integration, but go back to the early days of broadcasting" (Killmeier 2015: 19).

A final, notable category of adaptation in audio that will be considered here is the biographical and fact-based drama. Sometimes the 'docudrama' form in a visual context can be problematic. The formal strategy in television documentary wherein dramatic 're-enactments' are interpolated between interviews has become something of a cliché, almost self-parodic in nature. In contrast, in the context of radio, the strategy continues to be effective: BBC Radio works such as *The Presence* in 2009 (Dannie Abse's account of the car accident that killed his wife), *Black Roses: The Killing of Sophie Lancaster* in 2011 (about a 2007 hate crime with a cycle of poems by Simon Armitage), and *Well, He Would, Wouldn't He?* in 2013 (about Mandy Rice-Davies and her involvement in the 1963 Profumo affair) are all examples wherein authentic testimony from those directly involved is interwoven with dramatic (re)interpretation to great effect, creating an adaptation of fact that is arguably more profound than either standalone documentary or dramatisation. The genre of docudrama is not uncontroversial. Guy Starkey (2014: 229) writes that as soon as a docudrama strays beyond the personal experience of the contributors, "someone's imagination begins to take over, and 'actuality' begins to lose its authenticity." This was evident in the reaction to a pioneering example of radio docudrama: the BBC's 1946 production *The Man from Belsen*, an adaptation of Harold Le Druillenec's testimony as the only British survivor of the Belsen concentration camp. The play steps back and forth between autobiographical narration and dramatic action, a strategy which creates a powerful experience but upset some contemporary reviewers who were deeply troubled by the "mixing of the fact and the fictionalised," a strategy in which, in the words of *The Listener*'s Philip Hope-Wallace, "One always kills the other" (Hand 2014: 38).

As we have seen in this chapter, adaptation has had a central place in audio drama from its beginnings through to the present day. Whether it is dramatising prose fiction or actuality, audio adaptation can be efficient and economical, capable of creating and capturing narratives of all dimensions and demands. In this regard, there are obvious advantages to what we might call audio technology's 'visionless spectacles,' but there are major challenges too: audio drama relies on the listener to actively participate in co-creating these narratives and to hold them in their mind's eye. In some cases, adaptation can help to build and sustain a narrative: perhaps it is easier to follow a story already familiar to us, no matter how radical the audio interpretation of it is. Overall, the flexibility of audio and the wealth of technology that affords us multifarious ways in which to be listeners means that there has never been a richer time in which to experience 'sound stories.' At its best, audio adaptation has always been able to immerse and implicate us in its interiority; it has been able to span galaxies or place us in the claustrophobic confines of complete darkness. To hear narratives unfurl in the space between our ears can be a subjectively intense experience and an unparalleled way to experience an adaptation.

Works cited

The Adventures of Mark Twain (1944), dir. I. Rapper, Warner Bros. DVD.
The Adventures of Mark Twain (1944), writ. A. Oboler, WFBG. Radio.
All About Eve (1950), dir. J.L. Mankiewicz. 20th Century Fox. DVD.
All About Eve (1951), on *The Lux Radio Theater*, writ. G. Wells and S. Barnett. CBS. Radio.
All That Fall (1956), writ. S. Beckett. BBC. Radio.
Aura (2008), prod. ZBS Foundation. Digital download.
Berberian Sound Studio (2012), dir. P. Strickland. Film 4. DVD.
Black Roses: The Killing of Sophie Lancaster (2011), writ. S. Armitage. BBC. Radio.
Blatty, W.P. (2011) *The Exorcist*, 1971, London: Corgi.
Blue, H. (2002) *Words at War: World War II Era Radio Drama and the Postwar Broadcasting Blacklist*, Lanham, MD: Scarecrow Press.
Bolik, S. (1998) "Für ein unreines Hörspiel: Zur (nicht gestellten) Frage der Literaturadaptation im Radio," *Zeitschrift für Literaturwissenschaft und Linguistik*, 28(111): 154–61.
Briggs, A. (1985) *The BBC: The First Fifty Years*, Oxford: Oxford University Press.

Brontë, E. (1965) *Wuthering Heights*, 1847, London: Penguin.

Chignell, H. (2009) *Key Concepts in Radio Studies*, London: Sage.

Crook, T. (1999) *Radio Drama: Theory and Practice*, London: Routledge.

Dickens, C. (2003) *Great Expectations*, 1861, Harmondsworth, UK: Penguin.

The Exorcist (1973), dir. W. Friedkin. Warner Bros. DVD.

The Exorcist (2014), writ. R. Forrest. BBC. Radio.

Fuentes, C. (2010) *Aura*, 1962, Manchester, UK: Manchester University Press.

The Ghosts of Bush (2012), prod. Robin The Fog. The Fog Signals. Digital download.

Gough, L. (2013) "*Wuthering Heights*: A radio adaptation," *Journal of Adaptation in Film & Performance*, 6(2): 157–315.

Hammett, D. (2013) *The Maltese Falcon*, 1929, London: Macmillan.

Hand, R.J. (2006) *Terror on the Air! Horror Radio in America, 1931–1952*, Jefferson, NC: McFarland.

Hand, R.J. (2014) *Listen in Terror: British Horror Radio from the Advent of Broadcasting to the Digital Age*, Manchester, UK: Manchester University Press.

Hand, R.J. and Traynor, M. (2011) *The Radio Drama Handbook: Audio Drama in Practice and Context*, New York: Continuum.

The Hitchhiker's Guide to the Galaxy (1978–2005), writ. D. Adams. BBC. Radio.

Hutcheon, L. (2006) *A Theory of Adaptation*, London: Routledge.

Iles, F. (1932) *Before the Fact*, London: Arcturus.

Johnny Got His Gun (1939), writ. A. Oboler. NBC, Radio.

Johnny Got His Gun (1971), dir. D. Trumbo. World Entertainment. DVD.

Killmeier, M.A. (2015) "The (radio) adventures of Mark Twain: Arch Oboler's adaptations of Warners' Picture," *Journal of Adaptation in Film & Performance*, 8(1): 5–21.

Kingsley, C. (1992) *Westward Ho!*, 1855, Upper Saddle River, NJ: Prentice Hall.

The Lord of the Rings (1981), writ. B. Sibley and M. Bakewell. BBC. Radio.

The Maltese Falcon (1941), dir. J. Huston. Warner Bros. DVD.

The Maltese Falcon (1946), on Academy Award Theater, writ. F. Wilson. CBS. Radio.

The Man from Belsen (1946), writ. L. Cottrell. BBC. Radio.

Mrs Dalloway (2012), writ. M. Wandor. BBC. Radio.

Orr, M. (1946) "The Wisdom of Eve," *Cosmopolitan*, May: 72–5, 191–5.

The Presence (2009), writ. D. Abse. BBC. Radio.

Rand Smith, B. (1982) *Johnny Got His Gun*, New York: Mainstage.

The Ring (2015), writ. A. Sullivan. BBC. Radio.

Ringu (1998), dir. H. Nakata. Toho. DVD.

Rostand, E. (2008) *Cyrano de Bergerac*, 1897, New York: Barnes & Noble.

Sherlock Holmes (1989–98), writ. B. Coules et al. BBC. Radio.

"Sorry, Wrong Number" (1943), on *Suspense*, writ. L. Fletcher. CBS. Radio.

Starkey, G. (2014) *Radio in Context*, London: Palgrave Macmillan.

Sticks (1998), prod. ZBS Foundation. Digital download.

The Stone Tape (1972), dir. P. Sasdy. BBC. Television.

The Stone Tape (1972), writ. P. Strickland and M. Graham. BBC. Radio.

Suspicion (1941), dir. A. Hitchcock. RKO. DVD.

Suspicion (1946), on Academy Award Theater, writ. F. Wilson. CBS. Radio.

Suzuki, K. (2007) *Ringu*, 1991, New York: Harper.

Trumbo, D. (1939) *Johnny Got His Gun*, New York: Citadel.

Wagner, K.E. (1998) "Sticks," in *Tales of the Cthulhu Mythos: Golden Anniversary Anthology*, 1974, Sauk City, WI: Arkham House, 374–91.

Walter, E. (1908) *The Wolf*, New York: G.W. Dillingham.

"The War of the Worlds" (1938), on *Mercury Theatre on the Air*, writ. H. Koch. CBS. Radio.

Well, He Would, Wouldn't He? (2013), writ. C. Williams. BBC. Radio.

Wells, H.G. (2005) *The War of the Worlds*, Harmondsworth, UK: Penguin.

The Wizard of Oz (1939), dir. V. Fleming. Metro-Goldwyn-Mayer. DVD.

The Wizard of Oz (1950), on *The Lux Radio Theater*, writ. G. Wells and S. Barnett. CBS. Radio.

Woolf, V. (1925) *Mrs Dalloway*, London: Hogarth.

Wuthering Heights (1939), on *The Lux Radio Theater*, writ. G. Wells and S. Barnett. CBS. Radio.

Wuthering Heights (2003), writ. L. Gough. BBC. Radio.

33

ADAPTATION AND NEW MEDIA

Establishing the video game as an adaptive medium

Dawn Stobbart

On 26 April 2014, at a landfill site in New Mexico, thousands of copies of the 1983 video game *ET* were found, games that had been buried by Atari after its economic and critical failure. Made and released in just six weeks, the game was intended to be, according to makers Atari, "emotionally oriented" and "based on the film's sentimentality for the alien" (Guins 2014: 216). As an adaptation of one of the highest grossing films of 1982, it was an attempt to build on its success, but players quickly discovered that it had no ludic or narrative depth, that the graphics were inferior (even by 1983 standards), and began returning the game to retailers in droves. As the flaws became manifest, Atari buried the games in a secret location, which quickly became urban myth, until a documentary crew rediscovered it over thirty years later. *ET* cost Atari approximately twenty million dollars for the license alone, and its failure helped bring about the company's subsequent demise; it has since gone on to be described as "the worst game ever" (Guins 2014: 217).

In 2015, video game adaptations have become much more successful and popular than this unidirectional and unsuccessful beginning suggests: tie-in video games are frequently made to coincide with film releases, novelisations of video games are becoming increasingly popular, and films based on video games are now a common form. Video games, while still being considered a niche market by some, have fast become a leading and profitable business, and have, in recent years, seen changes in their reception and creation, including their ability to adapt and remediate other media. This chapter considers how the video game informs adaptation studies, arguing that while video games can be analysed using methodologies employed to study other media, they also require new methods of analysis to explicate their interactive and ludic content. The chapter also considers how and why the medium succeeds (or does not succeed) in adaptation, asking whether the interactivity of video game narratives extends to greater interactivity in adaptation across media, particularly within entertainment franchises.

As *ET* highlights, video games did not have a successful genesis as an adaptive medium. *Death Race* (1976), the first direct adaptation from film to video game, was also the first video game to incite controversy; players had to "drive cars so as to run over 'gremlins' (which looked like human stick figures) to score points" (Arsenault 2008: 277), a plot similar to that of the film, on which Roger Ebert (1975) comments, "the winner is determined, not merely by his speed, but also by the number of pedestrians he kills." What proved controversial was not the adaptation itself, or even the content that closely resembled the film, but the interactivity that playing brought to it – the requirement for a consumer to actively kill for the narrative/game to progress. It is this interactivity that carries adaptation in video games beyond forms in other media – through direct physical action, rather than solely through visual observation or auditory stimuli. Interactivity is a defining feature of the video game, and its presence, or lack thereof, can significantly change how an adaptation is received by players, critics, and the wider community, as was the case with *Death Race* as well as the first video game to be adapted to film: *Super Mario Bros.*

When it was released in 1993, *Super Mario Bros.* was derided by critics, viewers, and cast alike, and did not even manage to recoup its production budget of forty-eight million dollars. While there were many problems with the film's production, one reason that is especially pertinent to this discussion is that it captures none of the interactivity of the games. As already indicated above, video games, more so than any other medium, rely on interaction for their progress; players are able to influence the world of the game and this, as Mark Wolf (2001: 114) explains, "is an essential part of every game's structure." Just as interactivity was the catalyst for the controversial reception of *Death Race,* its absence was instrumental in the *Super Mario* film adaptation's lack of success. In narrative terms, the *Super Mario* games are not storytelling devices: they use a quest structure that involves attempting to save a princess while traversing a series of maps/terrains. The success of the franchise is primarily due to the player's ability to explore the game landscape as she attempts to carry out the quest to save the princess. The choice the player has in this exploration is not available in the film adaptation, in which it is instead determined by the director and cinematography. Furthermore, none of the characters in the *Mario* games have any narrative or psychological depth, and, for the games, this is perfectly acceptable, as players interact with the ludic elements, which provide psychological stimulation as players avoid traps and defeat monsters to conquer the landscape, over any narrative. When the game was adapted to film, however, the physical activity that gives the games their interest and dynamism was replaced by physically passive viewing of flat characters navigating the landscape. While the repetitive formula of the journey that Mario makes across the game landscape was valuable for learning ludic skills and conquering the game, it was tedious for the passive viewer who was no longer directly involved in the process. Recognising this, the film's production team attempted to impose an alternative narrative onto the Mario diegesis that diverged from the source material, lacked complexity and dynamism, and resulted in a poorly made film that bore little resemblance to the source 'text' or the excitement of that text. Subsequent film adaptations of video games, such as *Lara Croft: Tomb Raider* (2001), occasionally became financial successes due to casting, rather than through success as an adaptation. Indeed, the same criticisms were levelled at *Lara Croft: Tomb Raider* as had been at *Super Mario Bros.*, with critics on the film review aggregator website *Rotten Tomatoes* giving the film an average score of 19% (n.d.). However, the film industry has taken lessons from these adaptations, and later films have attempted to create adaptations that achieve greater critical success.

These early attempts at video game adaptation led game designers and filmmakers alike to consider the differences and similarities between video games and films more carefully, such as the level of control given to the player/consumer. In a video game, the player's ludic success is crucial to the success and completion of the narrative. The player makes choices in the game, which drive the narrative forward. If play ceases, then the game stops and the story stops: the played character will stand still on the screen until the game hardware shuts down or the player resumes play. A film, on the other hand, can (and does) carry on, regardless of any interaction by the viewer; once begun, a film will carry on until the end, whether anyone is watching it or not. Interactivity changes the way that a video game player interacts with a narrative through identification with and as the played character in that narrative, effectively placing the player in the role of the protagonist – akin, but not identical to, an actor playing a character – and asking her to walk a mile in his or her shoes.

Video games are by no means limited to being adapted to and from film. There is a growing number of video games that are based on or adapted to the printed page. Two such games are *The Walking Dead* (Telltale Games, 2012) and *The Wolf Among Us* (Telltale Games, 2013), both adapted from graphic novels. Both video games offer narrative-dominant gameplay based on decisions that the player makes on behalf of the character she is controlling, decisions that expand on the source texts. *The Walking Dead* is an episodic video game based on the zombie apocalypse graphic novel series of the same name (Kirkman and Moore 2011) and influenced by the AMC television series (AMC Networks, 2010–) that also adapts the novels. Set before the events of the graphic novels, the video game prequel introduces a new set of characters and plotlines, while incorporating familiar aspects of *The Walking Dead* mythology. Here, as elsewhere, franchises expand beyond Linda Hutcheon's definition of adaptation proper, and they offer a method by which to satisfy fans who demand fidelity by creating prequels, sequels, and spin-off texts. Both the video game and graphic novels "explore how people deal with extreme situations and how these events *change* them" (Kirkman and Moore 2011: Introduction, emphasis in original), with the game using a branching structure that allows the player to influence the narrative direction. These video games deal with adaptation not only as a cross-media concept, but more widely ask how narratives change their participants and require them to adapt. Both gameplay and narrative are delivered via 'point and click' gameplay (which requires the player to point a cursor at an object on the screen, usually with a computer mouse, and click the mouse buttons to interact with that object), generally precluding more animated gameplay such as fighting and warfare. Instead, the player makes decisions and reacts to situations by interacting with other characters and diegetic objects to influence events, becoming an adaptor of the narrative trajectory. The play here is decidedly narrative play, and at set points the player is required to make decisions that affect subsequent play, linking all the episodes and giving the illusion that the player has influenced the narrative, even though the narrative is strictly linear. While the events of the adaptation are pre-authored, it appears that the narrative is being created by the player as she engages with the adaptation.

Despite the negative beginnings, video games have become successful adaptations of other media. *Alice's Adventures in Wonderland* (Carroll 1865), for instance, was adapted – and remediated – to become *American McGee's Alice* (Electronic Arts, 2000) and its sequel *Alice: Madness Returns* (Electronic Arts, 2011). Cathlena Martin (2010: 134) remarks that a narrative such as *Alice in Wonderland* has "gone through an intermediary stage of film and other forms of adaptation, so that multiple incarnations of each have placed [it] into cultural consciousness," and considers that "video games offer the opportunity to analyse and revisit classic children's literature [and] to explore new and additional facets of the story." *American McGee's Alice* and *Alice: Madness Returns* follow the protagonist, Alice Liddell, as she makes her way through a dark, ruinous wonderland to reach and destroy the Red Queen. Although it is narratively a sequel to the novels written by Carroll, the games integrate much of the novels' original material to create a new version of *Alice*, one that is mature, dark, and fearless. This adaptation takes the narratives of the Carroll stories, and rewrites them, using cultural references and suggestions surrounding the writing of those stories (including claims regarding Carroll's paedophilic interest in Alice Liddell), to create its own narrative, adding puzzle solving and fighting elements that carry the player through Wonderland as an active, gaming participant. In terms of adaptation studies, this game not only reinterprets and adapts Carroll's *Alice* books; it further revises classic children's stories into a mature, twisted narrative that asks the player to re-evaluate her understanding of the original texts.

Narrative perspective, such as narrative distance or lack of narrative distance, determines the way that a consumer identifies with a narrative; adapting that perspective to another medium changes that identification. How a video game player views the character that she is controlling is integral to the interpretation of that narrative. Most video games have the player take on the role of their respective protagonists. *Tomb Raider* is a franchise that began as a video game (1996) and was subsequently adapted across a variety of media, including two film adaptations starring Angelina Jolie, which were subsequently novelised as *Lara Croft: Tomb Raider* (2001) and *Lara Croft: Tomb Raider – The Cradle of Life* (2003). These adaptations highlight the differences between film, video game, and the novel identification, both with and as characters. In the video game, the player controls Lara Croft from a viewpoint that combines first and third person, something I refer to as the decentred perspective. This perspective is dominant in contemporary video games, mixing the tracking camera with the over-the-shoulder shot, offering clearly defined and externally focalised perspectives, similar to those in film. Encompassing aspects of the third-person perspective, the character is frequently seen on the screen, creating narrative distance between player and protagonist. Structurally, however, the player is offered a visual and auditory perspective, similar to that of the first person, with the player positioned as if above the shoulder of the protagonist, rather than situated as if viewing the scene directly through the protagonist's eyes and ears. This creates a slightly decentred position that allows the player to see everything the protagonist sees, but slightly askance from the protagonist's view and, more importantly, with a wider field of vision than that of the character's. The decentred perspective, therefore, clearly identifies the character as a separate entity, akin to the third-person perspective, but places the player in much closer proximity to the character's view, allowing her to identify almost – but not quite – as the character. When coupled with the interactivity of controlling the character, this perspective functions much like the proverbial good/bad angel on the protagonist's shoulder, with the player in some cases acting in the role of the angel/devil – making moral choices that affect the character and/or subsequent events within the narrative.

This is not the case in either the film or the novelisation of the Lara Croft video games. In the film, the third-person perspective is the chief vantage point; although the video game has clearly influenced the perspective of camera angles and movement, the focalisation has been adjusted to work with the conventions of the film medium. The film does not use over-the-shoulder shots as often: the camera focuses on the protagonist as the object of the film. However, during the film's action sequences, the camera switches between over-the-shoulder shots and reverse angle shots, creating an "insistent and intimate shot-countershot technique" (Monaco 2009: 211) that occupies many points of view. Within the context of the film, switching between shots allows the viewer to both identify with the protagonist as a subject and to see the protagonist as a screen object. This unfolds in contrast to the decentred video game perspective, which predominantly shows actions from the viewpoint of the player; the film, therefore, allows the viewer to experience a relaxation from the relentless identification of the video game perspective, and to experience a more objective identification of the protagonist as observed object, rather than identified with subject. The novel combines a third-person external focalisation with an internal, character-bound focalisation. However, the novel also presents the reader with Lara's thoughts, and, thus, moves away from the objective view of her actions to the subjective view of her verbal thinking and emoting, so that the viewpoint becomes internally focalised on Lara. Neither the film nor the video game chooses to represent these thoughts, although both media are capable of doing so.

The relationship that the video game has with the printed word is by no means unidirectional, as it was so often in literature-to-film adaptation; there is a growing selection of novels beyond the Lara Croft franchise that are adapted from video games. Part of the *Assassin's Creed* franchise (Ubisoft, 2007–) has been adapted to novels, and these novels have, in turn, influenced subsequent games. In *Assassin's Creed: Brotherhood* (2010), when playing as the protagonist Ezio Auditore da Firenze, there is an opportunity to complete extra quests and missions that have no direct relevance to the game's completion, with one set of these missions being 'The Cristina Memories.' There are five of these missions to be completed, each resulting in a small cut-scene; these cut-scenes are based on *Assassin's Creed: Renaissance* (Bowden 2009), the novel adaptation of *Assassin's Creed 2* (2009), and chart the relationship between Ezio and his lover, Cristina, from his first meeting with her in 1476 to her death in his arms in 1489. These memories bear no relation to the ludic completion or the fictional time of the game itself; they are specifically narrative elements whose sole purpose is to bring a deeper narrative understanding of Ezio's character. Here the adaptation does not replace, or even replicate, the gameplay, but brings the novelisation into it to enrich the narrative with literary and filmic traditions.

Not only is narrative being adapted to and from video games to other media forms and conventions; ideologies are also being adapted across media and cultures. Once again, the active role that the player takes in a video game means that these ideologies can be explored in new ways, with more agency (or with more illusion of agency). The 2007 video game *BioShock* (2K Games) is an example of such a game, adapting Ayn Rand's 1957 novel *Atlas Shrugged* and, more explicitly, its ideological epistemology of Objectivism. It goes beyond adaptation to present a critique of Rand's philosophy and ties questions of social and political agency and free will to issues of agency in video gaming, creating a metacriticism of the medium via adaptation.

In *BioShock*, ludic play is inextricable from narrative ideology and, more than that, informs it in new ways. The setting, Rapture, although dystopian and filled with death, is full of movement and sound: water drips, voices re-echo, and ghostly images inhabit rooms. The player must move through these oppositional game spaces into the underwater city's darkest corners in order to discover the narrative of the game's backstory, mostly in the form of audio logs. As part of this exploration, Rand's ideology is didactically represented. Drawing on first-person narration in literature, theatre monologues, and voiceover conventions in film, during the introduction, NPC (i.e., non-player character) Andrew Ryan's voice narrates Rapture's ideology, telling the player that the city was born from his dissatisfaction with American left-wing politics in the Second World War. Finding that there was no place for "men who believed that work was sacred and property rights inviolate" (Fuller 2007: 42), he decided to create one. This adapts the creation of Galt's Gulch by its namesake, John Galt, in *Atlas Shrugged*, as a place where a man "hold[s] three things as the supreme and ruling values of his life: Reason – Purpose – Self-esteem" (Rand 2007: 1018), and which stands in contrast and opposition to a mainstream society that requires a producer, an individual who is "independent, rational, and committed to the facts of reality … and to their own happiness" (Younkins 2013: 168), or entrepreneur to not only be sacrificed by society for the greater good, but to accept this as fair and just. Both settings follow Rand's philosophy; however, where Galt's Gulch remains utopian, Rapture becomes dystopian. The freedom of the inhabitants to do as they wish engenders a society addicted to a drug named ADAM and its derivative, EVE. Just as Adam and Eve ate of the fruit of the Tree of Knowledge in Genesis, causing the Fall from divine grace, so too does the use of ADAM and EVE cause Rapture to become dystopian.

However, the adaptation shifts from translation to critique, and from adaptation to contesting the original's narrative ideologies. The game is not a simple remediation of Rand's philosophy: it serves as a critique of it; the player is shown both positive and negative views of Objectivism as opposed to solely Rand's position. In this regard, the adaptation is no different from texts or films that present multiple points of view. However, it goes beyond presenting different perspectives, requiring the player to explore them and make decisions within the narrative that change the narrative trajectory. To critique Objectivism, rather than simply endorse or damn it, the game positions the player as the protagonist, Jack. Predominantly a first-person shooter game (FPS), the player controls Jack as he seeks to escape Rapture. Unlike Rand's novel, but like most video games, *BioShock* is concerned primarily with the destruction of enemies, ranging from splicers to the main antagonist, Frank Fontaine. However, the combination of politics, ideology, narrative, and ludology creates a representational fictional form resembling that described by Kendall Walton in *Mimesis as Make-believe*. Walton (1990: 42) considers the role of "props," which include environmental objects, as well as weaponry with which the player interacts, to be "enormously important. They give fictional worlds and their contents a kind of objectivity … which contributes much to the excitement of our adventures with them." This objectivity extends from fictional worlds and their contents to their political and ideological contents. In the case of *BioShock,* these props bring "a kind of objectivity" (Walton 1990: 42) to the philosophy of Objectivism, and indeed, some of the props in *BioShock* are sentences of Objectivist philosophy. In this game, the ideology of the adapted text is thus implanted in the environment itself, creating an adaptation that functions on several levels: as a ludic game in its own right, a narrative in and of itself, and as an adaptation of Rand's novel.

Many video game adaptations immerse the player in a situation that is morally ambiguous, requiring her to question her actions both in the game world and outside it. *Spec Ops: The Line* (2K Games, 2012) offers the player the ability to do this, within a physically safe environment, while at the same time engaging with the themes and motifs of the source text. Self-reflexivity is an established feature of media (Stam 1992), including film and literature (Poulaki 2014; Huber, Middeke, and Zapf 2005). Within the context of this chapter, self-reflexivity is defined as any aspect of a video game that points towards it as a game, whether its creation, its conceptualisation, the processes by which it is constructed and consumed, and the methods by which it critiques itself. Self-reflexivity is frequently identified with postmodernism, a movement that embraces instability and is characterised by scepticism, the rejection of cultural progress, and the implementation of metanarratives (Sim 2011). The video game, still a young medium, is becoming increasingly postmodern and self-reflexive, particularly when it subverts traditional distinctions between reality and simulation. Simon Gottschalk's 1995 article "Videology: Video-games as postmodern sites/sights of ideological reproduction" offers a detailed study of the medium as postmodern. At first, *Spec Ops* appears to be a clone of shooter games such as the *Call of Duty* franchise, using the same tropes and ludic strategies. However, the intertextuality of the game to other shooter games is complicated and deepened by its relationship with philosophical literature and postmodern film. The two kinds of intertextuality lead it to question not only the legitimacy of violent occupation in the social world, but also to question the violent ludic structures of the mainstream video game tropes and conventions that it adopts and adapts. *Spec Ops: The Line* is a loose adaptation of both Joseph Conrad's *Heart of Darkness* (1899) and *Apocalypse Now* (1979), set in a speculative version of Dubai, which has been destroyed by a series of sandstorms that engulf the city, and as a result has become a dark place, reminiscent of Conrad's Africa and Coppola's Vietnam. *Spec Ops* offers a psychological exploration of its protagonist, Captain Martin Walker, as he makes his way through a ruined Dubai in search of the 'Damned' 33rd Battalion and their commanding officer, Konrad, who have gone missing, much as the protagonists of *Heart of Darkness* and *Apocalypse Now* undertake their actions against the backdrop of the atrocities of the ivory trade and the Vietnam War, respectively. The discomfort of playing this game serves to deconstruct player expectations of the shooter genre – its pleasures and its rewards – and to map her virtual violence onto real-world situations in which the innocent and powerless die along with the armed enemy. *Spec Ops* offers a commentary on choice in gaming and real life; repeatedly, pieces of dialogue feature Walker telling other characters that he has no choice in the actions he takes: for example, when he responds to Lugo's challenge about the morally dubious use of white phosphorus. Walker insists that there is no choice, despite Lugo's assertion that there is always a choice. The game sides with Walker diegetically, and with Lugo extradiegetically. While the game does not allow diegetic choice, at each point in the game where actions are immoral, even by the standards of FPS gaming, the player *is* offered a choice: to quit playing. Her continuation of the game, much like Walker's continued journey into Dubai, is a choice, and each time she makes the choice to continue, the game 'rewards' her with more death, more destruction, more discomfort, and more uneasy complicity in Walker's unethical actions. The game also demonstrates the growing sophistication with which video games can achieve photorealism and thereby become more cinematic, more closely resembling film visually and narratively.

In conclusion, while early video game adaptations were problematic, clumsy, and unsatisfying to consumers and unprofitable to producers, contemporary video game technologies mean that adaptations can simulate and engage a variety of media (gaming, text, cinematic scenes, music, and more) to create richly complex, intertextual forms of adaptation. They also offer an interactivity that changes the way in which consumers engage narrative and adaptation. Interaction allows a player to explore, discover, and shape narrative, both narratologically and ideologically, and alter how it unfolds and how it is consumed. While there is a growing body of work examining the cinematic capacities of video games, their functions as adaptations of other media has yet to receive the same scrutiny. Video games as adaptations is a rich field awaiting extensive future study. As its technologies continue to change and develop, it promises to continue to change the ways in which we understand, engage, and interpret adaptation specifically and intertextual narratives generally.

Works cited

Alice: Madness Returns (2011), dir. A. McGee. Spicy Horse. Electronic Arts. Video game.

American McGee's Alice (2000), dir. A. McGee. Rogue Entertainment. Electronic Arts. Video game.

Apocalypse Now (1979), dir. F. F. Coppola. Studio Canal. DVD.

Arsenault, D. (2008) "The video game as an object of controversy," in M. J. P. Wolf (ed.) *The Video Game Explosion: A History from PONG to PlayStation and Beyond*, Westport, CT: Greenwood Press, 277–82.

Assassin's Creed (2007–), dir. P. Désilets et al. Ubisoft Montreal. Ubisoft. Video game series.

Assassin's Creed 2 (2009), dir. P. Désilets et al. Ubisoft Montreal. Ubisoft.

Assassin's Creed Brotherhood (2010), dir. S. Bernard et al. Ubisoft Montreal. Ubisoft.

BioShock (2007), dir. K. Levine. 2K Boston/2K Australia. 2K Games. Video game.

Bowden, O. (2009) *Assassin's Creed: Renaissance*, London: Penguin.

Carroll, L. (1992) *Alice's Adventures in Wonderland, 1865*, Ware, UK: Wordsworth Editions.

Conrad, J. (1990) *Heart of Darkness*, 1899, Mineola, NY: Dover Thrift Editions.

Death Race (1976), dev. H. Ivy. Exidy. Exidy. Arcade game.

Ebert, R. (1975) "Death Race 2000," *Roger Ebert*, 27 April. Online. Available at: http://www.rogerebert.com/reviews/death-race-2000 (accessed 6 October 2015).

Fuller, B. (ed.) (2007) *Bioshock: A Game by Take Two & 2K Boston/Australia*. Script. Online. Available at: https://www.scribd.com/document/74201316/Bio-Shock.

Gottschalk, S. (1995) "Videology: Video-games as postmodern sites/sights of ideological reproduction," *Symbolic Interaction*, 18(1): 1–18. Online. Available at: http://onlinelibrary.wiley.com/doi/10.1525/si.1995.18.1.1/full (accessed 15 March 2016).

Guins, R. (2014) *Game After: A Cultural Study of Video Game Afterlife*, Cambridge, MA: MIT Press.

Huber, W., Middeke, M., and Zapf, H. (2005) *Self-Reflexivity in Literature*, Wiesbaden, Germany: Königshausen & Neumann.

Kirkman, R. and Moore, T. (2011) *The Walking Dead*, vol. 1, Berkeley, CA: Image Comics.

Lara Croft: Tomb Raider (2001), dir. S. West. Paramount Home Entertainment. DVD.

Martin, C. (2010) "'Wonderland's become quite strange': From Lewis Carroll's Alice to American McGee's Alice," in P. Frus and C. Williams (eds) *Beyond Adaptation: Essays on Radical Transformations of Original Works*, Jefferson, NC: McFarland, 133–43.

Monaco, J. (2009) *How to Read a Film: Movies, Media, and Beyond*, 4th edn, Oxford: Oxford University Press.

Poulaki, M. (2014) "Puzzled Hollywood and the return of complex films," in W. Buckland (ed.) *Hollywood Puzzle Films*, New York: Routledge, 35–54.

Rand, A. (2007) *Atlas Shrugged*, 1957, London: Penguin.

Rotten Tomatoes (n.d.) "Lara Croft: Tomb Raider," *Rotten Tomatoes*. Online. Available at: https://www.rottentomatoes.com/m/lara_croft_tomb_raider/ (accessed 15 March 2016).

Sim, S. (2011) "The modern, the postmodern, and the post-postmodern," in S. Sim (ed.) *The Routledge Companion to Postmodernism*, 2001, London: Routledge, vii–xiv.

Spec Ops: The Line (2012), dir. C. Davis. Yager Development. 2K Games. Video game.

Stam, R. (1992) *Reflexivity in Film and Culture: from* Don Quixote *to Jean Luc Godard*, New York: Columbia Univesity Press.

Stern, D. (2001) *Lara Croft: Tomb Raider*, London: Pocket Books.

Stern, D. (2003) *Lara Croft: Tomb Raider – The Cradle of Life*, London: Pocket Books.

Super Mario Bros. (1985), dir. S. Miyamoto. Nintendo R&D4. Nintendo. Video Game.

Super Mario Bros. (1993), dir. R. Morton and A. Jankel. Second Sight. DVD.

Tomb Raider (1996), dev. T Gard. Core Design. Eidos Interactive. Video game.

The Walking Dead (2010–), dev. Frank Darabont. AMC Networks. Television series.

The Walking Dead (2012), dir. Sean Vanaman et al. Telltale Games. Telltale Games. Video game.

Walton, K. L. (1990) *Mimesis as Make-Believe: On the Foundations of the Representational Arts*, Cambridge, MA: Harvard University Press.

The Wolf Among Us (2013), dir. Nick Herman et al. Telltale Games. Telltale Games. Video game.

Wolf, M. J. P. (2001) *The Medium of the Video Game*, Austin, TX: University of Texas Press.

Younkins, E. W. (2013) *Exploring Capitalist Fiction: Business through Literature and Film*, Lanham, MA: Lexington Books.

34

MEMES, GIFS, AND REMIX CULTURE

Compact appropriation in everyday digital life[1]

Eckart Voigts

Mashup, remix, and cultural appropriation

Translations, adaptations, appropriations, and remakes transform, change, and vary existing cultural material. So do other forms of 'chronological' or 'genealogical' *inter*textuality or *inter*mediality or *inter*arts in a diachronic perspective, such as allusions, parodies, pastiches, imitations, rewritings, mashups, remixes, samples, prequels, sequels, and – in the arts – montage, collage, and assemblage. Illegal intertextuality is called plagiarism or fake – for artistic purposes, we speak of fake art. Originality, it seems, is an innovative form of transforming existing material. Culture is produced via a reshaping of existing culture. We might address these cultural processes by using verb-derived nouns (adapting, transforming, remaking, remixing) in a sense quite distinct from the nouns denoting a rather more specific genre (adaptation, transformation, remake, remix). Whenever the synchronic dimension needs to be highlighted, researchers prefer the term *trans*mediality. In sum, we might speak of the poly-processes of artistic production, all of them adaptive, derivative, and appropriative in character.

Remixing and mashing, however, are special cases of transformation that can be distinguished from other forms of adaptation and appropriation. Only very wide definitions of 'adapting' (in diachronic perspective) or 'transmedia storytelling' (in synchronic view) would cover remixing, mashing, and sampling. Daniel Fischlin and Mark Fortier (2000: 4) see adaptation as "the general process of cultural recreation," and, more recently, they have argued against the 'bureaucratic' narrowing of adaptation terminologies because this less expansive notion would exclude the potentially limitless transformative activities at work in parodies, travesties, sequels, and so forth (Fischlin 2014: 22; Fortier 2014: 373).

First of all, remixes, mashups, and samples as well as collages, assemblages, and montages work with techniques of duplication – most likely with a multiplicity of duplications. Remixes and montages are rearrangements of existing cultural particles. Remixing, thus, is neither retelling, nor expanding, nor revisiting pre-existing material. In aesthetic terms, remixes – similar to montage – require a recognition of their separate components as distinct and heterogeneous. The term montage is in a way misleading, as we might better speak of 'demontage,' in that film or photomontage keeps the differences between separate units in view. The classic Eisensteinian montage, as well as jump cuts or split-screen techniques, makes editing visible rather than rendering it invisible, as in continuity editing or *découpage classique*. There is no such thing as a holistic, homogeneous collage or mashup. A sample is a recontextualised particle (sound recording, etc.) whereas a cover tends to derive from the initial version, but not use pre-existing material. Depending of the quantity of original material used, we tend to speak of either a sample or a remix version. The quantity of cited and recontextualised material is an important criterion in the analysis of remixes (as well as in distinguishing a mere allusion from an adaptation), but it is definitely not the only criterion in the aesthetic and legal consideration of remix cases. I would like to appropriate the terminology of narratology here and distinguish between 'kernels' (signature elements, not just events, but also characters, phrases and other writing, visual content, sounds, etc.) that are recognisable, attributable, citable, iterable, and necessary for a cultural work, and 'satellites,' not just events, but also characters, phrases and other writing, visual content, sounds, etc. that are non-mandatory for the recognisability of a fragment in a mix. In that sense, Lewis Carroll's original sketch of Alice for his iconic character (although more 'original' and 'authorial') is less of an Alice kernel than John Tenniel's illustration for the first published edition. A twenty-year-old copyright dispute about a two-second rhythm sequence sampled by Moses Pelham from German band Kraftwerk's song "Metall auf Metall" is currently being decided by the European Court of Justice. In 1997, Pelham sampled two seconds from the 1977 Kraftwerk song, and one of the core questions in this ongoing dispute is if the two-second sample shaped the song decisively – in other words, was it a kernel or merely a satellite? Histories of difficult cases of sample clearing in hip hop and elsewhere are legion. Another important category shifts the burden of clarifying what a cultural entity or kernel is to the author/curator and the recipient. Generic marking emerges in a contextual and paratextual negotiation between the producer and the consumer – increasingly maybe also in the circulation of meanings between pro-sumers and prod-users of culture.

Mashup and remix are arguably the most important emanations of cultural appropriation. It is fascinating to note that two almost contradictory attitudes towards cultural appropriation have emerged. On the one hand, from an aesthetic perspective within adaptation and remix studies, appropriating is praised for the creation of new cultural diversity. On the other hand, more recent concepts of cultural appropriation reverse this structure, attacking cases in which a dominant or hegemonic culture appropriates a practice or style (twerking, dreadlocks) from a position of cultural power and dominance. This paper addresses cases in which existing cultural material narratives, films, characters, etc. – is broken up and recombined in animated GIFs, thus generating a cultural practice I have called 'recombinant appropriation.'

Whereas the term 'adaptation' originally pertained to the transcoding of literary texts into the audiovisual medium of film, the terms 'remix' and 'mashup' tend to be used in the context of an everyday aesthetics. Remix and mashup are based on techniques of montage, collage, and assemblage and the constitutive aesthetic practices of 'editing,' 'cutting,' and 'rearranging.' In contrast to practices of adaptation, remixing is based on the persistent and demarcated textual contrast between the recombined *objets trouvés*. The terms 'remix' and 'mashup' encompass a variety of different techniques and subgenres. Film, of course, tends to be – technically speaking – always a montage, even if the 'suturing' of shots, for instance, along the lines of so-called continuity editing suggests unity, thus masking the processes of montage that inevitably occur in the composition of the aural and visual tracks of film. We can sum up that every film is a montage, but that very few films actually foreground this feature, marking the mixing and editing that make films edits or mixes. Hence, only if the formal features of a given film suggest it, it will be regarded as a remix – analogous to Linda Hutcheon's idea that only some adaptations will be read as adaptations (Hutcheon 2013: 6–9). These techniques are not necessarily bound to a specific function. They can have a satiric function or, in a more fundamental perspective, interrogate from a perspective of an avant-garde aesthetic the intentionalities, semantics, and aesthetics of traditional, bourgeois arts.

They may also very simply re-situate existing material without any political or aesthetic agenda. The increasing cultural relevance of remixes consists of the fact that they have become the standard form of low-threshold, creative participation in everyday culture in media-saturated, mobile societies of 'convergence culture.' They are "auditively, visually, audiovisually 'mixed' rearrangements, collages, bricolages in music, video, computer games, the visual arts (particularly contemporary media art), and in the net architecture" (Wilke 2015: 13; my translation). Just as the fragmentary aesthetics of the modernist avant-garde reacted to the incomprehensible, unintelligible fluidity of the industrialised, urban world, so have remix and mashup become the new paradigm of digital textuality. Mashup and remix, however, do not exclude non-digital genres, but have on the contrary influenced them – historically in the modernist montage novels, in the political montage of John Heartfield, the cut-ups of William Burroughs, the remix texts of Mark Amerika (2011), or the flood of mashup novels that inundated the market for popular fiction roughly ten years ago. Without a doubt, digital remix relies on established techniques that have become easier and more elegant and proficient in execution by applying digital tools, circulating and situating them in the networked platforms the digital world has provided.

Just how aesthetically advanced, politically progressive, and culturally innovative are these texts? This question can only be answered by analysing individual case studies. And can we really differentiate between remixes and mashups? According to which set of criteria? Is a remix lacking in semantic repositioning, not aiming at a new, different meaning-making via montage, collage, and alienation techniques (*Verfremdung*) (Wilke 2015: 14)? The discussion among remix and mashup scholars seems eerily analogous to the debates in adaptation studies. Here, the degree of creative repositioning of an adaptation in comparative interpretation with the adapted 'source' text (or, with Gérard Genette, 'hypotext') opens up the continuum between an adaptation and an appropriation.

Julie Sanders (2006: 26) has picked the spatial metaphor of journey and transition in her by now classic description of the space between the 'closer' adaptation and the 'wider' appropriation. Appropriations are recontextualised and repurposed texts that lead the new product into a completely new cultural sphere, "a more decisive journey away from the informing source into a wholly new cultural product and domain" (Sanders 2006: 26). As we have seen in Wilke's approach, remix scholarship has tended to opt for creative repurposing and refashioning as the key criterion for their typologies in ways quite similar to the emerging continua between adaptations and appropriations. Eduardo Navas (see 2012: 93), for instance, proposes a dichotomy between regressive and reflexive mashups: only reflexive mashups are able to constitute a new force field of meaning by re-composing cultural material.

All of the items that are available in digital code can become part of this game of reduplication and recontextualisation as long as the conditions outlined, for instance, by Felix Stalder (2009) are met: In short, there needs to be a cornucopia of mashable material in a data-saturated, connected culture; the material must be inexpensive or free and free of copyright; and, finally, remix and mashup must be conventionally accepted as a collective cultural praxis and technique.

Remixing and mashing have become a paradigmatic praxis of digital culture. A pioneering diagnosis of remix as an agent of cultural transformation is Lawrence Lessig's text on the read/write culture (Lessig 2008). The techniques of remix and mashup may not be new, but they have become a decisive, distinguishing trait of digital network culture, and their proponents frequently sound the horns of euphoria and panegyric. Dirk von Gehlen (2011: 19) praises the artistic potential of appropriation in a "creative reference culture" that enables content to be "easily adapted, remixed and parodied." Stalder (2009) compared this paradigmatic shift towards remixing and mashing to the emergence of bourgeois capitalism in the eighteenth century, which culminated in the cultural industries of the twentieth century. Maybe we are indeed witnessing a radical transformation of the public sphere and the end of Fordist production with its attendant phenomena such as the complete redefinition of copyright laws and the collapse of cultural distribution, exploding distinctions between consumers and producers.

Animated GIFs as appropriations in ordinary culture

Compact audiovisual clips are standard fare in web-based databases and on searchable platforms, and these short visual narratives shall serve as the prime example of appropriation in the remainder of this essay. The animated GIF is a small, shareable image file that combines several images, thus enabling mashers to produce a micronarrative within a single image file. Lacking sound, it is a visual micronarrative. The GIF file constitutes a 'cinematic' narration process within a single digital file. I argue, however, that the GIF is not so much a data format, but an indicator of contemporary cultural production. It is not only linked to the predominantly affective modes of early and contemporary cinema (Hesselberth and Poulaki 2017: 1), but also indicative of contents and modes of circulation fostered by networked social media.

GIFs merit our attention precisely because they are appropriation at its most ordinary and quotidian. Raymond Williams (2011: 53) argued in 1958 that "Culture is ordinary. That is where I must start … Culture is ordinary. In every society and in every mind." Williams's legacy suggests that cultural studies must continue to embrace the ordinary in its claim to move from the theoretical study of contemporary social contexts to an active participation of academia in current cultural debates. Cultural studies, too, must remain ordinary, and GIFs must be analysed not just in terms of how the media form determines what they are, but also in terms of the social contexts in which they appear, and in terms of which messages they are to convey. In the course of Brexit and the Trump election of 2016, a critical consensus has emerged which believes that social media are transforming the Habermasian public sphere – already damaged, according to Richard Sennett (in *The Fall of Public Man*, 1977) and others, through the onslaught of private media and publicised privacy. As Sebastian Sevignani (2016: 3) has summarised, "the capitalist nexus of commodification, between private property, surveillance, and privacy, is normatively challenged because it involves exploitation, social sorting, and exclusion, as well as alienation and heteronormy," and commodification "is described as ultimately contributing to individual and social unfreedom."

GIFs may also be described as the most participatory versions of adaptation and appropriation. A quick survey of the criticism of active audience concepts in cultural studies reveals the conflict between a predominantly culturalist view of participation and a predominantly political–economic model of assessing participation. Henry Jenkins's focus has long been on low thresholds to creativity and assertion – so that both the surrealist GIFs discussed below and the emergence of a post-factual echo chamber of truthiness with Donald Trump supporters on social media are two sides of the same coin (Tufekci 2016). John Fiske's notion of 'producerly' texts and his celebration of 'active' television viewing in the 1980s as part of his concept of 'audiencing' are central here. Fiske's terminology speaks of 'formations' rather than 'mass media audiences,' hence implying the 'prod-users' and nomadic interpretive communities of social media:

> Popular culture is made by the people, not produced by the culture industry. All the culture industry can do is produce a repertoire of texts or cultural resources for the various formations of the people to use or reject in the ongoing project to produce their popular culture.
>
> *Fiske 2010: 19*

Applying John Hartley's notion of 'Power Viewing' (Hartley 1992) from his *The Politics of Pictures*, we can describe the remixers in social media as both *pervasive* (ubiquitous) and *pervaded* (by other cultural practices). Applying Hartley's ideas to digital media, we could argue that to conspicuously consume and produce memes in social media is already transgressive: It is a non-functional activity that does not directly involve a commodified cultural exchange. On the other hand, the algorithmically established echo chamber is, of course, part and parcel of the pervasive commodification of all activities that capitalist societies strive for.

Fiske and Hartley's championing of 'active' consumption has generated numerous hostile responses – mainly from researchers who criticise the culturalist focus on creativity, cultural sharing, and mutual significance within communities. Jim McGuigan sought to re-establish critical cultural studies, castigating the celebratory model of consumer sovereignty as *Cultural Populism* (1992). Ten years later, he renewed and adapted his critique of *Cool Capitalism* as "the incorporation of disaffection into capitalism itself" (McGuigan 2012: 431). Trebor Scholz and Christian Fuchs have cast doubt on the transgressive potential of participatory culture on fundamentally Marxist grounds: the free and creative variations of 'cultural jazz' remixes are merely a playground fostered by Big Bad Media to better situate and flog the products of corporate intertextuality. Arguably, social media provide both a playground and a factory, in which fans are duped to supply their labour for free to capitalists such as Mark Zuckerberg (Scholz 2013: 8). As Christian Fuchs (2011) has argued, the primarily culturalist understanding of participation is flawed, as it excludes participation in "economic decision-making." Fuchs's critique of Jenkins harps on his narrowly culturalist notion of participation, as well as his technological determinism. Fuchs (2011) argues that "[s]tructures of control in the economy today and in the political system are based on power asymmetries. Although we produce information ourselves, this does not mean that all people benefit from it to the same extent." For Fuchs, there is no participation without a truly democratic society in which there is grassroots decision-making and common ownership of the means of production.

What all of the above critics agree on is the pivotal role of popular culture and its socio-economic framework for the contemporary world. As Oliver Marchart (2008: 14) holds in Foucauldian fashion, popular culture always implies relations of power. Terry Eagleton's point about the existential dimension of popular culture makes clear that even the most flippant web meme ought to be taken seriously and can, in fact, have grave consequences:

> In Bosnia or Belfast, culture is not just what you put on the cassette player; it is what you kill for. What culture loses in sublimity, it gains in practicality. In these circumstances, for both good and ill, nothing could be more bogus than the charge that culture is loftily remote from everyday life.
>
> *Eagleton 2000: 38*

Hence, when the encounter of Man and God on Michelangelo's "The Creation of Adam" (1511–2), the quintessentially high-cultural fresco on the ceiling of the Sistine Chapel, Vatican, is transformed into a ritual of streetwise interaction or a game of stone–paper–scissors in a recent meme remix, pop culture appropriates high art to its purposes; the Renaissance attempt to visualise a Christian condition is not just turned into a cheap and fleeting laugh, but is indicative of the circulation of cultural meanings in participatory culture. And, indeed, to share a GIF in social media can result in grave practical consequences, for instance when a Reddit user found that his anti-CNN GIF was picked up in a tweet by US President Donald Trump (on 2 July 2017).

Animated GIFs and meme culture

GIFs may be seen as sharable, compact, reduced, simple, or, even more normatively, atrophied narratives – but the term 'atrophied cinema,' suggesting a cinema that is wasted away, would be inadequate. Animated GIFs operate in a post-cinematic, post-TV world, outside the confines of the cinema and more likely in the context of shared contents and files encountered via social media on smartphones. What I am discussing here might be addressed as para-cinematic – a new cultural sphere of circulation (as short-circuited distribution and consumption). It is marked by new modes of engagement (mobilised and manipulable, spreadable and sharable, accelerated, compact, and integrated in day-to-day activities).

In the arena of the mobilised Web, the ephemeral materials of contemporary popular culture undergo a permanent remix, and remixers use and transform the available material, seeking to empower themselves by appropriating the work of others. In so doing, they pre-empt the inevitable threat of being appropriated themselves. In a performance arena of permanently supplanting styles and repertoires, these connoisseurs of the everyday seek distinction through participatory recognition.

Under conditions of low-threshold access, easy usability and findability, and information-rich media saturation, a meme culture emerges that has taken a particular liking to producing animated GIFs that frequently reference established art forms that utilise *objets trouvés*. Hence, "Internet memes can be treated as (post)modern folklore, in which shared norms and values are constructed through cultural artifacts such as Photoshopped images or urban legends" (Shifman 2014: 18). The Bardic role of the relatively homogeneous, programmed TV, as suggested by John Fiske or John Hartley in classic cultural studies analyses, has now been taken over by TV series on Netflix and circulated as user-generated content on mashed-up social media sites.

GIFs are memes, a particular type of public discourse. Limor Shifman (2014: 4) is caught between "(skeptic) academic and (enthusiastic) popular discourse about memes." With Shifman (2014: 18), Internet memes can be treated as (post)modern folklore, "in which shared norms and values are constructed through cultural artifacts such as Photoshopped images or urban legends." She defines memes as "(a) a group of digital items sharing common characteristics of content, form, and/or stance; (b) that were created with awareness of each other; and (c) were circulated, imitated, and/or transformed via the Internet by many users" (Shifman 2014: 41). Shifman isolates three dimensions of meme culture: popular, political, and global. There are pre-digital precursors such as the 'Kilroy was here' graffito, which emerged with American soldiers during World War II, and it is not just since the Facebook 'wall' or Twitter 'feed' that memes and graffiti share certain aspects, such as anonymous circulation.

In Shifman's distinction between viral, founder-based, and egalitarian memes, the degree of change, manipulation, and participation – that is, the generative and participatory aspect of meme culture – becomes the key criterion for differentiation, evident in the categories "user involvement" and "derivatives" (Shifman 2014: 82, 83). Manipulability is the measure of a meme's participatory and interactive quality. In terms of content, according to Shifman (2014: 74), vernacular activities are facilitated and marked by six features: "A focus on ordinary people, flawed masculinity, humour, simplicity, repetitiveness and whimsical content" – animated GIFs tend to tick many of these boxes. In terms of textual features, therefore, we need to look at the texts' quality of difference, generated by their whimsical and humourous qualities. The conditions of consumption necessitate simplicity and repetitiveness.

Any analysis of meme success, however, needs to go beyond mere textual analysis and address the technical and social dimensions of its low-threshold adaptability and spreadability, facilitating effortless circulation: "A GIF can be embedded directly in a webpage, where it loads immediately without plugins or third-party players, because it's an open format. And as simple files, GIFs are promiscuous and frictionless, with low barriers for viewing, possessing, and sharing" (Eppink 2014: 303). My initial argument, then, is that while the single GIF may be marginal and forgettable, the volume of shared GIFs makes them, along with other memes, "epitomize … the very essence of the so-called Web 2.0 era" (Shifman 2014: 15).

One reason for the GIF's ubiquity is that it is supported by nearly all Web browsers. Animated GIFs have become easy to transport, and ever since the earliest GIFs – such as the "Dancing Baby" (1996), which was etched in the memory of 1990s popular culture as it was picked up in the successful dramedy *Ally McBeal* (1997–2002) – they have proliferated in social media. The first clips using the GIF format, which appeared around 1995, were marked by simple, low-resolution, untextured 'clip-art' graphics (Eppink 2014: 300). With the advent of photographic GIFs around the 2000s, the popularity of the graphics format exploded (Eppink 2014: 300).

Animated GIFs as short, non-terminating, repeating narratives

Animated GIFs are marked by short, non-terminating, repeating processes. They repeat ad infinitum, or at least until the viewer's device runs out of battery, or until she terminates the application. They are "pretty silly—a few frames of video, endlessly looping in time," as one commentator put it in *Wired* (Thompson 2013). I suggest that all of these aspects deserve closer inspection – (1) short, (2) non-terminating, and (3) repeating narratives – and that the result is not necessarily silly. The non-terminating and repetitive qualities of the GIF are likely to result in a hypnotic, de-semanticised effect if a viewer is exposed to it over a longer stretch of time. The paradoxical permanence generated by the moving stasis of these "silly" loops, however, also lends itself to non-teleologic gestures and motifs – aesthetic ideals that seek formal expression in, for instance, modernism.

The GIF is hence a prime example of what a recent publication has termed compact cinematics. Pepita Hesselberth and Maria Poulaki (2017: 1, 6) link the GIF to Tom Gunning's cinematic attractions, as well as to Walter Benjamin's dispersed viewing. Their term, compact cinematics, suggest the brevity and usability of the emerging forms. The diversified, convergent forms and practices of visual circulation, beyond established genre boundaries such as 'film' or 'TV,' they argue, "challenge the concepts that have traditionally been used to understand the moving image, and call attention to complex and modular forms of expression and perception of which the cinematic partakes" (Hesselberth and Poulaki 2017: 3).

In the quality of being short, the animated GIF is indicative of a larger trend towards concise, compact, and portable narrativity that has recently generated much academic interest (see Gamper and Mayer 2017). Short textual forms, however, have been a trademark of modernity from the very beginning: miscellaneous notes, anecdotes, the funnies in newspapers, etc. Their fragmentation may result in both reduced and heightened complexity. Short duration may facilitate effective observation as a heuristic advantage. For instance, a GIF can be effortlessly integrated into a slideshow while the judicious presentation of either a feature film or a novel is impossible within the confines of a short visual presentation. Yet, it is almost impossible, without resorting to awkward ekphrasis, to represent a GIF in an essay like this, taking away its essential quality of repetitive motion; the quality of movement suggests that it transcends both traditional approaches to (still) photography and (more complex) film/cinema. The lack of sound accounts for the ambient quality of a GIF. One could imagine a GIF being projected on a living room wall in lieu of a non-animated image. Eppink (2014: 302) argues that its non-interactive qualities have, in fact, contributed to its success: "The format's lack of audio and playback control, frequently cited as shortcomings, enforce a silent and non-interactive form that doesn't demand as much attention as a full-featured video player."

As we have seen above, animated GIFs are short, visual narratives – minimal stories, sketches, and vignettes – that have always been heuristically useful. A standard minimal definition of narrative reads: "the semiotic representation of a series of events meaningfully connected" (Herman and Vervaeck 2005: 13). Certainly, there is a tendency among GIFs towards semiotic representation, more than one event, and a change of state in these GIF micro narratives. One can argue that the sequential images of the GIF have a higher potential for narrativity than the single representational picture, whereas it is restricted, albeit more ambient, due to its lack of sound. Frequently, GIFs remediate existing visual content from films and television. A purely abstract GIF, however, may be non-representational and, therefore, low on narrativity. The visuality of narratives has generated much interest that can be blamed on the visual/pictorial turn, declared by W.J.T. Mitchell, of merging art histories with other disciplines into visual studies. A narrative, however, is not necessarily – and, in fact, rarely – just visual, auditory, verbal, just oral, or written.

Let us again look at the brevity of the animated GIF. We could apply arguments that have been made for short fiction to the GIF: its brevity is particularly suited to effects of epiphany, and its suddenness lends itself to the comic or violent effect, according to the Kantian focus on the suddenness and absurdity of laughter: "Laughter is an affect that arises if a tense expectation is transformed into nothing."[2]

Beyond Hamlet's scolding of Polonius's verbosity, arguing simply that "brevity is the soul of wit," we might apply Adrian Hunter's idea and argue that the brief but repetitive, fragmentary GIF is particularly well suited to the modernist or postmodernist experience of life as rapid, disconnected events. In this view, the best GIFs achieve the modernist ideal of "a creative transaction between brevity and complexity – the art of saying less but meaning more" (Hunter 2007: 2). We might claim, with Charles May (1994: 131), that short forms are more primal, mythical, and natural: "the short story precedes the long story as the most natural means of narrative communication"; in addition, we might invoke the "rigor of brevity" with Henry James (1989: 190), who praises "the detached incident, single and sharp, as clear as a pistol-shot." Brevity, conciseness, and immediacy are qualities that are necessary when moving in the spaces of social media, but these qualities are to be found in classical rhetoric as well as in contemporary 'flash fiction' (less than 300 words).

The generic diversification of animated GIFs

The ways in which animated GIFs appropriate and remix existing cultural material may be studied by looking at the case of *The Gashleycrumb Tinies*, 26 alphabetically structured cartoons that initially appeared in comic panels. *The Gashleycrumb Tinies* was originally published in 1963 as an abecedarian book written by Edward Gorey. Gorey tells the tale of the 26 children (each representing a letter of the alphabet) and their untimely deaths in rhyming dactylic couplets, accompanied by the author's distinctive black and white illustrations. In print, *The Gashleycrumb Tinies* was a codex, and for its sequential effect the cartoon relied on the turning of new pages, each new leaf introducing new deaths. They may be described as polyphase pictures, and thus quite rich in narrativity if compared to one-panel, single, monophase pictures (Wolf 2008: 431). In terms of the classic distinction of Gotthold Ephraim Lessing's *Laokoon* (1766), *The Gashleycrumb Tinies* can be described as a spatial work of art, whereas the animation turns them into temporal art. The text is not only available as a book/codex, but also as a poster and as a scrollable pdf, which changes the sequentiality of the reception. These temporal and spatial manifestations of the text may support W.J.T. Mitchell's point that there is no categorical, fundamental distinction between temporal and spatial art (Mitchell 1986: 98). Werner Wolf (2008: 431) is able to address the problem of pictorial narrativity, even in representational, single, monophase pictures.

Just as any comic book, the animated GIF has both a spatial and a temporal dimension that can constitute its narrativity. However, comics on film are generally addressed as animation, so let us take seriously, for a moment, the qualifier 'animated' GIF.

From its beginnings in the mid-1990s, GIFs have also produced their very own brand of visual art that can be witnessed in various surrealist animations to be found in the works collected in Rhizome, the GIF art gallery, or the sections in giphy or other GIF database collections. Excellent examples can be found in recently circulating GIFs, such as Roy Lichtenstein-style cartoon faces remixed with the left screen swipe familiar from dating apps, or René Magritte's "La trahison des images" (1929) remixed with the clickbait GIF button – precisely not a GIF, but a non-animated image file.

My primary interest, however, is not in readings of GIFs as visual art, but as everyday, cultural circulation. For all their colourful diversity, GIFs most frequently express emotions. In fact, the reductive approach in the empirical study by Jou et al. (2014) starts from the assumption expressing the emotive is precisely their primary aim and *raison d'être*:

> Animated GIF images are a largely unexplored media in Multimedia and Computer Vision research. Their use for conveying emotions has become widely prevalent on the Web, and are now massively found on digital forums, message boards, social media, and websites of every genre … Meanwhile, animated GIFs have quickly become a channel for visually expressing emotion in our modern society. Their role in popular culture has even contributed to the rise of widely, rapidly spread cultural references called memes. This social function of GIFs today provides confidence that GIFs gathered from the Web for research are emotionally expressive.
>
> *Jou et al. 2014: 213*

A quality that is often taken for granted but rarely discussed is the humourous, recreational quality of many GIFs. Humourous GIFs may be most frequently shared as memes in popular culture. Regarding humour theory, animated GIFs portray unruly bodies under heteronomous control – bodies being controlled by the graphics interchange format. Sigmund Freud (1989: 437–8), of course, connects humour to artistic creativity, infantilism, and the pleasure principle, thus seeing laughter as an outlet for suppressed, unconscious desires, while Mikhail Bakhtin (1968) regards grotesque humour, in particular, as an articulation of revitalised materiality. Similarly, according to Henri Bergson (1911: 32, 35), humour can fulfil the important social function of reviving and reinvigorating society when we are laughing at inelasticity and constraining forces. All of these approaches might be usefully evoked to account for much of the GIF material on the Web. Humour is what is needed to generate attention for GIFs – an attention, however, that need not be sustained over long periods of time.

GIFs, then, do much more than visually expand and put into motion the expression of simple emotions. They are more than just elaborate emoticons. There is now plenty of generic diversification that complexifies the range of artistic expression to be found in GIFs. Consider, for instance, the cinemagraph: "Coined (and trademarked) by fashion photographer Jamie Beck and designer Kevin Burg in 2011, the cinemagraph is an animated GIF in which most of an image remains still while one element moves in a seamless loop" (Eppink 2014: 303). The cinemagraph makes full use of the perpetuity of the recurring GIF movement.

Another frequent variety of GIF is the so-called "reaction GIF" – a GIF subcategory in which the GIF expresses a physical or affective reaction. The "Confused Travolta" meme is such a reaction GIF series featuring a remixed, appropriated cutout of actor John Travolta, in the 1994 black comedy crime film *Pulp Fiction*, edited into other base images of various contexts. It was first uploaded to the image sharing platform Imgur by user karmafrappuccino in November 2015, going viral as another user mashed it with a background shot of a toy supermarket. The original GIF is composed of the Pulp Fiction character Vincent Vega (played by John Travolta) who looks around a room, expressing bewilderment. Frequently, the animated GIF consists of remixed visual material. It first appeared in a so-called Subreddit of the news aggregator Reddit (a portmanteau word combining 'read it' and 'edit').

The example exhibits the two qualities that the animated GIF shares with other memes, according to Shifman (2014: 20), those of "mimicry and remix. Mimicry involves the practice of 'redoing' … remixing, is a newer one. It involves technology-based manipulation, for instance by Photoshopping an image or adding a new soundtrack." Animated GIFs appear in content chains that mimic – that is, re-do – existing cultural material, and they frequently manipulate and mix existing cultural material. They appear in the context of a praxis, they develop through competitive processes of selection, and the praxis can be traced to the social 'meiosis' or 'cell division' suggested by the term 'meme.'

The "Confused Travolta" meme is an excellent example of the connectivity of meaning making in so-called 'content chains,' and the spreading of these memes is the consequence of both textual and circulatory features. The "Confused Travolta" meme has become productive in the most variegated and diversified contexts – as Brian Feldman (2015) praises it in *New York Magazine*, "It's very relatable, as is the best viral content." We might ask for indicators of productivity, or the ability to produce further parodic and non-parodic content chains. The 'citability' of these texts (a term cited from Jacques Derrida) or the 'relatability' of these miniature narratives (see Voigts-Virchow 2013: 76) are the key criteria for the success of GIFs such as the "Confused Travolta" meme, which is also available in compilation on various video-sharing websites.

The historical precedents of animated GIFs

Not only do the GIFs' immediacy enable ad-hoc reactions, but frequently they are also fragmentary, suggestive of non-present macro-narratives. More specifically, it has been associated with the 'cinema of attractions' (Gunning 1986), that is, the simple visual stimuli of early cinema. Indeed, both the scopophilic impetus of wanting to see (not necessarily to process narratively) and the narcissism associated with the pleasures of being seen link the GIF to early and contemporary cinematic practices. Considering early cinema, Tom Gunning (1986: 70) saw "effects" as "tamed attractions." Yet he argued that even "with the introduction of editing and more complex narratives, the aesthetic of attractions can still be sensed in periodic doses of non-narrative spectacle given to audiences" (Gunning 1989: 38). If this is true, then animated GIFs are examples of the essential core component one finds both in complex and non-complex visual narratives.

Eppink (2014: 298–9) lists a number of precursors for the pre-cinematic visual frenzy such as phenakistoscopes (1832), zoetropes (1834), and praxinoscopes (1877). The symmetrical and seamless loops, often portraying people or animals, seem to offer quite similarly simple pleasures to many 'cinematic' GIFs, that is, GIFs that are cut from digital film. With these digital GIFs, however, simple visual stimuli are disconnected from the shared presence of viewer and device. Appearing on the Web, the GIF has overcome certain limitations of the earlier technologies: the motion needed for the continuous loop need not be generated by the viewer, and the audience is largely expanded by the digital copying-machineries. Eppink suggests that linear flip books (1868) generated narratively more advanced formations than other early visual gimmicks because of their linear nature. He succinctly links the contents of pre-cinematic visual gimmickry and post-cinematic GIFs:

Both the electric kinetoscope (1894) and the hand-cranked mutosocope (1895) offered short, silent, photographed moving images as objects of entertainment. Early subjects included actualities (documentary-like footage of people and events) and loose, often sexually-charged narratives.

Eppink 2014: 298–9

Precisely the grotesque micronarratives we have seen in the current GIFs, therefore, also existed nearly 200 years ago. The phenakistoscope, for instance, is a device that enabled viewers to watch revolving pictures through a narrow slit, which in cognitive visual processing is translated as continuous motion. In 1825, Joseph Plateau described what he called the persistence of vision, a visual retinue in the spectator's mind that formed the basis of these visual contraptions explored not just for entertainment but also for science (for an overview on the expansion of Victorian visuality, see Brosch 2008).

To sum up, we have seen that the GIF is much more than an emoticon in motion. It is a simple and compact device to express an emotional response, true, but it can also be used for a great variety of humourous, satiric, parodic, and artistic purposes, particularly as it provokes a sudden reaction and requires the timing needed for comic effects. It is a dominant cultural mode of easy-access, compact, everyday appropriation. It is linked to the visual attractions at the core of many varieties of visual entertainment and tends to blur clear boundaries between established media genres. Its ambient, non-auditive quality makes it equally adequate for a number of different performative contexts. Low thresholds to generating GIFs as well as their evident citability, mashability, and remixability guarantee their potential for circulation. The GIF is generically versatile and eminently sharable – and a prime example of how culture circulates under conditions of the digital copy-paste culture.

Notes

1 This paper is a remix, that is, a revised and shortened version of material that was used in two publications – the online publication of the 'Mashup' section for the Anglistentag 2016 in Hamburg, and a volume on Adaptation and Transmedia Storytelling ('Appropriation and Parody in Remix Culture'), co-edited by Monika Pietrzak-Franger and Lucia Krämer.
2 Kant (1987: 203). In German: "der plötzliche Umschlag einer gespannten Erwartung ins nichts."

Works cited

Amerika, M. (2011) *Remixthebook*, Minneapolis, MN: University of Minnesota Press.
Bakhtin, M. (1968) *Rabelais and His World*, 1941, Cambridge, MA: MIT Press.
Bergson, H. (1911) *Laughter: An Essay on the Meaning of the Comic*, 1900, London: Macmillan.
Brosch, R. (2008) "Victorian challenges to ways of seeing: Everyday life, entertainment, images, and illusions," in R. Brosch (ed.) *Victorian Visual Culture*, Heidelberg, Germany: Winter, 21–64.
Eagleton, T. (2000) *The Idea of Culture*, Oxford: Blackwell.
Eppink, J. (2014) "A brief history of the Gif," *Journal of Visual Culture*, 13(3): 298–306.
Feldman, B. (2015) "'Confused John Travolta' meme is John Travolta's best role in years," *New York Magazine*, 9 December. Online. Available at: http://nymag.com/following/2015/12/confused-john-travolta-achieves-annual-relevance.html (accessed 20 May 2017).
Fischlin, D. (2014) "Introduction: OuterSpeares: Shakespeare, intermedia, and the limits of adaptation," in D. Fischlin (ed.) *OuterSpeares: Shakespeare, Intermedia, and the Limits of Adaptation*, Toronto: University of Toronto Press, 3–51.
Fischlin, D. (ed.) (2014) *OuterSpeares: Shakespeare, Intermedia, and the Limits of Adaptation*, Toronto: University of Toronto Press.
Fischlin, D. and Fortier, M. (eds) (2000) *Adaptations of Shakespeare: A Critical Anthology of Plays from the Seventeenth Century to the Present*, New York: Routledge.

Fiske, J. (2010) *Understanding Popular Culture*, 2nd edn, New York: Routledge.

Fortier, M. (2014) "Beyond adaptation," in D. Fischlin (ed.) *OuterSpeares: Shakespeare, Intermedia, and the Limits of Adaptation*, Toronto: University of Toronto Press, 372–86.

Freud, S. (1989) "Creative writers and day-dreaming," 1907, in P. Gay (ed.) *The Freud Reader*, New York: Norton, 436–43.

Fuchs, C. (2011) "Against Henry Jenkins: Remarks on Henry Jenkins' ICA talk 'Spreadable Media,'" *Christian Fuchs: Information − Society − Technology & Media*, 30 May. Blog. Online. Available at: http://fuchs.uti.at/570/ (accessed 20 June 2013).

Gamper, M. and Mayer, R. (eds) (2017) *Kurz & Knapp: Zur Mediengeschichte kleiner Formen vom 17. Jahrhundert bis zur Gegenwart*, Bielefeld, Germany: transcript.

Gunning, T. (1986) "The cinema of attractions: Early film, its spectator and the avant-garde," *Wide Angle*, 8(3/4): 63–70.

Gunning, T. (1989) "An aesthetic of astonishment: Early film and the (in)credulous spectator," *Art & Text*, 34: 31–45.

Hartley, J. (1992) *The Politics of Pictures. The Creation of the Public in the Age of Popular Media*, London: Routledge.

Herman, L. and Vervaeck, B. (2005) *Handbook of Narrative Analysis*, Lincoln, NE: University of Nebraska Press.

Hesselberth, P. and Poulaki, M. (2017) "Introduction: Screen − capture − attention," in P. Hesselberth and M. Poulaki (eds) *Compact Cinematics: The Moving Image in the Age of Bit-Sized Media*, London: Bloomsbury, 1–17.

Hunter, A. (2007) "Introduction," in *The Cambridge Introduction to the Short Story in English*, New York: Cambridge University Press, 1–4.

Hutcheon, L. (2013) *A Theory of Adaptation*. 2nd ed., London & New York: Routledge, 2013.

James, H. (1989) "The story-teller at large: Mr Henry Harland," 1898, in L. Edel (ed.) *The American Essays of Henry James*, Princeton, NJ: Princeton University Press, 186–96.

Jou, B., Bhattacharya, S., and Shih-Fu, C. (2014) "Predicting viewer perceived emotions in animated GIFs," in K.A. Hua et al. (eds) *MM'14: Proceedings of the 22nd ACM International Conference on Multimedia*, New York: ACM, 213–6. Online. Available at: http://dl.acm.org/citation.cfm?id=2656408 (accessed 16 July 2017).

Kant, I. (1987) *Critique of Judgment*, 1790, Indianapolis, IN: Hackett.

Know Your Meme (n.d.) "Confused Travolta," *Know Your Meme*. Online. Available at: http://knowyour-meme.com/memes/confused-travolta (accessed 20 May 2017).

Lessig, L. (2008) *Remix: Making Art and Commerce Thrive in the Hybrid Economy*, London: Bloomsbury.

McGuigan, J. (1992) *Cultural Populism*, London: Routledge

McGuigan, J. (2012) "The coolness of capitalism today," *tripleC*, 10(2): 425–38.

Marchart, O. (2008) *Cultural Studies*, Konstanz: UTB.

May, C.E. (1994) "The nature of knowledge in short fiction," in C.E. May (ed.) *The New Short Story Theories*, Athens, OH: Ohio University Press, 131–43.

Mitchell, W.J.T. (1986) *Iconology: Image, Text, Ideology*, Chicago, IL: University of Chicago Press.

Navas, E. (2012) *Remix Theory: The Aesthetics of Sampling*, Wien: Springer.

Sanders, J. (2006) *Adaptation and Appropriation*, London: Routledge.

Scholz, T. (2013) *Digital Labor: The Internet as Playground and Factory*, New York: Routledge.

Sennett, R. (1977) *The Fall of Public Man*, Cambridge, UK: Cambridge University Press.

Sevignani, S. (2016) *Privacy and Capitalism in the Age of Social Media*, New York: Routledge.

Shifman, L. (2014) *Memes in Digital Culture*, Cambridge, MA: MIT Press.

Stalder, F. (2009) "Neun Thesen zur Remix-Kultur," *iRights.info*. Projekt Arbeit 2.0. Online. Available at: https://irights.info/wp-content/uploads/fileadmin/texte/material/Stalder_Remixing.pdf (accessed 25 January 2016).

Thompson, C. (2013) "The animated GIF: Still looping after all these years," *Wired*, 1 March. Online. Available at: https://www.wired.com/2013/01/best-animated-gifs/ (accessed 13 February 2017).

Tufekci, Z. (2016) "Mark Zuckerberg is in denial," *The New York Times*, 15 November. Online. Available at: https://www.nytimes.com/2016/11/15/opinion/mark-zuckerberg-is-in-denial.html (accessed 13 February 2017).

Voigts-Virchow, E. (2013) "Anti-essentialist versions of aggregate Alice: A grin without a cat," in K. Krebs (ed.) *Translation and Adaptation in Theatre and Film*, New York: Routledge, 63–77.

Von Gehlen, D. (2011) *Mashup: Lob der Kopie*, Frankfurt: Suhrkamp.

Wilke, T. (2015) "Kombiniere! Variiere! Transformiere! Mashups als performative Diskursobjekte in populären Medienkulturen," in F. Mundhenke, F. Ramos Arenas, and T. Wilke (eds) *Mashups: Neue Praktiken und Ästhetiken in populären Medienkulturen*, Wiesbaden, Germany: Springer, 11–44.

Williams, R. (2011) "Culture is ordinary," 1958, in I. Szeman and T. Kaposy (eds) *Cultural Theory: An Anthology*, Oxford: Wiley-Blackwell, 53–9.

Wolf, W. (2008) "Pictorial narrative," in D. Herman et al. (eds) *Routledge Encyclopedia of Narrative Theory*, London: Routledge, 431–35.

INDEX